Statistics and Data with R

Statistics and Data with R:

An applied approach through examples

Yosef Cohen
University of Minnesota, USA.

Jeremiah Y. Cohen
Vanderbilt University, Nashville, USA.

WILEY
A John Wiley and Sons, Ltd., Publication

This edition first published 2008
© 2008 John Wiley & Sons Ltd.

Registered office
John Wiley & Sons Ltd, The Atrium, Southern Gate, Chichester, West Sussex, PO19 8SQ, United Kingdom

For details of our global editorial offices, for customer services and for information about how to apply for permission to reuse the copyright material in this book please see our website at www.wiley.com.

The right of the author to be identified as the author of this work has been asserted in accordance with the Copyright, Designs and Patents Act 1988.

All rights reserved. No part of this publication may be reproduced, stored in a retrieval system, or transmitted, in any form or by any means, electronic, mechanical, photocopying, recording or otherwise, except as permitted by the UK Copyright, Designs and Patents Act 1988, without the prior permission of the publisher.

Wiley also publishes its books in a variety of electronic formats. Some content that appears in print may not be available in electronic books.

Designations used by companies to distinguish their products are often claimed as trademarks. All brand names and product names used in this book are trade names, service marks, trademarks or registered trademarks of their respective owners. The publisher is not associated with any product or vendor mentioned in this book. This publication is designed to provide accurate and authoritative information in regard to the subject matter covered. It is sold on the understanding that the publisher is not engaged in rendering professional services. If professional advice or other expert assistance is required, the services of a competent professional should be sought.

Library of Congress Cataloging-in-Publication Data

Cohen, Yosef.
   Statistics and data with R : an applied approach through examples / Yosef Cohen, Jeremiah Cohen.
       p. cm.
   Includes bibliographical references and index.
   ISBN 978-0-470-75805-2 (cloth)
  1. Mathematical statistics—Data processing. 2. R (Computer program language)
I. Cohen, Jeremiah. II. Title.
   QA276.45.R3C64 2008
   519.50285′2133—dc22

                                                               2008032153

A catalogue record for this book is available from the British Library.

ISBN   978-0-470-75805-2

Typeset in 10/12pt Computer Modern by Laserwords Private Limited, Chennai, India

To the memory of **Gad Boneh**

Contents

Preface   xv

Part I Data in statistics and R

1   **Basic R**   3
    1.1   Preliminaries   4
       1.1.1   An R session   4
       1.1.2   Editing statements   8
       1.1.3   The functions `help()`, `help.search()` and `example()`   8
       1.1.4   Expressions   10
       1.1.5   Comments, line continuation and Esc   11
       1.1.6   `source()`, `sink()` and `history()`   11
    1.2   Modes   13
    1.3   Vectors   14
       1.3.1   Creating vectors   14
       1.3.2   Useful vector functions   15
       1.3.3   Vector arithmetic   15
       1.3.4   Character vectors   17
       1.3.5   Subsets and index vectors   18
    1.4   Arithmetic operators and special values   20
       1.4.1   Arithmetic operators   20
       1.4.2   Logical operators   21
       1.4.3   Special values   22
    1.5   Objects   24
       1.5.1   Orientation   24
       1.5.2   Object attributes   26
    1.6   Programming   28
       1.6.1   Execution controls   28
       1.6.2   Functions   30
    1.7   Packages   33

1.8	Graphics	34
	1.8.1 High-level plotting functions	35
	1.8.2 Low-level plotting functions	36
	1.8.3 Interactive plotting functions	36
	1.8.4 Dynamic plotting	36
1.9	Customizing the workspace	36
1.10	Projects	37
1.11	A note about producing figures and output	39
	1.11.1 `openg()`	39
	1.11.2 `saveg()`	40
	1.11.3 `h()`	40
	1.11.4 `nqd()`	40
1.12	Assignments	41

2 Data in statistics and in R 45
2.1	Types of data	45
	2.1.1 Factors	45
	2.1.2 Ordered factors	48
	2.1.3 Numerical variables	49
	2.1.4 Character variables	50
	2.1.5 Dates in R	50
2.2	Objects that hold data	50
	2.2.1 Arrays and matrices	51
	2.2.2 Lists	52
	2.2.3 Data frames	54
2.3	Data organization	55
	2.3.1 Data tables	55
	2.3.2 Relationships among tables	57
2.4	Data import, export and connections	58
	2.4.1 Import and export	58
	2.4.2 Data connections	60
2.5	Data manipulation	63
	2.5.1 Flat tables and expand tables	63
	2.5.2 Stack, unstack and reshape	64
	2.5.3 Split, unsplit and unlist	66
	2.5.4 Cut	66
	2.5.5 Merge, union and intersect	68
	2.5.6 `is.element()`	69
2.6	Manipulating strings	71
2.7	Assignments	72

3 Presenting data 75
3.1	Tables and the flavors of `apply()`	75
3.2	Bar plots	77
3.3	Histograms	81
3.4	Dot charts	85
3.5	Scatter plots	86
3.6	Lattice plots	88

3.7	Three-dimensional plots and contours	90
3.8	Assignments	90

Part II Probability, densities and distributions

4 Probability and random variables — 97
- 4.1 Set theory — 98
  - 4.1.1 Sets and algebra of sets — 98
  - 4.1.2 Set theory in R — 103
- 4.2 Trials, events and experiments — 103
- 4.3 Definitions and properties of probability — 108
  - 4.3.1 Definitions of probability — 108
  - 4.3.2 Properties of probability — 111
  - 4.3.3 Equally likely events — 112
  - 4.3.4 Probability and set theory — 112
- 4.4 Conditional probability and independence — 113
  - 4.4.1 Conditional probability — 114
  - 4.4.2 Independence — 116
- 4.5 Algebra with probabilities — 118
  - 4.5.1 Sampling with and without replacement — 118
  - 4.5.2 Addition — 119
  - 4.5.3 Multiplication — 120
  - 4.5.4 Counting rules — 120
- 4.6 Random variables — 127
- 4.7 Assignments — 128

5 Discrete densities and distributions — 137
- 5.1 Densities — 137
- 5.2 Distributions — 141
- 5.3 Properties — 143
  - 5.3.1 Densities — 144
  - 5.3.2 Distributions — 144
- 5.4 Expected values — 144
- 5.5 Variance and standard deviation — 146
- 5.6 The binomial — 147
  - 5.6.1 Expectation and variance — 151
  - 5.6.2 Decision making with the binomial — 151
- 5.7 The Poisson — 153
  - 5.7.1 The Poisson approximation to the binomial — 155
  - 5.7.2 Expectation and variance — 156
  - 5.7.3 Variance of the Poisson density — 157
- 5.8 Estimating parameters — 161
- 5.9 Some useful discrete densities — 163
  - 5.9.1 Multinomial — 163
  - 5.9.2 Negative binomial — 165
  - 5.9.3 Hypergeometric — 168
- 5.10 Assignments — 171

6 Continuous distributions and densities — 177
- 6.1 Distributions — 177
- 6.2 Densities — 180
- 6.3 Properties — 181
  - 6.3.1 Distributions — 181
  - 6.3.2 Densities — 182
- 6.4 Expected values — 183
- 6.5 Variance and standard deviation — 184
- 6.6 Areas under density curves — 185
- 6.7 Inverse distributions and simulations — 187
- 6.8 Some useful continuous densities — 189
  - 6.8.1 Double exponential (Laplace) — 189
  - 6.8.2 Normal — 191
  - 6.8.3 χ^2 — 193
  - 6.8.4 Student-t — 195
  - 6.8.5 F — 197
  - 6.8.6 Lognormal — 198
  - 6.8.7 Gamma — 199
  - 6.8.8 Beta — 201
- 6.9 Assignments — 203

7 The normal and sampling densities — 205
- 7.1 The normal density — 205
  - 7.1.1 The standard normal — 207
  - 7.1.2 Arbitrary normal — 210
  - 7.1.3 Expectation and variance of the normal — 212
- 7.2 Applications of the normal — 213
  - 7.2.1 The normal approximation of discrete densities — 214
  - 7.2.2 Normal approximation to the binomial — 215
  - 7.2.3 The normal approximation to the Poisson — 218
  - 7.2.4 Testing for normality — 220
- 7.3 Data transformations — 225
- 7.4 Random samples and sampling densities — 226
  - 7.4.1 Random samples — 227
  - 7.4.2 Sampling densities — 228
- 7.5 A detour: using R efficiently — 230
  - 7.5.1 Avoiding loops — 230
  - 7.5.2 Timing execution — 230
- 7.6 The sampling density of the mean — 232
  - 7.6.1 The central limit theorem — 232
  - 7.6.2 The sampling density — 232
  - 7.6.3 Consequences of the central limit theorem — 234
- 7.7 The sampling density of proportion — 235
  - 7.7.1 The sampling density — 236
  - 7.7.2 Consequence of the central limit theorem — 238
- 7.8 The sampling density of intensity — 239
  - 7.8.1 The sampling density — 239

		7.8.2 Consequences of the central limit theorem	241

	7.9	The sampling density of variance	241
	7.10	Bootstrap: arbitrary parameters of arbitrary densities	242
	7.11	Assignments	243

Part III Statistics

8 Exploratory data analysis — 251
   8.1 Graphical methods — 252
   8.2 Numerical summaries — 253
      8.2.1 Measures of the center of the data — 253
      8.2.2 Measures of the spread of data — 261
      8.2.3 The Chebyshev and empirical rules — 267
      8.2.4 Measures of association between variables — 269
   8.3 Visual summaries — 275
      8.3.1 Box plots — 275
      8.3.2 Lag plots — 276
   8.4 Assignments — 277

9 Point and interval estimation — 283
   9.1 Point estimation — 284
      9.1.1 Maximum likelihood estimators — 284
      9.1.2 Desired properties of point estimators — 285
      9.1.3 Point estimates for useful densities — 288
      9.1.4 Point estimate of population variance — 292
      9.1.5 Finding MLE numerically — 293
   9.2 Interval estimation — 294
      9.2.1 Large sample confidence intervals — 295
      9.2.2 Small sample confidence intervals — 301
   9.3 Point and interval estimation for arbitrary densities — 304
   9.4 Assignments — 307

10 Single sample hypotheses testing — 313
   10.1 Null and alternative hypotheses — 313
      10.1.1 Formulating hypotheses — 314
      10.1.2 Types of errors in hypothesis testing — 316
      10.1.3 Choosing a significance level — 317
   10.2 Large sample hypothesis testing — 318
      10.2.1 Means — 318
      10.2.2 Proportions — 323
      10.2.3 Intensities — 324
      10.2.4 Common sense significance — 325
   10.3 Small sample hypotheses testing — 326
      10.3.1 Means — 326
      10.3.2 Proportions — 327
      10.3.3 Intensities — 328

xii   Contents

	10.4 Arbitrary statistics of arbitrary densities	329
	10.5 p-values	330
	10.6 Assignments	333
11	**Power and sample size for single samples**	**341**
	11.1 Large sample	341
	11.1.1 Means	342
	11.1.2 Proportions	352
	11.1.3 Intensities	356
	11.2 Small samples	359
	11.2.1 Means	359
	11.2.2 Proportions	361
	11.2.3 Intensities	363
	11.3 Power and sample size for arbitrary densities	365
	11.4 Assignments	365
12	**Two samples**	**369**
	12.1 Large samples	370
	12.1.1 Means	370
	12.1.2 Proportions	375
	12.1.3 Intensities	379
	12.2 Small samples	380
	12.2.1 Estimating variance and standard error	380
	12.2.2 Hypothesis testing and confidence intervals for variance	382
	12.2.3 Means	384
	12.2.4 Proportions	386
	12.2.5 Intensities	387
	12.3 Unknown densities	388
	12.3.1 Rank sum test	389
	12.3.2 t vs. rank sum	392
	12.3.3 Signed rank test	392
	12.3.4 Bootstrap	394
	12.4 Assignments	396
13	**Power and sample size for two samples**	**401**
	13.1 Two means from normal populations	401
	13.1.1 Power	401
	13.1.2 Sample size	404
	13.2 Two proportions	406
	13.2.1 Power	407
	13.2.2 Sample size	409
	13.3 Two rates	410
	13.4 Assignments	415
14	**Simple linear regression**	**417**
	14.1 Simple linear models	417
	14.1.1 The regression line	418
	14.1.2 Interpretation of simple linear models	419

		Contents	xiii

	14.2 Estimating regression coefficients	422
	14.3 The model goodness of fit	428
	14.3.1 The F test	428
	14.3.2 The correlation coefficient	433
	14.3.3 The correlation coefficient vs. the slope	434
	14.4 Hypothesis testing and confidence intervals	434
	14.4.1 t-test for model coefficients	435
	14.4.2 Confidence intervals for model coefficients	435
	14.4.3 Confidence intervals for model predictions	436
	14.4.4 t-test for the correlation coefficient	438
	14.4.5 z tests for the correlation coefficient	439
	14.4.6 Confidence intervals for the correlation coefficient	441
	14.5 Model assumptions	442
	14.6 Model diagnostics	443
	14.6.1 The hat matrix	445
	14.6.2 Standardized residuals	447
	14.6.3 Studentized residuals	448
	14.6.4 The RSTUDENT residuals	449
	14.6.5 The DFFITS residuals	453
	14.6.6 The DFBETAS residuals	454
	14.6.7 Cooke's distance	456
	14.6.8 Conclusions	457
	14.7 Power and sample size for the correlation coefficient	458
	14.8 Assignments	459
15	**Analysis of variance**	**463**
	15.1 One-way, fixed-effects ANOVA	463
	15.1.1 The model and assumptions	464
	15.1.2 The F-test	469
	15.1.3 Paired group comparisons	475
	15.1.4 Comparing sets of groups	484
	15.2 Non-parametric one-way ANOVA	488
	15.2.1 The Kruskal-Wallis test	488
	15.2.2 Multiple comparisons	491
	15.3 One-way, random-effects ANOVA	492
	15.4 Two-way ANOVA	495
	15.4.1 Two-way, fixed-effects ANOVA	496
	15.4.2 The model and assumptions	496
	15.4.3 Hypothesis testing and the F-test	500
	15.5 Two-way linear mixed effects models	505
	15.6 Assignments	509
16	**Simple logistic regression**	**511**
	16.1 Simple binomial logistic regression	511
	16.2 Fitting and selecting models	519
	16.2.1 The log likelihood function	519
	16.2.2 Standard errors of coefficients and predictions	521
	16.2.3 Nested models	524

xiv   Contents

    16.3 Assessing goodness of fit   525
           16.3.1 The Pearson χ^2 statistic   526
           16.3.2 The deviance χ^2 statistic   527
           16.3.3 The group adjusted χ^2 statistic   528
           16.3.4 The ROC curve   529
    16.4 Diagnostics   533
           16.4.1 Analysis of residuals   533
           16.4.2 Validation   536
           16.4.3 Applications of simple logistic regression to 2×2 tables   536
    16.5 Assignments   539

17 Application: the shape of wars to come   541
    17.1 A statistical profile of the war in Iraq   541
           17.1.1 Introduction   542
           17.1.2 The data   542
           17.1.3 Results   543
           17.1.4 Conclusions   550
    17.2 A statistical profile of the second Intifada   552
           17.2.1 Introduction   552
           17.2.2 The data   553
           17.2.3 Results   553
           17.2.4 Conclusions   561

References   563

R Index   569

General Index   583

Preface

For the purpose of this book, let us agree that Statistics[1] is the study of data for some purpose. The study of the data includes learning statistical methods to analyze it and draw conclusions.

Here is a question and answer session about this book. Skip the answers to those questions that do not interest you.

- Who is this book intended for?

The book is intended for students, researchers and practitioners both in and out of academia. However, no prior knowledge of statistics is assumed. Consequently, the presentation moves from very basic (but not simple) to sophisticated. Even if you know statistics and R, you may find the many many examples with a variety of real world data, graphics and analyses useful. You may use the book as a reference and, to that end, we include two *extensive* indices. The book includes (almost) parallel discussions of analyses with the normal density, proportions (binomial), counts (Poisson) and bootstrap methods.

- Why "Statistics *and* data with R"?

Any project in which statistics is applied involves three major activities: preparing data for application of some statistical methods, applying the methods to the data and interpreting and presenting the results. The first and third activities consume by far the bulk of the time. Yet, they receive the least amount of attention in teaching and studying statistics. R is particularly useful for any of these activities. Thus, we present a balanced approach that reflects the relative amount of time one spends in these activities.

The book includes over 300 hundred examples from various fields: ecology, environmental sciences, medicine, biology, social sciences, law and military. Many of the examples take you through the three major activities: They begin with importing the data and preparing it for analysis (that is the reason for "and data" in the title), run the analysis (that is the reason for "Statistics" in the title) and end with presenting the results. The examples were applied through R scripts (that is the reason for "with R" in the title).

---

[1] From here on, we shall not capitalize Statistics.

Our guiding principle was "what you see is what you get" (WYSIWYG). Thus, whether examples illustrate data manipulation, statistical methods or graphical presentation, accompanying scripts allow you to produce the results as shown. Each script is presented as code snippets or as line-numbered statements. These enhance explanation of the scripts. Consequently, some of the scripts are not short and there are plenty of repetitions. Adhering to our goal, a wiki website, http://turtle.gis.umn.edu includes all of the scripts and data used in the book. You can download, cut and paste to reproduce all of the tables, figures and statistics that are shown. You are invited to modify the scripts to fit your needs.

Albeit not a database management system, R integrates the tasks of preparing data for analysis, analyzing it and presenting it in powerful and flexible ways. Power and flexibility do not come easily and the learning curve of R is steep. To the novice—in particular those not versed in object oriented computer languages—R may seem at times cryptic. In Chapter 1 we attempt to demystify some of the things about R that at first sight might seem incomprehensible.

To learn how to deal with data, analysis and presentation through R, we include over 300 examples. They are intended to give you a sense of what is required of statistical projects. Therefore, they are based on moderately large data that are not necessarily "clean"—we show you how to import, clean and prepare the data.

- What is the required knowledge and level of presentation?

No previous knowledge of statistics is assumed. In a few places (e.g. Chapter 5), previous exposure to introductory Calculus may be helpful, but is not necessary. The few references to integrals and derivatives can be skipped without missing much. This is not to say that the presentation is simple throughout—it starts simple and becomes gradually more advanced as one progresses through the book. In short, if you want to use R, there is no way around it: You have to invest time and effort to learn it. However, the rewards are: you will have complete control over what you do; you will be able to see what is happening "behind the scenes"; you will develop a good-practices approach.

- What is the choice of topics and the order of their presentation?

Some of the topics are simple, e.g. parts of Chapters 9 to 12 and Chapter 14. Others are more advanced, e.g. Chapters 16 and 17. Our guiding principle was to cover large sample normal theory and in parallel topics about proportions and counts. For example, in Chapter 12, we discuss two sample analysis. Here we cover the normal approach where we discuss hypotheses testing, confidence intervals and power and sample size. Then we cover the same topics for proportions and then for counts. Similarly, for regression, we discuss the classical approach (Chapter 14) and then move on to logistic regression (Chapter 16). With this approach, you will quickly learn that life does not end with the normal. In fact, much of what we do is to analyze proportions and counts. In parallel, we also use Bootstrap methods when one cannot make assumptions about the underlying distribution of the data and when samples are small.

- How should I teach with this book?

The book can be covered in a two-semester course. Alternatively, Chapters 1 to 14 (along with perhaps Chapter 15) can be taught in an introductory course for seniors and first-year graduate students in the sciences. The remaining Chapters, in a second course. The book is accompanied by a solution manual which includes solutions to most of the exercises. The solution manual is available through the book's site.

- How should I study and use this book?

To study the book, read through each chapter. Each example provides enough code to enable you to reproduce the results (statistical analysis and graphics) exactly as shown. Scripts are explained in detail. Read through the explanation and try to produce the results, as shown. This way, you first see how a particular analysis is done and how tables and graphics may be used to summarize it. Then when you study the script, you will know what it does. Because of the adherence to the WYSIWYG principle, some of the early examples might strike you as particularly complicated. Be patient; as you progress, there are repetitions and you will soon learn the common R idioms.

If you are familiar with both R and basic statistics, you may use the book as a reference. Say you wish to run a simple logistic regression. You know how to do it with R, but you wish to plot the regression on a probability scale. Then go to Chapter 16 and flip through the pages until you hit Figure 16.5 in Example 16.5. Next, refer to the example's script and modify it to fit your needs. There are so many R examples and scripts that flipping through the pages you are likely to find one that is close to what you need.

We include two indices, an R index and a general index. The R index is organized such that functions are listed as first entry and their arguments as sub-entries. The page references point to applications of the functions by the various examples.

- Classical vs. Bayesian statistics

This topic addresses developments in statics in the last few decades. In recent years and with advances in numerical computations, a trend among natural (and other) scientists of moving away from classical statistics (where large data are needed and no prior knowledge about it is assumed) to the so-called Bayesian statistics (where prior knowledge about the data contributes to its analysis) has become fashionable. Without getting into the sticky details, we make no judgment about the efficacy of one as opposed to the other approach. We prescribe to the following:

- With large data sets, statistical analyses using these two approaches often reach the same conclusions. Developments in bootstrap methods make "small" data sets "large".
- One can hardly appreciate advances in Bayesian statistics without knowledge of classical statistics.
- Certain situations require applications of Bayesian statistics (in particular when real-time analysis is imperative). Others, do not.
- The analyses we present are suitable for both approaches and in most cases should give identical results. Therefore, we pursue the classical statistics approach.

- A word about typography, data and scripts

We use monospaced characters to isolate our code work with R. Monospace lines that begin with > indicate code as it is typed in an R session. Monospace lines that begin with line-number indicate R scripts. Both scripts and data are available from the books site (http://turtle.gis.umn.edu).

- Do you know a good joke about statistics?

Yes. Three statisticians go out hunting. They spot a deer. The first one shoots and misses to the left. The second one shoots and misses to the right. The third one stands up and shouts, "We got him!"

Yosef Cohen and Jeremiah Cohen

Part I

Data in statistics and R

1

Basic R

Here, we learn how to work with R. R is rooted in S, a statistical computing and data visualization language originated at the Bell Laboratories (see Chambers and Hastie, 1992). The S interpreter was written in the C programming language. R provides a wide variety of statistical and graphical techniques in a consistent computing environment. It is also an *interpreted* programming environment. That is, as soon as you enter a statement, or a group of statements and hit the **Enter** key, R computes and responds with output when necessary. Furthermore, your whole session is kept in memory until you exit R. There are a number of excellent introductions to and tutorials for R (Becker et al., 1988; Chambers, 1998; Chambers and Hastie, 1992; Dalgaard, 2002; Fox, 2002). Venables and Ripley (2000) was written by experts and one can hardly improve upon it (see also The R Development Core Team, 2006b). An excellent exposition of S and thereby of R, is given in Venables and Ripley (1994). With minor differences, books, tutorials and applications written with S are relevant in R. A good entry point to the vast resources on R is its website, http://r-project.org.

R is not a panacea. If you deal with tremendously large data sets, with millions of observations, many variables and numerous related data tables, you will need to rely on advanced database management systems. A particularly good one is the open source PostgreSQL (http://www.postgresql.org). R deals effectively—and elegantly—with data that include several hundreds of thousands of observations and scores of variables. The bulkier your system is in memory and hard disk space, the larger the data set that R can handle. With all these caveats, R is a most flexible system for dealing with data manipulation and analysis. One often spends a large amount of time preparing data for analysis and exploring statistical models. R makes such tasks easy. All of this at no cost, for R is free!

This chapter introduces R. Learning R might seem like a daunting task, particularly if you do not have prior experience with programming. Steep as it may be, the learning curve will yield much dividend later. You should develop tolerance of ambiguity. Things might seem obscure at first. Be patient and continue. With plenty of repetitions and examples, the fog will clear. Being an introductory chapter, we are occasionally compelled to make simple and sweeping statements about what R does or does not do. As you become familiar with R, you will realize that some of our statements may be slightly (but not grossly) inaccurate. In addition to R, we discuss some abstract concepts such as classes and objects. The motivation for this seemingly unrelated material is this: When you start with R, many of the commands (calls to functions) and the results they produce may mystify you. With knowledge of the general concepts introduced here, your R sessions will be clearer. To get the most out of this chapter, follow this strategy: Read the whole chapter lightly and do whatever exercises and examples you find easy. Do not get stuck in a particular issue. Later, as we introduce R code, revisit the relevant topics in this chapter for further clarification.

It makes little sense to read this chapter without actually implementing the code and examples in a working R session. Some of the material in this chapter is based on the excellent tutorial by Venables et al. (2003). Except for starting R, the instructions and examples apply to all operating systems that R is implemented on. System specific instructions refer to the Windows operating systems.

1.1 Preliminaries

In this section we learn how to work with R. If you have not yet done so, go to http://r-project.org and download and install R. Next, start R. If you installed R properly, click on its icon on the desktop. Otherwise, click on Start | Programs | R | R x.y.z where x.y.z refers to the version number you installed. If all else fails, find where R is installed (usually in the Program Files directory). Then go to the bin directory and click on the Rgui icon. Once you start R, you will receive a welcome statement that ends with a prompt (usually >).

1.1.1 An R session

An R session consists of starting R, working with it and then quitting. You quit R by typing q() or (in Windows) by selecting File | Exit.[1] R displays the prompt at the beginning of the line as a cue for you. This is where you interact with the system (you do not type the prompt). The vast majority of our work with R will consist of executing functions. For now, we will say that a function is a small, single-purpose executable program that R can be instructed to run. A function is identified by a name followed by parentheses. Often the parentheses are separated by words. These are called arguments. We say that we execute a function when we type the function's name, with the parentheses (and possibly arguments) and then hit the Enter key.

Let us go through a simple session. This will give you a feel for R. At any point, if you wish to quit, type q() to end the session. To quit, you can also use the File | Exit menu. At the prompt, type (recall that you do not type the prompt):

```
> help.start()
```

---

[1] As a rule, while working with R, we avoid menus and graphical interface and stick with the command line interface.

R as a calculator

Here are some ways to use R as a simple calculator (note the # character—it tells R to treat everything beyond it to the end of the line as is):

```
> 5 - 1 + 10   # add and subtract
[1] 14
> 7 * 10 / 2   # multiply and divide
[1] 35
> pi           # the constant pi
[1] 3.141593
> sqrt(2)      # square root
[1] 1.414214
> exp(1)       # e to the power of 1
[1] 2.718282
```

The order of execution obeys the usual mathematical rules.

Assignments

Assignments in R are directional. Therefore, we need a way to distinguish the direction:

```
> x <- 5    # The object (variable) x holds the value 5
> x         # print x
[1] 5
> 6 -> x    # x now holds the value 6
> x
[1] 6
```

Note that to observe the value of x, we type x and follow it with Enter. To save space, we sometimes do this:

```
> (x <- pi)  # assign the constant pi and print x
[1] 3.141593
```

Executing functions

R contains many function. We execute a function by entering its name followed by parentheses and then Enter. Some functions need information to execute. This information is passed to the function by way of arguments.

```
> print(x)  # print() is a function. It prints its argument, x
[1] 6
> ls()      # lists the objects in memory
[1] "x"
```

```
> rm(x)      # remove x from memory
> ls()       # no objects in memory, therefore:
character(0)
```

A word of caution: R has a large number of functions. When you create objects (such as x) avoid function names. For example, R includes a function c(); so do not name any of your variables (also called objects) c. To discover if a function name may collide with an object name, before the assignment, type the object's name followed by **Enter**, like this:

```
> t
function (x)
UseMethod("t")
<environment: namespace:base>
```

From this response you may surmise that there is a function named t(). Depending on the function, the response might be some code.

Vectors

We call a set of elements a vector. The elements may be floating point (decimal) numbers, integers, strings of characters and so on. Many common operations in R require sequences. For example, to generate a sequence of 20 numbers between 0 and 19 do:

```
> x <- 0 : 19
```

To see the values stored in the vector x we just created, type x and hit **Enter**. R prints:

```
> x
 [1]  0  1  2  3  4  5  6  7  8  9 10 11 12 13 14 15 16 17
[19] 18 19
```

Note the way R lists the values of the vector x. The numbers in the square brackets on the left are not part of the data. They are the index of the first element of the vector in the listed row. This helps you identify the order of the data. So there are 18 values of x listed in the first row, starting with the first element. The value of this element is 0 and its index is 1. This index is denoted by [1]. The second row starts with the 19th element and so on. To access the value of the fifth element, do

```
> x[5]
[1] 4
```

Behind the scenes

To demystify R, let us take a short detour. Recall that statements in R are calls to functions and that functions are identified by a name followed by parentheses (e.g. help()). You may then ask: "To print x, I did not call a function. I just entered the name of the object. So how does R know what to do?" To fully answer this question we need to discuss objects (soon). For now, let us just say that a vector (such as x) is an object. Objects belong to object-types (sometimes called classes). For example x belongs to the type vector. Objects of the same type have (inherit) the properties of their type. One of these properties might be a print function that knows how to print the object. For example, R knows how to print objects of type vector. By entering the name of the vector only, you are calling a function that knows how to print the vector object x. This idea extends to other types of objects. For example, you can create objects of type data.frame. Such objects are versatile. They hold different kinds of data and have a comprehensive set of functions to manipulate the data. data.frame objects have a print function different from that of vector objects. The specific way that R accomplishes the task of printing, for example, may be different for different object-types. However, because all objects of the same type have similar structure, a print function that is attached to a type will work as intended on any object of that type.

In short, when you type x <- 0 : 19 you actually create an object and enter the data that the object stores. Other actions that are common to and necessary for such objects are known to R through the object-type.

Plots

Back to our session. Let us create a vector with some random numbers and plot it. We create a vector y from x. To each element of x, we add a random value between 10 and 20. Here is how:

```
> y <- x + runif(20, min = 10, max = 20)
```

Here we assign each value of x + (a random value) to the vector y. When you assign an object to a named object that does not exist (y in our example), R creates the object automatically. The function runif(), standing for "random uniform," produces random numbers. The call to it includes three arguments. The first argument, 20, tells R to produce 20 random numbers. The second and third, min = 10 and max = 20, ask that each number between the values of 10 and 20 have the same probability of occurring. The addition of the random numbers to x is done element by element. Thus, the statement above produces

```
y[1] <- x[1] + the first random number
y[2] <- x[2] + the second random number
...
y[20] <- x[20] + the 20th random number
```

To plot the values of y against x, type

```
> plot(x, y)
```

In response, you should get something like Figure 1.1.

Statistics

R is a statistical computing environment. It contains a large number of functions that apply statistical analyses to data. Two familiar statistics are the mean and variance of a set (vector) of numbers:

```
> mean(x)
[1] 9.5
> var(x)
[1] 35
```

8   Basic R

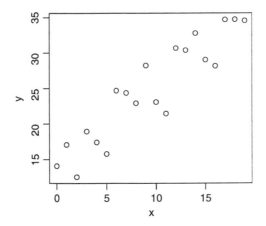

Figure 1.1   plot(x,y).

The function mean() takes—in this case—a single argument, a vector of the data (x). It responds with the mean of the data. The function var() responds with the variance.

1.1.2 Editing statements

You can recall and edit statements that you have already typed. R saves the statements in a session in a buffer (think of a buffer as a file in memory). This buffer is called history. You can use the ↑ and the ↓ keys to recall previous statements or move down from a previous statement. Use the ← and → arrows to move within a statement in the active line. The active line is the last line in the session window. It is the line that is executed when you hit Enter. The Delete or Backspace keys can be used to delete characters on the command line. There are other commands you can use. See the Help | Console menu.

1.1.3 The functions help(), help.search() and example()

Suppose that you need to generate uniform random numbers. You remember that there is such a function, named runif(), but you do not exactly remember how to call it. To see the syntax for the call and other information related to runif(), call the function help() with the argument set to runif(), like this:

> help(runif)

(or help('runif')). In response, R displays a window that contains the following (some output was omitted and line numbers added):

```
1  Uniform      package:base     R Documentation
2
3  The Uniform Distribution
4
5  Description:
6        These functions provide information about ,,,
7
```

```
 8  Usage:
 9          ...
10          runif(n, min=0, max=1)
11
12  Arguments:
13          ...
14          n: number of observations. ....
15    min,max: lower and upper limits of the distribution.
16      ...
17
18  Details:
19          If 'min' or 'max' are not specified they assume ...
20          '0' and '1' respectively.
21          ...
22
23  References:
24          ....
25
26  See Also:
27          '.Random.seed' about random number generation, ....
28
29  Examples:
30          u <- runif(20)
31          ...
```

You can access help for all of the functions in R the same way—type help(*function-name*), where *function-name* is a name of a function you need help with. If you are sure that help is available for the requested topic and you get no response, enclose the topic with quotes. Functions' help is standard, so let us go through the help text for runif() in detail. In line 1, the header of the help window for runif() declares that the function resides in a package, named base—we will talk about packages soon. The Description section starts in line 5. Since this help window provides help for all of the functions that relate to the so-called uniform distribution, the description mentions all of them.

The Usage section begins on line 8. It explains how to call this function. In line 10, it says that runif() is called with the arguments n, min and max. Because n represents the number of random numbers you wish to generate and because it is not written as argument-name = *argument-value*, you should realize that this is a required argument. That is, if you omit it, R will respond with an error message. On line 12 begins a section about the Arguments. As it explains, n is the number of observations. min and max are the limits between which the random numbers will be generated. Note that in the call (line 10), they are specified as min = 0 and max = 1. This means that you do not have to call these arguments explicitly. If you do not, then the default values will be 0 and 1. In other words, runif(n) will produce n random numbers between 0 and 1. Because it is the uniform distribution, each of the numbers between 0 and 1 is equally likely to occur. We say that n is an *unnamed argument* while min and max are *named* arguments. Usually, named arguments have default values and unnamed arguments are required.

The `Details` section explains other issues related to the functions on this help page. There is usually a `Reference` section where more details about the functions may be found. The `See Also` section names relevant functions and finally there is an `Examples` section. All functions are documented according to this template. The documentation is often terse and it takes time to get used to.

Suppose that you forgot the exact function name. Or, you may wish to look for a concept or a topic. Then use the function `help.search()`. It takes a single argument. The argument is a string and strings in R are delineated by single or double quotes. So you may look for a topic such as

```
> help.search('random')
```

This will bring up a window with a list such as this (line numbers were added and the output was edited):

```
1  Help files with alias or concept or title matching 'random'
2  using fuzzy matching:
3
4  REAR(agce)           Fit a autoregressive model with ...
5  r2dtable(base)       Random 2-way Tables with ...
6  Random.user(base)    User-supplied Random Number ...
7  ...
```

Line 4 is the beginning of a list of topics relevant to "random." Each topic begins with the name of a function and the package it belongs to, followed by a brief explanation. For example, line 6 says that the function `Random.user()` resides in the package `base`. Through it, the user may supply his or her own random number generator.

Once you identify a function of interest, you may need to study its documentation. As we saw, most of the functions documented in R have an `Examples` section. The examples illustrate various applications of the documented function. You can cut and paste the code from the example to the console and study the output. Alternatively, you can run all of the examples with the function `example()`, like this:

```
> example(plot)
```

The statement above runs all the examples that accompany the documentation of the function `plot()`. This is a good way to quickly learn what a function does and perhaps even modify an example's code to fit your needs. Copy as much code as you can; do not try to reinvent the wheel.

1.1.4 Expressions

R is case sensitive. This means that, for example, `x` is different from `X`. Here `x` or `X` refer to object names. Object names can contain digits, characters and a period. You may be able to include other characters in object names, but we will not. As a rule, object names start with a letter. Endow ephemeral objects—those you use within, but not between, sessions—with short names and save on typing. Endow persistent object—those you wish to use in future sessions—with meaningful names.

In object names, you can separate words with a period or underscore. In some cases, you can names objects with spaces embedded in the name. To accomplish this, you will have to wrap the object name in quotes. Here are examples of *correct* object names:

a A hello.dolly the.fat.in.the.cat hello.dOlly Bush_gore

Note that `hello.dolly` and `hello.dOlly` are two different object names because R is case sensitive.

You type expressions on the current (usually last) line in the console. For example, say x and y are two vectors that contain the same number of elements and the elements are numerical. Then a call to

```
> plot(x, y)
```

is an expression. If the expression is complete, R will execute it when you hit **Enter** and we get a plot of y vs. x. You can group expressions with braces and separate them with semicolons. All statements (separated by semicolons) on a single line are grouped. For example

```
> x <- seq(1 : 10) ; x # two expressions in one line
[1]  1  2  3  4  5  6  7  8  9 10
```

creates the sequence and prints it.

1.1.5 Comments, line continuation and Esc

You can add comments almost anywhere. We will usually add them to the end of expressions. A comment begins with the hash sign # and terminates at the end of the line. For example

```
# different from seq(1, 10, by = 2); try it
> (x <- seq(1, 10, length = 5))
[1]  1.00  3.25  5.50  7.75 10.00
> # also different from seq(1 : 10)
```

If an expression is incomplete, R will allow you to complete it on the next line once you hit **Enter**. To distinguish between a line and a continuation line, R responds with a +, instead of with a > on the next line. Here is an example (the comments explain what is going on):

```
> x <- seq(1 : 10 # no ')' at the end
+ ) #now '(' is matched by ')'
> x
[1]  1  2  3  4  5  6  7  8  9 10
```

If you wish to exit a continuation line, hit the **Esc** key. This key also aborts executions in R.

1.1.6 `source()`, `sink()` and `history()`

You can store expressions in a text file, also called a script file and execute all of them at once by referring to this file with the function `source()`. For example, suppose we want to plot the normal (popularly known as the bell) curve. We need to create a vector that holds the values of x and then calculate the values of y based on the values of x using the formula:

$$y(x) = \frac{1}{\sigma\sqrt{2\pi}} e^{-(x-\mu)^2/2\sigma^2} . \tag{1.1}$$

Here σ and μ (the Greek letters sigma and mu) are fixed, so we choose $\sigma = 1$ and $\mu = 0$. With these values, $y(x)$ is known as the standard normal probability density. Suppose we wish to create 101 values of x between -3 and 3. Then for each value of x, we calculate a value for y. Then we plot y vs. x. To accomplish this, open a new text file using Notepad (or any other text editor you use) and save it as normal.R (you can name it anything). Note the directory in which you saved the file. Next, type the following (without the line numbers) in the text editor:

```
x <- seq(-3, 3, length = 101)
y <- dnorm(x)           # assign standard normal values to y
plot(x, y, type = 'l')  # 'l' stands for line
```

and save it. Click on the File | Change dir... menu in the console and change to the directory where you saved normal.R. Next, type

```
> source('normal.R')
```

or click on File | Source R code.... Either way, you should obtain Figure 1.2. Note the statement in line 2. We use the 101 values of x to generate 101 values of y according to (1.1). The function dnorm()—for density normal—does just that. The function source() treats the content of the argument—a string that represents a file name—as a sequence of commands. It will read one line at a time and execute it. Because the argument to source() must be a constant string of characters representing a file name, you must delineate the string with single or double quotes.

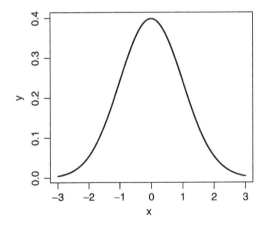

Figure 1.2 The standard normal (bell) curve.

The complement of source() is the function sink(). Occasionally you may find it useful to divert output from the console to a file. Let us create a vector of 100 elements, all initialized to zero and write the vector to a file. Type the following (without the line numbers) in your console:

```
1  > x <- vector(mode = 'numeric', length = 100) #create the vector
2  > sink('x.txt')                #open a text file named x.txt
3  > x                            #write the data to the file
4  > sink()                       #close the file
```

Go to the directory in which R is working now and open the file x.txt. You should see the output exactly as if you did not redirect the output away from the console.

In line 1 of the code, the function vector() creates a vector of length 100.[2] If you do not specify mode = 'numeric', vector() will create a vector of 100 elements all set to FALSE. The latter are logical elements that have only two values, TRUE or FALSE. Note that TRUE and FALSE are not strings (you do not enclose them in quotes and when R prints them, it does not enclose them in quotes). They are special values that are simply represented by the tokens TRUE or FALSE. You will later see that vector() and logical variables are useful.

history() is another useful housekeeping function. Now that we have worked a little in the current session, type

```
> history(50)
```

In response, a new text window will pop up. It will include your last 50 statements (not output). The argument to history is any number of lines you wish to see. You can then copy part or all of the text to the clipboard and then paste to a script file. This is a good way to build scripts methodically. You try the various statements in the console, make sure they work and then copy them from a history window to your script. You can then edit them. It is also a good way to document your work. You will often spend much time looking for a particular function, or an idiom that you developed only to forget it later. Document your work.

1.2 Modes

In R, a simple unit of data occupies the same amount of memory. The type of these units is called mode. The modes in R are:

LOGICAL   Has only two values, TRUE or FALSE.
character   A string of characters.
numeric   Numbers. Can be integers or floating point numbers, called double.
complex   Complex numbers. We will not use this type.
raw   A stream of bytes. We will not use this type.

You can test the mode of an object like this:

```
> x <- integer() ; mode(x)
[1] "numeric"
> mode(x <- double())
[1] "numeric"
> mode(x <- TRUE)
```

---

[2] You can use x <- numeric(100) instead of a call to vector().

```
[1] "logical"
> mode(x <- 'a')
[1] "character"
```

1.3 Vectors

As we saw, R works on objects. Objects can be of a variety of types. One of the simplest objects we will use is of type vector. There are other, more sophisticated types of objects such as matrix, array, data frame and list. We will discuss these in due course. To work with data effectively, you should be proficient in manipulating objects in R. You need to know how to construct, split, merge and subset these objects. In R, a vector object consists of an ordered collection of elements of the *same* mode.

The length of a vector is the number of its elements. As we said, all elements of a vector must be of the same mode. There is one exception to this rule. We can mix with any mode the special token NA, which stands for Not Available. It will be worth your while to remember that NA is *not* a value! Therefore, to work with it, you must use specific functions. We will discuss those later. We call a vector of dimension zero the empty vector. Albeit devoid of elements, empty vectors have a mode. This gives R an idea of how much additional memory to allocate to a vector when it is needed.

1.3.1 Creating vectors

Vectors can be constructed in several ways. The simplest is to assign a vector to a new symbol:

```
> v <- 1 : 10       # assign the vector of elements 1...10 to v
> v <- c(1, 5, 3) # c() concatenates its elements
```

You can create a vector with a call to

```
> v <- vector()   # vector of length 0 of logical mode
> (v <- vector(length = 10))
 [1] FALSE FALSE FALSE FALSE FALSE FALSE FALSE FALSE FALSE
[10] FALSE
```

From the above we conclude that vector() produces a vector of mode logical by default. If the vector has a length, all elements are set to FALSE. To create a vector of a specific mode, simply name the mode as a function, with or without length:

```
> v <- integer()
> (v <- double(10))
 [1] 0 0 0 0 0 0 0 0 0 0
> (v <- character(10))
 [1] "" "" "" "" "" "" "" "" "" ""
```

You can also create vectors by assigning vectors to them. For example the function c() (for concatenate) returns a vector. When the result of c() is assigned to an object, the object becomes a vector:

```
> v <- c(0, -10, 1000)
[1] 0 -10 1000
```

Because a vector must include elements of a simple mode, if you concatenate vectors with c(), R will force all modes to the simplest mode. Here is an example:

```
1  > (x <- c('Alice', 'in', 'lalaland'))
2  [1] "Alice"   "in"      "lalaland"
3  > (y <- c(1, 2, 3))
4  [1] 1 2 3
5  > (z <- c(x,y))
6  [1] "Alice"   "in"      "lalaland" "1"       "2"
7  [6] "3"
```

Study this code snippet. It will pay valuable dividends later. In line 1 we assign three strings to x. Because the function c() constructs the vector and returns it, the assignment in line 1 creates the vector x. Its type is character and so is its mode. In line 3, we construct the vector y with the elements 1, 2 and 3. The mode(y) is numeric and the typeof(y) is double. In line 5 we concatenate (with c()) the vectors x and y. Because all elements of a vector must be of the same mode, the mode of z is reduced to character. Thus, when we print z (lines 6 and 7), the numbers are turned into their character representation. They therefore are enclosed with quotes. This may be confusing, but once you realize that R strives to be coherent, it is no longer so.

1.3.2 Useful vector functions

Working with data requires manipulating vectors frequently. R provides functions that allow such manipulations.

length()

The length of a vector is obtained with length():

```
> (t.f <- c(TRUE, TRUE, FALSE, TRUE)) ; length(t.f)
[1] TRUE  TRUE FALSE  TRUE
[1] 4
```

sum()

Here is an example where we need the length() of a vector and the sum() of its elements:

```
> (a <- 1 : 11)
 [1]  1  2  3  4  5  6  7  8  9 10 11
> (average <- sum(a) / length(a))
[1] 6
```

As you might expect, mean() does this and then some (see help(mean)).

1.3.3 Vector arithmetic

The binary (in the sense that they need two objects to operate on) arithmetic operations addition, subtraction, multiplication and division, +, -, * and /, respectively, can be applied to vectors. In such cases, they are applied element-wise:

```
1  > (x <- 1.2 : 6.4)
2  [1] 1.2 2.2 3.2 4.2 5.2 6.2
3  > x * 2
```

```
[1]  2.4  4.4  6.4  8.4 10.4 12.4
> x / 2
[1] 0.6 1.1 1.6 2.1 2.6 3.1
> x - 1
[1] 0.2 1.2 2.2 3.2 4.2 5.2
```

In line 1 we create a sequence from 1.2 to 6.4 incremented by 1 and print it. In line 2 we multiply x by 2. This results in multiplying each element of x by 2. Similarly (lines 5 and 7), division and subtraction are executed element-wise.

When two vectors are of the same length and of the same mode, then the result of x / y is a vector whose first element is x[1] / y[1], the second is x[2] / y[2] and so on:

```
> (x <- seq(2, 10, by = 2))
[1]  2  4  6  8 10
> (y <- 1 : 5)
[1] 1 2 3 4 5
> x / y
[1] 2 2 2 2 2
```

Here, corresponding elements of x are divided by corresponding elements of y. This might seem a bit inconsistent at first. On the one hand, when dividing a vector by a scalar we obtain each element of the vector divided by the scalar. On the other, when dividing one vector by another (of the same length), we obtain element-by-element division. The results of the operations of dividing a vector by a scalar or by a vector are actually consistent once you realize the underlying rules.

First, recall that a single value is represented as a vector of length 1. Second, when the length of x > the length of y and we use binary arithmetic, R will cycle through the elements of y until its length equals to that of x. If x's length is not an integer multiple of y, then R will cycle through the values of y until the length of x is met, but R will complain about it.

In light of this rule, any vector is an integer multiple of the length of a vector of length 1. For example, a vector with 10 elements is 10 times as long as a vector of length 1. If you use a binary arithmetic operation on a vector of length 10 with a vector of length 3, you will get a warning message. However, if one vector is of length that is an integer multiple of the other, R will not complain. For example, if you divide a vector of length 10 by a vector of length 5, then the vector of length 5 will be duplicated once and the division will proceed, element by element. The same rule applies when the length of x < the length of y: x's length is extended (by cycling through its elements) to the length of y and the arithmetic operation is done element-by-element. Here are some examples.

First, to avoid printing too many digits, we set

```
> options(digits = 3)
```

This will cause R to print no more than 3 decimal digits. Next, we create x with length 10 and y with length 3 and attempt to obtain x / y:

```
> (x <- 1 : 10) ; (y <- 1 : 3) ; x / y
 [1]  1  2  3  4  5  6  7  8  9 10
 [1] 1 2 3
 [1]  1.0  1.0  1.0  4.0  2.5  2.0  7.0  4.0  3.0 10.0
Warning message:
longer object length
         is not a multiple of shorter object length in: x/y
```

Note that because x is not an integer multiple of y, R complains. Here is what happens:

```
    x y  x/y
1   1 1  1.0
2   2 2  1.0
3   3 3  1.0
4   4 1  4.0
5   5 2  2.5
6   6 3  2.0
7   7 1  7.0
8   8 2  4.0
9   9 3  3.0
10 10 1 10.0
```

As expected, y cycles through its values until it has 10 elements (the length of x) and we get an element by element division. The same rule applies with functions that implement vector arithmetic. Consider, for example, the square root function `sqrt()`. Here R follows the rule of operating on one element at a time:

```
> (x <- 1 : 5) ; sqrt(x)
[1] 1 2 3 4 5
[1] 1.00 1.41 1.73 2.00 2.24
```

1.3.4 Character vectors

Data, reports and figures require frequent manipulation of characters. R has a rich set of functions to deal with character manipulations. Here we mention just a few. More will be introduced as the need arises. Character strings are delineated by double or single quotes. If you need to quote characters in a string, switch between double and single quotes, or preface the quote by the escape character \. Here is an example:

```
> (s <- c("Florida; a politician's",'nightmare'))
[1] "Florida; a politician's"
[2] "nightmare"
```

The vector s has two elements. Because we need a single quote in the first element (after the word `politician`), we add '. To create a single string from s[1] and s[2], we `paste()` them:

```
> paste(s[1], s[2])
[1] "Florida; a politician's nightmare"
```

By default, `paste()` separates its arguments with a space. If you want a different character for spacing elements of characters, use the argument `sep`:

```
> paste(s[1], s[2], sep = '-')
[1] "Florida; a politician's-nightmare"
```

The examples above demonstrate how to include a single quote in a string and how to `paste()` character strings with different separators (see `help(paste)` for further information).

1.3.5 Subsets and index vectors

One of the most frequent manipulations of vectors (and other classes of objects that hold data) is extracting subsets. To extract a subset from a vector, specify the indices of the elements you wish to extract or exclude. To demonstrate, let us create

```
> x <- 15 : 30
```

Here x is a vector with elements 15, 16, ..., 30. To create a new vector of the second and fifth elements of x, we do

```
> c(x[2], x[5])
[1] 16 19
```

A more succinct way to extract the second and fifth elements is to execute

```
> x[c(2, 5)]
[1] 16 19
```

Here `c(2, 5)` is an index vector. The index vector specifies the elements to be returned. Index vectors must resolve to one of 4 modes: logical, positive integers, negative integers or strings. Examples are the best way to introduce the subject.

Index vector of logical values and missing data

Here is a typical case. You collect numerical data with some cases missing. You might end up with a vector of data like this:

```
> (x <- c(10, 20, NA, 4, NA, 2))
[1] 10 20 NA  4 NA  2
```

You wish to compute the mean. So you try:

```
> sum(x) / length(x)
[1] NA
```

Because operations on `NA` result in `NA`, the result is `NA`. So we need to find a way to extract the values that are *not* `NA` from the vector. Here is how. x above has six elements, the third and the fifth are `NA`. The following statement identifies the `NA` elements in i:

```
> (i <- is.na(x))
[1] FALSE FALSE  TRUE FALSE  TRUE FALSE
```

The function `is.na(x)` examines each element of the vector x. If it is `NA`, it assigns the element the value `TRUE`. Otherwise, it assigns it the value `FALSE`. Now we can use i as an *index vector* to extract the *not* `NA` values from x. Recall that i is a vector.

If an element of x is a value, then the corresponding element of the vector i is FALSE.
If an element of i has no value, then the corresponding element of i is TRUE. So i
and *not* i, denoted by !i give:
```
> i ; !i
[1] FALSE FALSE  TRUE FALSE  TRUE FALSE
[1]  TRUE  TRUE FALSE  TRUE FALSE  TRUE
```
Therefore, to extract the values of x that are not NA, we do
```
> (y <- x[!i])
[1] 10 20  4  2
```
Now to calculate the mean of x, a vector with some elements containing missing values, we do:
```
> n <- length(x[!i]) ; new.x <- x[!i] ; sum(new.x) / n
[1] 9
```
The same result can be achieved with
```
> mean(x, na.rm = TRUE)
[1] 9
```
na.rm is a named argument. Its default value is FALSE. If you set it to TRUE, mean()
will compute the mean of its unnamed argument (x in our example) even though the
latter contains NA.

Index vector of positive integers

Here is another way to exclude NA data:
```
> (x <- c(160, NA, 175, NA, 180))
[1] 160  NA 175  NA 180
> (no.na <- c(1, 3, 5))
[1] 1 3 5
> x[no.na]
[1] 160 175 180
```
Again, we use no.na as an index vector. Another example: we wish to extract elements
20 to 30 from a vector x of length 100:
```
> x <- runif(100) # 100 random numbers between 0 and 1
> round(x[20 : 30], 3) # print rounded elements 20 to 30
 [1] 0.249 0.867 0.946 0.593 0.088 0.818 0.765 0.586 0.454
[10] 0.922 0.738
```
First we create a vector of 100 elements, each a random number between 0 and 1. To
interpret the second statement, we go from the inner parentheses out: 20 : 30 creates
an index vector of positive integers 20, 21, ..., 30. Then x[20 : 30] extracts the
desired elements. We wish to see only the first 3 significant digits of the elements. So
we call round() with two arguments—the data and the number of significant digits.
Here is a variation on the theme:
```
> (i <- c(20 : 25, 28, 35 : 40)) # 20 to 25, 28 and 35 to 40
 [1] 20 21 22 23 24 25 28 35 36 37 38 39 40
> round(x[i], 3)
 [1] 0.249 0.867 0.946 0.593 0.088 0.818 0.454 0.675 0.834
[10] 0.131 0.281 0.636 0.429
```

Index vector of negative integers

The effect of index vectors of negative integers is the mirror of positive. To extract all elements from x except elements 20 to 30, we write x[-(20 : 30)]. Here we must put 20 : 30 in parenthesis. Otherwise R thinks that we want to extract elements -20 to 30. Negative indices are not allowed.

Index vector of strings

We measure the weight (x) and height (y) of 5 persons:

```
> x <- c(160, NA, 175, NA, 180)
> y <- c(50, 65, 65, 80, 70)
```

Next, we associate names with the data:

```
> names(x) <- c("A. Smith", "B. Smith",
+    "C. Smith", "D. Smith", "E. Smith")
```

The function names() names the elements of x. To see the data, we column-bind it with the function cbind():

```
> cbind(x, y)
           x   y
A. Smith 160  50
B. Smith  NA  65
C. Smith 175  65
D. Smith  NA  80
E. Smith 180  70
```

Now that the indices are named, we can extract elements by their names:

```
> x[c('B. Smith', 'D. Smith')]
B. Smith D. Smith
      NA       NA
```

Observe this:

```
> y[c('A. Smith', 'D. Smith')]
[1] NA NA
```

Because y has no elements named A. Smith and D. Smith, R assigns NA to such elements.

1.4 Arithmetic operators and special values

R includes the usual arithmetic operators and logical operators. It also has a set of symbols for special values, or no values at all.

1.4.1 Arithmetic operators

Arithmetic operators consist of $+, -, *, /$ and the power operator ^. All of these operate on vectors, element by element:

```
> x <- 1 : 3 ; x^2
[1] 1 4 9
```

1.4.2 Logical operators

Logical operators include "and" and "or", denoted by & and |. We compare values with the logical operators >, >=, <, <=, == and != standing for greater than, greater equal than, less than, less equal than, equal and not equal. ! is the negation operator. Upon evaluation, logical operators return the logical values TRUE or FALSE. If the operation cannot be accomplished, then NA is returned. With vectors, logical operators work as usual—one element at a time. Here are some examples:

```
> 5 == 4 & 5 == 5
[1] FALSE
> 5 != 4 & 5 == 5
[1] TRUE
> 5 == 4 | 5 == 5
[1] TRUE
> 5 != 4 | 5 == 5
[1] TRUE
> 5 > 4 ;   5 < 4 ;   5 == 4 ;   5 != 4
[1] TRUE
[1] FALSE
[1] FALSE
[1] TRUE
```

Like any other operation, if you operate on vectors, the values returned are element by element comparisons among the vectors. The rules of extending vectors to equal length still stand. Thus,

```
> x <- c(4, 5) ; y <- c(5, 5)
> x > y ; x < y ; x == y ; x != y
[1] FALSE FALSE
[1]  TRUE FALSE
[1] FALSE  TRUE
[1]  TRUE FALSE
```

Here is an example that explains what happens when you compare two vectors of different lengths:

```
> x <- c(4,5) ; y <- 4
> cbind(x, y)
     x y
[1,] 4 4
[2,] 5 4
> y
[1] 4
> x == y
[1]  TRUE FALSE
> x < y
[1] FALSE FALSE
> x > y
[1] FALSE  TRUE
```

In line 1 we create two vectors, with lengths of 2 and 1. We column bind them. So R extends y with one element. When we implement the logical operations in lines 8, 10 and 12, R compares the vector 5, 4 to 4, 4. To make sure that you get what you want, when comparing vectors, always make sure that they are of *equal* length.

1.4.3 Special values

Because R's orientation is toward data and statistical analysis, there are features to deal with logical values, missing values and results of computations that at first sight do not make sense. Sooner or later you will face these in your data and analysis. You need to know how to distinguish among these values and test for their existence. Here are the important ones.

Logical values

Logical values may be represented by the tokens TRUE or FALSE. You can specify them as T or F. However, you should avoid the shorthand notation. Here is an example why. We wish to construct a vector with three logical elements, all set to TRUE. So we do this

```
> T <- 5
...
> (x <- c(T, T, T))
[1] 5 5 5
```

Some time earlier during the session, we happened to assign 5 to T. Then, forgetting this fact, we assign c(T, T, T) to x. The result is not what we expect. Because TRUE and FALSE are reserved words, R will not permit the assignment TRUE <- 5. The tokens TRUE and FALSE are represented internally as 1 or 0. Thus,

```
> TRUE == 0 ; TRUE == 1
[1] FALSE
[1] TRUE
> TRUE == -0.1 ; FALSE == 0
[1] FALSE
[1] TRUE
```

NA

This token stands for "Not Available" It is used to designate missing data. In the next example, we create a vector x with five elements, the first of which is missing. To test for NA, we use the function is.na(). This function returns TRUE if an element is NA and FALSE otherwise.

```
> (x <- c(NA, 2 : 5))
[1] NA  2  3  4  5
> (test <- is.na(x))
[1]  TRUE FALSE FALSE FALSE FALSE
```

It is important to realize that is.na() returns FALSE if the element tested for is *not* NA. Why? Because there are other values that are not numbers. They may result from computations that make no sense, but they are not NA.

NaN and Inf

These designate "Not a Number" and infinity, respectively. Division by zero does not result in a number; it therefore returns NaN. You may wish to assign Inf to a vector (for example when you wish any vector to be smaller than Inf in a comparison). In both cases, these are *not* NA; they are NaN and Inf, respectively. Furthermore, Inf is a number (you can verify this with the function is.numeric()); NaN is not. To distinguish among these possibilities, use the function is.nan().

Distinguishing among NA, NaN and Inf

Distinguishing among these in data can be confusing. Unless interested, you may skip this topic. Consider the following vector:

```
> (x <- c(NA, 0 / 0, Inf - Inf, Inf, 5))
[1]  NA NaN NaN Inf   5
```

Here 0/0 is undefined and therefore not a number. So is Inf-Inf. Albeit not a real number, Inf is part of the set of numbers called extended real numbers. We need to distinguish among vector elements that are a number, NA, NaN and Inf in x. First, let us test for NA:

```
> is.na(x)
[1]  TRUE  TRUE  TRUE FALSE FALSE
```

As you can see, NA and NaN are undefined and therefore the test returns TRUE for both. Now let us test x with is.nan():

```
> is.nan(x)
[1] FALSE  TRUE  TRUE FALSE FALSE
```

The first element of x is NA. It is distinguishable from NaN and we get FALSE for it. Finally, because Inf is a value, we test it as usual with the logical operator ==. This operator returns TRUE if the left equals the right hand side:

```
> x == Inf
[1]    NA    NA    NA  TRUE FALSE
```

Note what happens. Because NA and NaN are undefined, comparing them to a defined value (Inf), we get NA. We therefore expect to get the similar result of the test

```
> x == 5
[1]    NA    NA    NA FALSE  TRUE
```

The next table summarizes these results.

	x	is.na(x)	is.nan(x)	Inf == x	5 == x
1	NA	TRUE	FALSE	NA	NA
2	NaN	TRUE	TRUE	NA	NA
3	NaN	TRUE	TRUE	NA	NA
4	Inf	FALSE	FALSE	TRUE	FALSE
5	5	FALSE	FALSE	FALSE	TRUE

1.5 Objects

We discussed objects on numerous occasions before. That was necessary because we introduced other topics that required the notion of objects (learning R cannot be linear). Here we discuss these and additional object-related topics in more detail. Understanding objects is key to working with R effectively.

In the next few statements, we assign values to x. We also explore the type of object created by the assignments:

```
> x <- 2                    # x is a vector of length 1
> x <- vector()             # x is a vector of 0 length
> x <- matrix()             # x is a matrix of 1 column, 1 row
> x <- 'Hello Dolly'        # x is a vector containing 1 string
> x <- c('Hello', 'Dolly')  # x is a vector with 2 strings
> x <- function(){}         # x is a function that does nothing
```

As we have seen, vectors are atomic objects—all of their elements must be of the same mode. In most cases, we work with vectors of modes logical, numeric or character. Most other types of objects in R are more complex than vectors. They may consist of collections of vectors, matrices, data frames and functions. When an object is created (for example with the assignment <-), R must allocate memory for the object. The amount of memory allocated depends on the mode of the object. Beside their mode and length, objects have other properties which we will learn about as we progress.

1.5.1 Orientation

The following is a general exposition of the idea of objects. This section is not related to R directly. Rather, it is conceptual. It is intended to demystify some of the baffling aspects of R.

Usually, computer software that deals with data (e.g. Excel, Oracle, other database management systems, programming languages) distinguish between what we call data types. For example, in Excel, you can format a column so that it is known to contain numbers, or text, or dates. In the programming language C, you distinguish between data that represent integers, floating (decimal point) numbers, single characters, collections of characters (called strings) and so on. "Why do we need to make these distinctions?" you might ask. The short answer is because of efficiency and error checking. If the software knows the intended use of data, it will allocate as much memory as is needed for it and no more. For example, the amount of memory that is needed to represent an integer is less than the amount needed to represent a string that contains 100 characters. So if you tell the software that x is intended to represent integers and y strings, computations will be more efficient than otherwise. Other reasons for specifying data types are consistency, ability to check for errors, pointer arithmetic and so on. For example, if the software knows that x and y represent numbers, then it will take special actions if you ask it to compute x/y when y = 0.

This leads to the definition of simple data types. These are types that cannot be broken into simpler data types.[3] An integer, a decimal number and a

---
[3]Unless you are ready to deal with bits.

character are examples. From these, more elaborate data types can be constructed. For example, a string is a collection of characters and a collection of integers is a vector.

This gives rise to the idea of structures. Instead of defining simple data types, such as integers, floating point numbers and characters, we can define data structures. For example, we can define a structure named *vector* and specify that such a structure contains a set of numbers. Then we can tell the software that x is a vector and assign data to it with a statement like x <- c(1, 2, 3). Better yet, we can define a structure named *matrix*, for example, that contains two or more *vector*s of the same data type and same length. We can then tell the software that y is a matrix and write

```
> (y <- cbind(letters[1 : 4], LETTERS[1 : 4]))
     [,1] [,2]
[1,] "a"  "A"
[2,] "b"  "B"
[3,] "c"  "C"
[4,] "d"  "D"
```

(cbind() is a function that binds vectors as columns). Structures do not need to be atomic. For example, a structure may contain a numeric and a character vector. In short, structures are user-defined data types. But why should we stop with structures? After all, we often apply similar actions to similar structures. Consider, for example, printing. All matrices are printed in the same way: numbers arranged in columns and rows. The only difference in printing matrix objects is their number of rows and columns. This leads to the idea of object types (also called classes). An object type is a definition of a collection of structures (data) and actions (functions) that we may apply to these structures.

Viewing a vector as a type, we can define it as a collection of elements (data) and a collection of actions (functions), such as printing and multiplying one vector by another.

An object type is a specification. As such, it is an abstract definition. It simply says what kinds of data and actions an object that is declared to be of that type can have. An object is a realization of a type. When we say that x is an object of type vector, we are creating a concrete object of type vector. By concrete we mean that R actually assigns memory to the object and we can assign data to it.

Suppose that we define a function print() for the object type *vector*. We also define objects of type *matrix* and a print() for it. Next, we say that x is an object of type *vector* and y is an object of type *matrix*. When we say print(x), the software knows that we are calling print() for vectors by context; that is, it knows we are asking for print() for vectors because x is of type *vector*. If we type print(y) then print() for matrices is invoked.

As you may guess, the whole approach can become much more syntactically involved, but we will not pursue it further. Instead, let us get back to R and see how all of this applies. Say we define a vector to be a collection of numbers:

```
> x <- 1 : 10
```

and a matrix

```
> y <- cbind(letters[1 : 4], LETTERS[1 : 4])
```

We can print x and y by simply saying

```
> x
 [1]  1  2  3  4  5  6  7  8  9 10
```

and

```
> y
     [,1] [,2]
[1,] "a"  "A"
[2,] "b"  "B"
[3,] "c"  "C"
[4,] "d"  "D"
```

By the assignments above, R knows that we wish to create a vector x and a matrix y. When we say y, R knows that we wish to print y and it invokes the matrix `print()` function because y is an object of type *matrix*. To convince yourself that this in fact is the case, try this:

```
> x <- 1 : 10 ; x
 [1]  1  2  3  4  5  6  7  8  9 10
> print(x)
 [1]  1  2  3  4  5  6  7  8  9 10
```

Observe that x and `print(x)` produce identical results; in other words, the statement x and the function-call `print(x)` are one and the same.

Of course we can have object types that are more complicated than the atomic types vector and matrix. Both are atomic because they must contain a single mode—strings of character only, numbers, or logical values. Lists and data frames are complex objects. Lists, for example, may consist of a collection of objects of any type (mode), including lists.

This, then, is the story of objects—behind every object lurks a type.

1.5.2 Object attributes

Object attributes can be examined and set with various functions: `mode()`, `attributes()`, `attr()`, `typeof()`, `dim()` (for dimension) and `dimnames()` (for dimension names). Instead of defining object attributes, we shall discuss these functions.

Here we discuss `mode()`, `is.x()` and `as.x()` where x is the object type. The other functions to set and explore object attributes will be discussed when needed.

The functions `mode()`, `is.object()` and `as.object()`

The mode attribute of an object is obtained with the function `mode()`:

```
> x <- 1 : 5 ; mode(x)
[1] "numeric"
> x <- c('a', 'b', 'c') ; mode(x)
[1] "character"
> x <- c(TRUE, FALSE) ; mode(x)
[1] "logical"
```

Here are the modes we will deal with:

```
> mode(mean) ; mode(1) ; mode(c(TRUE, FALSE))
[1] "function"
[1] "numeric"
[1] "logical"
> mode(letters)
[1] "character"
```

Any of these can be created, tested and set (coerced) with the functions "mode name", "is" and "as". Setting a mode from one to another is called coercion. Beware of coercion. If the coercion is not well defined (for example, attempting to change the mode of a vector from character to numeric), R will go through the coercion but will set all the elements to NA. Keep in mind that objects of types other than vector also have functions mode(), is.x() and as.x(), where x stands for the object type. For example, x may be matrix, list or data.frame. Then, mode(), is.list() and as.list() parallel mode(), is.vector() and as.vector(). All of these functions take the object name as an argument.

We follow with some examples. Generalizing these examples to R's rules of coercion and naming is immediate. First, we create vectors of various modes:

```
> logical(3)           # a vector of 3 logical elements
[1] FALSE FALSE FALSE
> (x <- numeric(3))    # a vector of 3 numeric values
[1] 0 0 0
> (x <- integer(3))    # a vector of 3 integers
[1] 0 0 0
> (x <- character(3))  # a vector of 3 empty strings
[1] "" "" ""
```

Here we test a vector of mode logical for its mode:

```
> x <- c(TRUE, FALSE, TRUE, FALSE)
> # test for mode:
> is.logical(x) ; is.numeric(x) ; is.integer(x)
[1] TRUE
[1] FALSE
[1] FALSE
> is.character(x)
[1] FALSE
```

Here we coerce a logical vector to numeric and character modes:

```
> as.numeric(x) ; as.character(x)
[1] 1 0 1 0
[1] "TRUE"  "FALSE" "TRUE"  "FALSE"
```

Here we test a numeric vector for its mode:

```
> (x <- runif(3, 0, 20))
[1]  0.97567  0.14065 12.31121
> is.numeric(x) ; is.integer(x) ; is.character(x)
```

```
[1] TRUE
[1] FALSE
[1] FALSE
> is.logical(x)
[1] FALSE
```

Here we coerce a numeric vector (x above) to integer, character and logical modes:

```
> as.integer(x) ; as.character(x) ; as.logical(x)
[1]  0  0 12
[1] "0.975672248750925" "0.140645201317966" "12.3112069442868"
[1] TRUE TRUE TRUE
> # integer is numeric but numeric is not an integer
> is.numeric(as.integer(x))
[1] TRUE
```

The code indicates that TRUE is coerced to 1 and FALSE to 0. Note that in the coercion from numeric to character, R attempts to produce x in its internal representation, hence the added decimal digits to the numeric strings above. Exact internal representation is not guaranteed. So you may lose precision in the process of

```
> as.numeric(as.character(x))
```

(where x is originally numeric) due to rounding errors.

Length

This object attribute is obtained with the function length():

```
> (x <- c(1 : 5, 8)) ; length(x)
[1] 1 2 3 4 5 8
[1] 6
```

length() applies to matrix, data.frame and list objects as well.

1.6 Programming

Like other programming languages, R includes the usual conditional execution, loops and such constructs. In this section, we discuss these constructs briefly. Because of its rich collection of functions and packages and because of its object oriented approach, we will avoid programming in R as much as possible. There are, however, situations where we will need to rely on programming.

1.6.1 Execution controls

Occasionally, we need to execute some statements based on some condition. On other occasions we need to repeat execution.

Conditional execution

Conditional execution is accomplished with the `if else` idiom. It has the following syntax

```
if (test) {
   executes something
} else {
   executes something else
}
```

test must return a logical value. If the result of *test* is TRUE then R *executes something* (a collection of zero or more statements), otherwise, R *executes something else* (also a collection of zero or more statements). Here is an example

```
> x <- TRUE
> if(x) y <- 1 else y <- 0
> y
[1] 1
```

Note that because there are single statements following `if` and `else`, we do not need to group them with braces {}. If you do not need the alternative for `if`, you can drop the `else`.

```
> x <- FALSE ; if(!x){y <- 1 ; z <- 2} ; y ; z
[1] 1
[1] 2
```

You can use & and | or && and || with `if` to accomplish more elaborate tests than shown thus far. The operators & and | apply to vectors element-wise. The operators && and || apply to the first element of vectors.

Repetitive execution and `break`

To repeat execution, you can use the loop statements `for`, `repeat` and `while`. While you are within a repetitive execution you can break out of the loop with the `break` statement. Here is an example:

```
1  > x <- as.logical(as.integer(runif(5, 0, 2))) ; x
2  [1] FALSE FALSE FALSE FALSE TRUE
3  > y <- vector() ; y
4  logical(0)
5  > for(i in 1 : length(x)){if(x[i]){y[i] <- 1}
6  +    else {y[i] <- 0}}
7  > y
8  [1] 0 0 0 0 1
```

The first line produces a 5-element logical vector with randomly dispersed TRUE FALSE values. To see this, we parse the innermost statement and then move out (always follow this approach to analyze code). First, we use `runif(5, 0, 2)` to produce 5 random numbers between 0 and 2. All the numbers between 0 and 2 have the same probability of occurrence. It so happens that the first 4 were less than 1 and the last one was greater than 1. Once these numbers are produced, they are coerced into integers. So the first 4 are turned into 0 and the last into 1. Next, the integers are coerced into logical values. By now we know that 0 is turned into FALSE and 1 into TRUE. Finally we assign these 5 numbers to x. The assignment generates a vector of mode logical.

In the third line, we create an empty vector y. By default, its mode is logical. If we assign data of any other mode to y, then y will be coerced into the appropriate mode automatically. In the fifth line we use both for and if. We add the braces to clarify the execution groupings. We repeat the loop for i in the sequence 1 : 5 where 5 is the length of x. Now inside the loop, if x[i] is TRUE, then y[i] is set to 1. Otherwise, it is set to zero. Because the first 4 elements of x are FALSE, the first four elements of y are set to 0. The result can be achieved with fewer statements, but here we do not intend to be unduly terse.

As we progress with our study of statistics and R, we shall meet loops and execution controls again. Please be aware that because R is object oriented, you can accomplish many tasks without having to resort to loops. Avoid loops whenever you can. The execution will be faster and less prone to errors. Here is how the previous example is done with vectors.

```
> x
[1] FALSE FALSE FALSE FALSE TRUE
> (y <- vector())
logical(0)
[1] 0
> ifelse(x, y <- 1, y <- 0)
[1] 0 0 0 0 1
```

The function ifelse(a, b, c) executes, element by element, b[i] if a[i] is TRUE and c[i] if a[i] is FALSE.

To use R efficiently, you should avoid using loops. There are numerous functions that help, but without motivation, it makes little sense to talk about them now. We shall meet these functions when we need them.

1.6.2 Functions

R has a rich set of functions. Before deciding to write a function of your own, see if one that does what you need already exists (refer to Section 1.7 for more details). Occasionally, you may need to write your own functions.

A function has a name and zero or more arguments. It has a body and often returns values. So the general form of a function is

```
function.name <- function(arguments){
    body and return values
}
```

Here is a simple example:

```
> dumb <- function(){1}
> dumb()
[1] 1
```

dumb() takes no arguments and when called it returns 1. Another example:

```
> dumber <- function(x){x + 1}
> dumber
function(x){x + 1}
> dumber()
Error in dumber() : Argument "x" is missing, with no default
> x <- runif(2) ; dumber(x)
[1] 1.1064 1.0782
```

dumber() takes one argument, adds 1 to it and returns the result. To see its code, type the function name without the parentheses. Because no default value is specified for the argument, you must call dumber() with one argument. Yet another example:

```
> dumbest <- function(x = 1){x}
> dumbest()
[1] 1
> dumbest() ; dumbest(2) ; dumbest(vector(length = 3))
[1] 1
[1] 2
[1] FALSE FALSE FALSE
```

dumbest() takes one optional argument. It is optional because it has a default value of 1.

A word about scope. When you create an object, say x, it is stored in memory and is accessible during your session. We then say that the scope of the variable is the workspace. When you define a function, like this:

```
> a.function <- function(z){
+     y <- 2 * z
+     y
+ }
```

Then function objects have a scope within the function only. Thus, calling a.function like this:

```
> a.function(2)
[1] 4
> y
Error: Object "y" not found
```

Note the error message above. Because y is defined inside the function, its scope is inside the function. When you try to access y from the workspace, you receive an error message. To elaborate slightly on the issue, consider this:

```
> (y <- 4)
[1] 4
> a.function(y) ; y
[1] 8
[1] 4
```

Here we assign 4 to y. Then we call a.function() with the argument y. The function multiples the value of y by 2 and returns 8. Once the function returned that value, we are back to the workspace. The y in the function is out of scope. Therefore, the recognized value of y is 4.

If you want to assign an object to be globally available (global scope) then use <<-:

```
> a.function <- function(z){
+     y <<- 2 * z
+     y
+ }
> y
[1] 4
> a.function(y)
[1] 8
> y
[1] 8
```

From the examples above, we glean the following rules about function arguments:

1. If you do not specify a default value, then the function argument is required. Therefore,
2. if you specify more than one argument with no default value, the arguments are required and the order they appear in the function argument list identifies them.
3. Arguments with default values are not required.
4. Arguments with no default values must appear first. After that, the order of named arguments is arbitrary.
5. Unless <<- or assign() are used for an assignment, the scope of variables in the function's body is local.

Lest you think all functions are dumb or dumber, here is a useful example. The example is from Venables et al. (2003). It is designed to address the following problem. Matrices and other structures in R have dimensions (1 for vector, 2 for matrix). These dimensions have assigned or default names. When you print such structures, the dimension names are printed as well. Here is an example:

```
> x <- as.integer(runif(5, 10, 20))
> y <- x + 2
> cbind(x, y)
      x  y
[1,] 11 13
[2,] 19 21
[3,] 16 18
[4,] 19 21
[5,] 14 16
```

The rows of the matrix created by cbind() are not named. Therefore, the row numbers [1,], ..., [5,] are printed. The columns are named x and y. We do not wish to print these so-called dimnames. Therefore, we need to define empty dimension names. The function no.dimnames() accomplishes this task:

```
> no.dimnames <- function(a) {
+     d <- list()
+     l <- 0
```

```
+     for(i in dim(a)) {
+             d[[l <- l + 1]] <- rep("", i)
+     }
+     dimnames(a) <- d
+     a
+ }
```

We did not discuss lists, dimensions and dimension indexing with [[]]. So we will not explain no.dimnames() here. However, this should not deter you from using the function. It is quite useful. With no.dimnames() defined, we get:

```
> no.dimnames(cbind(x, y))

  11 13
  19 21
  16 18
  19 21
  14 16
```

At this time, this is all we say about functions. When the need to write functions arises later, we will linger on the syntax and explain things in detail.

1.7 Packages

R is modular. Modularity is achieved by implementing the idea of packages. These are cohesive units that provide functions, data and other facilities to implement specific topics that might not be of interest to many R users. For example, many of the tables used in this book were formatted in R. To accomplish this task, we used a package named xtable. This package provides functions that allow one to format R's output according to LaTeX (a typesetting software) specifications. One can then paste the output directly into a LaTeX file (this book was written in LaTeX). Other topic-specific packages relate to time series analysis (ts), survival analysis (survival), spatial analysis (splancs is but one of them) and many others. Click on the Help | Html menu. Then, in the Html page that is loaded to your browser, click on Packages to see what packages are installed with R on your system. Study the R's Packages menu for further options. In the ensuing chapters, we will use packages frequently. We will then explain what they do.

Except for a few core ones, packages are not loaded automatically when you invoke R (otherwise R will consume much of your memory). You need to load them manually. To load a package, use

```
> library(package-name)
```

where *package-name* is the name of the package you wish to load. All of the package's functions and data are then available for the remainder of the session or until you detach the package. It is a good idea to detach loaded packages as soon as you are done with them for two reasons. First, packages consume memory and therefore may slow down computations. Second, some packages have functions with names that conflict with similarly named functions in other packages. For example, date includes functions that allow you to work with dates. Using functions in date, you can add, subtract and use other date-related operations:

```
> library(date)

Attaching package: 'date'

        The following object(s) are masked from
package:survival :
        as.date date.ddmmmyy date.mdy date.mmddyy
date.mmddyyyy ...
```

Note that when date loads, it prints information about function names that may be masked in other packages (the shown output was edited). For example, if you load date, a call to date.mdy will execute this function from the date package, not from the survival package. You can load packages that you frequently use automatically by calling for them in .First() or in the Rprofile file (see Section 1.9). Once done working with a package, unload it with

```
> # ... work with the package and then detach it:
> detach(package:date)
```

Along with R's installation, you can choose to install various packages. If you have plenty of hard disk space and memory, install all of them. You have two choices: you can download the packages and then install them from your download directory, or you can install them directly from the Web. You can also update these packages from the Web. To accomplish any of these tasks, use the menu **Packages**.

If you wish to install directly from the Web, use the menu **Packages | Install package(s) from CRAN**.... The other menus under **Packages** are self-explanatory.

1.8 Graphics

In addition to its rich statistical procedures and data-handling facilities, R excels in its graphics facilities. With few exceptions, all of the graphs in this book were produced with R. We will talk about these when they occur. For now, we shall just introduce the subject. Graphical display of data is an integral (and important) part of data analysis. With versatile graphics, your expertise in data and statistical analysis become versatile as well.

All executable graphics statements (calls to graphics functions) are directed to an active graphics window (unless you explicitly specify not to) called the graphics device or graphics driver. You can have several graphics windows open in a single session, but only one is active at a time. You start a graphics window with the command

```
> windows()
```

in the Windows environment. In Unix, you start a graphics window with the call to x11(). Some functions start the device driver on their own. Once a graphics device is open, most plotting commands will be directed to it. Thus, you can, for example, create a plot and add lines, points, annotations and so on to it.

Plotting functions are categorized into high-level, low-level and interactive. Dynamic plotting is in a category by itself. The first category produces complete plots from data that you pass to the high-level plotting functions as arguments. Low-level functions allow you to add information to plots. Here you annotate the plots, add lines and points and so on. With interactive graphics functions you identify data on the plot, add or remove data and further annotate the plot. Dynamic plotting provides facilities such as three dimensional rotation of plots.

1.8.1 High-level plotting functions

High-level plotting functions produce complete plots. If a graphics window is active, these functions erase whatever is displayed in it and plot into it. Otherwise, they open a new graphics window and plot into it. The most frequently used plotting function is plot(). The type of plot produced by plot() depends on the type of object that is given as arguments to it. We have already seen plot() in action (Figure 1.1).

Example 1.1. A so-called scatter plot, where the values of y are plotted against the values of x is common. So let us create vectors with 20 points of random data, between 0 and 1 and plot the data (Figure 1.3—we will learn how to improve upon figures later):

> x <- runif(20) ; y <- runif(20) ; plot(x, y)   □

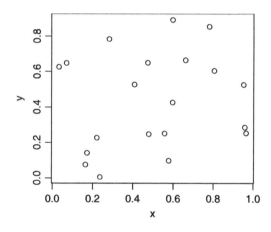

Figure 1.3   A scatter plot.

Other high-level plotting functions are:

pairs()   plots all possible pairs of matrix or data frame columns.[4]
coplot() plots pairs of vectors for fixed values of a third.
hist()   plots histograms.
perspective() produces three dimensional plots.

We shall discuss these and other plotting functions when we use them.

---
[4]We will talk about data frame objects later.

1.8.2 Low-level plotting functions

We use low-level plot functions to modify and enhance plots. Among the commonly used such functions are `points()`, `lines()`, `polygon()` and `legend()`.

Example 1.2. If you still have the graphics window with Figure 1.3 active, type

```
> lines(x, y)
```

and you will get something like Figure 1.4. □

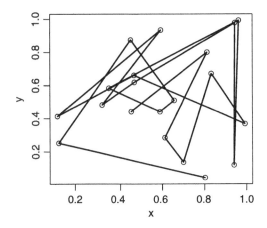

Figure 1.4   Lines added to Figure 1.3.

1.8.3 Interactive plotting functions

`locator()` and `identify()` are two commonly used functions to identify specific data in a plot and add annotation. We shall have the opportunity to use them later.

1.8.4 Dynamic plotting

Rotating three-dimensional plots is extremely useful. Often a cloud of points might not reveal relationships among variables. When rotated or viewed from the right angle, trends may become obvious. Another useful feature of dynamic plots is highlighting pairs of data points simultaneously in different scatter plots. To access such facilities, you need to install the extensive dynamic graphics facilities available in the system XGobi. You may download and install the system from http://www.research.att.com/areas/stat/xgobi/. Once xgobi is installed, you can access its facilities directly from R.

1.9 Customizing the workspace

You can customize your workspace (or environment) in ways that suit your work habits. For example, if you work with this book, you may wish to create a different project for each chapter (see Section 1.10). Then if you want to remind yourself which chapter you are working on, you can do something like this:

```
> options(prompt = 'ch1> ', continue = "+   ")
ch1> #this is the new prompt
```

Here you tell R to use "ch1> " as a prompt. A continuation line will begin with "+ " (note the 3 spaces after +). `prompt` and `continue` are but two named arguments to the function `options()`. It has many other arguments (see `help(options)`). The problem with this approach is that every time you start the ch1 project, you will need to type the `options()` command.

To avoid this extra step, you can type the line above (without the prompt) in special files that R executes every time it starts. If you wish to set the same options for all projects—for example the `continue` option may be applied for all projects—then type your options in the `Rprofile` file. This file resides in R's installation directory, in the `etc` subdirectory. The default R's installation directory in the Windows environment is `C:\Program files \R\rwxxxx` where xxxx stands for the version number of R. Here is one possible setup in `Rprofile`:

```
options(prompt = '$ ', continue = "+   ",
    digits = 5, show.signif.stars = FALSE)
```

The `prompt` and `continue` arguments are set to $ and +, respectively. The number of significant digits is set to 5. The effect of setting `show.signif.stars` to `FALSE` is that no extra stars are printed to indicate significance (we shall talk about significance later).

If R finds an `Rprofile` file in the working directory, it executes it next. So in a directory named ch1, you can place the following in your Rprofile file:

```
options(prompt = 'ch1> ')
```

Now every time you start a different chapter's project, you will be reminded by the prompt where you are.

We already talked about the function `.First()` in Example 1.3. You can place your options in `.First()`. It is executed at the environment's initialization after `Rprofile`. `.First()` is then saved in `.RData` when you save the workspace. Upon starting R in the appropriate project (workspace), R will run it first. Another function, `.Last()`, can be coded and saved in `.RData`. It will be executed upon exiting the workspace.

Keep in mind that R executes `Rprofile` and `.First()` when it starts in a particular workspace, *not* when you switch to a workspace. So if you wish to switch form ch1 project to ch2 project, use the File | Load workspace... menu. However, your prompt will change only after you issue `.First()` once you load the workspace. Also note that switching workspace is different from switching the active directory with the File | Change dir... menu. The latter simply changes the directory from which R reads and to which it writes files. These distinctions might be confusing at first. Experiment and things will become clear.

1.10 Projects

In this section, we will learn how to organize our work. As you will see, working with R requires special attention to organization. In R, anything that has a name, including functions, is an object. When objects are created, they live in memory for the remainder of the session. If you assign a new value to an already named object, the old value vanishes.

To see what objects are currently stored in memory, use the function

```
> ls()
[1] "x" "y" "z"
```

The function ls(), short for "list", lists the objects in memory. You can click the menu Misc | List objects to achieve the same result. All the objects comprise the workspace. Upon exit, R asks if you wish to save the workspace. If you click Yes, then all objects (your workspace) will be saved as an image (binary) file named .RData in the current directory. When you restart R from this directory (by, say, double-clicking on .RData), the image of the workspace is loaded into memory. As you keep working and saving your workspace, the number of objects increases. You will soon forget which object holds which data. Also, objects consume memory and slow down execution. To keep your work organized, follow these rules:

1. Isolate your work into well defined projects and create a different workspace for each project. To create a workspace for a project
   (a) Click on the File | Change dir... menu and change to a directory in which you wish to work (or create one).
   (b) Click on the File | Save workspace... menu. This will create a .RData file in the directory. The file stores your current workspace.
2. Next time you start R, Click on File | Load workspace... menu to load the project's workspace.
3. Occasionally, use rm() to remove from the workspace objects which are no-longer needed.
4. If you want R to load a particular workspace every time you start it, do this:
   (a) Right click on R's shortcut (on the Programs menu or on the desktop) and choose Properties.
   (b) In the Properties window, specify in which directory you wish to start R in the space to the right of Start in:.

Let us implement these suggestions anticipating further work with R. Follow the spirit of the example on your computer.

Example 1.3. Because we anticipate working with R throughout the book, we create a directory named Book somewhere in the directory tree. Book will be our root directory. Next, we create a directory named ch1 for Chapter 1. All the work that relates to this chapter will reside in the ch1 directory. Next, to start with a clean workspace, we remove all objects and then list whatever is left:

```
> rm(list = ls())
> ls()
```

Let us analyze the first line, from inside out. The function ls() lists all the objects. The list is assigned to list. The function rm() removes whatever is stored in list and we end up with a workspace devoid of objects. You can achieve the same result by clicking on the Misc | Remove all objects menu. Use this feature carefully. Everything, whether you need it or not, is removed!

To have something to save in the workspace, type (including the leading period):

```
> .First <- function(){options(prompt = 'ch1> ')}
```

The effect of this statement is to create a special function, named .First(). Every time you start a session, R executes .First() automatically. Here, we use .First() to modify the appearance of the workspace. From within the body of the function .First()—we use braces, {}, to group statements—we call the function options() with the argument prompt = 'ch1> '. This has the effect of changing the prompt to ch1>. Next, we click on the File | Save workspace... menu and save the workspace in the ch1 directory. The function options() is useful. It takes time to learn to take full advantage of it. We will return to options() frequently.

Because we anticipate working for a while in ch1, we next set up R to start in ch1. So, we type q() to quit R. Alternatively, select File | Exit and answer Yes to save the workspace. Next, we right click on the shortcut to R on the desktop and choose Properties. In the Properties window we instruct R to start in ch1. Now we start R again. Because it starts in ch1 and because .First() is stored in .RData—the latter was created when we saved the workspace—the workspace is loaded and we get the desired prompt. □

To see the code that constitutes .First(), we type .First without the parenthesis. You can view the code of many R functions by typing their name without the parenthesis. Use this feature liberally; it is a good way to copy code and learn R from the pros.

1.11 A note about producing figures and output

Nearly all the figures in this book were produced by the code that is explained in detail with the relevant examples. However, if you wish to produce the figures exactly as they are scaled here and save them in files to be included in other documents, you will need to do the following.

Use the function openg() before plotting. When done, call saveg(). To produce some of the histograms, use the function h() and to produce output with no quotes and no dimension names, use nqd(). These functions are explained next.

1.11.1 openg()

This function opens a graphics device in Windows. If you do not set the width and the height yourself, the window will be 3×3 inches. Here is the code

```
openg <- function (width = 3, height = 3, pointsize = 8)
{
    windows(width = width, height = height,
        pointsize = pointsize)
}
```

and here are examples of how to use it

```
> openg() # 3in by 3in window with font point size set to 8
> openg(width = 4, height = 5, pointsize = 10)
> openg(4, 5, 10) # does the same as the above line
```

The second call draws in a window 4×5 inches and font size of 10 point.

1.11.2 saveg()

This function saves the graphics device in common formats.

```
saveg <- function (fn, width = 3, height = 3, pointsize = 8)
{
   dev.copy(device = pdf, file = paste(fn, ".pdf", sep = ""),
       width = width, height = height,
       pointsize = pointsize)
   dev.off()
   dev.copy(device = win.metafile, file = paste(fn, ".wmf",
       sep = ""), width = width, height = height,
       pointsize = pointsize)
   dev.off()
   dev.copy(device = postscript, file = paste(fn, ".ps",
       sep = ""),
       width = width, height = height,
       pointsize = pointsize)
   dev.off()
}
```

The first (and required) argument is fn, which stands for file name. The function saves the plotting window in PDF, WMF and PS formats. You can use these formats to import the graphics files into your documents. PDF is a format recognized by Adobe Acrobat, WMF by many Windows applications and PS by application that recognize postscript files. Here are a couple of examples of how to use saveg():

```
> saveg('a-plot')
> saveg('b-plot', 5, 4, 11)
```

The first statement saves the current graphics device in three different files, named a-plot.pdf, a-plot.wmf and a-plot.ps. The second statement saves three b-plot files, each with width of 5 in, height of 4 in and font size of 11 point. To avoid distortions in the graphics files, always use the same width, height and pointsize in both openg() and saveg().

1.11.3 h()

This function is a modification of hist(). Its effect is self-explanatory.

```
h <- function(x, xlab = '', ylab = 'density', main = '', ...){
  hist(x, xlab = xlab, main = '', freq = FALSE,
     col = 'grey90', ylab = ylab, ...)
}
```

1.11.4 nqd()

This function prints data with no quotes and no dimension names.

```
nqd <- function(x){
  print(noquote(no.dimnames(x)))
}
```

1.12 Assignments

Exercise 1.1. Answer briefly:

1. What is the difference between `help()` and `help.search()`?
2. Show an output example for each of these two functions.
3. List and explain the contents of each section in the `help()` window.
4. When you type the command `help.search(correlation)` you get an error message. Why?
5. How would you correct this error?
6. In response to `help.search('variance')` you get a window that shows two items: `var(base)` and `Var(car)`. Explain the difference between these. If you do not see these two items, explain why they do not show up.
7. When you type `example(plot)`, you end up with a single plot. Yet, if you look at the `Examples` section in the help window for `plot()`, you will see that there is code that produces more than one plot. Why do you end up with only one plot?

Exercise 1.2.

1. Give the command that creates the sequence 0, 2, 4, ..., 20.
2. Give the command that creates the sequence 1, 0.99, ..., 0. Do not use the `by` argument. Use the `length` argument.
3. Create a sequence that includes the first 5 and last 5 of the English lower case letters. Use the symbol `:`, `c()` and `length()` in a single statement to create this sequence.
4. Do the same, but the last 5 letters should be in upper case.

Exercise 1.3. Answer the following briefly:

1. What prompt do you get following the statement `seq(1 : 10, by?`
2. Why?
3. How would you restart typing the statement above by getting out of the continuation prompt?

Exercise 1.4.

1. Create a file, named `script.R`. The file should include statements that plot 100 uniform random values of x against 100 uniform random values of y, both between 0 and 100. Attach a printout of the file to your answers.
2. Attach the plot you produced. The plot should be embedded in your favorite word processor.
3. Print the values of x and y that you created to two separate files, `x.txt` and `y.txt`. Attach a printout of these files with your answers.
4. Attach an unadulterated history file of the last 50 statements that you used in producing the answers to this exercise.

Exercise 1.5.

1. Create two projects in a root directory named `book` on your computer. Call one project `ch2` and the other `ch3`.

42   Basic R

2. In each project directory, save a .First() function in .RData. Set in .First() the prompt to 'ch1> ' and 'ch2> ' (without the single quotes).
3. Your answer should include instructions about how to accomplish these tasks.

Exercise 1.6.

1. How would you prove that the assignment x <- 1 produces a vector?
2. Will the following addition work? Why?

    ```
    x <- c(1, '2', 3) ; y <- 5
    x + y
    ```

3. If it did not work, how would you fix the code above such that x + y would work?
4. What is the mode and length of x in the statement x <- vector()?
5. How would you create a zero length vector of numeric mode?
6. Let x <- c(1 : 1000, by = 2). What is the value of $x[1]^2 + \cdots + x[n]^2$ divided by the number of elements of x?
7. Let x <- 1 : 6 and y <- 1 : 3. What is the length of the vector x * y? What are the values of its elements? Why?
8. Let x <- 1 : 7 and y <- 1 : 2. What is the length of the vector x + y? What are the values of its elements? Why?
9. Will the assignment x <- c(4 < 5, 'a' < 'b') work? What do you get?
10. What are the return values of the following statements? Explain!
    (a) 4 == 4 & 5 == 5
    (b) 5 != 5 | 6 == 6
    (c) 5 == 5 | 6 != 6
    (d) x <- 5 & y <- 6
    (e) x <- NA ; y <- 5 ; x == NA & y == 5
11. Even though it is not as terse, you should insist on using TRUE instead of T in expressions. Why?
12. Discuss the difference between the functions is.na() and is.nan().
13. How would you show that R treats Inf as a number?
14. What will be the modes of x under the following assignments? Explain.
    (a) x <- c(TRUE, 'a')
    (b) x <- c(TRUE, 1)
    (c) x <- c('a', 1)
15. Give examples (with code) of how to subset vectors using the following index types:
    (a) The index is a vector of logical values.
    (b) The index is a vector of positive integers.
    (c) The index is a vector of negative integers.
    (d) The index is a vector of strings.
16. Explain the following result:

    ```
    > x <- c(160, NA, 175, NA, 180)
    > y <- c(NA, NA, 65, 80, 70)
    > cbind(x = x[!is.na(x) & !is.na(y)],
    +       y = y[!is.na(x) & !is.na(y)])
           x   y
    [1,] 175  65
    [2,] 180  70
    ```

Exercise 1.7.

1. Execute x <- letters in R. What did you get?
2. Execute x <- LETTERS in R. What did you get?
3. Use your discovery of letters and LETTERS to create a vector x of the first 10 lower case alphabet letters.
4. What is the mode of x?
5. What happens when you coerce x to logical?
6. What happens when you coerce x to numeric?
7. Now let x <- 0 : 10. What happens when you coerce x to logical?

Exercise 1.8.

1. What would be the results of the following statements? Why?

   ```
   > x <- c(TRUE, TRUE, FALSE) ; y <- c(0, 0, 0)
   > x & y
   > x && y
   ```

2. Write a short script that uses a for loop to create a vector x of length 10. Each element of x must be a uniform random number between 0 and 1.

Exercise 1.9. Write a function that takes a vector x as input and returns $\sqrt{x}+1$. Show the code and the results of calling the function with the sequence $-1:10$. Call the R function sqrt() from within your function.

Exercise 1.10. Find the package to which the function cor.test() belongs. Run cor.test() on x <- runif(10) and y <- runif(10) and display the results. If you cannot find the package on your system, install it from the Web.

Exercise 1.11. Customize your environment so that no more than 60 characters per line are written on the console. Show the content of the appropriate file or function that accomplishes this task for every R project.

2

Data in statistics and in R

You cannot use statistics without data. Different statistical methods are appropriate for different types of data. Moreover, different statistical analyses require different representations of the same data. This means that we have to know something about how data are categorized, represented, manipulated and managed. Database management is a vast field that is independent of statistics. To analyze data effectively, some knowledge of database concepts is helpful. Statistical analysis requires a significant amount of time preparing data for analysis.

Here we introduce basic ideas about data: What types we recognize, how to organize them and some principles of manipulating them.

2.1 Types of data

Data are either provided to you or you collect them yourself. In the latter case, it will be worth your while to think about how you enter (key in) the data. For example, counts are represented as nonnegative integers while measurements are real numbers. Like any other computer language, R has what one might call basic data types. Furthermore, when it comes to analyzing and presenting data, the same method will display data differently based on their type.

2.1.1 Factors

A factor is the most general data type. Factors are also called *categories* or *enumerated types*. Think of a factor as a set of category names. Factors are qualitative classification of objects. Categories do not imply order. A black snake is different from a brown snake. It is neither larger nor smaller.

Example 2.1. Here are some examples of categorical data:

- a division of a population into males and females
- the number of dots that appear on the face of a die
- head or tail in flipping a coin
- species
- color of flowers □

Categorical data may be presented in graphs. However, the location of categories along the x or the y axes does not imply order.

Example 2.2. The results of the 2000 presidential election in the U.S. were controversial. The vote count for Gore and Bush in Florida was close and the winner was to become the next president (Adams and Fastnow, 2000). Figure 2.1 shows the vote count results. The fact that Gore appears first on the x-axis and Nader last does not mean that Gore got more votes than Bush or that Nader got fewer votes than Buchanan. The following script produces Figure 2.1.

```
e <- read.csv('elections2000.csv')
barplot(sapply(e[, 2 : 5] / 1000, sum), las = 2,
  main = 'elections 2000, Florida', ylab = 'in thousands',
  col = 'gray90')
```

In the book's site, the script is stored in a file named `elections-2000-barplot.R`. To run it, we

```
> source('elections-2000-barplot.R')
```

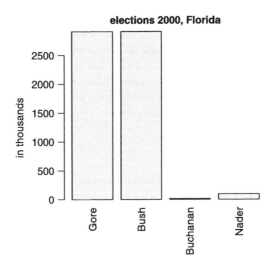

Figure 2.1 Florida vote counts in the 2000 U.S. presidential election. Votes for only 4 candidates are shown.

In line 1 we read the data from a comma separated values text file. `read.csv()` returns a data frame. We name it e. Here are the first few lines of the data frame:

```
> head(e)
     County    Gore    Bush Buchanan Nader
1   ALACHUA   47365   34124      263  3226
2     BAKER    2392    5610       73    53
3       BAY   18850   38637      248   828
4  BRADFORD    3075    5414       65    84
5   BREVARD   97318  115185      570  4470
6   BROWARD  386561  177323      788  7101
```

The function `head()` prints the first few lines of the data. A related function, `tail()` prints the last few lines of the data. We need to sum the total votes for each candidate. The number of votes by candidate and county appear in columns 2 to 5. To extract these columns, we use `e[, 2 : 5]`. Nothing followed by a comma refers to all the rows. On the right side of the comma we use the index notation to extract the needed columns. The sum is large, so we divide each county votes by 1 000. To sum the columns in one stroke, we use the function `sapply()`. The function takes two unnamed (required) arguments: the data (`e[, 2 : 5]` in our case) and a function to be applied to the elements of the data (columns in our case). The function we apply is `sum()`. So the effect of `sapply(e[, 2 : 5],sum)` is to apply `sum` to each column and return them in an array, like this:

```
> sapply(e[, 2 : 5], sum)
    Gore     Bush Buchanan    Nader
 2909117  2910078    17465    97416
```

Now `barplot()` puts the column name (`candidate`) on the x-axis and uses the data to scale the heights of the bars. The heights of these bars reflect the number of votes per candidate, divided by 1 000. `main` and `ylab` set the main title and the label of the y-axis. The named argument `las` is set to 2. This sets the ticks' text perpendicular to the axes. The named argument `col` sets the bars' color to light gray (see Figure 2.1). □

A factor is said to have *levels*. Calling the different values that a factor can take levels is somewhat misleading because we usually think of levels as reflecting order. In the context of factors, this is not always the case. In Example 2.2, a candidate is a factor variable. It has four levels, labeled `Gore`, `Bush`, `Buchanan` and `Nader`. These levels do not imply order. To create factors in R, use the function `factor()`. However, many operations on data in R create factors by default. If you ever grade exams, you may find the next example useful.

Example 2.3. There are 65 students in your class. You score (in %) the final test and wish to assign a letter grade to the score. You used to work with Excel and decided it is time to switch to R. First, you save the Excel file as a comma separated values file (`.csv`) and then import it like this:

```
> grades <- read.table('score.csv', sep = ',', header = TRUE)
```

This creates a data frame named `grades` from the file `score.csv`. The named argument `sep` tells R that columns are separated by commas. Use the comma separator even though `score.csv` has one column with no commas. Otherwise, R will use space as field separator and you might get undesired results. The named argument `header` tells R that the first row in the data file contains the names of the data columns. Next, we use the first four upper case letters for the grade:

```
> (grade <- LETTERS[4 : 1])
[1] "D" "C" "B" "A"
```

Next, we want to cut the scores into categories: D = [60, 70), C = [70, 80), B = [80, 90) and A ≥ 90. The symbol $[x, y)$ says "an interval between x and y, including x, but not y." This is accomplished like this:

```
> (letter <- cut(grades$score,
+       breaks = c(60, 70, 80, 90, 101),
+       labels = grade, include.lowest = TRUE, right = FALSE))
 [1] A D D C D D B C C A C A A C A C D C C B B B B D B C B B
[29] D C D B C C C D B D C C B D C D C C B B C B C C C A B C
[57] D C C D C C B B C
Levels: D C B A
```

Here are the first few lines of `grades`:

```
> head(grades, 5)
  score letter
1  97.9      A
2  63.0      D
3  68.1      D
4  70.9      C
5  65.3      D
```

With the letter grade as a factor, it makes working with the data easy. For example,

```
> table(grades[, 2])

 D  C  B  A
14 28 17  6
```

counts the number of students receiving each letter grade. □

2.1.2 Ordered factors

Factors have levels. Sometimes we use the levels to indicate order, but not necessarily magnitude. For example, we can define the label of presidential candidates as implying order from the most popular (having the most number of votes) to the least popular. Then in the U.S. elections, we might have the factor variable named `candidate`, with 4 levels such that Gore > Bush > Nader > Buchanan.[1] One candidate might have gotten 10 million votes and the other 1 vote. Ordinal data do not reveal this kind of information. For example, we generally agree that rabbits are faster than turtles. We rarely know by how much. To order factors, use

---

[1] Gore *was* first in the number of votes, but did not win the election.

```
> (grade <- LETTERS[1 : 4])
[1] "A" "B" "C" "D"
> (grade.factor <- factor(grade))
[1] A B C D
Levels: A B C D
> (grade.ordered <- factor(grade, ordered = TRUE))
[1] A B C D
Levels: A < B < C < D
```

You can check if a factor is ordered with

```
> is.ordered(grade.factor)
[1] FALSE
> is.ordered(grade.ordered)
[1] TRUE
```

2.1.3 Numerical variables

Numerical variables reflect magnitude and, as such, order. Numerical variables can be *discrete* or *continuous*. Counts, for example, are discrete variables that can take only nonnegative integer values. Other variables can take on any value (real numbers). Examples are:

- height of trees (continuous);
- concentration of a pollutant in the air in units of parts per million (discrete);
- weight of an animal (continuous);
- number of birds in a flock (discrete);
- average number of birds per flock (continuous);
- density of animal population (continuous).

From a strictly mathematical point of view, the distinction we make here between continuous and discrete is not correct. For our purpose, the distinction is useful. In R, numbers can be either integer or decimal. Decimal numbers are stored in what is called double-precision. Here is an example:

```
> x <- 1
> is.numeric(x) ; is.integer(x) ; is.double(x)
[1] TRUE
[1] FALSE
[1] TRUE
```

By default, x is stored as a decimal number. Therefore, x is numeric; it is not an integer and it is stored in memory as a double. If you want x to be an integer, do this:

```
> x <- as.integer(1)
> is.numeric(x) ; is.integer(x) ; is.double(x)
[1] TRUE
[1] TRUE
[1] FALSE
```

Now x is numeric, it is an integer and it is not a double.

2.1.4 Character variables

In addition to numbers and factors, we can store data as strings of characters. Here is an example:

Example 2.4. We create a vector of strings and store it in a data frame.

```
> v <- c('The', 'rain', 'in', 'Spain')
> df <- data.frame(factors = v, strings = v)
```

By default, R will convert the strings to factors:

```
> c(is.character(df$factors), is.character(df$strings))
[1] FALSE FALSE
```

We turn the second column of `df` to characters:

```
> df[, 2] <- as.character(df[, 2])
> c(is.character(df$factors), is.character(df$strings))
[1] FALSE  TRUE
```

You can change the default conversion of strings to factors with

```
> options(stringsAsFactors = FALSE)
```

This will result in creating data frames without converting characters to factors. ☐

R includes a rich set of functions that manipulate character strings. We will discuss them as needed.

2.1.5 Dates in R

Dates are not easy to deal with. They are written in different order (month-day-year in the U.S., day-month-year in most of the rest of the world), different number of digits (01-01-01, 1-1-01, 1-01-2001 or any other combination you like), or mixed digit-character format (June-20-2002 or any other combination you like). Representing dates in R (or any other system) and conversion from different formats is tedious. We will discuss dates when we need them. For now, just notice this seemingly esoteric behavior:

```
> Sys.Date()
[1] "2008-04-11"
> Sys.time()
[1] "2008-04-11 17:33:41 Central Daylight Time"
> c(Sys.Date(), Sys.time())
[1] "2008-04-11" "9233-09-07"
> c(Sys.time(), Sys.Date())
[1] "2008-04-11 17:34:35 Central Daylight Time"
[2] "1969-12-31 21:53:00 Central Standard Time"
```

All of these relate to the current computer system date and time.

2.2 Objects that hold data

In addition to vectors, matrices, lists and data frames are object types that hold data. Learning to work with these objects is essential to working with data in R.

2.2.1 Arrays and matrices

Arrays generalize the concept of vectors. Recall that a vector has a dimension 1 and a length of at least 0. The ith element of a vector is accessed via the subscript notation; e.g. v[i]. Matrices are two-dimensional arrays. They are rectangular. The element in the intersection of the ith row and jth column is accessed with m[i, j]. Arrays have k dimensions. Each element of an array is accessed with k indices, a[i1, i2, ..., ik].

An array object of dimension 1 differs from a vector object by virtue of having a dimension vector. The dimension vector is a vector of positive integers. The length of this vector gives the dimension of the array. The dimension vector is an attribute of an array. The name of the attribute is dim. Here are some statements that clarify these ideas:

```
> (v <- 1 : 10) # a vector
 [1]  1  2  3  4  5  6  7  8  9 10
> c('vector?' = is.vector(v), 'array?' = is.array(v))
vector?  array?
   TRUE   FALSE
> dim(v) <- c(10) # endow v with dim and it is an array
> c('vector?' = is.vector(v), 'array?' = is.array(v))
vector?  array?
  FALSE    TRUE
> v # array of dimension 1 prints like a vector
 [1]  1  2  3  4  5  6  7  8  9 10
```

A matrix object is a two-dimensional array. It therefore has a dim attribute. Its dimension vector has a length of 2. The first element indicates the number of rows and the second the number of columns. Here is an example of how to create a matrix with 3 columns and 2 rows with matrix():

```
> (m <- matrix(0, ncol = 3, nrow = 2))
     [,1] [,2] [,3]
[1,]    0    0    0
[2,]    0    0    0
```

Next, we verify that m is in fact an array with is.array():

```
> c('matrix?' = is.matrix(m), 'array?' = is.array(m))
matrix?  array?
   TRUE    TRUE
```

As is.matrix() illustrates, m is both a matrix and an array object. In other words, every matrix is an array, but not every array is a matrix. Like any other object, you get information about the attributes of a matrix with

```
> attributes(m)
$dim
[1] 2 3
```

Just like vectors, you index and extract elements from arrays with index vectors (see Section 1.3.5). In the next example, we create a matrix of 5 columns and 4 rows and extract a submatrix from it:

```
> (m <- matrix(1 : 20, ncol = 5, nrow = 4))
     [,1] [,2] [,3] [,4] [,5]
[1,]    1    5    9   13   17
```

```
[2,]     2    6   10   14   18
[3,]     3    7   11   15   19
[4,]     4    8   12   16   20
> i <- c(2, 3) ; j <- 2 : 4 # index vectors
> m[i, j] # extract rows 2,3 and columns 2 to 4
     [,1] [,2] [,3]
[1,]   6   10   14
[2,]   7   11   15
```

Note how the matrix is created from a sequence of 20 numbers. The first column is filled in first, then the second and so on. This is a general rule. Matrices are filled column-wise because the leftmost index runs the fastest. This rule applies to dimensions higher than 2 (i.e., to arrays). You can fill matrix by row by using the named argument byrow in your call to matrix(). You can also name the matrix dimensions by using the named argument dimnames.

Arrays are constructed with array():

```
> v <- 1 : 24 ; (a <- array(v, dim = c(3, 5, 2)))
, , 1

     [,1] [,2] [,3] [,4] [,5]
[1,]   1    4    7   10   13
[2,]   2    5    8   11   14
[3,]   3    6    9   12   15

, , 2

     [,1] [,2] [,3] [,4] [,5]
[1,]  16   19   22    1    4
[2,]  17   20   23    2    5
[3,]  18   21   24    3    6
```

The printing pattern follows the array filling rule: from the slowest running index (depth of 2), to the next slowest (5 columns) to the fastest (3 rows). Note the cycling— v is not long enough to fill the array, so after 24 elements, its values start recycling.

We have already seen how to construct matrices with matrix(). Matrices can also be constructed with the cbind() (column bind) and rbind() (row bind) functions. We discussed them in Section 1.3.5.

2.2.2 Lists

Lists are objects that can contain arbitrary objects. The elements of a list constitute an ordered collection of objects. To construct a list, use list(). In the next example, we make a list of a character vector, integer vector and a matrix. Each component of the list is named during construction:

```
> ch.v <- letters[1 : 5]                            #character vector
> int.v <- as.integer(1 : 7)                        # integer vector
> m <- matrix(runif(10), ncol = 5, nrow = 2)        # matrix
> (hodge.podge <- list(integers=int.v,              # list
+       letter = ch.v, floats = m))
```

```
$integers
[1] 1 2 3 4 5 6 7

$letter
[1] "a" "b" "c" "d" "e"

$floats
          [,1]      [,2]      [,3]      [,4]      [,5]
[1,] 0.5116554 0.3470034 0.2139750 0.3776336 0.3646456
[2,] 0.5246382 0.8092359 0.4230139 0.7846506 0.7316200
```

When the components of the list are named, they can be accessed in two ways—by name or by index:

```
> rbind(hodge.podge$letter, hodge.podge[[2]])   # row bind
     [,1] [,2] [,3] [,4] [,5]
[1,] "a"  "b"  "c"  "d"  "e"
[2,] "a"  "b"  "c"  "d"  "e"
```

Single list components are accessed by double square brackets, *not* by single square brackets: We use single brackets to access elements of an array or a vector. Here we extract the second and third elements from the second component of hodge.podge:

```
> cbind(hodge.podge$letter[2 : 3],     # column bind
+       hodge.podge[[2]][2 : 3])
     [,1] [,2]
[1,] "b"  "b"
[2,] "c"  "c"
```

The length attribute of a list is the number of its components. Here are the lengths of various parts of hodge.podge: Try to decipher the following

```
> length(hodge.podge)              # no. of list components
[1] 3
> length(hodge.podge$floats)       # no. of elements in floats
[1] 10
> length(hodge.podge$floats[, 1])  # no. of rows in floats
[1] 2
> length(hodge.podge$floats[1, ])# no. of columns in floats
[1] 5
> length(hodge.podge[[3]][1, ])  # no. of columns in floats
[1] 5
```

Another way to access named list components is using the name of the component in double square brackets. Compare the following:

```
> (x <- hodge.podge$integers)
[1] 1 2 3 4 5 6 7
> (y <- hodge.podge[['integers']])
[1] 1 2 3 4 5 6 7
```

Like any other object in R, lists can be concatenated with c().

2.2.3 Data frames

As you will quickly find out, we do much of our work with objects of type data.frame. These objects fit somewhere in between matrices and lists. They are not as rigid as matrices—they can contain columns of different modes—but they are not as loose as lists—they are required to have a rectangular structure.

Data frames are closest to what we think of as data tables (see Example 2.7 and Figure 2.2). You refer to their objects as you do in matrices. Many functions in R use data frames as the starting point for analysis. For this reason alone, you should put your data into data frames before analysis. We shall use the convenience that data frames provide frequently.

We construct data frames with data.frame() from appropriate objects. They can be of almost any mode. But they all must have equal length:

```
> composers <-c ('Sibelius', 'Wagner', 'Shostakovitch')
> grandiose <- c(1, 3, 2)
> (music <- data.frame(composers, grandiose))
        composers grandiose
1         Sibelius         1
2           Wagner         3
3    Shostakovitch         2
```

We can also construct a data frame with read.table() which reads appropriately saved data from a text file directly into a data frame (see Section 2.4). Appropriate objects can be coerced into data.frames with as.data.frame():

```
> as.data.frame(matrix(1 : 24, nrow = 4, ncol = 6))
  V1 V2 V3 V4 V5 V6
1  1  5  9 13 17 21
2  2  6 10 14 18 22
3  3  7 11 15 19 23
4  4  8 12 16 20 24
```

In the case above, data.frame() will also work, except that it will name the columns as X1, ..., X6, instead of V1, ..., V6. This (at the time of writing) small inconsistency might get you if such calls are embedded in scripts that refer to columns by name.

You can refer to columns of a data frame by index or by name. If by name, associate the column name to the data frame with $. For example, for the music data frame above, you access the composer column in one of three ways:

```
> noquote(cbind('BY NAME' = music$composer,
+   '|' = '|', 'BY INDEX' = music[, 1],
+   '|' = '|', 'BY NAMED-INDEX' = music[, 'composers']))
     BY NAME       | BY INDEX      | BY NAMED-INDEX
[1,] Sibelius      | Sibelius      | Sibelius
[2,] Wagner        | Wagner        | Wagner
[3,] Shostakovitch | Shostakovitch | Shostakovitch
```

To access a row, use, for example, music[1,]. Index vectors work the usual way on rows and columns, depending on whether they come before or after the comma in the square brackets. Instead of accessing `composer` with `music$composer`, you can `attach()` the data frame and then simply indicate the column name:

```
> attach(music) ; composer
[1] "Sibelius"     "Wagner"         "Shostakovitch"
```

Attaching a data frame can also be by position. That is, `attach(music, pos = 1)` attach `music` ahead of other objects in memory. So if you have another vector named `composer`, for example, then after attaching `music` to position 1, `composer` refers to `music`'s `composer`. If you change the attached column's data by their name, instead of with the data frame name followed by $ and the column name, then the data in the data frame do not change. Once done with your work with a data frame you can `detach()` it with

```
> detach(music)
```

`attach()` and `detach()` work on objects of just about any type that you can name: lists, vectors, matrices, packages (see Section 1.7) and so on. Judicious use of these two functions allows you to conserve memory and save on typing. Data frames have many functions that assist in their manipulation. We will discuss them as the need arises.

2.3 Data organization

Data that describe or measure a single attribute, say height of a tree, are called *univariate*. They are composed of a set of observations of objects about which a single value is obtained. *Bivariate* data are represented in pairs. *Multivariate* data are composed of a set of observations on objects. Each observation contains a number of values that represent this object.

Statistical analysis usually involves more than one data file. Often we use several files to store different data that relate to a single analysis. We then need to somehow relate data from different files. This requires careful consideration of how the data are to be organized. Once you commit the data to a particular organization it is difficult to change. The way the data are organized will then dictate how easy they are to prepare for different types of statistical analyses.

Data are organized into tables and tables are related to each other. The tables, their relationship and other auxiliary information form a *database*. For example, you may have data about air pollution. The pollution is measured in numerically labeled stations and the data are stored in one table. Another table stores the correspondence between the station number and the name of the closest town (this is how the U.S. Environmental Protection Agency saves many of its pollution-related data). Tables are often stored in separate files.

2.3.1 Data tables

We arrange data in columns (variables) and rows (observations). In the database vernacular, we call variables *fields* and observations *cases* or *rows*. The following example demonstrates multivariate data. It is a good example of how data should be reported succinctly and referenced appropriately.

Example 2.5. R comes with some data frames bundled. You can access these data with data() and use its name as an argument. If you give no argument, then R will print a list of all data that are installed by default. The table below shows the first 10 observations from a data set with 6 variables

```
> data(airquality)
> head(airquality)
  Ozone Solar.R Wind Temp Month Day
1    41     190  7.4   67     5   1
2    36     118  8.0   72     5   2
3    12     149 12.6   74     5   3
4    18     313 11.5   62     5   4
5    NA      NA 14.3   56     5   5
6    28      NA 14.9   66     5   6
```

We view the data frame by first bringing it into our R session with data() and then printing its head(). The data represent daily readings of air quality values in New York City for May 1, 1973 through September 30, 1973. The data consist of 154 observations on 6 variables:

1. Ozone (ppb) – numeric values that represent the mean ozone in parts per billion from 1300 to 1500 hours at Roosevelt Island.
2. Solar.R (lang) – numeric values that represent solar radiation in Langleys in the frequency band 4000–7700 Angstroms from 0800 to 1200 hours at Central Park.
3. Wind (mph) – numeric values that represent average wind speed in miles per hour at 0700 and 1000 hours at La Guardia Airport.
4. Temp (degrees F) – numeric values that represent the maximum daily temperature in degrees Fahrenheit at La Guardia Airport.
5. Month – numeric month (1–12)
6. Day – numeric day of month (1–31)

The data were obtained from the New York State Department of Conservation (ozone data) and the National Weather Service (meteorological data). The data were reported in Chambers and Hastie (1992). □

The output in Example 2.5 illustrates typical arrangement of data and reporting:

- the data in a table with observations in rows and variables in columns;
- the variable names and their type;
- the units of measurement;
- when and where the data were collected;
- the source of the data;
- where they were reported.

This is a good example of how data should be documented. Always include the units and cite the source. Give variables meaningful names and you will not have to waste time looking them up. The distinction between variables and observations need not be rigid. They may even switch roles, based on the questions asked.

Example 2.6. The data on vote counts in Florida were introduced in Example 2.2. In Table 2.1, the candidates are variables. Each column displays the number of votes

Table 2.1 Number of votes by county and candidate. U.S. 2000 presidential elections, Florida counts.

County	Gore	Bush	Buchanan	Nader
ALACHUA	47 365	34 124	263	3 226
BAKER	2 392	5 610	73	53
BAY	18 850	38 637	248	828
BRADFORD	3 075	5 414	65	84
BREVARD	97 318	115 185	570	4 470

for the candidate. The counties are the observations (rows). In Table 2.2, the counties are the variables. The columns display the number of votes cast for different candidates in a county. Now if you want to compute the total votes cast for Gore, you might have to present the data in Table 2.1 to your statistical package. If you want the total number of votes cast in a county, you might have to produce the data in Table 2.2 to your statistical package. Contrary to appearances, switching the roles of rows and columns may not be a trivial task. We shall see that R is particularly suitable for such switches. □

Table 2.2 Number of votes by candidate and county. U.S. 2000 presidential elections, Florida counts.

Candidate	ALACHUA	BAKER	BAY	BRADFORD	BREVARD
Gore	47 365	2 392	18 850	3 075	97 318
Bush	34 124	5 610	38 637	5 414	115 185
Buchanan	263	73	248	65	570
Nader	3 226	53	828	84	4 470

2.3.2 Relationships among tables

Many tables may be part of a single project that requires statistical analysis. Creating these tables may require data entry—a tedious and error-prone task. It is therefore important to minimize the amount of time spent on such activities. Sometimes the tables are very large. In epidemiological studies you might have hundreds of thousands of observations. Large tables take time to compute and consume storage space. Therefore, you often need to minimize the amount of space occupied by your data.

Example 2.7. The World Health Organization (WHO) reports vital statistics from various countries (WHO, 2004). Figure 2.2 shows a few lines from three related tables from the WHO data. One table, named who.ccodes stores country codes under the variable named code and the country name under the variable named name. Another table, named who.pop.var.names stores variable names under the column var and description of the variable under the column descr. For example the variable Pop10 stores population size for age group 20 to 24. The third table, who.pop.2000, stores population size for country (rows) by age group (columns).

If you wish to produce a legible plot or summary of the data, you will have to relate these tables. To show population size by country and age group, you have to read the country code from who.pop.2000 and fetch the country name from who.ccodes. You

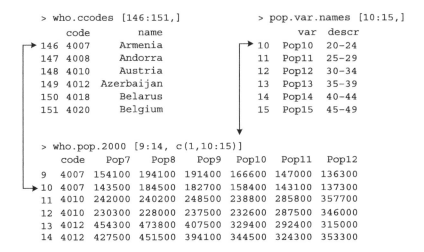

Figure 2.2   Sample of WHO population data from three related tables.

will also have to read population size for a variable and fetch its description from pop.var.names. In the figure, the population size in 2000 in Armenia for age group 20–24 was 158 400. □

We could collapse the three tables in Example 2.7 into one by replacing the country code by the country name and variable name by its description. With a table with thousands of records, the column code would store names instead of numbers. If, for example, some statistical procedure needs to repeatedly sort the table by country, then sorting on a string of characters is more time-consuming than sorting by numbers. Worse yet, most statistical software and database management systems cannot store a variable name such as 20–24. The process of minimizing the amount of data that needs to be stored is called *normalization* in database "speak." If the tables are not too large, you can store them in R as three distinct data frames in a list.

2.4   Data import, export and connections

Unless you enter data directly into R (something you should avoid), you will need to know how to import your data to R. To exchange data with those poor souls that do not use R, you also need to learn how to export data. If you routinely need to obtain data from a database management system (DBMS), it may be tedious to export the data from the system and then import it to R. A powerful alternative is to access the data directly from within R. This requires connection to the DBMS. Connecting directly to a DBMS from within R has three important conveniences: Automation (thus minimizing errors), working with a remote DBMS (that is, data that do not reside on your computer) and analysis in real time. R comes with an import/export manual (The R Development Core Team, 2006a). It is well written and you should read it for further details. We will discuss some of these R's capabilities when we need them.

2.4.1   Import and export

Exporting data from R is almost a mirror to importing data. We will concentrate on importing. There are numerous functions that allow data imports. Some of them access binary files created with other software. You should always strive to import text data. If the data are in another system and you cannot export them as text from that system, then you need to import the binary files as written by the other system without conversion to text first.

Text data

The easiest way to import data is from a text file. Any self-respecting software allows data export in a text format. All you need to do is make sure you know how the data are arranged in the text file (if you do not, experiment). To import text data, use one of the two: `read.table()` or `read.csv()`. Both are almost identical, so we shall use them interchangeably.

Example 2.8. We discussed the WHO data in Example 2.7. Here is how we import them and the first few columns and rows:

```
> who <- read.table('who.by.continents.and.regions.txt',
+   sep = '\t', header = TRUE)
> head(who[, 1 : 3], 4)
         country continent          region
1         Africa    Africa            <NA>
2 Eastern Africa    Africa  Eastern Africa
3        Burundi    Africa  Eastern Africa
4        Comoros    Africa  Eastern Africa
```

We tell `read.table()` that columns are separated by the tab character (`sep = '\t'`). The first row of the text file holds the headers of the data columns. The data were obtained as an Excel spreadsheet and then saved as a text file with tab as the separating character. □

A useful function to import text data is `scan()`. You can use it to read files and control various aspects of the file to be read. We find it particularly useful in situations like this: You browse to a web page that contains a single column data; a string of numbers; something like:

10
50
120
.
.
.

Then copy the numbers to your clipboard and in R do this:

```
> new.data <- scan()
1:
```

The number prompt indicates that you are in input mode. Paste the data you copied to the clipboard and enter a return (extra blank line) when done. You can also use `scan()` to enter data manually.

Data from foreign systems

The package `foreign` includes functions that read and write in formats that other system recognize. At the time of writing, you can import data from SAS, DBF, Stata, Minitab, Octave, S, SPSS and Systat. Let us illustrate import from a Stata binary file.

Example 2.9. The data were published in Krivokapich et al. (1999) and downloaded from the University of California, Los Angeles Department of Statistics site at http://www.stat.ucla.edu/. Import cardiac.dta like this:

```
> library(foreign)
> cardiac <- read.dta('cardiac.dta')
> head(cardiac[, 1:4])
  bhr basebp basedp pkhr
1  92    103   9476  114
2  62    139   8618  120
3  62    139   8618  120
4  93    118  10974  118
5  89    103   9167  129
6  58    100   5800  123
```

Importing from other systems is done much the same way. □

2.4.2 Data connections

Here we discuss one way to connect to data stored in formats other than R. Such connections allow us to both import data and manipulate it before importing. Often, we need to import—or even manipulate—data stored in a variety of formats. For example, Microsoft's Excel is widely used to both analyze and store data. Another example would be cases where tremendous amounts of data are stored in a DBMS such as Oracle and dBase and large statistical software such as SAS, SPSS and Stata. R is not a DBMS and as such, is not suitable to hold large databases.

Open Data Base Connectivity (ODBC) is a protocol that allows access to database systems (and spreadsheets) that implement it. The protocol is common and is implemented in R. Among others, the advantages of connecting to a remote database are: Data safety and replication, access to more than one database at a time, access to (very) large databases and analysis in real time of changing data. In the next example, we connect to a worksheet in an Excel file. The package `RODBC` includes the necessary functions.

Example 2.10. An Excel file was obtained from WHO (2004). The file name is who-population-data-2002.xls. Minor editing was necessary to prepare the data for R. These include, for example, changing the spreadsheet notation for missing data from .. to NA. So we created a new worksheet in Excel, named **MyFormat**. Here we connect to this worksheet via RODBC and import the data. The task is divided into two steps. First, we make a connection to the spreadsheet at the operating system level. Then we open the connection from within R.

If you are using a system other than Windows, read up on how to name an ODBC connection. A connection is an object that contains information about the data location, the data file name, access rights and so on. To name a connection, go to your systems **Control Panel** and locate a program that is roughly named **Data Sources (ODBC)**.[2] Next, activate the program. A window, named **ODBC Data Source Administrator** pops up. We are adding a connection, so click on **Add**.... A window, named **Create a New Data Source** shows up. From the list of ODBC drivers, choose **Microsoft Excel Driver (*.xls)** and click on **Finish**. A new window, named **ODBC Microsoft Excel Setup** appears. We type who in the **Data Source Name** entry and something to describe the connection in the **Description** entry. Then we click on **Select Workbook...** button. Next, we navigate to the location of the Excel file and click on it. We finally click on OK enough times to exit the program. So now we have an ODBC connection to the Excel data file. The connection is named who and any software on our system that knows how to use ODBC can open the who connection.

Next, we connect and import the data with

```
> library(RODBC) ; con <- odbcConnect('who')
> sqlTables(con) ; who <- sqlFetch(con, 'MyFormat')
> odbcClose(con)
```

Here, we load the RODBC package (by Lapsley and Ripley, 2004) and use odbcConnect() with the argument 'who' (a system-wide known ODBC object) to open a connection named con. sqlTables() lists all the worksheet names that the connection con knows about. In it, we find the worksheet MyFormat. Next, we access the MyFormat worksheet in the data source with sqlFetch(). This function takes two unnamed arguments, the connection (con in our case) and the worksheet name (MyFormat). sqlFetch() returns a data frame object and we assign this object to who. Now who is a data.frame. When done, we close the connection with odbcClose(). □

In the next example, we show how to access data from a MySQL DBMS that resides on another computer. We use MySQL not because it is the best but because it is common (and yes, it is free). We recommend the more advanced *and* open source DBMS from http://archives.postgresql.org.

Example 2.11. We will use a MySQL database server installed on a remote machine. The database we use is called rtest. Before accessing the data from R, we need to create a system-wide Data Source Name (DSN). To create a DSN, follow these steps:

1. Download the so-called MySQL ODBC driver from http://MySQL.org and install it according to the instructions.
2. Read the instructions that come with the driver on how to create a DSN under a Windows system.

Now in R, we open a connection to the data on the remote server:

```
> library(RODBC)
> (con <- odbcConnect('rtest', case = 'tolower'))
RODB Connection 13
```

---
[2] In Windows XP, the program resides in the Control Panel, under Administrative Tools.

```
Details:
  case=tolower
  DATABASE=rtest
  DSN=rtest
  OPTION=0
  PWD=xxxxxx
  PORT=0
  SERVER=yosefcohen.org
  UID=root
```

The argument `case` causes all characters to be converted to lower case. The details tell us that `rtest` is open, the server has been located and the user ID is `root` (a user known to the remote DBMS).

We communicate with the data via the Simple Query Language (SQL)—a standard language that provides database facilities. To see what data tables are available in the database, we do

```
> (sqlTables(con))
  TABLE_CAT TABLE_SCHEM TABLE_NAME TABLE_TYPE      REMARKS
1     rtest                   who      TABLE MySQL table
```

Next, we import the data in `who` into a data frame:

```
> who.from.MySQL <- sqlQuery(con, 'select * from who')
```

Let us see some data and close the connection:

```
> head(who.from.MySQL[, 1 : 3])
          country continent         region
1          Africa    Africa           <NA>
2  Eastern Africa    Africa Eastern Africa
3         Burundi    Africa Eastern Africa
4         Comoros    Africa Eastern Africa
5        Djibouti    Africa Eastern Africa
6         Eritrea    Africa Eastern Africa
> (odbcClose(con))
[1] TRUE
```

Be sure to close a connection once you are done with it. □

Let us upload data from R to the `rtest` database.

Example 2.12. The data for this example are from United States Department of Justice (2003). It lists all of the 7 658 capital punishment cases in the U.S. between 1973 and 2000 (data prior to 1973 were collapsed into 1973). We load the data, open a connection and use `sqlSave()`:

```
> load('capital.punishment.rda')
> con <- odbcConnect('rtest')
> (sqlTables(con))
  TABLE_CAT TABLE_SCHEM          TABLE_NAME TABLE_TYPE      REMARKS
1     rtest                             who      TABLE MySQL table
> sqlSave(con, capital.punishment,
+     tablename = 'capital_punishment')
```

Next, we check that all went well and close the connection:

```
> (sqlTables(con))
  TABLE_CAT TABLE_SCHEM           TABLE_NAME TABLE_TYPE    REMARKS
1     rtest             capital_punishment      TABLE MySQL table
2     rtest                            who      TABLE MySQL table
> odbcClose(con)
```

Because we cannot use dots for names in MySQL, we specify `tablename = 'capital_punishment'`. □

In addition to the mentioned connections, you can open connections to files and read and write directly to them and to files on the Web.

2.5 Data manipulation

The core of working with data is the ability to subset, merge, split and perform other such data operations. Applying various operations to subsets of the data wholesale is as important. These are the topics of this section. Unlike traditional programming languages (such as C and Fortran), executing loops in R is computationally inefficient even for mundane tasks. Many of the functions we discuss here help avoid using loops. We shall meet others as we need them. We discussed how to subset data with index vectors in Section 1.3.5.

2.5.1 Flat tables and expand tables

If you chance upon data that appear in a contingency table format, you may read (or write) them with `read.ftable()` (or `write.ftable()`). If you use `table()` (we will meet it again later), you can `expand.table()`, a function in the package `epitools`.

Example 2.13. Back to the capital punishment data (Example 2.12). First, we load the data and view the unique records from two columns of interest:

```
> load('capital.punishment.rda')
> cp <- capital.punishment
> unique(cp[, c('Method', 'Sex')])
          Method Sex
1              9   M
27   Electrocution   M
130      Injection   M
1022           Gas   M
1800   Firing squad   M
2175       Hanging   M
7521             9   F
7546     Injection   F
7561 Electrocution   F
```

(9 stands for unknown or NA). unique() returns only unique records. This reveals the plethora of execution methods, including the best one, NA. Next, we want to count the number of executions by sex:

```
> (tbl <- table(cp[, c('Method', 'Sex')]))
              Sex
Method          F    M
  9           133 6842
  Electrocution 1  148
  Firing squad  0    2
  Gas           0   11
  Hanging       0    3
  Injection     4  514
```

Let us expand rows 3 to 5 of the table. This should give us $2 + 11 + 3$ records:

```
> library(epitools)
> (e.tbl <- expand.table(tbl[3 : 5, ]))
        Method Sex
1  Firing squad  M
2  Firing squad  M
3           Gas  M
4           Gas  M
5           Gas  M
6           Gas  M
7           Gas  M
8           Gas  M
9           Gas  M
10          Gas  M
11          Gas  M
12          Gas  M
13          Gas  M
14      Hanging  M
15      Hanging  M
16      Hanging  M
```

You may need such an expansion for further analysis. For example, you can now do

```
> tapply(e.tbl$Method, e.tbl$Method, length)
Firing squad          Gas      Hanging
           2           11            3
```

which is another way of counting records. tapply() takes three unnamed arguments. In our case, the first is a vector, the second must be a factor vector and the third is a function to apply to the first based on the factor levels. You can use most functions instead of length(). □

2.5.2 Stack, unstack and reshape

From the R help page on stack() and unstack(): "Stacking vectors concatenates multiple vectors into a single vector along with a factor indicating where each observation originated. Unstacking reverses this operation." The need for stack() can best be explained with an example.

Example 2.14. Let us stack the 2000 U.S. presidential elections in Florida (see Example 2.2). First, we import the data and look at their head:

```
> e <- read.csv('elections-2000.csv')
> head(e)
    County    Gore   Bush Buchanan Nader
1  ALACHUA   47365  34124      263  3226
2    BAKER    2392   5610       73    53
3      BAY   18850  38637      248   828
4 BRADFORD    3075   5414       65    84
5  BREVARD   97318 115185      570  4470
6  BROWARD  386561 177323      788  7101
```

A box plot is a way to summarize data (we will discuss it in detail later). The data on the x-axis are factors. So we stack() the data and, for posterity, add a column for the counties:

```
> stacked.e <- cbind(stack(e), county = e$County)
> head(stacked.e)
   values  ind   county
1   47365 Gore  ALACHUA
2    2392 Gore    BAKER
3   18850 Gore      BAY
4    3075 Gore BRADFORD
5   97318 Gore  BREVARD
6  386561 Gore  BROWARD
```

Next, we do

```
> plot(stacked.e[, 2 : 1])
```

(see Figure 2.3)

reshape() works much like stack().

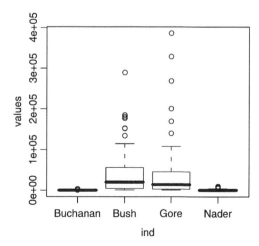

Figure 2.3   Candidates and county votes for the Florida 2000 U.S. presidential elections.   □

2.5.3 Split, unsplit and unlist

Occasionally, we need to split a data frame into a list based on some factor. A case in point might be when the frame is large and we wish to analyze part of it. Here is an example.

Example 2.15. The data refer to a large study about fish habitat in streams, conducted by the Minnesota Department of Natural Resources. The data were collected from two streams, coded as OT and YM under the factor CODE. We wish to split the data into two components, one for each of the streams. So we do this:

```
> load('fishA.rda')
> f.s <- split(fishA, fishA$CODE)
> (is.list(f.s))
[1] TRUE
> (names(f.s))
[1] "OT" "YM"
```

is.list() verifies that the result of split() is a list and names() gives the names of the list components. □

The functions unsplit() and unlist() do the opposite of what split() does.

2.5.4 Cut

From the R's help page for cut(): "cut divides the range of x into intervals and codes the values in x according to which interval they fall. The leftmost interval corresponds to level one, the next leftmost to level two and so on." The need for cut is illustrated with the next example.

Example 2.16. The site http://icasualties.org maintains a list of all U.S. Department of Defense confirmed military casualties in Iraq. To import the data, we save the HTM page, open it with a spreadsheet and save the date column. We then cut dates into 10-day intervals and use table() to count the dead. First, we import the data and turn them into "official" date format:

```
> casualties <- read.table('Iraq-casualties.txt', sep = '\t')
> casualties$V1 <- as.Date(casualties$V1, '%m/%d/%Y')
> head(casualties, 5)
          V1
1 2007-01-04
2 2006-12-30
3 2006-12-27
4 2006-12-26
5 2006-12-26
```

as.Date() turns the data in the only column of casualties to dates according to the format it was read: month, day and four-digit year ('%m/%d/%Y'). Next, we sort

the dates by increasing order, add a Julian day column that corresponds to the date
of each casualty and display the first few rows of the data frame:

```
> casualties$V1 <- sort(casualties$V1)
> jd <- julian(casualties$V1)
> casualties <- data.frame(Date = casualties$V1, Julian = jd)
> head(casualties)
        Date Julian
1 2003-03-21  12132
2 2003-03-21  12132
3 2003-03-21  12132
4 2003-03-21  12132
5 2003-03-21  12132
6 2003-03-21  12132
```

Julian date is a count of the number of days that elapsed since some base date. We now determine the number of 10-day intervals, cut the Julian dates into these intervals and count the number of deaths within each interval:

```
> (b <- ceiling((jd[length(jd)] - jd[1]) / 10))
[1] 139
> cnts <- table(cut(jd, b))
```

`ceiling()` returns the smallest integer larger than a decimal number and `cut()` cuts the data in b equal intervals and returns a vector, like this:

```
> head(cut(jd, b))
[1] (1.213e+04,1.214e+04] (1.213e+04,1.214e+04]
[3] (1.213e+04,1.214e+04] (1.213e+04,1.214e+04]
[5] (1.213e+04,1.214e+04] (1.213e+04,1.214e+04]
```

(the intervals are factors named after the Julian date). Finally, we count the number of occurrences of each interval, i.e. the number of reported deaths during 10-day intervals. So the counts look like this:

```
> head(cnts, 5)

(1.213e+04,1.214e+04] (1.214e+04,1.215e+04]
                   60                    57
(1.215e+04,1.216e+04] (1.216e+04,1.217e+04]
                   14                     8
(1.217e+04,1.218e+04]
                    4
```

During the first 10-day interval, there were 60 reported dates (which refer to 60 casualties). Finally we plot the data (Figure 2.4). We shall see how the plot was produced later. □

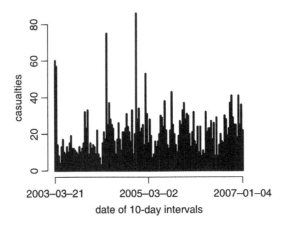

Figure 2.4 U.S. military casualties in Iraq. Counts are shown in 10-day intervals.

2.5.5 Merge, union and intersect

These operations are best explained with an example.

Example 2.17. Let us create a vector of the first six upper case letters, the corresponding integer code of these letters and a data frame of these two:

```
> a <- data.frame(letter = LETTERS[1 : 6])
> a <- data.frame(a, code = apply(a, 1, utf8ToInt))
```

LETTERS (for upper case) and letters (for lower case) are data vectors supplied with R that hold the English letters. utf8ToInt() is a function that returns the integer code of the corresponding letter in the so called UTF-8 format. apply() applies to the data frame a, by rows (1) the function utf8ToInt(). Next, we create a similar data frame b and display both data frames:

```
> b <- data.frame(letter = LETTERS[4 : 9])
> b <- data.frame(b, code = apply(b, 1, utf8ToInt))
> cbind(a, '|' = '|', b)
  letter code | letter code
1      A   65 |      D   68
2      B   66 |      E   69
3      C   67 |      F   70
4      D   68 |      G   71
5      E   69 |      H   72
6      F   70 |      I   73
```

cbind() binds the columns of a and b and a column of separators between them. Now here is what merge() does:

```
> merge(a, b)
  letter code
1      D   68
2      E   69
3      F   70
```

Contrast this with `union()`:

```
> union(a, b)
[[1]]
[1] A B C D E F
Levels: A B C D E F

[[2]]
[1] 65 66 67 68 69 70

[[3]]
[1] D E F G H I
Levels: D E F G H I

[[4]]
[1] 68 69 70 71 72 73
```

which creates a list of the columns in a and b. Note the asymmetry of `intersect()`:

```
> cbind(intersect(a, b), '|' = '|', intersect(b, a))
  letter | letter
1      D |      A
2      E |      B
3      F |      C
4      G |      D
5      H |      E
6      I |      F
```

The data frames do not have to have equal numbers of rows. □

2.5.6 `is.element()`

This is a very useful functions that in "Data Speak" relates many records in one, say, data frame, to many records in another, based on common values. The function `is.element()` takes two vector arguments and checks for common elements in the two. It returns an index vector (`TRUE` or `FALSE`) that gives the common argument values (as `TRUE`) in its first argument. You can then use the returned logical vector as an index to extract desired elements from the first vector. Thus, the function is *not* symmetric. The next example illustrates one of the most commonly encountered problems in data manipulation. Its solution is not straightforward.

Example 2.18. In longitudinal studies, one follows some units (subjects) through time. Often, such units enter and leave the experiment after it began and before it ends. A two-year imaginary diet study started with six patients:

```
> begin.experiment
      name weight
1 A. Smith    270
2 B. Smith    263
3 C. Smith    294
```

```
4 D. Smith    218
5 E. Smith    305
6 F. Smith    261
```

After one year, three patients joined the study:

```
> middle.experiment
     name weight
1 G. Smith   169
2 H. Smith   181
3 I. Smith   201
```

Five patients, some joined at the beginning and some joined in the middle, finished the experiment:

```
> end.experiment
     name weight
1 C. Smith   107
2 D. Smith   104
3 A. Smith   104
4 H. Smith   102
5 I. Smith   100
```

Imagine that each of these data frames contains hundred of thousands of records. The task is to merge the data for those who started and finished the experiment. First, we identify all the elements in end.experiment that are also in begin.experiment:

```
> (m <- is.element(begin.experiment$name, end.experiment$name))
[1]  TRUE FALSE  TRUE  TRUE FALSE FALSE
```

Next, we create a vector of those patient names that started in the beginning and ended in the end:

```
> (begin.end <- begin.experiment[m, ])
     name weight
1 A. Smith   270
3 C. Smith   294
4 D. Smith   218
> (p.names <- begin.experiment[m, 1])
[1] "A. Smith" "C. Smith" "D. Smith"
```

We merge the data for the weights at the beginning and end of the experiment:

```
> (patients <- cbind(begin.experiment[m, ],
+     end.experiment[is.element(end.experiment$name, p.names), ]))
     name weight    name weight
1 A. Smith   270 C. Smith   107
3 C. Smith   294 D. Smith   104
4 D. Smith   218 A. Smith   104
```

patients is still not very useful. Our goal is to obtain this:

```
   name  time weights
1 A. Smith begin    270
2 C. Smith begin    294
3 D. Smith begin    218
4 C. Smith   end    107
5 D. Smith   end    104
6 A. Smith   end    104
```

(you will see in a moment why). To achieve this goal, we stack names and then weights:

```
> (p.names <- stack(patients[, c(1, 3)]))
    values    ind
1 A. Smith   name
2 C. Smith   name
3 D. Smith   name
4 C. Smith name.1
5 D. Smith name.1
6 A. Smith name.1
> (weights <- stack(patients[, c(2, 4)])[, 1])
[1] 270 294 218 107 104 104
```

Now create the data frame:

```
> (experiment <- data.frame(p.names, weights))
    values    ind weights
1 A. Smith   name     270
2 C. Smith   name     294
3 D. Smith   name     218
4 C. Smith name.1     107
5 D. Smith name.1     104
6 A. Smith name.1     104
```

This is it. All that is left is to rename columns and factor levels:

```
> levels(experiment$ind) <- c('begin', 'end')
> names(experiment)[1 : 2] <- c('name', 'time')
```

and we achieved our goals. Why do we want this particular format for the data frame? Because

```
> tapply(experiment$weights, experiment$time, mean)
   begin      end
260.6667 105.0000
```

is so easy. To handle data with hundreds of thousands of subjects, all you have to do is change the indices in this example. □

2.6 Manipulating strings

R has a rich set of string manipulations. Why are these useful? Consider the next example.

Example 2.19. One favorite practice of database managers is to assign values to a field (column) that contain more than one bit of information. For example, in collecting environmental data from monitoring stations, the European Union (EU) identifies the stations with code such as DE0715A. The first two characters of the station code give the country id (DE = Germany). For numerous reasons, we may wish to isolate the two first characters. We are given

```
> stations
 [1] "IT15A"  "IT25A"  "IT787A" "IT808A" "IT452A" "DE235A"
 [7] "DE905A" "DE970A" "DE344A" "DE266A"
```

and we want to rename the stations to their countries:

```
> stations[substr(stations, 1, 2) == 'DE'] <- 'Germany'
> stations[substr(stations, 1, 2) == 'IT'] <- 'Italy'
> stations
 [1] "Italy"   "Italy"   "Italy"   "Italy"   "Italy"
 [6] "Germany" "Germany" "Germany" "Germany" "Germany"
```

Here substr() returns the first two characters of the station id string. □

We have already seen how to use paste(), but here is another example.

Example 2.20.

```
> paste('axiom', ':', ' power', ' corrupts', sep = '')
[1] "axiom: power corrupts"
```
□

Sometimes, you might get a text file that contains data fields (columns) separates by some character. Then strsplit() comes to the rescue.

Example 2.21.

```
> (x <- c('for;crying;out;loud', 'consistency;is;not;a virtue'))
[1] "for;crying;out;loud"
[2] "consistency;is;not;a virtue"
> rbind(strsplit(x, ';')[[1]], strsplit(x, ';')[[2]])
     [,1]          [,2]      [,3]  [,4]
[1,] "for"         "crying"  "out" "loud"
[2,] "consistency" "is"      "not" "a virtue"
```
□

In cases where you are not sure how character data are formatted, you can transform all characters of a string to upper case with toupper() or all characters to lower case with tolower(). This often helps in comparing strings with ==. We shall meet these again.

2.7 Assignments

Exercise 2.1. Classify the following data as categorical or numerical. If numerical, classify into ordinal, discrete or continuous.

1. Number of trees in an area
2. Sex of a trapped animal

3. Carbon monoxide emission per day by your car
4. Order of a child in a family

Exercise 2.2. For the following questions, use the literature cited in the article you found as a template to cite the article you found. Also, list the data as given in the paper. If you do not wish to type the data, then you may attach a copy of the relevant pages.

Cite one paper from a scientific journal of your choice where the data are:

1. Categorical
2. Numerical
3. Univariate
4. Bivariate
5. Multivariate

Exercise 2.3.

1. Create a vector of factors with the following levels: Low, Medium, High.
2. Create the same vector, but with the levels ordered.

Exercise 2.4. In the following, show how you verify that your answer is correct.

1. Create a vector of double values 1, 2, 3.
2. Create the same vector, but now with integer values.

Exercise 2.5.

1. Create a vector of strings that hold the following data: "what" "a" "shame".
2. How would you test that the vector you created is in fact a vector of strings?
3. By default, when R reads a character input vector into a data frame, it converts it to a factor variable. How would you override this default?

Exercise 2.6. What is the difference in the output between sys.Date() and Sys.time()

Exercise 2.7.

1. Prove that x produced with x <- 1:10 is not an array.
2. Turn x into an array.
3. Is x produced with matrix(0, ncol=3, nrow=2) an array? Is it a matrix? Prove it!

Exercise 2.8. You are given a list of 10 "names" and 10 test scores:

names <- c(A, B, C, D, E, F, G, H, I, J)
scores <- c(59, 51, 72, 79, 79, 83, 69, 81, 51, 87)

Show the code and the result for the following:

1. Make a data frame with the first column named "score" and the second named "names."
2. Make a list with the first element named "names" and the second named "score."
3. Show two ways to access names in the list you just created.

Exercise 2.9. Pick 10 people at random (5 males and 5 females) and create a data frame with the following columns: Gender—a factor with two levels, M and F, Height—a numeric variable holding the height of each person (in cm), First Name—a character variable and Last Name—a character variable.

Exercise 2.10. Use scan() to create a vector of 10 integers

Exercise 2.11. Import data directly to R from an Excel file of your choice.

Exercise 2.12. Given i, which was produced as follows:

```
> set.seed(1)
> (i <- as.integer(runif(10, 1, 5)))
 [1] 2 2 3 4 1 4 4 3 3 1
```

how would you use R to tally the number of occurrences of each digit?

3
Presenting data

In this chapter we will learn how to present data in tables and graphics. Often, tables require a good deal of data manipulations. Graphics is an important tool not only in presenting data but also in cleaning them and later analyzing them.

3.1 Tables and the flavors of `apply()`

Strictly speaking, `table()`s in R produce counts of categorical variables. They are useful in exploring associations among factors in data (e.g. *contingency tables*). We used `table()` in Example 2.3, where we discussed how to create letter grades from exam scores, Example 2.13, where we saw how to use `expand.table()` and Example 2.16, where we counted the number of U.S. casualties in Iraq in 10-day intervals.

Often, we need to produce marginal values for tables (e.g. totals at the bottom or the right end of a table). The various flavors of `apply()` are extremely useful. Let us learn about these through examples.

Example 3.1. The following data pertain to high school graduation rates (%) and number of graduates (n) in the U.S. for the academic years 2000–01, 2001–02 and 2002–03, by state. Data were obtained from the U.S. Department of Education Web site. We import the data, name the columns and view the first three rows and six columns of the data frame:

```
> graduation <- read.table('graduation.txt',
+    header = TRUE, sep = '\t')
> names(graduation) <- c('region', 'state', '% 00-01',
+    'n 00-01', '% 01-02', 'n 01-02',
+    '% 02-03', 'n 02-03')
> (head(graduation[, 1 : 6]))
  region     state % 00-01 n 00-01 % 01-02 n 01-02
```

Statistics and Data with R: An applied approach through examples   Y. Cohen and J.Y. Cohen
© 2008 John Wiley & Sons, Ltd.

```
1   S    Alabama     63.7    37082   62.1   35887
2   NW   Alaska      68.0     6812   65.9    6945
3   SW   Arizona     74.2    46733   74.7   47175
4   S    Arkansas    73.9    27100   74.8   26984
5   W    California  71.6   315189   72.7  325895
6   M    Colorado    73.2    39241   74.7   40760
```

Next, we want to spell fully the region names in graduation. So we use the R internal data frame, called state.region:

```
> region <- c(as.character(state.region[1 : 8]),
+    as.character(state.region[4]),
+    as.character(state.region[9 : 50]))
> graduation$region <- as.factor(region)
> (head(graduation[, 1 : 6], 3))
  region       state % 00-01 n 00-01 % 01-02 n 01-02
1 South       Alabama    63.7    37082    62.1   35887
2 West        Alaska     68.0     6812    65.9    6945
3 West        Arizona    74.2    46733    74.7   47175
```

We want to obtain the average graduation rate (%) and the number that graduated. So we do:

```
> (round(apply(graduation[, 3 : 8], 2, mean), 1))
% 00-01 n 00-01 % 01-02 n 01-02 % 02-03 n 02-03
  73.0  50376.5    73.9 51402.6   74.9 53332.3
```

apply() applies to the data graduation[, 3 : 8], by column (hence the unnamed argument 2) the function mean(). To obtain the average by state, we do:

```
> pct <- round(apply(graduation[, c(3, 5, 7)], 1, mean), 1)
> n <- round(apply(graduation[, c(4, 6, 8)], 1, mean), 1)
> by.state <- data.frame(graduation[, 'state'], pct, n)
> names(by.state) <- c('state', '%', 'n')
> head(by.state)
       state    %        n
1    Alabama 63.5  36570.0
2     Alaska 67.3   7018.0
3    Arizona 74.9  47964.7
4   Arkansas 75.1  27213.0
5 California 72.8 327393.7
6   Colorado 74.8  40793.3
```

Here we use the unnamed argument 1 to apply() mean() to the appropriate rows. □

Next, let us examine a few related functions: tapply(), sapply(), lapply() and mapply().

Example 3.2. Continuing with the U.S. high school graduation rate data (Example 3.1), we wish to compute means by region:

```
> pct.region <- round(tapply(by.state[, 2],
+   graduation$region, mean), 0)
> n.region <- round(tapply(by.state[, 3],
+   graduation$region, mean), 0)
> rbind('%' = pct.region, n = n.region)
  North Central Northeast  South  West
%            80        77     68    73
n         54713     51717  52702 47612
```

Next, we want to calculate the average graduation rate and the average number of graduates by year and by region. So we first split graduation by region:

```
> grad.split <- split(graduation[, 3 : 8],
+   graduation$region)
> names(grad.split)
[1] "North Central" "Northeast"     "South"         "West"
```

and then sapply() means to the list components (the regions):

```
> round(sapply(grad.split, mean), 0)
        North Central Northeast South West
% 00-01            79        77    67   73
n 00-01         53731     50849 50982 46161
% 01-02            80        77    68   74
n 01-02         54303     51275 52391 47521
% 02-03            81        78    69   74
n 02-03         56104     53027 54734 49152
```

Let us double check that we are getting the right results for, say, the Western region of the U.S.:

```
> round(apply(grad.split$West, 2, mean), 0)
% 00-01 n 00-01 % 01-02 n 01-02 % 02-03 n 02-03
    73   46161      74   47521      74   49152
```
□

From the R help page, "sapply is a user-friendly version of lapply by default returning a vector or matrix if appropriate." mapply() gives results identical to those obtained from sapply() in Example 3.2. See Example 7.18 for an application of mapply().

3.2 Bar plots

Bar plots are the familiar rectangles where the height of the rectangle represents some quantity of interest. Each bar is labeled by the name of that quantity. Bar plots are particularly useful when you have two-column data, one categorical and the other numerical (usually counts). For example, you may have data where the first column holds species names and the second the number of individuals.

Example 3.3. The WHO data were introduced in Example 2.7. Figures 3.1 and 3.2 show data about the distribution of the population over age in two countries with very different cultures, economies and histories. The following script produces Figures 3.1 and 3.2.

```
1  load('who.pop.2000.rda')                  # population data
2  load('who.ccodes.rda')                    # country codes
3  load('who.pop.var.names.rda')#variable names in who.pop.2000
4
5  cn <- 'Austria'                           # country name
6  #cn <- 'Armenia'          # uncomment for Armenia bar plot
7  cc <- who.ccodes$code[who.ccodes$name == cn] # country code
8
9  par(mfrow = c(2, 1))
10 bl <- as.character(pop.var.names$descr[2 : 26]) # bar labels
11 gender <- 1                               # males
12 rows <- who.pop.2000$code == cc &         # row to be plotted
13    who.pop.2000$sex == gender
14 columns <- 5 : 29                         # columns to be plotted
15
16 barplot(t(who.pop.2000[rows, columns])[, 1]/1000,
17    names.arg = bl, main = paste(cn, ', males'),
18    las = 2, col = 'gray90')
19 gender <- 2                               # females
20 rows <- who.pop.2000$code == cc &
21    who.pop.2000$sex == gender
22 barplot(t(who.pop.2000[rows, columns])[, 1]/1000,
23    names.arg = bl, main = paste(cn, ', females'),
24    las = 2, col = 'gray90')
```

The script illustrates several important features of R. In particular, linking data from different data frames and using `barplot()`. It merits a detailed examination.

To produce the annotation in Figures 3.1 and 3.2, we need three data frames: `who.ccodes` contains the country codes which match a country code to a country name. `pop.var.names` matches variable names in `who.pop.2000` to meaningful names. For example, in `who.pop.2000`, there is a variable named Pop10. This variable holds the population of age group 20 to 24. Using `pop.var.names`, we can display the variable description (the string 20-24 that corresponds to the variable Pop10). `who.pop.2000` holds the data—population by age and countries. Figure 2.2 shows typical observations (rows) for each frame and the links that we need to display the bar plot properly.

In lines 1 to 3 we load the data. In line 5 we assign `Austria` to `cn`. If you wish to produce the bar plot for `Armenia`, comment line 5 and uncomment line 6. In line 7 we extract country codes from `who.ccodes`. This is how it is done: The statement inside the square brackets,

`who.ccodes$names == cn`

creates an unnamed logical vector. The length of this vector is the length of `who.ccodes$code`. All elements of this vector are set to `FALSE` except those elements whose value is `Austria`. These elements are set to `TRUE`. Because this unnamed logical vector appears in the square brackets, the index of the `TRUE` elements is used to extract the desired values from `who.ccodes$code`. These extracted values are stored

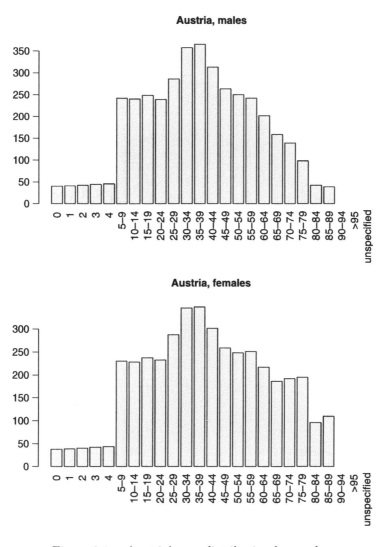

Figure 3.1   Austria's age distribution by gender.

in the vector cc. Thus, using the country name we extract the country code. Country names and country codes are unique. Therefore, cc contains one element only.

In line 9 we divide the graphics window into two rows and one column, so that we can plot both sexes on the same graphics window. Preparing a graphics window (also called a device) to accept more than one plot is common. It is done by specifying the number of rows and columns with the argument mfrow. In our case, we specify 2 rows and 1 column with c(2,1). We then set the argument mfrow with a call to par(). In line 10, we assign labels to the bars we are going to produce. The labels we need are for variables 2 to 26. The labels reside in the descr column of the pop.var.names data frame. In line 11 we set the gender to males. In lines 12 to 14 we prepare the logical vectors that will be used to extract the necessary row and columns from who.pop.2000. We need to extract the row whose code value corresponds to the

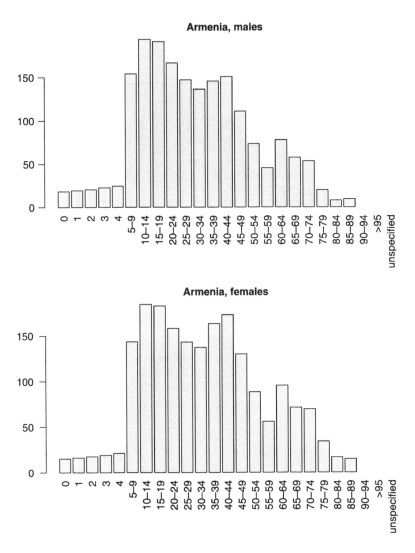

Figure 3.2   Armenia's age distribution by gender.

country code of Austria. We have this value in cc. The row we need to extract is for males, so the row we choose must have a value of gender = 1 in the sex column in who.pop.2000. The condition for extraction is stored in rows. Columns 5 to 29 in who.pop.2000 contain the age group populations.

We are now ready to call barplot(). In line 16, we extract the row and needed columns from the data frame. Before plotting, we must transpose the data because now the columns' populations must be represented as data (rows) to barplot(). This is done with a call to the transpose function t(). Note the division by 1 000. In line 17 we set the labels for the bars with the named argument names.arg. We also create the main title for the bar plot with paste(). In line 18 we set las to 2. This plots the tick labels perpendicular to the axes. The named argument col is set to gray90. This color is known to R as light gray. To find out color names in R, type colors(). Lines 19 to 24 repeat the bar plot, this time for females.   □

In Example 3.3, the x-axis shows the age categories into which the populations are divided. For example, in both countries, ages 10–14 and 15–19 are the most prevalent in the population. The example reveals interesting differences between and within countries with respect to gender. Think about answers to these questions:

- Why is there a dip in both male and female populations at the ages of 25–40 in Armenia compared to Austria?
- Why are there more older females than older males in Austria?
- Why is there a big jump in the age group from age 4 to ages 5–10 in both countries?

3.3 Histograms

Histograms are close relatives of bar plots. The main difference is that in histograms we are interested in the distribution of data. In other words, we wish to know if there is regularity in the number of observations that fall within a category. This means that how the data are binned takes on an additional importance.

Example 3.4. One of the most important activities that field ecologists pursue is estimating population densities by recording distances to observed organisms (see Buckland et al., 2001). It turns out that the way distance data are binned affects the way the data are adjusted and later used to infer population densities. When it comes to endangered and rare species, the decision on how to bin the data can influence decision making—about conservation actions, court rulings, etc. A circular plot is used to census birds. The observer sits at the center of an imagined disc with radius r and records distance and species for spotted individuals. In a study of bird density in the Sierra Nevada, such data were recorded for the Nashville warbler. Here are the 84 observations of distances, recorded from 20 such plots (personal data), each with a radius of 50 m:

```
15 16 10  8  4  2 35  7  5 14 14  0 35 31  0
10 36 16  5  3 22  7 55 24 42 29  2  4 14 29
17  1  3 17  0 10 45 10  9 22 11 16 10 22 48
18 41  4 43 13  7  7  8  9 18  2  5  6 48 28
 9  0 54 14 21 23 24 35 14  4 10 18 14 21  8
14 10  6 11 22  1 18 30 39
```

Figure 3.3 summarizes the data for different numbers of binning categories. `breaks = 11` is the default chosen by R. For `breaks = 4`, the data clearly indicate a regular (monotonic) decay in detectability of Nashville warblers as distance increases. This is not so for the other binned histograms.

The following script produces Figure 3.3.

```
load('distance.rda')
par(mfrow = c(2, 2))
hist(distance, xlab = '', main = 'breaks = 11',
    ylab = 'frequency', col = 'gray90')
hist(distance, xlab = '', main = 'breaks = 20',
    ylab = '' , breaks = 20, col = 'gray90')
```

```
hist(distance, xlab = 'distance (m)', main = 'breaks = 8',
    ylab = 'frequency', breaks = 8, col = 'gray90')
hist(distance, xlab = 'distance (m)', main = 'breaks = 4',
    ylab = '', breaks = 4, col = 'gray90')
```

We use `load()` to load the R data (a vector) `distance` in line 1. In line 2 we instruct the graphics device to accept four figures in a 2 by 2 matrix with a call to `par()` and with the named argument `mfrow` set to a 2 × 2 matrix of plots. The matrix is filled columns first. If we draw more than four, they will recycle in the graphics window.

In lines 3 and 4 we call `hist()` to plot the data with the default number of `breaks`, (which happens to be 11) with our own y-axis label (`ylab`) and with color (`col`) set to `gray90`. In lines 5 and 6 we plot the same data. But now we ask to break them into 20 categories of distances. In lines 7–10 we do the same for different numbers of breaks. □

Figure 3.3 is revealing. You may arrive at different conclusions about the distribution of the data based on different numbers of breaks. This provides an opportunity to

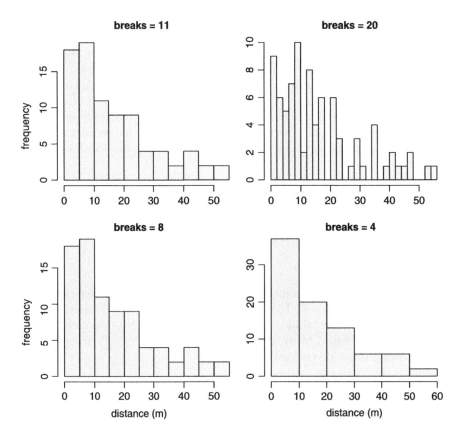

Figure 3.3 Histograms of distances to 84 observed Nashville warblers in twenty 50 m circular plots. The histograms are shown for different numbers of binning categories (*breaks*) of the data.

question conclusions from data. You should strive to have some theoretical (mechanistic) idea about what the distribution of the data should look like. The fact that there are some "holes" in observations when `breaks = 20` indicates that perhaps there are too many of them. Histograms are useful in exploring differences among treatments in experiments. Here is an example.

Example 3.5. The data, included with R's distribution, are about plant growth (Dobson, 1983). The data set compares yields—as measured by dry weight of plants—from a control and two treatments. There are 30 observations on 2 variables: weight (g) and treatment with three levels: `ctrl`, `trt1` and `trt2`. From Figure 3.4 it seems that the most frequent weight under the control experiment was between 5 and 5.5 g. In treatment 1, it was between 4 and 5 and in treatment 2 between 5.25 and 5.75. Note the insistence on consistent scales among the histograms of the different treatments. The following code was used in this example to produce Figure 3.4.

```
1  data(PlantGrowth) ; attach(PlantGrowth)
2  par(mfrow = c(1, 3))
3  xl <- c(3, 6.5) ; yl <- c(0, 4)
4  a <- hist(weight[group == 'ctrl'], xlim = xl, ylim = yl,
5     xlab = '', main = 'control',
6     ylab = 'frequency', col = 'gray90')
7  b <- hist(weight[group == 'trt1'], xlim = xl, ylim = yl,
8     xlab = 'weight', ylab = '', main = 'treatment 1',
9     col = 'gray90')
10 c <- hist(weight[group == 'trt2'], xlim = xl, ylim = yl,
11    xlab = '', ylab = '', main = 'treatment 2',
12    col = 'gray90')
```

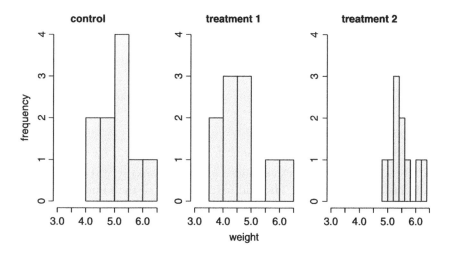

Figure 3.4 Control and two treatments in a plant growth experiment. Weight refers to dry weight.

The `PlantGrowth` data come with R. To load them, we call `data()` in line 1. To avoid extra typing, we `attach()` the data frame (also in line 1). In line 2 we tell the graphics device to accept one row of 3 plots. Because we wish all the plots to scale identically for all figures, we set `xl` and `yl` in line 3 and then in line 4 we specify the x- and y-axis limits with the `xlim` and `ylim` arguments. We do the same for the other 2 histograms.

In line 4, we choose a subset of the weight data that corresponds to the values of group = 'ctrl'. We do it similarly for the other two histograms in lines 7 and 10. We also set the x label to `xlab = 'weight'` in line 8. The y label (`ylab`) is set to frequency. Because we do not wish to clutter the graphs, we set `ylab = ''` for the other two histograms in lines 8 and 11. We distinguish between the histograms by specifying different `main` titles to each in lines 5, 8 and 11.

Note the assignment of the histograms to a, b and c. These create lists that store data about the histograms. This allows us to examine the breakpoints (`breaks`) and frequencies that `hist()` uses. We often use the data stored in the histogram list for further analysis. Let us see what a, for example, contains:

```
> a
$breaks
[1] 4.0 4.5 5.0 5.5 6.0 6.5

$counts
[1] 2 2 4 1 1

$intensities
[1] 0.4 0.4 0.8 0.2 0.2

$density
[1] 0.4 0.4 0.8 0.2 0.2

$mids
[1] 4.25 4.75 5.25 5.75 6.25

$xname
[1] "weight[group == \"ctrl\"]"

$equidist
[1] TRUE

attr(,"class")
[1] "histogram"
```

a stores vectors of the `breaks`, their `counts` and their `density`. `intensities` give the same information as `density`. The mid (`mids`) values of the binned data are listed as well. If you do not specify `xlab`, `hist()` will label x with `xname`. In this case the label will be `weight[group == "ctrl"]`. The extra backlashes are called escape characters. They ensure that the quotes are treated as characters and not as quotes. Another piece of information is whether the histogram is equidistant or not. Finally, we see that the attribute (`attr()`) of a is a `class` and the classname is `histogram`. You can use this information to later build your own graphs or tables. □

3.4 Dot charts

Dot charts are a good way to examine simple data-derived statistics such as means. Whenever you think "pie chart," think dot chart. Here is a direct citation from R's help page for `pie()`:

> "Pie charts are a very bad way of displaying information. The eye is good at judging linear measures and bad at judging relative areas. A bar chart or dot chart is a preferable way of displaying this type of data.
>
> Cleveland (1985), page 264: 'Data that can be shown by pie charts always can be shown by a dot chart. This means that judgments of position along a common scale can be made instead of the less accurate angle judgments.' This statement is based on the empirical investigations of Cleveland and McGill as well as investigations by perceptual psychologists."

One then wonders why pie charts are so popular in corporate financial reports.

Example 3.6. We discussed the WHO data in Example 2.7. The death rates, grouped by the WHO classified regions, are instructive (Figure 3.5). Africa stands apart from other regions in its high death rate. So does (to a lesser extent) Eastern Europe. The following script produces Figure 3.5.

```
who <- read.csv('who.by.continents.and.regions.txt',
  sep = '\t')
m.region <- tapply(who$dr, who$region, mean, na.rm = TRUE)
dotchart(m.region, xlab = 'death rate')
```

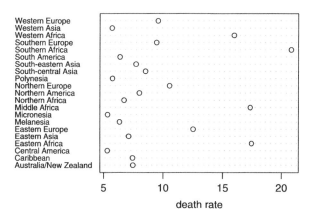

Figure 3.5 Death rate (per 1000). Based on WHO 2003 data, pertaining to 1995–2000.

The data are stored in a comma separated value (csv) text file. The file was saved from the WHO Excel file. In line 1 we import the data with `read.csv()`. The data

86   Presenting data

columns in the file are separated by tabs. Therefore, we set the Argument. sep = '\t'
In line 3 we compute the means by region with tapply(). Note how the argument to
mean(), na.rm (remove NA data), is specified. The call to tapply() returns an array
object m.region. In line 4 we call dotchart().   □

3.5 Scatter plots

In scatter plots we assign one variable to the x-axis and the other to the y-axis. We
then plot the pairs (x_i, y_i) of the data.

Example 3.7. In the WHO data (first introduced in Example 2.7), birth rate is
defined as the number of births per 1 000 people in the population in a year. Death
rate is defined similarly. Figure 3.6 shows the scatter plot for 222 nations. There
seems to be some regularity—nations with high birth rate also have high mortality
rate. Nations with very low birth rate seem to have higher mortality than nations
with moderate birth rate. Some nations of interest are identified by name.

Figure 3.6 was obtained thus:

```
1  who.fertility.mortality <- read.table(
2      'who.by.continents.and.regions.txt', sep = '\t',
3      header = TRUE)
4  names(who.fertility.mortality) <- c('country', 'continent',
5      'region', 'population', 'density', '% urban', '% growth',
6      'birth rate', 'death rate', 'fertility',
7      'under 5 mortality')
8
9  save(who.fertility.mortality,
10     file = 'who.fertility.mortality.rda')
11 d <- who.fertility.mortality
12 plot(d[, 'birth rate'], d[, 'death rate'],
13     xlab = 'birth rate', ylab = 'death rate')
14 unusual <- identify(d[, 8], d[, 9], labels = d[, 1])
15 points(d[unusual, 8], d[unusual, 9], pch = 19)
```

The script demonstrates the low level plotting function points() and other fancy plot
enhancements. In lines 1–3 we import the data and in lines 4–8 we name the columns
and save the data. We plot the data in lines 11–12. In line 13, we use identify()
to click on the points we wish to annotate. As labels for these points, we use the
country names in the first column of the data frame. identify() returns the index
of the points we clicked on. We store these indices in unusual. To differentiate these
from the remaining points, we plot the unusual with points() again in line 14, with
the plot character pch = 19 (solid circles).   □

The variable on the x-axis can be and often is, time. In such cases, it makes sense
to create a time series object, as opposed to a data frame. Here is an example.

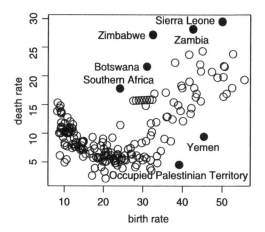

Figure 3.6   Death rate vs. birth rate for 222 nations.

Example 3.8. A celebrated data set is the one that raised the suspicion of global climate change (Keeling et al., 2003). As stated in their report,

> "Monthly values are expressed in parts per million (ppm) and reported in the 1999 SIO manometric mole fraction scale. The monthly values have been adjusted to the 15th of each month. Missing values are denoted by −99.99. The 'annual' average is the arithmetic mean of the twelve monthly values. In years with one or two missing monthly values, annual values were calculated by substituting a fit value (4-harmonics with gain factor and spline) for that month and then averaging the twelve monthly values."

The data are reported for the years 1958 through 2002. First, we import the data. This time, we shall do it with scan():

```
> manua <- scan()
1:
```

The data are in a single-column text file. So we copy all of it to the clipboard and then paste it into the work space. Here are the first few lines as they paste themselves:

```
1: NA
2: 315.98
3: 316.91
4: 317.65
5: 318.45
6: 318.99
7: NA
8: 320.03
```

When pasting is done, we hit the enter for an empty line. We now have the vector manua with the yearly CO_2 measurements. Next, we turn manua into a time series object like this:

```
> manua.ts <- ts(manua, start = 1958, end = 2002)
```

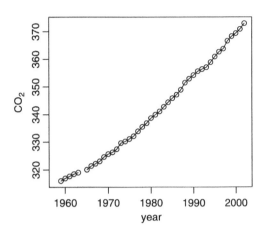

Figure 3.7 Atmospheric CO_2 (ppm) from Manua-Loa, Hawaii.

Now that we have the time series, plot() will display it appropriately:

```
> plot(manua.ts, xlab = 'year', ylab = expression(CO[2]))
```

(Figure 3.7). To show how time series plots deal with NA (not available) data, we plot the data with lines (the default) and add points. We assign a label to the y-axis with expression(). This function allows you to specify mathematical (TEX-like) expressions. Thus, CO[2] appears as CO_2. See plotmath() for further details on how to typeset mathematics in plots. expression() returns an expression object. Such objects can be later evaluated with a call to eval(). These two functions are very useful in cases where you wish to evaluate expression that you may build as strings. □

A good way to display potential paired interactions among variables in a multivariate data frame is the pairs() scatter plot.

Example 3.9. Let us continue with the WHO fertility and mortality data (see Example 3.7). We load the data and then plot columns 7–9 (% growth, birth rate and death rate) in pairs

```
> load('who.fertility.mortality.rda')
> pairs(who.fertility.mortality[, 7 : 9])
```

(Figure 3.8). Note the seemingly positive relationship between birth rate and % annual growth and negative relationship between death rate and % annual growth. In the latter, there are numerous countries that float above the seemingly negative relationship. We shall examine this in a moment. □

3.6 Lattice plots

So far, we discussed mostly univariate and bivariate data (bar, scatter and paired plots). Let us see how we may present trivariate data where one or two of the variables are factors.

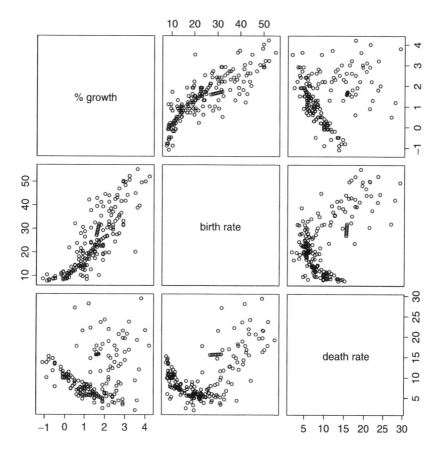

Figure 3.8   Growth, death and birth in 222 nations.

Example 3.10. In Example 3.9, we examined pairs of variables where the relationship between % growth (x-axis) and death rate were seemingly negative with a cloud of countries overhanging most others (bottom left, Figure 3.8). Let us see if we can isolate these countries by region. To that end, we

```
> load('who.fertility.mortality.rda')
> d <- who.fertility.mortality
> library(lattice)
```

and plot the data with

```
> xyplot(d[, 9] ~ d[, 7] | d[, 3], xlab = '% growth',
+    ylab = 'death rate', par.strip.text = list(cex = 0.6))
```

(Figure 3.9). Here we use R's *formula* syntax for the plot. Anything on the left side of ~ is a dependent variable and on the right an independent variable. In our case, we have death rate and % growth (columns 9 and 7 in d). The vertical bracket, |, indicates conditioning. We condition the xyplot() on the region factor (column 3 in d). Thus, we obtain a separate scatter plot for each level of region. Because the names of the regions are too long to fit in their space, we use

90   Presenting data

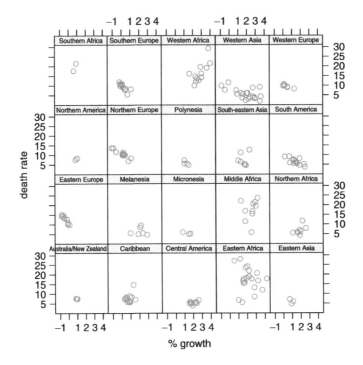

Figure 3.9   Death rate vs. % growth by region for 222 nations.

the strip-text argument. This argument reads a list(). One of the list components it knows about is cex—an argument that specifies the relative size of text in the graphics. Hence the list(cex = 0.6). Compare Figure 3.9 to the bottom left corner of Figure 3.8: The cloud of countries above the main trend (of negative relationship) is produced mostly by countries from Western, Middle and Eastern Africa.   □

3.7 Three-dimensional plots and contours

Three-dimensional plots and contour plots are used to represent relief, scatter plots and surfaces. The main functions and their corresponding packages are listed in the data frame graphics.3d.rda (available at the book website). We will use 3D plots here and there and illustrate them when the need arises.

3.8 Assignments

Exercise 3.1. The following data appeared in the World Almanac and Book of Facts, 1975 (pp. 315–318). It was also cited by McNeil (1977) and is available with R. It lists the number of discoveries per year between 1860 and 1959.

```
Start = 1860
End = 1959
5 3 0 2 0  3 2  3 6 1 2 1 2 1 3 3 3 5 2 4
4 0 2 3 7 12 3 10 9 2 3 7 7 2 3 3 6 2 4 3
5 2 2 4 0  4 2  5 2 3 3 6 5 8 3 6 6 0 5 2
2 2 6 3 4  4 2  2 4 7 5 3 3 0 2 2 2 1 3 4
2 2 1 1 1  2 1  4 4 3 2 1 4 1 1 1 0 0 2 0
```

1. Load the data. Then, of the reported data, in how many cases were there between 0 and less than 2 discoveries? Between 2 and less than 4 discoveries, between 4 and less than 6 and so on up to 12?
2. Based on the results in (1), plot a histogram of discoveries.
3. Do the data remind you of some regular curve that you may be familiar with? What is that curve?

Exercise 3.2.

1. Compare the age distribution of males to females in Austria (Figure 3.1). Speculate about the reasons for the difference in the survival of females and males.
2. Compare the age distribution of males to females in Armenia (Figure 3.2). Speculate about the reasons for the difference in the survival of females and males.
3. Compare the age distribution of males in Austria (Figure 3.1) and Armenia (Figure 3.2). Speculate about the differences in the survival of males in these two countries.
4. Compare the age distribution of females in Austria (Figure 3.1) and Armenia (Figure 3.2). Speculate about the differences in the survival of females in these two countries.
5. What information would you need to verify that your speculations are reasonable?

Exercise 3.3. Go to http://www.google.com. In the search box, enter the following string exactly as shown (including the quotes) "wind energy in X" where X stands for a state name. Spell the state names fully, including upper case letters. For example for X = New York, you enter (including the quotes) "wind energy in New York". Once you enter the string, click on the Google Search button. Under the search button, you will see how many items were found in the search. Record the number of items found and the state name. Repeat the search for X = all of the contiguous states in the U.S. Using the data you thus gathered (state name vs. the number of search items that came up):

1. Plot a histogram of the data.
2. What is the most common number of items found per state? How many states belong to this number?
3. What is the least common number of items found per state?
4. Using a histogram, identify a region in the U.S. (e.g. Northeast, Northwest, etc.) where most of the found items show up.
5. Why this particular region compared to others?

Exercise 3.4. Sexual dimorphism is a phenomenon where males and females of a species differ with respect to some trait. Among species of spiders, sexual dimorphism is widespread. Females are usually much larger than males (so much so, that they often eat the male after mating). Plot—by hand or with R—an imaginary graph that reflects the histogram of weights of individuals from various spider species. Explain the plot. If you choose R to generate the data, you can use `rnorm()` (look it up in Help).

Exercise 3.5. Use the discoveries data shown in Exercise 3.1:

1. Introduce a new factor variable named period. The variable should have 20 year periods as levels. So the first level of period is "1860–1879," the second is "1880–1899" and so on.
2. Compute the mean number of discoveries for each of these periods.
3. Construct a dot chart for the data.
4. Draw conclusions from the chart.

Exercise 3.6. For this exercise, you will need to use the function `rnorm()` (see Help). You have a vector that contains data about tree height (m). The first 30 observations pertain to aspen, the next 25 to spruce and the last 34 to fir.

1. Use a single statement to create imagined data from a normal distribution with means and standard deviations set to aspen: 5, 2; spruce: 8, 3; fir: 10.4.
2. Use a single statement to create an appropriate factor vector
3. Use a single statement to create a data frame from the two vectors.

No need to report the data. Just the code.

Exercise 3.7. Use a single statement to compute the mean height for aspen, spruce, fir and spruce from the `data.frame` created in Exercise 3.6

Exercise 3.8. Continuing with Exercise 3.6:

1. Use a single statement to reorder the levels of the species column in the data.frame you created in Exercise 3.6 such that species is an ordered factor with the levels aspen > spruce > fir.
2. Use an appropriate printout to prove that the factor is ordered.
3. Use a single statement to compute the means of the species height. The printout should arrange the means according to the ordered levels.

Exercise 3.9. In this exercise, before every call to a function that generates random numbers, call the function `set.seed(1)` exactly as shown. This will have the effect of getting the same set of random numbers every time you answer the exercise. You will also need to use the functions `runif()` and `round()`.

1. Create a matrix with 30 columns and 40 rows. Each element is a random number from a normal distribution with mean 10 and standard deviation 2. Show the code, not the data.
2. Create a submatrix with 6 rows and 6 columns. The rows and columns are chosen at random from the matrix. Show the code, not the data.
3. Print the submatrix with 3 decimal digits and without the row and column counters shown in the printout (i.e. without the dimension names; see `no.dimnames()` on page 32). Show the printout; it should look like this:

```
 9.220  9.224 10.122 10.912  9.671 10.445
13.557  8.984 10.508  4.006 11.976 11.321
 7.905  8.579 11.953 10.150 10.004 10.389
 8.688  9.950  7.087 12.098 11.141 10.305
11.890  9.409  6.299  5.620 11.083 13.288
10.198 11.002 12.363  8.662  4.222  9.074
```

Exercise 3.10. In 2003, there were 35 students in my statistics class. To protect their identity, they are labeled S1, ..., S35.

1. With a single statement, including calls to `factor()` and `paste()`, create a factor vector that contains the student labels.
2. Here are the results of the midterm exam

   ```
   68 76 66 90 78 66 79 82 80 71 90 78 68 52 86
   74 74 84 83 80 84 82 75 55 81 74 73 60 70 79
   88 73 78 74 61
   ```

   and final

   ```
   67 76 87 65 74 76 80 73 90 73 78 82 71 66 89
   56 82 75 83 78 91 65 87 90 75 55 78 70 81 77
   80 77 83 72 68
   ```

   Both results above are sorted by student label. Save the data as text files, named `midterm.txt` and `final.txt`. Import the data to R.
3. Create a data frame, named `exams`, with student labels and their grades on the midterm and final. Name the columns `student`, `midterm` and `final`. You may need to use `dimnames()` to name the columns.
4. Create a vector, named `average` that holds the mean of the midterm and final grade.
5. Add this vector to the `exams` data frame. When done with this part of the exercise, the `exams` data frame should look like this (only the first 5 records are shown; your frame should have 35 records).

   ```
   > exams[1 : 5,]
     student midterm final average
   1      S1      68      67    67.5
   2      S2      76      76    76.0
   3      S3      66      87    76.5
   4      S4      90      65    77.5
   5      S5      78      74    76.0
   ```

6. Create a data frame named `class`. Your data frame should look as follows (your data frame should include values for grades instead of `NA`):

   ```
   > class
        exam grade
   1 midterm    NA
   2   final    NA
   3   total    NA
   ```

7. Create a list, named class.03. The list has two components, class and exams; both are the data frames you created. The list should look as shown next (your list should include data instead of NA). Only the first 5 rows of the second component are shown.

```
> class.03
$class.mean
    exam grade
1 midterm NA
2   final NA
3   total NA
$student.grades
  student midterm final average
1      S1      68    67    67.5
2      S2      76    76    76.0
3      S3      66    87    76.5
4      S4      90    65    77.5
5      S5      78    74    76.0
```

8. Show two ways to access the first 5 records of the students.grades data frame in the class.03 list.

Exercise 3.11. Download the file elections-2000.csv from the book's website and:

1. Create a data frame named Florida.
2. How many counties are present in the data file?
3. In how many counties did the majority vote for Gore? For Bush?
4. Suppose that all the votes for Buchanan were to go to Bush and all the votes for Nader were to go to Gore. Who wins the election? By how many votes?

Part II

Probability, densities and distributions

4

Probability and random variables

Probability theory involves the study of uncertainty. It is a branch of applied mathematics and statistics. The latter permeates every aspect of human endeavor: science, engineering, behavior and games. In building safe cars, airplanes, trains and bridges, engineers address the issue of safety with probability. Statistical inference is closely related to probability. To introduce the subject, we begin with an example. The example demonstrates how often uncertainty influences our everyday perceptions.

Example 4.1. This example was cited in Dennett (1995) in reference to evolution. I claim that I can produce a person who guesses correctly ten consecutive flips of a coin. I can produce such a person even if the coin is not fair; but let us assume it is. With flipping a fair coin, heads are as likely to show up as tails. Your initial reaction would be disbelief. "After all," you might reason, "the probability that a person guesses right a single flip of a fair coin once is $1/2$, twice $1/2 \times 1/2 = 2^{-2}$, ..., 10 times in a row $2^{-10} \approx 0.00098$." Certainly an unlikely event. But imagine the following experiment. We pick $2^{10} = 1\,024$ people for a tournament. Of each pair of players, the one who guesses the outcome of a coin-flip wrong is eliminated. In the first round, we start with 1 024 contestants and end with 512 winners. In the second round we end with 256 winners; all of them guessed right two flips in a row. After 10 rounds, we are guaranteed to have a single winner who guessed correctly 10 flips in a row.

When evolution is regarded as practically an infinite number of elimination contests, with few winners, it's no wonder that things look so unlikely to have happened without divine intervention! ☐

What do we mean when we say probability? Intuitively, we mean the chance that a particular event (or a set of events) will occur. This chance is quantified with a real number between 0 and 1. The closer the number is to 1, the more likely the event. To deal with probability, we need to start with set theory.

4.1 Set theory

In this section we first discuss sets and algebra of sets. Next, we discuss applications of R to set theory.

4.1.1 Sets and algebra of sets

A *set* is a collection of objects. These objects are called *elements*. We say that B is a *subset* of A if all the elements of B are also elements of A. We write it as $B \subset A$. Accordingly, any set is a subset of itself. We denote sets with upper-case letters and elements of sets with lower-case letters. To indicate that a collection of elements form a set we enclose the elements, or their description, in braces. The expression

$$A = \{a_1, a_2, \ldots, a_n\} \tag{4.1}$$

says that the set A consists of a collection of n elements, a_1 through a_n. A set can be characterized by its individual elements—as in (4.1)—or by the properties of its elements. For example,

$$A = \{a : a > 1\} \tag{4.2}$$

says that the set A consists of elements a such that $a > 1$. Often we need to be more explicit about the set from which the elements are drawn. For example, (4.2) describes two different sets when a is an integer, or when it is a real number. In the present context, the integers or the real numbers are the underlying *spaces* or the *universe*. We often denote the universal set by S. We call a set with no elements the *empty* or the *null* set and denote it by $\varnothing$.

Example 4.2. An organism is either alive (a) or dead (d). Therefore we write

$$S = \{a, d\} \; .$$

The set of all possible subsets of S is

$$\mathbb{P} = \{\varnothing, \{a\}, \{d\}, S\} \; .$$

The set of all subsets, $\mathbb{P}$, consists of $2^2 = 4$ subsets. □

Elements of sets can be sets. We call the set of all possible subsets the *power set* and denote it by $\mathbb{P}$. It can be shown that the power set of any set with n elements consists of 2^n subsets.

To understand how probabilities are constructed and manipulated, we need to be familiar with set operations. We introduce these operations with the help of Venn diagrams. In the following diagrams, squares represent the universal set S.

Transitivity If $A \subset B$ and $B \subset C$ then $A \subset C$ (Figure 4.1). Thus, for any set A,

$$A \subset A, \quad \varnothing \subset A, \quad A \subset S \, .$$

Example 4.3. Let C be the set of all people in South Africa, B be the set of all black people in South Africa and A the set of all black men in South Africa. Then A is a subset of B and B is a subset of C. Obviously, A is a subset of C. Here we may identify S with C. □

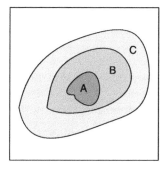

Figure 4.1   Transitivity.

Equality Set A equals set B if and only if every element of A is an element of B and every element of B is an element of A. That is,

$$A = B \text{ if and only if } A \subset B \text{ and } B \subset A.$$

Union The union of two sets A and B is a set whose elements belong to A, or to B, or to both (Figure 4.2). The union is denoted by $\cup$ and $A \cup B$ reads "A union B."

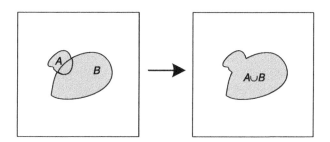

Figure 4.2   Union.

Example 4.4. Let A be the set of all baseball, basketball and football players and B the set of all basketball and hockey players. Then $A \cup B = \{$baseball, basketball, football, hockey players$\}$. Note that basketball players are *not* counted twice. In unions, common elements are never counted twice. □

Associativity For any sets A, B and C

$$A \cup B \cup C = (A \cup B) \cup C = A \cup (B \cup C).$$

Example 4.5. Consider a small hospital with 3 wards. Let A, B and C be the sets of patients in each of these wards. Then, $A \cup B \cup C$ is the set of all patients in the hospital. Now take $D := A \cup B$. Obviously, $D \cup C$ is the set of all patients in the hospital. Also, for $D := B \cup C$, we have that $A \cup D$ is the set of all patients in the hospital. □

Commutativity Using the same reasoning as for associativity, you can easily verify that

$$A \cup B = B \cup A.$$

Also,
$$A \cup \emptyset = A, \quad A \cup S = S,$$
and if $B \subset A$ then $A \cup B = A$.

Example 4.6. Let
$$A = \{\text{oranges, tomatoes}\},$$
$$B = \{\text{bananas, apples}\},$$
$$S = \{\text{fruits}\}.$$

Then
$$A \cup B = \{\text{oranges, tomatoes, bananas, apples}\}$$
and
$$B \cup A = \{\text{bananas, apples, oranges, tomatoes}\}.$$

Because order is not important, we can rearrange the items in $A \cup B$ such that $A \cup B = B \cup A$. Also, the union of A with nothing gives A. And the union of A with S gives fruits because all member of the union are fruits. □

Intersection The intersection of two sets, A and B, is a set consisting of all elements belonging to both A and B. This is written as $A \cap B$. From Figure 4.3,
$$(A \cap B) \cap C = A \cap (B \cap C) = A \cap B \cap C.$$

It then follows that

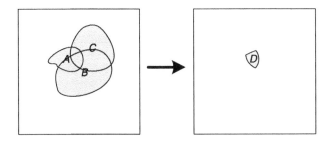

Figure 4.3   Intersection.

$$A \cap A = A, \quad A \cap \emptyset = \emptyset, \quad A \cap S = A.$$

Example 4.7. Let
$$A = \{1, 2, 3, 4\},$$
$$B = \{3, 4, 5, 6\},$$
$$C = \{4, 5, 6, 7\},$$
$$S = \text{the set of all integers}.$$

The elements 1, 2, 3 and 4 are members of the set A. The common elements of $A \cap A$ are also these elements. Therefore, $A \cap A = A$. There are no elements common to A and $\varnothing$. Therefore, $A \cap \varnothing = \varnothing$. Finally, the elements common to A and S are the elements of A. Therefore, $A \cap S = A$.

Note that
$$A \cap B = \{3, 4\} \ .$$
Then
$$(A \cap B) \cap C = \{3, 4\} \cap \{4, 5, 6, 7\} = \{4\} \ .$$
Similarly,
$$B \cap C = \{4, 5, 6\} \ .$$
Then
$$A \cap (B \cap C) = \{1, 2, 3, 4\} \cap \{4, 5, 6\} = \{4\} \ .$$
Again, note that common elements are not counted twice. □

If two sets have no elements in common then
$$A \cap B = \varnothing \ .$$
A and B are then said to be *disjoint* sets (or *mutually exclusive* sets).

Distribution As Figure 4.4 illustrates, the distributive law for sets is

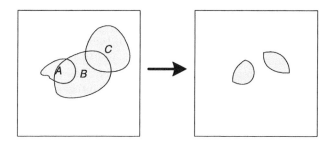

Figure 4.4   Distribution.

$$A \cap (B \cup C) = A \cap B \cup A \cap C \ .$$

Example 4.8. Returning to Example 4.7, we have
$$D := B \cup C = \{3, 4, 5, 6, 7\}$$
and
$$A \cap D = \{1, 2, 3, 4\} \cap \{3, 4, 5, 6, 7\}$$
$$= \{3, 4\} \ .$$
Similarly,
$$D := A \cap B = \{3, 4\} \ , \quad E := A \cap C = \{4\} \ .$$
Therefore,
$$D \cup E = \{3, 4\} \cup \{4\} = \{3, 4\} \ .$$
□

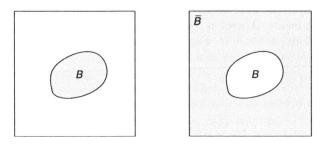

Figure 4.5 Complement.

Complement The complement of the set A, denoted by $\overline{A}$, is the set of all elements of S that are not in A (Figure 4.5). Thus,

$$\overline{\varnothing} = S, \quad \overline{S} = \varnothing,$$
$$\overline{\overline{A}} = A, \quad A \cup \overline{A} = S,$$
$$A \cap \overline{A} = \varnothing,$$
$$\text{if } B \subset A \text{ then } \overline{B} \supset \overline{A},$$
$$\text{if } A = B \text{ then } \overline{A} = \overline{B}.$$

Example 4.9. Let S be the set of all integers. Then in the defined space, $\varnothing$ has no elements. All of the elements that are not in $\varnothing$ are integers and they constitute S. In notation, $\overline{\varnothing} = S$. Similarly, because S includes all of the integers, the set that has no integers in the space of all integers is empty. In notation, $\overline{S} = \varnothing$.

Let $A = \{1, 2\}$. Then, $\overline{A}$ is the set of all integers except 1 and 2. The set of all elements that are not in $\overline{A}$ is $\overline{\overline{A}} = \{1, 2\} = A$.

Let $B = \{1\}$. Then, $B \subset A$. Also, $\overline{B}$ is the set of all integers except 1 and $\overline{A}$ is the set of all integers except 1 and 2. Therefore, $\overline{A} \subset \overline{B}$.

Finally, let $C = \{1, 2\}$. Then obviously $\overline{A} = \overline{C}$. □

Difference The set $A - B$ consists of all of the elements of A that are not in B. Similarly, the set $B - A$ consists of all elements of B that are not in A (Figure 4.6). To distinguish the "$-$" operation on sets from the usual subtraction operation, we sometimes write $A - B := B \backslash A$. Note that

$$A - B = A \cap \overline{B} = A - A \cap B.$$

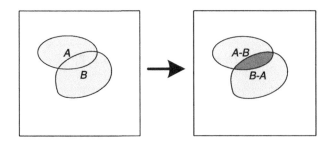

Figure 4.6 Difference.

To convince yourself that this indeed is the case, trace the sets $\overline{B}$, $A \cap \overline{B}$ and $A \cap B$ in Figure 4.6. In general, $(A - B) \cup B$ does not equal A. Furthermore,

$$(A \cup A) - A = \varnothing \quad \text{while} \quad A \cup (A - A) = A.$$

Further reflection will convince you that for any A,

$$A - \varnothing = A, \quad A - S = \varnothing, \quad S - A = \overline{A}.$$

Example 4.10. Let $A = \{1, 2, 3, 4\}$, $B = \{3, 4, 5, 6\}$ and S the set of all integers. Then

$$A - B = \{1, 2\}, \quad B - A = \{5, 6\}.$$

Also,

$$\begin{aligned} A \cap \overline{B} &= \{1, 2, 3, 4\} \cap \{\text{all integers except } 3, 4, 5, 6\} \\ &= \{1, 2\} = A - B \end{aligned}$$

and

$$\begin{aligned} A - A \cap B &= \{1, 2, 3, 4\} - \{3, 4\} \\ &= \{1, 2\} = A - B. \end{aligned}$$

□

Sum The sum of two sets is a new set with all the elements of both. Common elements are counted twice. Thus,

$$A + B = A \cup B + A \cap B.$$

Therefore,

$$A \cup B = A + B - A \cap B.$$

Example 4.11. The sum of sets A and B in Example 4.4 is

$$A \cup B = \{\text{Baseball, basketball, football, basketball, hockey players}\}.$$

Basketball players are counted twice! □

4.1.2 Set theory in R

The most obvious applications of set theory ideas in R relate to data manipulation and spatial analysis. Examples of common tasks are union and intersection of polygons and tests for whether points are within a polygon. R includes several packages that make such work easy.

Some of the spatially related packages are geoR, gstat, splancs and gpclib. Review the help for these packages and you will probably find functions that do what you need. We discussed merge(), union() and intersect() in Example 2.17.

4.2 Trials, events and experiments

It is beyond our scope to define probabilities, events, probability spaces, sample spaces and chance experiments rigorously. The exposition below is heuristic and therefore not entirely correct because (technical) details must be omitted. However, you should get a feel for what these concepts mean. The following sections are based on Papoulis (1965) and Evans et al. (2000).

We start with the following definition:

Outcome Any observable phenomenon is said to be an outcome.

In the context of probability theory, we define a set of outcomes from the description of an experiment. The outcomes may not be unique, so we must agree upon their definition to avoid ambiguity. We associate uncertainty with outcomes. The uncertainty is measured with probability. The latter ranges from 0 to 1. The probability of an outcome that is certain to occur is 1 and the probability of an outcome that never occurs is 0. An experiment here does not necessarily mean some activity that we undertake. It refers to anything we wish to observe.

Example 4.12. Table 4.1 illustrates some experiments and their outcomes. Note that the outcomes can be factors, integers, real numbers or anything else you wish to define as an outcome. □

Table 4.1   Experiments and outcomes.

Experiment	Outcomes
Observing a sick person	sick, healthy, dead
Treating a sick person	sick, healthy, dead
Rolling a die	number of dots facing up
Earthquake	magnitude
Weighing an elephant	the elephant's weight

The definition of outcome leads to another important concept in probability theory, namely the definition of

Sample space The set of all possible outcomes, denoted by S, is called the sample space. A sample space is also known as an event space, possibility space or simply the space.

Example 4.13. The sample space of the state of two organisms (dead (d) or alive (a)) is

$$S = \{aa, ad, da, dd\} \ .$$

The sample space of the magnitude of an earthquake is

$$S = \text{ the set of all real numbers} \ .$$

□

With the concept of sample space, we have the definition of

Event An event is a subset of the sample space.

In notation, if S is a sample space, then $E \subset S$ is an event. Because sets are subsets of themselves, S is also an event.

Example 4.14. In the context of an experiment, we may define the sample space of observing a person as

$$S = \{\text{sick, healthy, dead}\} \ .$$

Therefore, the following are all events:

{sick} , {healthy} , {dead} ,
{sick, healthy} , {sick, dead} , {healthy, dead} ,
{sick, healthy, dead} , {none of the above} .

The sample space of elephant weights is

$$S = \text{real numbers} .$$

Therefore, the following are all events:

the set of all real numbers between $-\infty$ and ∞ ,

the set of all real numbers between -10 and 5.32 ,

any real number .

□

You may object to some event definitions for elephant weights. However, we can assign 0 probability to events. Therefore, negative weights are acceptable as weights, as long as we assign zero probability to them. A special kind of event is an

Elementary (or simple) event An event that cannot be divided into subsets.

Example 4.15. Consider a study of animal movement. We classify behaviors as a - standing, b - walking and c - running. Then $A := \{a\}$, $B := \{b\}$ and $C := \{c\}$ are all elementary events. a, b and c are not events at all because they are not sets. We may consider an observation as a complete sequence (in any order) of A, B and C. Then $\{a, b, c\}$ is an elementary event, but $\{A, B, C\}$ is not. □

Because events are subsets of S, they can include more than one outcome. If one of these outcomes occurred, we say that the event occurred.

Example 4.16. In a study of animal behavior, we classify the following events: $A =$ standing, $B =$ walking, $C =$ running, $D =$ lying and $E =$ other. We are interested in two events: F - moving, G - not moving. Then if we observe the animal walking, we say that event F occurred. Similarly, if we observe the animal laying, then we say that event G occurred. □

Recall that we defined disjoint sets as those sets whose intersection is empty. For events, we define

Disjoint events A and B are said to be disjoint events if $A \cap B = \emptyset$.

Example 4.17. Let $A =$ pneumonia, $B =$ gangrene and $C =$ dead be possible outcomes of the observation that a person is alive. Then A and C are disjoint events. A and B are not. □

The ideas of outcome, sample space and event lead to the following definition:

Trial A single performance of an experiment whose outcome is in S. The following are examples of trials: flipping a coin, rolling a die, treating a pond with rotenone, treating a patient with a particular drug and recording the magnitude of an earthquake. The simplest trial is defined as

Bernoulli trial A trial with only two possible outcomes, one arbitrarily named a success and the other a failure.

Example 4.18. A single flip of a coin is one example of a Bernoulli trial. It may consists of an idealized coin—a circular disk of zero thickness. When flipped, it will come to rest with either face up ("heads", H, or "tails", T) with equal probability. A regular coin is a good approximation of the idealized coin. □

Chance experiment A chance experiment (or experiment for short) is a trial with more than one possible outcomes where the amount of uncertainty of different outcomes and their combinations is known or deducible.

Example 4.19. Flipping a fair coin, with outcomes defined as H and T is a chance experiment. The sample space is

$$S = \{H, T\} \ .$$

and the amount of uncertainty of any outcome or their combinations is known.

In a medical study, giving patients a drug and observing the outcome is a chance experiment. The outcome is uncertain and we assign hypothetical probability to outcomes (e.g. healthy, sick, or dead). When the experiment is over, we may use the results to test if our hypotheses about the probabilities—of being healthy, sick, or dead—were justified. □

Here is an example where a chance experiment is defined and the sample space is determined.

Example 4.20. You observe deer crossing the highway. The experiment consists of observing the sex of two consecutive deer. Let M be the event that a male crossed the highway and F the event that a female did. To define the sample space, we use a tree diagram (Figure 4.7). Here, the set of all possible outcomes is a female crossed the road and then a female or a male, a male crossed the road and then a female or a male. Therefore,

$$S = \{FF, FM, MF, MM\} \ .$$

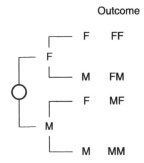

Figure 4.7 A tree diagram.

To create the sample space with R, we use `combn()`.

```
> no.dimnames(t( combn(c('F', 'M', 'F', 'M'), 2)))
 "F" "M"
 "F" "F"
 "F" "M"
 "M" "F"
 "M" "M"
 "F" "M"
```

From the innermost parentheses out: We create a vector that labels the possible outcomes in the first and second pair of observations. Next, `combn()` creates a matrix of all possible combinations of two from the vector. Next, we transpose the matrix (i.e. rows become columns) with `t()`. Finally, we print this matrix with `no.dimnames()` (see page 32). □

When the number of possible outcomes is small, we can present the possible outcomes with a tree diagram. In Example 4.20, an elementary event consists of two (*not one*) crossings. Here is another example.

Example 4.21. Mist nets are used to catch birds. They are made of fine nylon mesh so birds do not see them. The nets hang somewhat loosely and when a bird flies into one, it gets tangled. Different meshes are used to catch different sizes of birds. Suppose you have four mist nets, each of a different mesh. Call them nets 1, 2, 3 and 4. You wish to allocate each of these nets to one of four study areas named a, b, c and d. Here are the possible outcomes:

```
> p <- t(expand.grid(letters[1 : 4], 1 : 4))
> no.dimnames(noquote(p))

 a b c d a b c d a b c d a b c d
 1 1 1 1 2 2 2 2 3 3 3 3 4 4 4 4
```

The function `expand.grid()` creates a data frame from all combinations of the supplied vectors. We give it a vector of the `letters` a through d and the sequence of numbers 1 through 4. With `t()` we switch columns and rows in the resulting data frame p. To see the combinations without quotes, we use `noquote()` and to see it without the dimension names we use `no.dimnames()`. The latter is discussed on page 32.

The output from the calls above produces the sampling space S. Here we label an elementary event by a pair of one letter and one digit—for example, $(a, 1)$ means that net 1 was assigned to area a. Because they are subsets of S, the following are events. The set of all events such that the chosen area is a is

```
> (A <- no.dimnames(noquote(p[, p[1, ] == 'a'])))

 a a a a
 1 2 3 4
```

From the inside out, the expression `p[1, ] =='a'` returns TRUE for all those columns in the first row of p that have the value a. Next, `p[, p[1, ] == 'a']` returns the subset of the data frame that has a in its first row. To avoid clutter, we print A with no dimension names. The events where mist net 1 is allocated to all areas are obtained with

```
> (B <- no.dimnames(noquote(p[, p[2, ] == 1])))

  a b c d
  1 1 1 1
```

Again, from the inside out, the expression p[2,] == 1 returns TRUE for all those columns in the second row of p that have the value 1. Next, p[, p[2,] == 1] returns the subset of the data frame that has 1 in its second row. Again, to avoid clutter, we print B with no dimension names. Now the intersection of the events A and B gives the single event

```
> noquote(intersect(A, B))
[1] a 1
```

You could accomplish all of this without R. However, if your sets are more complex and the combinations and subsets more elaborate, deriving the results by hand may be tedious. □

One of the most fundamental and pervasive experiments is the Bernoulli experiment.

Bernoulli experiment We call an experiment with a single event and two outcomes a Bernoulli experiment.

Example 4.22. The simplest example of a Bernoulli experiment is a flip of a coin. The event is a side of a coin landing face up. The two possible outcomes are heads (H) or tails (T). □

A Bernoulli experiment is important in its own right. In many situations in life we face binary choices with no certain outcomes. We drive a car and may or may not get into an accident. Any aspect of computer logic and computations whose outcome may not be certain involves binary operations. In decision making we often reduce choices to binary operations—to act or not to act. Bernoulli trials serve as the starting point for many other probability models which we shall meet as we proceed.

4.3 Definitions and properties of probability

There are several competing interpretations—and therefore definitions—of probability. Yet, they all agree that a probability is a real number between 0 and 1 and that the larger its value, the more likely the event. We write $P(E)$ to mean the probability of event E.

4.3.1 Definitions of probability

As a prelude to the definition of probability, let us consider two examples. These illustrate two different approaches to the definition of probability.

Example 4.23. Consider a global disease epidemic that infects one quarter of the world's population. You work for the World Health Organization, so you travel the world and record whether every person you meet (independent of any other person) is uninfected or infected. Assign the value 1 to an infected person and 0 to an uninfected person. What would be the running proportion of infected persons you record? Let n_S (success) be the event that a person is infected and n the total number of persons you encounter. The first person you meet may be infected or uninfected. Suppose they turn out to be infected. Then the current proportion is $p_1 = 1$. The next person you encounter is uninfected and the current proportion is $p_2 = 1/2$. After n encounters, you have $p_n = n_S / n$. As n increases, you expect that $p_n \approx 1/4$. Instead of physically counting the infected, let us simulate the process.

```
1  set.seed(100)
2  n.S <- ifelse(runif(1, 0, 1) < 0.25, 1, 0)
3  p <- vector()
4  for(n in 2 : 10000){
5    n.S <- n.S + ifelse(runif(1, 0, 1) < 0.25, 1, 0)
6    p[n] <- n.S / n
7  }
```

In line 1, we set.seed(). This allows us to repeat the same sequence of random numbers every time we run the script. Thus, the numbers we generate are called pseudo-random numbers. In line 2, we set our first "success" to zero or one. We generate a single random number that has equal probability of being between zero and one. Hence the call to runif(1,0,1). Accordingly, ifelse() assigns 1 if the random number is < 0.25 and 0 otherwise. If we runif() many times with these argument values, then approximately 1/4 of them will be less than 0.25.

In line 3, we initiate the vector of proportions with vector(). Then, we "encounter" 10 000 individuals with for(). We accumulate the number of infected persons we meet in n_S in line 5. In line 6, we accumulate the proportion of infected persons to the total number of persons we met thus far. This is an inefficient way of doing things in R, but we do it for heuristic reasons. The "process" above can be generated with the single line

```
p <- cumsum(ifelse(runif(10000, 0, 1) < 0.25, 1, 0)) / (1:10000)
```

To see our experiment, we plot the accumulating proportions with this:

```
plot(p, type = 'l', ylim = c(0, 0.3),
  xlab = expression(italic(n)), ylab = expression(italic(p)))
```

The type = 'l' argument ensures that a line (as opposed to the default points) is drawn. To zoom in on the results, we set the limits of the y-axis between 0 and 0.3 with ylim = c(0, 0.3). To annotate the axes with xlab and ylab with italics, we call the function italic() and run the function expression() on the result. Finally, we add a horizontal line using abline() with the argument h (for horizontal) at 0.25:

```
abline(h = .25)
```

Figure 4.8 illustrates the results of our experiment. As we accumulate samples, the proportion of infected persons approaches the true proportion of 0.25. The fact that this happens seems trivial. The mathematical proof that this will happen every time you run a similar experiment is not. □

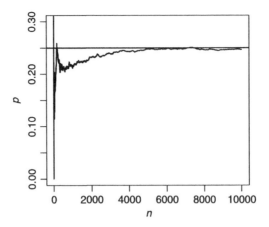

Figure 4.8   A simulation of a disease epidemic.

The result in Example 4.23 is based on an experiment. Next, we obtain results based on reasoning.

Example 4.24. Select a card from a well-mixed deck. Let C be the event that the selected card's suit is clubs. There is a 25% chance that the selected card's suit is clubs. How are we to interpret this statement? Imagine that we draw a card many times. In approximately 25% of the draws the card's suit will be clubs. After all, a single draw must be of a card with only one suit. Thus we write

$$\text{relative frequency of } C = \frac{\text{number of times } C \text{ occurs}}{\text{number of draws}}.$$

Therefore,
$$P(C) = 0.25 \,.$$

Also,
$$\text{not } C = 1 - P(C) = 0.75 \,.$$

Now draw five cards. Let F be the event that all five cards are of the same suit. Dealing 5 cards repeatedly, we write

$$\text{relative frequency of } F = \frac{\text{number of times } F \text{ occurs}}{\text{number of draws}}.$$

Therefore,
$$P(F) = \frac{13}{52} \times \frac{12}{51} \times \frac{11}{50} \times \frac{10}{49} \times \frac{9}{48} = 0.000\,495 \,. \qquad \square$$

By the same token, instead of tossing a coin and determining the probability of heads, we could have assumed that the coin is perfect, there is no wind, it falls on a flat, horizontal surface, etc. In other words, under ideal conditions, we *reason* that the probability of heads is 1/2.

As you can see, the perception of probability in Examples 4.23 and 4.24 differs. The former relies on experimentation and limit arguments, the latter on logical consequences of the assumptions. Both examples reflect the view of probability as a frequency. We call this the *frequentist* view of probability. The distinction between the two is that in Example 4.23, relying on the law of large numbers, probability is defined as a limit:

$$p = \lim_{n \to \infty} \frac{n_A}{n}$$

where n_A is the number of times the event A occurs in n trials. The notation indicates that p equals the right hand side in the limit as $n \to \infty$. According to Example 4.24,

$$p = \frac{n_A}{n}$$

is a logical consequence of the assumptions.

There is a different view of probability called the *Bayesian* view. Here probability is treated somewhat subjectively. Probability is presumed to have a distribution. A statistical procedure is then applied to estimate the parameters of the underlying distribution based on the observed distribution. We will not discuss this view at all (for further details, see Jaynes, 2003).

Our working definition of probability corresponds to the frequentist view. Therefore, we define probability as follows:

Probability The probability of event E, denoted by $P(E)$, is the value approached by the relative frequency of occurrences of E in a long series of replications of a chance experiment.

4.3.2 Properties of probability

From the above discussion, we conclude the following:

1. For any event E, $0 \leq P(E) \leq 1$.
2. The probabilities of all elementary events must sum to 1.
3. Let E be composed of a set of elementary events. Then, $P(E)$ is the sum of the probabilities of all elementary events contained in E.
4. For any event E, $P(E) + P(\text{not } E) = 1$.

A more general version of these propertiesis called the probability axioms. These were first articulated by Kolmogoroff (1956).

Example 4.25. Assume that a certain forest tract is visited by individual birds randomly and independent of each other. Based on a few days of observations you conclude that 30% of the birds visiting the tract are birds of prey and the rest are song birds. Of the song birds, 25% are finches, 18% are warblers, 15% are vireos and 12% are chickadees. Define the event E as the next visiting bird is a song bird and D as the event that the next visiting bird is a bird of prey. Then from the second property of probabilities, we have

$$P(E) + P(D) = 1 \quad \text{or} \quad P(E) = 1 - 0.3 = 0.7 \ .$$

Another way of achieving the same result is

$$P(E) = 0.25 + 0.18 + 0.15 + 0.12 = 0.7 \ .$$

112 Probability and random variables

The probability that the next visiting bird is not a song bird is obviously 0.3. This probability can also be obtained from

$$P(\text{not } E) = 1 - P(E) = 1 - 0.7 = 0.3 \,. \qquad \square$$

From the properties of probability, we conclude the following. Recall that elementary events are disjoint and when calculating probabilities of their combinations, we simply add the probabilities of the elementary events. Some combined events can be disjoint as well. Thus, let $E_1, E_2, \ldots, E_n$ be disjoint events. Then[1]

$$P(E_1 \text{ or } E_2 \text{ or } \cdots \text{ or } E_n) = P(E_1) + P(E_2) + \cdots + P(E_n) \,.$$

4.3.3 Equally likely events

Equally likely events refer to events with equal probability of occurrence. When we say an individual is chosen randomly, we mean that all individuals have equal probability of being chosen.

Example 4.26. In a field research project, a crew of four students, two males and two females, is selected to set traps in a certain area. They decide to divide themselves randomly into two pairs. Denote the female students by f_1 and f_2 and the male students by m_1 and m_2. The possible pairs are

$$\{(f_1, f_2), (f_1, m_1), (f_1, m_2), (f_2, m_1), (f_2, m_2), (m_1, m_2)\}$$

and therefore

$$P\{(f_1, f_2)\} = \cdots = P\{(m_1, m_2)\} = \frac{1}{6} \,.$$

Let $E =$ both members of a pair are of the same gender. Then

$$E = \{(f_1, f_2), (m_1, m_2)\} \,, \quad P(E) = \frac{2}{6} \,.$$

Let F be the event that at least one of the members of a pair is a female. Then

$$P(F) = \frac{5}{6} \,. \qquad \square$$

When the number of elementary events is large, finding various combinations of events is difficult. We then rely on counting rules (see Section 4.5.4).

4.3.4 Probability and set theory

Let us reflect upon the connection between probability and set theory. Recall that a set is a collection of elements. The elements themselves can be sets. To establish the connection between sets and events we shall use a specific example. Generalization is immediate.

Suppose that 24 people are interviewed as potential jurors for a trial. One of them is to be chosen randomly. Of the 24, 6 are black, 6 are Asian and 12 are white. Label

---

[1] The equation needs a proof, which we skip.

the blacks by $a_1, \ldots, a_6$, the Asians by $b_1, \ldots, b_6$ and the whites by $c_1, \ldots, c_{12}$. The set of all elements is

$$S = \{a_1, \ldots, a_6, b_1, \ldots, b_6, c_1, \ldots, c_{12}\} \ .$$

There are three obviously disjoint subsets,

$$A = \{a_1, \ldots, a_6\} \ ,$$
$$B = \{b_1, \ldots, b_6\} \ ,$$
$$C = \{c_1, \ldots, c_{12}\} \ .$$

Let E_i be the event that the ith person is chosen as a juror randomly. Then, $P(E_i) = 1/24$. The sets A, B and C correspond to the events E_A, that one of the selected jurors is black, E_B, that one is Asian and E_C, that one is white. Then the correspondence between probabilities and sets is detailed in Table 4.2. Events are mapped to probabilities. Therefore, set operations apply to the corresponding probabilities. For example, corresponding to the set elements we have simple events, each with equal probability. Also,

$$P(E_A) = P(E_1) \cup \cdots \cup P(E_6) \ .$$

Table 4.2 Correspondence between set theory and probability.

Set theory	Events	Probabilities
Elements	Elementary, E_i	$P(E_i) = 1/24$
Disjoint subsets	Mutually exclusive	$P(E_A), P(E_B), P(E_C)$
S	Sample space	$P(S) = 1$

Because elementary events are disjoint,

$$\begin{aligned} P(E_A) &= P(E_1) \cup \cdots \cup P(E_6) \\ &= P(E_1) + \cdots + P(E_6) \\ &= \frac{6}{24} = \frac{1}{3} \ . \end{aligned}$$

$P(E_B)$ and $P(E_C)$ are computed similarly. We conclude that the probability of choosing a black juror is $1/3$. The probability that the selected juror is Asian or white is

$$\begin{aligned} P(E_B) \cup P(E_C) &= P(E_B) + P(E_C) - P(E_B) \cap P(E_C) \\ &= \frac{6}{24} + \frac{12}{24} - 0 \\ &= \frac{2}{3} \ . \end{aligned}$$

4.4 Conditional probability and independence

Conditional probability and independence deal with how to compute probabilities and the meaning of probability. They are directly related to the description of Bayesian probability that we alluded to in Section 4.3.1.

4.4.1 Conditional probability

The occurrence of one event might change the likelihood of another. For example you may be asked about the likelihood that a day was cloudy when you know it rained during that day. In such a case, you would say that the likelihood is 100%. If you are asked about the likelihood of rain given that it was cloudy during the day, your answer will be less than 100%.

Example 4.27. Suppose that in a population, 0.1% have a certain disease. A diagnostic test is available, but it is correct in only 80% of the cases—; diagnosing the disease when a person is actually infected. The other 20% show false positives. Now choose a person from the population randomly and consider the following events: E, an individual carries the disease and F, an individual's diagnostic test result is positive. Let $P(E|F)$ denote the probability of E given F. Then from the data

$$P(E) = 0.001, \quad P(E|F) = 0.8.$$

From this, we conclude that before having the test result, the probability of E is unlikely. Once we have the test result, the probability that the person is infected has increased several folds. □

Here is another example of how conditional probabilities are calculated.

Example 4.28. For reasons one might guess, identical drugs are more expensive in the U.S. than in Canada. Some states that border Canada decided to offer their residents the option to buy drugs from Canada. As you might expect, the drug companies oppose this effort voraciously. They claim that drugs from Canada may be tainted—more so than drugs bought in the U.S. The data for this example are imagined. Yet, it can serve as a model to resolve the drug manufacturers claim.

You buy 25 pills, manufactured by a single company, from Canada and from the U.S. Some of the pills are tainted according to the following data:

	Not tainted	Tainted	Total
Canada	11	4	15
US	7	3	10
Total	18	7	25

Select a pill randomly for analysis and let E be the event that the chosen pill is from Canada and F the event that the chosen pill is tainted. From the table above we conclude:

$$P(E) = \frac{15}{25} = 0.60, \quad P(F) = \frac{7}{25} = 0.28, \quad P(E \text{ and } F) = \frac{4}{25} = 0.16.$$

Now suppose that the analysis revealed that the pill is tainted. How likely is it that the pill came from Canada? Again, based on the table

$$P(E|F) = \frac{4}{7} \approx 0.571.$$

This is smaller than the original $P(E)$ because there is a lower percentage of tainted pills from Canada than from the U.S. The same conditional probability can also be calculated according to:

$$P(E|F) = \frac{4}{7} = \frac{4}{25} \div \frac{7}{25} = \frac{P(E \text{ and } F)}{P(F)}.$$

In other words, $P(E|F)$ is the ratio of the probability that both events occur divided by the probability of the conditioning event F.

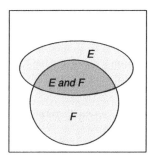 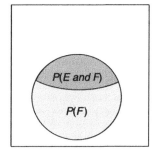

Figure 4.9 Conditional probability.

The idea of conditional probability can be represented with a Venn diagram (Figure 4.9). We know that the outcome was F. The likelihood that E also occurred is the size of E and F relative to the size of F. □

We thus arrive at the definition of:

Conditional probability Let E and F be two events with $P(F) > 0$. Then the conditional probability of E given that F has occurred is

$$P(E|F) = \frac{P(E \text{ and } F)}{P(F)}.$$

Example 4.29. The seed banks in two prairie areas, labeled A and B, were studied. The data consist of the relative number of seeds of three major grass species; call them a, b and c. The table below shows the fraction of the seeds from each species and area.

	Species			
Area	a	b	c	Total
A	0.40	0.21	0.09	0.70
B	0.10	0.09	0.11	0.30
Total	0.50	0.30	0.20	1.00

Tables such as this are called *joint probability tables*. Examples from the table: 70% of all seeds were from area A, 50% of the seeds came from species a. Denote the following events: E, a selected seed is from A and F, a selected seed is from a. Now select a seed at random and identify it. It turns out to be from a. What is the probability that the seed was collected from A?

$$P(E|F) = \frac{P(E \text{ and } F)}{P(F)} = \frac{0.40}{0.50} = 0.80 \ .$$

In other words, 80% of the seeds from species a came from area A. Here

$$P(E|F) = 0.80 > P(E) = 0.70 \ .$$

Furthermore,

$$P(F|E) = \frac{P(E \text{ and } F)}{P(E)} = \frac{0.40}{0.70} = 0.571 > 0.5 = P(F) \ .$$

It is not always the case that conditional probability improves chances that an event will occur. For example, let C be the event that the selected seed came from b. Then

$$P(E|C) = \frac{P(E \text{ and } C)}{P(C)} = \frac{0.21}{0.30} = 0.70 = P(E) \ .$$

In other words, $P(E|C) = P(E)$. That is, if we are told that the seed belongs to a, the likelihood that it came from A remains unchanged. □

Throughout the preceding (and future) discussion of probability, we always interpret it in frequentist terms: "If we repeat the experiment many times, then the probability is obtained from the ratio of the number of times an event occurred to the total number of repetitions."

4.4.2 Independence

If the occurrence of one event does not change the probability that another event will occur, we say that the events are independent.

Independent events (first definition) Events E and F are said to be independent if

$$P(E|F) = P(E) \ .$$

If E and F are not independent then we say that they are *dependent*.

Similarly, if $P(F|E) = P(F)$ then E and F are said to be independent. Independence implies the following:

$$P(\text{not } E|F) = P(\text{not } E) ,$$
$$P(E|\text{not } F) = P(E) ,$$
$$P(\text{not } E|\text{not } F) = P(\text{not } E) \ .$$

Another way to define independent events is:

Independent events (second definition) The events E and F are independent if and only if

$$P(E \text{ and } F) = P(E)P(F) \ .$$

Conditional probability and independence

This identity is called the *multiplicative rule*. The "if and only if" statement above implies that if E and F are independent, then the multiplicative rule is true and if the multiplicative rule is true, then the events are independent.

Example 4.30. Let E be the event that a statistics class begins on time and F the event that an ornithology class begins on time. The professors of both classes are unaware of each other's behavior. We therefore assume that E and F are independent. Suppose that $P(E) = 0.9$ and $P(F) = 0.6$. Then

$$P(E \text{ and } F) = P(\text{both classes begin on time})$$
$$= P(E)P(F)$$
$$= 0.9 \times 0.6 = 0.54 .$$

Also

$$P(\text{not } E \text{ and not } F) = P(\text{neither class begins on time})$$
$$= P(\text{not } E)P(\text{not } F)$$
$$= 0.1 \times 0.4 = 0.04 .$$

The probability that exactly one of the two classes begins on time is

$$P(\text{exactly one class begins on time}) = 1 - (0.54 + 0.04)$$
$$= 0.42 .$$

□

Independence applies to more than two events. If $E_1, E_2, \ldots, E_n$ are independent then

$$P(E_1 \text{ and } E_2 \text{ and } \ldots \text{ and } E_n) = P(E_1)P(E_2)\cdots P(E_n) . \qquad (4.3)$$

Note that the relations are *not* if and only if. In other words, if (4.3) is true, it does not necessarily mean that the events are independent. We say that n events are independent if and only if (4.3) is true *and* all possible pairs of events are independent *and* all possible triplets are independent and so on. To see this, consider the following example.

Example 4.31. In Figure 4.10 the proportion of the size of the rectangles A, B and C to the rectangular space S reflects their probability and the inset darker rectangle represents $A \cap B \cap C$. Given that

$$P(A) = P(B) = P(C) = \frac{1}{6}$$

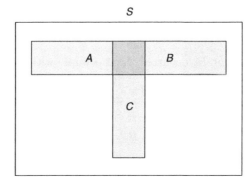

Figure 4.10   Seemingly independent events that are not.

we have
$$P(A \cap B) = P(A \cap C) = P(B \cap C) = P(A \cap B \cap C) = \frac{1}{36}.$$
Also
$$P(A)P(B) = P(A)P(C) = P(B)P(C) = \frac{1}{36},$$
so the events appear to be independent. However,
$$P(A \cap B \cap C) \neq P(A)P(B)P(C) = \frac{1}{216}.$$
In other words, pairs of events are independent, but their triplet is not. Therefore, the events A, B and C are not independent. □

4.5 Algebra with probabilities

As we have seen, we combine probabilities in different ways for dependent and independent events. When the number of outcomes is small, we can enumerate all outcomes and compute probabilities. This is not possible when the number of outcomes is large. Thus, we need to have some (when possible) rules about how to deal with addition, subtraction and in general, how to combine events and obtain their probabilities.

4.5.1 Sampling with and without replacement

Sampling with replacement refers to drawing a sample from a population and then putting the sample units back in the population before drawing another sample. Sampling without replacement refers to removing the sample from the population after it is drawn. When the population is small, sampling without replacement may change probabilities significantly. In other words, for a small population, sampling without replacement introduces noticeable dependency among the probabilities of events. Here is an example with a small population.

Example 4.32. Last semester, there were 35 students in my statistics class, 20 females and 15 males. Of the females, 15 had blond hair. Of the males, 10 were blond. Consider the following experiment: select a student at random with replacement and record gender and hair color. Let B_1 denote the event that the first chosen student is a male blond, B_2 the second chosen student is a male blond and B_3 the third chosen student is a male blond. Then
$$P(B_3) = \frac{10}{35} = 0.28\,571$$
regardless of whether B_1 or B_2 occurred. Next, sample without replacement. Then
$$P(B_3|B_1 \text{ and } B_2) = \frac{10-2}{35-2} = 0.24\,242.$$
Also
$$P(B_3|\text{not } B_1 \text{ and not } B_2) = \frac{10}{35-2} = 0.30\,303.$$
Thus, probabilities may be noticeably different, depending on whether we sample with or without replacement. □

Next, consider a large population.

Example 4.33. The Minnesota Vikings and the Green Bay Packers are two football teams with a long history of rivalry. Of the 10 000 people that show up to a game between these teams, 2 500 are Packers fans. The rest are Vikings fans. Choose 3 fans without replacement and define the following events: E_1, the first choice is a Packers fan, E_2, the second choice is a Packers fan and E_3, the third choice is a Packers fan. Then
$$P(E_3|E_1 \text{ and } E_2) = \frac{2\,498}{9\,998} = 0.24\,985$$
and
$$P(E_3|\text{not } E_1 \text{ and not } E_3) = \frac{2\,500}{9\,998} = 0.25\,005 \ .$$
Thus, for all practical purposes, E_1, E_2 and E_3 are independent. □

A rule of thumb If at most 5% of the population is sampled without replacement, then we may consider the sample as if it is with replacement.

4.5.2 Addition

As we already saw in Section 4.3.4, adding dependent events is like adding sets that are not disjoint. Consequently, the probabilities of adding two events when they are dependent or independent are different. We need to subtract the common outcomes of the events. For two events, we have the

Addition rule For any two events, E and F,
$$P(E \text{ or } F) = P(E) + P(F) - P(E \text{ and } F) \ .$$

In words, the probability that the events E or F occur equals the sum of the probabilities that each event occurs, minus the probability that both E and F occur.

Example 4.34. Of the students in an Ecology class, 60% took statistics, 40% took calculus and 25% took both. Select a student randomly. What is the probability that the student took at least one of these two courses? Let E be the event that the selected student took statistics, F that the selected student took calculus and G that the selected student took at least one of the courses. Then
$$P(E) = 0.6 \ , \quad P(F) = 0.4 \ , \quad P(E \text{ and } F) = 0.25 \ .$$
Therefore,
$$\begin{aligned} P(G) &= P(E \text{ or } F) \\ &= P(E) + P(F) - P(E \text{ and } F) \\ &= 0.60 + 0.40 - 0.25 \\ &= 0.75 \ . \end{aligned}$$
Now let H be the event that the selected student took none of the courses. Then
$$\begin{aligned} P(H) &= P(\text{not } (E \text{ or } F)) \\ &= 1 - P(E \text{ or } F) \\ &= 0.25 \ . \end{aligned}$$

Let I be the event that the selected student took exactly one of the courses. Then
$$P(I) = P(E \text{ or } F) - P(E \text{ and } F)$$
$$= 0.75 - 0.25$$
$$= 0.50 \ .$$

□

4.5.3 Multiplication

Recall that for conditional probability with $P(F) > 0$, we have
$$P(E|F) = \frac{P(E \text{ and } F)}{P(F)} \ .$$
Therefore,
$$P(E \text{ and } F) = P(E|F)P(F) \ . \quad (4.4)$$
When E and F are independent, $P(E|F) = P(E)$ and the last equation reduces to
$$P(E \text{ and } F) = P(E)P(F) \ . \quad (4.5)$$

Multiplication rule If two events E and F are dependent, then (4.4) holds. If the events are independent, then (4.5) holds.

Example 4.35. You are told that in a certain area, 70% of the birds are song birds. Of these, 20% are sparrows. Also, 30% of the birds in the area are hummingbirds and of these, 40% are Calliope hummingbirds. Suppose that your best guess of what bird you will be seeing next as you walk in the forest is that it will be a random individual among the birds in the area. What is the probability that you will see a sparrow next? First, we write the data in a convenient format:

	% of birds	Of these
Songbirds	70%	20% sparrows
Hummingbirds	30%	40% Calliope

Let D be the event that a song bird is observed and E a sparrow is observed. From the data
$$P(D) = 0.70 \ ,$$
$$P(E|D) = 0.20 \ .$$
Therefore,
$$P(D \text{ and } E) = P(E|D) \times P(D) = 0.70 \times 0.20 = 0.14 \ . \quad □$$

4.5.4 Counting rules

So far, we have dealt with a small number of events. Drawing tree diagrams and computing probabilities of events with and without replacement was relatively easy. When the number of outcomes is large, computing all possible outcomes becomes impossible. We have to be a bit more clever in calculating probabilities. We deal with these issues next.

Multiplication

Multiplication applies to sampling with replacement. When we have two experiments, the first with n_1 possible outcomes and the second with n_2 possible outcomes, then the total number of outcomes, n, is

$$n = n_1 \times n_2 .$$

Example 4.36. Ten single males and 12 single females are invited to a party. In how many ways can they be paired? Identify each male as M_i, $i = 1, \ldots, 10$ and each female as F_i, $i = 1, \ldots, 12$. Here $n_1 = 10$ and $n_2 = 12$. Now M_1 can be paired in 12 different ways (with $F_1, \ldots, F_{12}$). M_2 can be paired in 12 different ways and so on up to M_{10}. Therefore, the number of possible pairs is

$$n = 10 \times 12 = 120 .$$

Next suppose that 3 of the males are from England and 4 of the females are from France. Let C be the event that the male member of a pair is from England and the female from France. Assume that pairs are formed randomly. Then

$$P(C) = \frac{\text{(number of pairs in } C\text{)}}{n} = \frac{3 \times 4}{120} = 0.10 .$$

□

When we combine k experiments, each with n_i outcomes, the number of possible outcomes, n, is

$$n = n_1 \times n_2 \times \cdots \times n_k .$$

Example 4.37. In a group of suspected terrorists, 10 speak English fluently, 15 are combat trained and 12 can fly planes. In how many ways can the group divide itself into subgroups of three? Here

$$n = 10 \times 15 \times 12 = 1\,800 .$$

Suppose that 3 of the English speaking suspected terrorists, 4 of the combat trained and 3 of those who can fly planes are from Saudi Arabia. Triplets are chosen randomly. What is the probability that all members of a triplet are from Saudi Arabia?

$$P(\text{all are from Saudi Arabia}) = \frac{3 \times 4 \times 3}{1\,800} = 0.02 .$$

□

Permutations

In the previous section, we chose members of a pair or a triplet with replacement. When the order of choosing is important and when it is done without replacement, the rule for finding the number of possible events is different.

Example 4.38. Twelve students apply for summer field work that requires 5 different tasks: trapping (s_1), mist-netting (s_2), collecting vegetation data (s_3), entering data (s_4) and analyzing data (s_5). All students are equally skilled at these tasks. How many different teams can be formed? We start with $n = 12$ students. For s_1 we can choose one student out of 12. Therefore, there are 12 possibilities. For a particular choice of a student for task s_1, there are 11 students to choose from for task s_2 and so on. Therefore, the number of different teams that can be chosen is

$$n(n-1)\cdots(n-4) = 12 \times 11 \times 10 \times 9 \times 8$$
$$= 95\,040 \,.$$
□

To generalize the example, we consider n objects. They are to be arranged into an ordered subset of k objects. Thus, for the first slot in k we can choose one of the n objects in n different ways. For the second slot in k we can choose one of the objects in $n-1$ ways. Therefore the first and the second slots in k can be chosen in $n(n-1)$ ways and so on. For the last kth slot, we have $n-(k-1)$ objects to choose from. We thus have the following definition:

Permutation The number of permutations of k objects selected randomly from a population of n objects, denoted by $P_{n,k}$ is

$$P_{n,k} = n(n-1)(n-2)\cdots(n-k+1) \,. \tag{4.6}$$

Instead of writing the permutations explicitly, we use a short hand notation, called factorial.

Factorials

$$n! := n \times (n-1) \times \cdots \times 2 \times 1 \,,$$
$$0! := 1 \,.$$

Therefore, we can write equation (4.6) as

$$P_{n,k} := \frac{n!}{(n-k)!} \,. \tag{4.7}$$

To see this, expand the numerator and denominator and cancel equal elements. Equation (4.7) is called the *permutations equation*. It gives the number of permutations (ways to arrange) of k objects taken from a population of size n, when the order of selecting the objects is important.

Example 4.39. In a random mating experiment, there are 10 females available for mating. A male mates with 6 of them. The mating order is important because the male's viability deteriorates with more matings. In how many ways can that male mate with 6 females?

$$P_{10,6} = \frac{10!}{(10-6)!} = 151\,200 \,.$$

Now suppose that another male chooses the 6 females in the same order. In other words, both males chose the same permutation out of 151 200. We will then conclude that it is highly unlikely that the choice of mating order is random. □

Often, we wish to produce the permutations themselves, instead of counting the number of permutations. Here is how we do it in R.

Example 4.40. We wish to produce a list of all permutations of the last five letters. First, we create a vector of these letters and print it:

```
> x <- letters[22 : 26]
> nqd(array(x))
```

v w x y z

The function nqd() prints an array with no quotes and no dimension names. Its code is

```
nqd <- function(x) print(noquote(no.dimnames(x)))
```

Next, we create a list of all permutations of the letters.

```
> library(combinat)
> px <- unlist(permn(x))
```

permn() returns a list of all permutations. We collapse the list into a vector with unlist(). Here are the first two permutations:

```
> nqd(array(px[1 : 10]))
```

v w x y z v w x z y

To display all permutations compactly, we cut the matrix pmx into two and then column bind them with space between them. Finally, we print the results:

```
> pmx <- matrix(px, ncol=4, byrow=TRUE)
> pmx <- cbind(pmx[1 : 6, ], '    ', pmx[7 : 12, ],
+              '    ', pmx[13 : 18, ], '    ', pmx[19 : 24, ])
> nqd(pmx)
```

```
w x y z      w y z x      z y x w      x z y w
w x z y      w y x z      y z x w      z x y w
w z x y      y w x z      y x z w      z x w y
z w x y      y w z x      y x w z      x z w y
z w y x      y z w x      x y w z      x w z y
w z y x      z y w x      x y z w      x w y z
```

□

Combinations

Recall that in the case of permutations, order was important. When order is not important we deal with combinations. It should be clear that the number of possibilities when order is not important is smaller than otherwise.

Example 4.41. Consider five genes: A, B, C, D and E. In how many ways can you arrange three of these genes when their order is important?

$$P_{5,3} = \frac{5!}{(5-3)!} = 60 \ .$$

If order is not important, the first gene in the arrangement can be in one of three positions (first, second or third), the second gene in one of two remaining positions and the last in the one remaining position. This means that we count fewer choices because different orders represent the same choice if the same three genes were selected. How many fewer choices? As many as the number of permutations of the three genes. In other words, by a factor of 3!. Therefore, when order is not important, the number of distinct arrangements of three out of five genes is

$$\frac{5!}{3!\,(5-3)!} = 10\,.$$

With R, we do this:

```
> library(combinat)
> nc <- nCm(5, 3)
> np <- nCm(5, 3) * fact(3)
> (c(comb = nc, perm = np))
comb perm
  10   60
```

Note that the number of permutations (np) = the number of combinations (nc) × 3!. To list the combinations, do

```
> x <- LETTERS[1 : 5]
> nqd(combn(x, 3))

A A A A A A B B B C
B B B C C D C C D D
C D E D E E D E E E
```

□

Thus we have the following definition:

Combination An unordered subset of k objects chosen from among n objects is called a combination. The number of such combinations is computed with

$$C_{n,k} := \binom{n}{k} := \frac{P_{n,k}}{k!} = \frac{n!}{k!\,(n-k!)}\,.$$

Example 4.42. In Example 4.39, we had 10 females and a male was to mate with 6 of them. We found that the male could mate with the 6 females in

$$P_{10,6} = \frac{10!}{(10-6)!} = 151\,200$$

different ways when order was important. Suppose that the male's viability does not deteriorate with more matings. Then the order of mating is not important. The number of possible matings with 6 females out of 10 now becomes

$$C_{10,6} = \binom{10}{6} = \frac{10!}{6!\,(10-6)!} = 210\,.$$

This represents a large number of choices. If another male mates with the same combination of females, we conclude that the choice of mates is not random.

Suppose that after the mating, the male shows a particular preference for 2 of the females. Now the male is presented with 8 females and is going to choose 4 of them randomly. What is the probability that his 2 preferred females will be included in the male's choice? Let E be the event that both preferred females are chosen. Assume that all possible choices (4 out of 8) are equally likely. Then

$$P(E) = \frac{\text{number of outcomes in } E}{\text{number of ways to select 4 out of 8 females}}$$
$$= \frac{\binom{6}{2}}{\binom{8}{4}} = \frac{6!}{2!(6-2)!} \times \frac{4!(8-4)!}{8!} \approx 0.214$$

□

Let us contrast permutations and combinations with and without replacement using R.

Example 4.43. The genome of an organism is carried in its DNA. Genes code for RNA, which in turn codes for amino acids. Genes that code for amino acids are composed of codons. When strung together (in a specific order), these amino acids form a protein. In RNA, each codon sub-unit consists of three of the following four nucleotide bases: adenine (A), cytosine (C), guanine (G) and uracil (U).

Imagine a brine with billions of A, C, G and U. In how many ways can a codon be arranged? The first slot of the sequence of three can be filled with four different nucleotides, the second with four and the third with four. Therefore, we have $4^3 = 64$ different codon combinations. These 64 permutations are called the RNA Codon Table. Here is how we make the table in R:

```
1  nucleotide <- c('U', 'C', 'A', 'G')
2  library(gtools)
3  RNA.codons <- permutations(4, 3, v = nucleotide,
4    repeats.allowed = TRUE)
5  RNA.table <- data.frame(RNA.codons[ 1 : 16, ], ' ')
6  for(i in 2 : 4){
7    RNA.table <- data.frame(
8      RNA.table, RNA.codons[ (16 * (i - 1) + 1) : (16 * i),
9      ], ' ')
10 }
11 nqd(as.matrix(RNA.table))
```

To use the function `permutations()`, we first load the package `gtools` with a call to `library()` (line 2). In lines 3 and 4 we call `permutations()` with the appropriate arguments v (a vector) and allowing repetition (`repeats.allowed`). We then create the RNA table as a `data.frame()` and print it `as.matrix()`. The output from this script is

```
A A A     C A A     G A A     U A A
A A C     C A C     G A C     U A C
A A G     C A G     G A G     U A G
A A U     C A U     G A U     U A U
A C A     C C A     G C A     U C A
```

A C C	C C C	G C C	U C C
A C G	C C G	G C G	U C G
A C U	C C U	G C U	U C U
A G A	C G A	G G A	U G A
A G C	C G C	G G C	U G C
A G G	C G G	G G G	U G G
A G U	C G U	G G U	U G U
A U A	C U A	G U A	U U A
A U C	C U C	G U C	U U C
A U G	C U G	G U G	U U G
A U U	C U U	G U U	U U U

Next, suppose that the brine is limited in the supply of A, C, G and U. Then change all 16 to 6 in the script above and change the call to permutations() to

```
RNA.codons <- permutations(4, 3, v = nucleotide,
    repeats.allowed=FALSE)
```

Now you get fewer permutations (because replacement is not allowed):

A C G	C A G	G A C	U A C
A C U	C A U	G A U	U A G
A G C	C G A	G C A	U C A
A G U	C G U	G C U	U C G
A U C	C U A	G U A	U G A
A U G	C U G	G U C	U G C

Order is important here because every codon sequence produces a different amino acid. Suppose that identical amino acids were produced with the same three nucleotides, regardless of their sequence with unlimited supply of nucleotides. Then the script

```
nucleotide <- c('U', 'C', 'A', 'G')
library(gtools)
RNA.codons <- combinations(4, 3, v = nucleotide,
    repeats.allowed = TRUE)
RNA.codons <- data.frame(RNA.codons[1 : 10, ], '   ',
    RNA.codons[11 : 20, ])
nqd(as.matrix(RNA.codons))
```

produces

A A A	C C C
A A C	C C G
A A G	C C U
A A U	C G G
A C C	C G U
A C G	C U U
A C U	G G G

```
A G G      G G U
A G U      G U U
A U U      U U U
```

and if the supply of the nucleotides is limited (combination is without replacement), then change the call to combinations()

```
RNA.codons <- combinations(4, 3, v = nucleotide,
  repeats.allowed = FALSE)
```

and you you get the possible combinations

```
A C G
A C U
A G U
C G U
```

Observe that sampling with and without replacement for permutations and combinations result in a different number of outcomes! □

4.6 Random variables

Events are associated with probabilities. A function that assigns real values to events associates the events' probabilities with those real values. Such functions are appropriately called random variables. From here on, we will use rv to denote both *random variable* and *random variables*.

Assigning real values to events lead to rv. The values of the rv inherit the probabilities (and operations on these probabilities) of their corresponding events. The links between the values that a rv takes and the probabilities assigned to these values then lead to densities and distributions. These links are illustrated in Figure 4.11. We discussed the link $P(E)$ throughout this chapter—it corresponds to the definition of probability. The remaining links are discussed here.

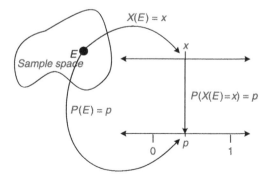

Figure 4.11   A random variable is a mapping of events to the real line.

Throughout, we use the concept of a real line. The real line is the familiar line that extends from $-\infty$ to ∞. It has an origin at 0 and each point on the line has a

128 Probability and random variables

value. The latter reflects the distance of the point from the origin. These values are called *real numbers*. We will agree that the (extended) real line includes both $-\infty$ and ∞.

We define rv thus:

Random variable A function that assigns real numbers to events, including the null event.

We usually denote rv with upper case letters, such as X and Y. As Figure 4.11 illustrates, a rv is a mapping of events to values on the real line. In this context, we say that the sample space, S, is the domain and the real line, $\mathbb{R}$, is the range. We write this as

$$X(E) : S \to \mathbb{R}.$$

The definition of a rv implies that the assignment of real numbers to events can be arbitrary. Because the definition includes the null event, we are free to assign any real value or range of values to the null event.

4.7 Assignments

Exercise 4.1. Males and females cross a street in no particular order. We note the gender of the first and second people who cross the street. The possible outcomes consist of male first and male second, male first and female second, and so on. Let F and M be the events that a female or a male crossed the street, respectively. Then $S = \{MM, MF, FM, FF\}$. Verify that the power set consists of $2^4 = 16$ subsets.

Exercise 4.2. This exercise demonstrates DeMorgan's Laws. Draw a Venn diagram picturing A and B that partially overlap.

1. Shade *not* $(A \text{ or } B)$. On a separate diagram, shade $(not\ A)\ and\ (not\ B)$. Compare the two diagrams.
2. Shade *not* $(A \text{ and } B)$. On a separate diagram shade $(not\ A)\ or\ (not\ B)$. Compare the two diagrams.

Exercise 4.3. An experiment consists of rolling a die and flipping a coin.

1. What is the sample space S? How many outcomes are in the sample space?
2. What are the outcomes of the event E that the side of the die facing up shows an even number of dots?
3. What are the outcomes of the event F that the coin lands on H?
4. What are the outcomes of $E \cup F$?
5. What are the outcomes of $E \cap F$?
6. Suppose that outcomes are equally likely. Compute:
    (a) $P(E)$
    (b) $P(F)$
    (c) $P(E \cup F)$
    (d) $P(E \cap F)$

Exercise 4.4. What is the sample space of an experiment that consists of drawing a card from a standard deck and recording its suit?

Exercise 4.5. Last semester, I took note of students late to class. The results from 22 students were:

0 2 5 0 3 1 8 0 3 1 1 9 2 4 0 2 9 3 0
1 9 8

1. What proportion of the students was never late?
2. What proportion of the students was late to class at most 8 times? At least 8 times?
3. What proportion of the students was late between 3 and 6 times during the semester?

Exercise 4.6. We record the fate of patients who arrive at a hospital emergency room. The two possible outcomes are the patient is admitted for further treatment or released. To test the hypothesis (we shall do that later) that the fate of two consecutive patients is independent, we choose the first patient at random and record his/her fate. Then we record the fate of the next arrival.

1. What is the sample space, i.e. the set of all possible outcomes?
2. Show the sample space in a tree diagram.
3. List the outcomes of the event B that at least one patient was released.
4. List the outcomes of the event C that exactly one patient was released.
5. List the outcomes of the event D that none of the patients were released.
6. Which of the events B, C and D is elementary?
7. List the outcomes in the events B and C.
8. List the outcomes in the events B or D.

Exercise 4.7. To test the efficacy of admissions or release, an emergency room embarks on a controlled experiment. Each experiment (there are many of them, but we shall examine only one) consists of choosing 4 patients at random on a particular night. The patients are selected from a group where the doctors are not certain whether they should be admitted or not. Name the patients P_1, P_2, P_3 and P_4. Of these 4 patients, we choose 2 at random. The first patient will be released and the second admitted.

1. Display a tree diagram of the possible outcomes.
2. Denote by A the event that at least one of the patients has an even numbered index (P_2 and P_4 have even numbered indices). Which outcomes are included in A?
3. Suppose that P_1 and P_2 are over 50 years old and P_3 and P_4 are less than 40 years old. Denote by B the event that exactly one of the patients selected is over 50 years old. Which outcomes are included in B?

Exercise 4.8. Starting at a certain time, you observe deer crossing a road and record their sex (M = male, F = female). The experiment terminates as soon as a male is observed.

1. Give 5 possible experimental outcomes.
2. How many outcomes are there in the sample space?
3. Let E = number of deer observed is even. What outcomes are in E?

Exercise 4.9. The following is a subset of the vital statistics data obtained from WHO (see Example 2.7). The data were collected during 1995 to 2000 and reported in 2003. Data include the death rate (per 1000 per year) for Eastern African countries only.

```
                           country   dr
2                   Eastern Africa 18.8
3                          Burundi 20.6
4                          Comoros  8.4
5                         Djibouti 17.7
6                          Eritrea 11.9
7                         Ethiopia 17.7
8                            Kenya 16.7
9                       Madagascar 13.2
10                          Malawi 24.1
11                       Mauritius  6.7
12                      Mozambique 23.5
13                         Reunion  5.5
14                          Rwanda 21.8
16                         Somalia 17.7
17                          Uganda 16.7
18      United Republic of Tanzania 18.1
19                          Zambia 28.0
20                        Zimbabwe 27.0
```

A person is picked at random from Eastern Africa.

1. Which country (or countries) could the person have come from if you are told that his probability of dying during the next year is greater than 0.0 167?
2. Which country (or countries) could the person have come from if you are told that his probability of dying during the next year is smaller than 0.0 067?
3. Which country (or countries) could the person have come from if you are told that his probability of dying during the next year is larger than 0.00 167 and smaller than 0.0 181?

Exercise 4.10. All of the terrorists in the 9/11 attack on the Twin Towers came from Middle Eastern Arab countries. The populations of Middle Eastern Arab countries (from the WHO data, see Example 2.7) are as follows (in 1 000):

```
                            country   pop
1                           Bahrain   724
2                             Egypt 71931
3         Iran (Islamic Republic of) 68919
4                              Iraq 25174
5                            Jordan  5472
6                            Kuwait  2521
7                           Lebanon  3652
8             Libyan Arab Jamahiriya  5550
9       Occupied Palestinian Territory 3557
10                             Oman  2851
```

11	Saudi Arabia	24217
12	Syrian Arab Republic	17799
13	United Arab Emirates	2994
14	Yemen	20010

Suppose that these terrorists were assembled independently.

1. What is the probability that one of the terrorists came from Saudi Arabia?
2. What is the probability that one of the terrorists came from Saudi Arabia or Egypt?
3. What is the probability that one of the terrorists came from neither Saudi Arabia nor from Egypt?

Exercise 4.11. A single card is randomly selected from a well-mixed deck.

1. How many elementary events are there?
2. What is the probability of an elementary event?
3. What is the probability that the selected card is a diamond? A face card (Jack, Queen or King)?
4. What is the probability that the selected card is both a diamond and a face card?
5. Let A be the event that the selected card is a face and B the event that the selected card is a diamond. What is $P(A \text{ or } B)$?

Exercise 4.12. Based on a questionnaire, a matching service finds 4 men and 4 women that match perfectly and are predicted to have a happy marriage. Any incorrect matching is predicted to result in a failed marriage. In their infinite wisdom, the matching service pairs the customers completely randomly. That is, all outcomes are equally likely. Label the males as A, B, C and D. To simplify the notation, consider one possible outcome: A is paired with B's perfect match, B is paired with C's perfect match, C is paired with D's perfect match and D is paired with A's perfect match. We write this outcome as $\{B, C, D, A\}$.

1. List the possible outcomes.
2. Consider the event that exactly two of the matchings result in a happy marriage. List the outcomes contained in this event.
3. What is the probability of this event?
4. What is the probability that exactly one matching results in a happy marriage?
5. What is the probability that exactly three matchings result in happy marriages?
6. What is the probability that at least two of the four matches result in happy marriages?

Exercise 4.13. Five drug addicts are shooting heroin in a crack house. Name them A, B, C, D and E. Each of them is equally likely to die from overdose. Two of them will die by the end of the evening.

1. List the possible outcomes.
2. What is the probability of each elementary event?
3. What is the probability that one of the dead addicts is A?

Exercise 4.14. Of five people in the emergency room (ER) of a certain hospital, A and B are first time patients. For patients C, D and E, it is their second visit to the ER. Two of the five are chosen randomly for treatment by the ER intern.

1. What is the probability that both selected patients are first-time visitors?
2. What is the probability that both selected patients are second-time visitors?
3. What is the probability that at least one of the selected patients is a first-time visitor?
4. What is the probability that of the selected patients, one is a "first-timer" and the other is a "second-timer?"

Exercise 4.15. A patient is seen at a clinic. A recent epidemic in town shows that the probability that the patient suffers from the flu is 0.75. The probability that he suffers from walking pneumonia is 0.55. The probability that he suffers from both is 0.50. Denote by F the event that the patient suffers from the flu and by M that he suffers from pneumonia.

1. Interpret and compute $P(F|M)$.
2. Interpret and compute $P(M|F)$.
3. Are F and M independent? Explain.

Exercise 4.16. The probability that a randomly selected student on a typical university campus showered this morning is 0.15. The probability that a randomly selected student on the campus had breakfast this morning is 0.05. The probability that a randomly selected student on the campus both took a shower and had breakfast is 0.009.

1. Given that the student took a shower, what is the probability that he had breakfast as well?
2. If a randomly selected student had breakfast, what is the probability that she also took a shower?
3. Are the events "took a shower" and "had breakfast" independent? Explain.

Exercise 4.17. In the U.S., racial profiling describes the practice of law enforcement agencies to search, stop and sometimes arrest people of a particular ethnic group more than their relative number in the population. Suppose that a population in a certain city is composed of 30% belonging to ethnic group 1 and 70% to ethnic group 2. Members of the ethnic groups are visibly different. Court records reveal that crime rate in group 1 is 25% and in group 2 it is 10%. A police officer stops a person at random. Let E_1 be the event that the person belongs to group 1, E_2 the event that the person belongs to group 2 and E_3 the event that the person is a criminal.

1. What is the probability that the person is a criminal?
2. What is the probability that the person is from A if he is a criminal?
3. What is the probability that the person is from B if she is a criminal?
4. In your opinion, do the results justify racial profiling?

Exercise 4.18. A small pond has 12 fish in it. Seven of them are walleye and five are Northern pike. On a particular day, only two fish are caught. Suppose that the two fish are caught randomly.

1. What is the probability that the first fish caught is a walleye?
2. What is the probability that the second fish caught is a walleye given that the first is a walleye?
3. What is the probability that the first and the second fish caught are walleye?
4. Explain the difference in the probabilities between case 2 and case 3.

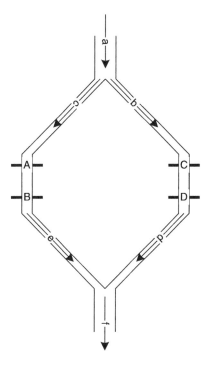

Figure 4.12 Dam gates.

Exercise 4.19. A series of gated dams along two parallel streams is shown in Figure 4.12. Denote E_1 as the event that gate A functions properly, E_2 as the event that gate B functions properly and so on. Suppose that $P(E_i) = 0.95$, $i = 1, 2, 3, 4$ and that gates function independently. A closed gate is considered to be functioning improperly.

1. What is the probability that water will flow uninterrupted through branch c?
2. What is the probability that water will flow uninterrupted through branch b?
3. What is the probability that water will flow through both branches uninterrupted?
4. What is the probability that water will flow through the system uninterrupted?

Exercise 4.20. To successfully treat a disease, a patient goes through a two-step treatment with clear criteria for successful treatment at each step. Let E denote the event that the first step of the treatment succeeds and F the event that the second step succeeds. The respective probabilities are $P(E) = 0.45$ and $P(F) = 0.25$. The probability that the two-step treatment succeeds is $P(E \text{ and } F) = 0.20$.

1. What is the probability that at least one step of the treatment succeeds?
2. What is the probability that neither step succeeds?
3. What is the probability that exactly one of the two steps succeeds?
4. What is the probability that only the first step succeeds?

Exercise 4.21. Tuberculosis is becoming a global health problem. There are strains of the bacillus that are resistant to antibiotics. Suppose that 0.2% of individuals in a population suffer from tuberculosis. Of those who have the disease, 98% test positive when administered a diagnostic test. Of those who do not have the disease, 85% test negative when the test is applied. Choose an individual at random and administer the test. Let E be the event that a person has tuberculosis and F the event that the test is positive.

1. Construct a tree diagram with two branches: infected with tuberculosis and not infected. From each of these branches, show two branches: test positive and test negative. Show the appropriate probabilities on each of the four branches.
2. What is $P(E \text{ and } F)$?
3. What is $P(F)$?
4. What is $P(E|F)$?

Exercise 4.22. At the time of writing, the Minneapolis - St. Paul metropolitan area has 4 major-sport teams: the Vikings (football), the Timberwolves (basketball), the Twins (baseball) and the Wild (hockey). When the teams are successful (they usually are not), game tickets are hard to come by. A scalper (a person who buys tickets at their box-office price and then sells them to the highest bidder—oddly deemed illegal in a capitalistic society) buys 5 tickets to 5 different Vikings games, 4 to 4 different Timberwolves games, 3 to 3 different Twins games and no Wild tickets. He then lets you select 3 tickets randomly.

1. In how many ways can you select one ticket for each team game?
2. In how many ways can you select 3 tickets without regard to the team?
3. If 3 tickets are selected completely randomly, what is the probability that the 3 are for different team games?

Exercise 4.23. You are shopping for a computer system. You have a choice of monitor from 6 manufactures, main unit (CPU) from 4 manufacturers and printer from 7 manufacturers. All are about equally priced. How many different system combinations can you assemble?

Exercise 4.24. You suspect that 1 of a 25-cow herd is sick with mad cow disease. The remaining 24 are healthy. You select one cow at a time and test for the disease. Once you detect the disease, you stop the experiment. What is the probability that you must examine at least 2 cows?

Exercise 4.25. You are admitted to a hospital for brain surgery. Before submitting to the operation, you wish to have an opinion from two physicians. You obtain a list of 5 physicians, along with their years of practice. The list says that the 5 physicians have been in practice for 2, 5, 7, 9 and 12 years. You choose two physicians randomly. What is the probability that the chosen two have a total of at least 13 years of practice experience?

Exercise 4.26. Each mouse entering a maze in an experiment can turn left (L), right (R), or go straight (S). The experiment terminates as soon as a mouse goes straight. Let Y denote the number of mice observed.

1. What are the possible values of Y?
2. List 5 different outcomes and their associated Y values.

Exercise 4.27. The deepest point in a lake is 100 ft. A point is randomly selected on the surface of the lake. $Y = $ the depth of the lake at the randomly selected point. What are the possible values of Y?

Exercise 4.28. A box contains four chocolate bars marked 1, 2, 3 and 4. Two bars are selected without replacement. Once you select a bar, you receive as many additional bars as the numbers that appear on the bars you select. List the possible values for each of the following random variables:

1. $X = $ the sum of the numbers on the first and second bar
2. $Y = $ the difference between the numbers on the first and second bar
3. $Z = $ the number of bars selected that show an even number
4. $W = $ the number of bars selected that show a 4

Exercise 4.29. During its 6 hour trans-Atlantic flight, it takes an airplane 15 minutes to reach a cruising altitude of 25 000 ft. It takes the airplane 15 minutes to descend from the cruising altitude until landing. Select a random time, T, between take-off and landing. Let $X(T)$ be the altitude of the plane at T.

1. What are the possible values of T?
2. What are the possible values of X?
3. Is X a rv? Justify your answer.
4. What is the probability that $X = 25\,000$?
5. In answering (4), do we have to assume that the speed of the plane is approximately constant throughout the flight? Explain.

5

Discrete densities and distributions

In this chapter, we define discrete densities and distributions and learn how to construct them. Our goal is to develop an understanding of what these mean and their relation to rv. Consequently, we will discuss specific densities and distributions that we will find useful later. There are numerous distributions (and their densities) that describe natural phenomena. Refer, among others, to Ross (1993), Johnson et al. (1994), McLaughlin (1999), Evans et al. (2000) and Kotz et al. (2000).

The notation $P(X = x)$ reads "the probability that the rv X takes on a value x." Similarly, the notation $P(X \leq x)$ reads "the probability that the rv X takes on any value $\leq x$." If we do not state otherwise, x is a real number (a point on the real line). To emphasize that $P(X = x)$ may depend on some given values that we call *parameters*, we write $P(X = x | \boldsymbol{\theta})$, where $\boldsymbol{\theta}$ is a vector of given constants.

The following distinctions will allow us to be succinct in our narrative. Let $\mathbb{R}$ denote the set of real numbers. Then an alternative way to saying that x is a real number is $x \in \mathbb{R}$, or in words, x is a member of ($\in$) the set $\mathbb{R}$. Similarly, we identify the nonnegative integers 0, 1, ... with the symbol $\mathbb{Z}_{0+}$. Thus, $n \in \mathbb{Z}_{0+}$ means that n takes on any value that is a nonnegative integer. We denote the set of positive integers with $\mathbb{Z}_+$. We distinguish between sets whose elements are countable or not countable. For example $\mathbb{R}$ is a *noncountable* set and $x \in \mathbb{R}$ can take an infinite number of values. A countable set is a set whose elements can be counted. For example $\mathbb{Z}_{0+}$ is a *countable* set and $n \in \mathbb{Z}_{0+}$ can take an infinite number of values that can be counted. Other examples are: $x \in [0, 1]$ is not countable while $n \in A := \{0, 1, \ldots, 10\}$ is countable with a finite number of values (A has a finite number of elements).

5.1 Densities

Let
$$A := \{E_1, E_2, \ldots\}$$

be a countable (finite or infinite) set of simple events. A includes only those simple events with a *positive* probability. Each of these events is identified by a unique nonnegative integer i and corresponds to a probability π_i. Let $X \in \mathbb{R}$ be a rv such that $P(X(E_i) = x_i) = \pi_i > 0$. Therefore, $X(E_i)$ (or equivalently x_i) are countable subsets of $\mathbb{R}$ with $P(X(A)) = 1$. Here each event E_i corresponds to a real value x_i. Let $\overline{A}$ be the complement of A and S be the event space. Then $S = A \cup \overline{A}$. Define $P(X(\overline{A})) = 0$. Thus, $P(X(S)) = 1$ for $X \in \mathbb{R}$. Note that $X(A)$ is countable while $X(\overline{A})$ is not necessarily.

We define a discrete probability density (discrete density for short) thus:

Discrete probability density Let $X \in \mathbb{R}$ and $x \in \mathbb{R}$. Define the countable (finite or infinite) set of events $A := \{E_1, E_2, \ldots\}$ and assign $P(X(E_i) = x_i) = \pi_i > 0$. Then the function

$$P(X = x | \boldsymbol{\pi}) = \begin{cases} \pi_i & \text{for } x = x_i \\ 0 & \text{otherwise} \end{cases} \tag{5.1}$$

(where $\boldsymbol{\pi} := [\pi_1, \pi_2, \ldots]$) is a discrete density.

Defining discrete densities this way conforms to the requirement that rv are real numbers. Furthermore, as we shall see, the R functions that provide discrete densities and distributions treat the rv X as real for both discrete and continuous densities.

Example 5.1. Consider the following experiment: An observer sits at the corner of a busy intersection and records whether a person crossing the street made it. Each person has the same probability, π, of being hit by a car and these are independent. The experiment continues until the first person is hit. Each trial (a person crossing the street) has one of two outcomes: success with probability π and failure (the person is not hit) with probability $1 - \pi$. Define the event E_i as the number of people that crossed the street successfully by the time the experiment ends. Then

$$A = \{E_0, E_1, \ldots\} \ .$$

Here E_0 is the event that the first person to cross the street was hit (i.e. no person crossed the street successfully), E_1 is the event that one person crossed the street successfully and second was hit and so on. Define the rv X as the number of people that crossed the street successfully. So $X(E_i) := i$. Then

$$P(X(E_0) = 0) = \pi \ , \quad P(X(E_1) = 1) = (1 - \pi) \pi \ , \quad \ldots \ ,$$
$$P(X(E_n) = n) = (1 - \pi)^n \pi \ , \quad \ldots \ , \quad n \in \mathbb{Z}_{0+} \ .$$

Simplifying the notation, we write the density as

$$P(X = x | \pi) = \begin{cases} (1 - \pi)^x \pi & \text{for } x = x_i \\ 0 & \text{otherwise} \end{cases} \tag{5.2}$$

where π is the probability of accident and $x_i \in \mathbb{Z}_{0+}$. □

Equation (5.2) is known as the *geometric density*. It is constructed from a sequence of independent Bernoulli trials where $(1 - \pi)$ is the probability of failure and π is the probability of success. The density describes the number of failures until the first success. Figure 5.1 illustrate two geometric densities, for $\pi = 0.3$ and for $\pi = 0.7$. Here is the script that produces Figure 5.1:

```
1   PI <- c(0.3, 0.7) ; x <- 0 : 10
2   xlab <- expression(italic(x))
3   par(mfrow = c(1, 2))
4   for(i in 1 : 2){
5     density <- dgeom(x, PI[i])
6     ylab <- bquote(italic(P(X==x))~~~~~pi) == .(PI[i]))
7     plot(x, density, type = 'h', lwd = 2,
8        xlab = xlab, ylab = ylab)
9     abline(h = 0, lwd = 2)
10  }
```

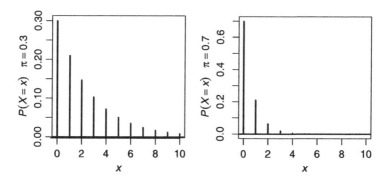

Figure 5.1   Geometric densities.

In line 1 we create a vector (with a call to c()) for the two π values of the geometric densities and a vector of x values (with :). In line 2 we prepare the label for the x-axis. The function expression() creates an R expression from its argument. In R, mathematical notations (including italic()) are produced with expression(). When the argument to expression() invokes text plotting functions (such as xlab = 'something'), R treats the argument as a mathematical expression. So italic(x) causes the function that draws the label for the x-axis (xlab in line 8) to draw x in italic. In line 3 we set the graphics device to accept two plots with a call to par() and the argument mfrow set to a matrix of one row and two columns. The densities are produced for $\pi = 0.3$ or 0.7 in line 5. We ask dgeom() (for geometric density) to produce one value for each x with the probability PI[i]. For example, for x = 2 and PI[1] $= 0.3$ we obtain

$$P(X = 2 | \pi = 0.3) = 0.7^2 0.3 = 0.147 \ .$$

Line 6 produces the mathematical notation for the label of the y-axis. The function bquote() (for back quote) quotes its argument (that is, it produces a string) with one exception. A term wrapped in .() is evaluated. So the effect of bquote() here is to produce the string "$P(X = x)$ $\pi = 0.3$" (or 0.7). The effect of ~~~~~ is to produce five spaces before PI (see help for plotmath()). We then call plot() with

140 Discrete densities and distributions

plot type type = 'h' and line width lwd = 2 in lines 7 and 8. This produces the stick plot. In line 9 we add a horizontal zero-line of width 2 with the arguments h and lwd given to abline(). We do this to emphasize that the rv X can be defined for any real value x with $P(X = x) \geq 0$ for some isolated (countable) values of x and zero everywhere else.

From Example 5.1 we derive the following definition:

Geometric density Equation (5.2) is known as the family of geometric densities. For a specific value of π, it is known as the geometric density.

We think of a density as a family. Densities have parameters that determine their shape, dispersion and location. A parameter is assigned a fixed value. For each parameter value we have a member of the family. Because the geometric has a single parameter (π), we sometimes say that the geometric is a *single parameter density*. With this definition of the geometric density, a full (and correct) representation of a density should show $P(X = x|\pi)$ for all values of x.

In research, we are often interested in empirical (experimentally based) densities. Such densities are constructed from data. Empirical densities can be presented with histograms (see Section 3.3).

Example 5.2. Beginning on 9/27/2000, the Israeli Foreign Ministry has posted information of incidents of terrorist attacks on Israelis (MFA, 2004). The data referred to a war dubbed the Second Intifada, including the date and a short description of the incident. The description includes the number killed, the number injured and the organization or organizations that claimed responsibility for the attack. Occasionally, the description details the number of children and women that were killed and injured in the attack (see Chapter 17 for further details). Here we are interested in the density of the number of Israelis that were killed per attack by Hamas (Figure 5.2).

An "experiment" is an attack. An event is the number of people that were killed in the attack. Because the distance between breaks in the histogram is 5, to compute the

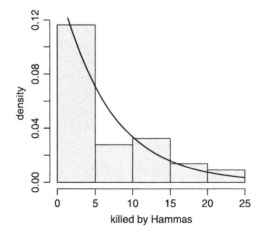

Figure 5.2   The density of the number of people killed per attack by Hamas. An exponential function fits the data.

probabilities, we multiply the densities shown in Figure 5.2 by 5. Thus we construct the density:

$$
\begin{array}{llllllll}
\text{Event:} & 0\text{-}4 & 5\text{-}9 & 10\text{-}14 & 15\text{-}19 & 20\text{-}24 & \varnothing \\
\text{Dead } (X)\text{:} & [0,5) & [5,10) & [10,15) & [15,20) & [20,25) & \text{otherwise} \\
P(X = x|\pi)\text{:} & 0.581 & 0.140 & 0.163 & 0.070 & 0.0\,465 & 0
\end{array}
$$

The following script was used in this example.

```
1  load('terror.by.Hamas.rda')
2  terror <- terror.by.Hamas
3  lambda <- 1 / mean(terror$Killed) ;
4  x <- 0 : 25
5  h(terror$Killed, xlab = 'killed by Hamas')
6  lines(x, dexp(x, lambda))
```

In line 1, we load the data frame `terror.by.Hamas` from a file. Here are the first three rows of the data:

```
   Julian    Date   Killed Injured Org.1 Org.2 Org.3
44 15038  3/4/2001      3         60 Hammas
47 15062 3/28/2001      2          4 Hammas
52 15087 4/22/2001      1         60 Hammas
```

`Julian` refers to Julian day, `Date` refers to the date of the attack, `Killed` and `Injured` give the number of people killed or injured by the attack and `Org.1`, `Org.2` and `Org.3` refer to the organization that claimed responsibility for the attack (in few cases there were more than one).

To save on typing, we assign the data frame to `terror` in line 2. In line 3, we compute an estimate of a parameter named λ of the exponential density (we will discuss it soon). In line 5, we plot the histogram of the number of Israelis killed per attack. We use a modified `hist()`, as detailed on page 40. We fit a curve to the histogram with a call to `dexp()` with the parameter `lambda`, embedded in `lines()`.   □

5.2 Distributions

Corresponding to each discrete density $P(X = x|\pi)$ there exists a

Discrete probability distribution If $P(X = x|\pi)$ is a discrete density, as defined in (5.1), then $P(X \leq x|\pi)$ is a discrete distribution.

What do we mean by $P(X \leq x|\pi)$? From the definition of rv, we have $X(E) = x$. Let A be the set of all events such that $X(A) \leq x$. Then the distribution function is $P(X(A) \leq x|\pi)$. Here is an example of how to construct a distribution.

142 Discrete densities and distributions

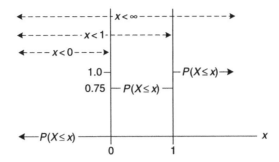

Figure 5.3 The distribution of a Bernoulli experiment of rain vs. no rain with $X(R) = 0.75$ and $X(\overline{R}) = 0.25$.

Example 5.3. Consider a day in a tropical rainy season. Let R be the event that it rained during the day and $\overline{R}$ the event that it did not. Rain occurs with $\pi = 0.75$ and rains on different days are independent. So we have $X(R) = 0.75$ and $X(\overline{R}) = 0.25$. The density of R is

$$P(X = x | \pi = 0.75) = \begin{cases} 0.75 & x = 0 \\ 0.25 & x = 1 \\ 0 & \text{otherwise} \end{cases}.$$

The corresponding distribution is illustrated in Figure 5.3. From the figure, we conclude:

$$P(X \leq x | x < 0, 0.75) = 0,$$
$$P(X \leq x | 0 \leq x < 1, 0.75) = 0.75,$$
$$P(X \leq x | x \geq 1, 0.75) = 1$$

or

$$P(X \leq x | 0.75) = \begin{cases} 0 & x < 0 \\ 0.75 & 0 \leq x < 1 \\ 1 & x \geq 1 \end{cases}. \tag{5.3}$$

Therefore, $P(X \leq x)$ is defined for all real x. □

Equation (5.3) is a distribution. It describes the outcome of Bernoulli trials and is known as the *Bernoulli distribution* with parameter π where π is the probability of success. In the next example, we construct the geometric distribution.

Example 5.4. In Example 5.1 we constructed the geometric density. To obtain the geometric distribution, we need to establish $P(X \leq x | \pi)$ for all values of x. For $x < 0$, no accident can occur because no one crossed the street. Therefore, $P(X < 0 | \pi) = 0$. At $x = 0$, we have no successful crossing and an accident on the first crossing. Therefore, $P(X \leq 0 | \pi) = \pi$. For $0 < x < 1$ no event can occur (we are counting integers). Therefore, $P(X < 1 | \pi) = P(X \leq 0 | \pi) + P(0 < X < 1 | \pi) = \pi$. At $x = 1$ we have one successful crossing and then an accident. Therefore,

$$P(X \leq 1 | \pi) = P(X < 0 | \pi) + P(X = 0 | \pi) + P(0 < X < 1 | \pi) + P(X = 1 | \pi)$$
$$= 0 + \pi + 0 + (1 - \pi)\pi$$
$$= (1 - \pi)^0 \pi + (1 - \pi)^1 \pi.$$

Continuing in this manner, we find that

$$P(X \leq x|\pi) = \int_{-\infty}^{x} P(X = \xi|\pi) \, d\xi \tag{5.4}$$

$$= \int_{-\infty}^{x} \sum_{i=0}^{\underline{x}} (1-\pi)^{i-1} \pi \delta(\xi - i) \, d\xi$$

where $i \in \mathbb{Z}_{0+}$ and $\underline{x}$, the floor of x, is the largest integer $\leq x$. The function $\delta(x)$ is zero for any value of $x \neq 0$. Also,

$$\int_{-\infty}^{\infty} f(a)\delta(x-a) \, dx = f(a) \, .$$

$\delta(x)$ is called the *Dirac delta function*. Because $P(X \leq x|\pi) = 0$ for any $x \notin \mathbb{Z}_{0+}$, (5.4) simplifies to

$$P(X \leq x|\pi) = \sum_{i=0}^{\underline{x}} P(X = x_i|\pi) \, . \tag{5.5}$$

To produce a value for of $P(X \leq x|\pi)$ for any x, with, say $\pi = 0.1$, use

```
> pgeom(10, 0.1)
[1] 0.6861894
> pgeom(10.5, 0.1)
[1] 0.6861894
```

and note this

```
> sum(dgeom(0 : 10, 0.1))
[1] 0.6861894
```

`pgeom()` and `dgeom()` are the geometric distribution and density, respectively. Also note this:

```
> dgeom(0.1, 0.1)
[1] 0
Warning message:
non-integer x = 0.100000
```

In other words, it is *not* an error to ask for `dgeom()` of a number other than an integer. However, R wants to remind you that you provided a discrete density with a value for x that is not an integer. □

Equation (5.4) is known as the *geometric distribution*. Figure 5.4 illustrates what it looks like. To produce the figure, follow the script on p. 138. However, instead of using `dgeom()`, use `pgeom()`.

5.3 Properties

From the definitions of discrete densities and distributions and the discussion above, we deduce their properties.

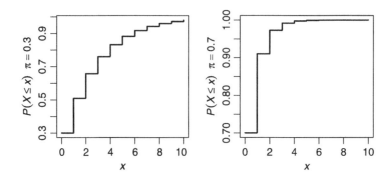

Figure 5.4   Geometric distributions (compare to Figure 5.1).

5.3.1 Densities

$P(X = x|\pi) \geq 0$

This property is a direct consequence of the definition of probability.

$\sum_i P(X = x_i|\pi) = 1$

Here x_i is a subset of x at which $P(X = x|\pi) > 0$. This property is a consequence of the fact that x_i for i indexing all possible x is a map of all events with $P(X > x|\pi)$ to the real numbers.

5.3.2 Distributions

$P(X \leq -\infty|\pi) = 0$

This is a consequence of the definition of X. Specifically, that $X > -\infty$.

$P(X \leq \infty|\pi) = 1$

This is a consequence of the definition of X. Specifically, that $X < \infty$.

$P(X \leq x_i|\pi) \leq P(X \leq x_j|\pi)$ for $i \leq j$

The distribution at x_j, $P(X \leq x_j|\pi)$, is the union of all density values at x_i. These density values are ≥ 0. Therefore, $P(X \leq x_i|\pi) \leq P(X \leq x_j|\pi)$.

$P(x_i < X \leq x_j|\pi) = P(X \leq x_j|\pi) - P(X \leq x_i|\pi)$ for $i < j$

This property can be observed from the distributions illustrated in Figures 5.1, 5.3 and 5.4.

In the last two properties, x_i and x_j are members of subset of x for which $P(X = x|\pi) > 0$. We discuss these properties in more detail in Section 6.3.

5.4 Expected values

Because a rv takes on certain values with certain probabilities, to obtain a mean value, we must sum over all the values that the rv might take, each value weighed by its probability.

Example 5.5. Say somebody offers you the following gamble: You are given a biased coin with probability of head $= 0.9$. You win \$1 when head (H) shows up and lose \$10 when tail (T) shows up. Should you take the gamble expecting to win? Our rv is

$X(H) := \$1$ and $X(T) := -\$10$. After many experiments, you should expect to win $\$1 \times 0.9 - \$10 \times 0.1 = -\$0.1$. □

Recall that we defined $A := \{E_1, E_2, \ldots\}$ as the countable set of all events such that $P(X(E_i)) = x_i) = \pi_i > 0$. Using the same argument that leads to (5.5), we define

Expected value The expected value of a discrete density $P(X = x|\pi)$ is

$$E[X] := \sum_i x_i P(X = x_i|\pi) .$$

Note that because $P(X = x|\pi) = 0$ for $x \neq x_i$ (for all i), we simply sum over the discrete values of $x_i \times P(X = x_i|\pi)$ and thus avoid integration for the remaining (real) values of x. The expected value of a discrete density is *not* necessarily a typical value. In fact, it may even be a value that the rv from the density might take with probability zero. Furthermore, the expected value of any density is *not* a rv.

Example 5.6. We arbitrarily assign values to X based on the number of dots that show on the face of a die:

Event:	1	2	3	4	5	6	∅
x_i:	100	−1.2	10	12.4	1 000	−5.24	
$P(X = x\|\pi)$:	1/6	1/6	1/6	1/6	1/6	1/6	0

Therefore, the expected value of X is

```
> x <- c(100, -1.2, 10, 12.4, 1000, -5.24)
> PI <- rep(1/6, 6)
> (E.x <- round(sum(x * PI), 2))
[1] 185.99
```

or in vector notation

```
> round(x %*% PI, 2)
      [,1]
[1,] 185.99
```

In our notation,

$$E[X] = \sum_{i=1}^{6} x_i P(X = x_i|\pi) \approx 185.99 .$$

In R, the sum of element by element multiplication of two vectors can be achieved in one of two ways

```
> x <- c(1 : 3) ; y <- c(4 : 6)
> sum(x * y)
[1] 32
> x %*% y
     [,1]
[1,]   32
```

Although we get the same answer, the objects returned from these two operations are different. Both operations correspond to the so-called vector *dot-product*. □

146 Discrete densities and distributions

The computation above is identical to the intuitive definition of the mean because each value of the rv is equally probable. This is not the case when probabilities of events are not equal. For some densities, it is possible to derive the expected value of the rv in a closed form.

Example 5.7. Using (5.2), we write

$$E[X] = \sum_{i=1}^{\infty} i(1-\pi)^{i-1}\pi$$

In Exercise 5.15, you are asked to prove that

$$E[X] = \frac{1}{\pi}$$

is the expectation of the geometric density. □

5.5 Variance and standard deviation

Intuitively, the variance of a rv from a known density reflects our belief that a particular value of a rv will be within some range. With the notation preceding the definition of expected values in mind and again, using the same argument that leads to (5.5), we define

Variance The variance of a discrete density $P(X = x|\pi)$ is

$$V[X] = \sum_i (x_i - E[X])^2 P(X = x_i|\pi) \ .$$

Like the expected value, the variance of a density is *not* a rv.

Example 5.8. Yellowstone was the first national park to be established in the U.S. A total of 3,019,375 people visited the park in 2003 (NPS, 2004). Two of the busiest entrances to the park are the Western and Northern. You obtain a summer job at the park and are asked to record the number of passengers in a car entering the park. You find that it ranges between 1 and 5. The densities of the number of passengers in a car (X and Y for the Western and Northern entrances) are shown in Table 5.1. We have

```
> Yellowstone <- cbind(passengers = c(1, 2, 3, 4, 5),
+    p.west = c(.4, .3, .2, .1,0),
+    p.north = c(.2, .6, .2, 0, 0))
> (E.west <- sum(Yellowstone[, 1] * Yellowstone[, 2]))
[1] 2
> (E.north <- sum(Yellowstone[, 1] * Yellowstone[, 3]))
[1] 2
```

or in our notation,

$$E[X] = 2 \ , \quad E[Y] = 2 \ .$$

Table 5.1 Passengers in cars entering Yellowstone National Park.

Passengers	Probabilities	
	West entrance	North entrance
1	0.400	0.200
2	0.300	0.600
3	0.200	0.200
4	0.100	0.000
5	0.000	0.000

For the variance, we obtain

```
> sum((Yellowstone[, 1] - E.west)^2 * Yellowstone[, 2])
[1] 1
> sum((Yellowstone[, 1] - E.north)^2 * Yellowstone[, 3])
[1] 0.4
```

or in our notation
$$V[X] = 1, \quad V[Y] = 0.4.$$

In passing, we note that because we are talking about the expectation and variance of the density, we must assume that the probabilities in Table 5.1 represent the proportions for all cars entering the park. Such proportions are sometimes called the true (or population) proportions. □

As was the case for the expected value, the variances of some distributions are known in a closed form.

Example 5.9. In Exercise 5.17 you are asked to show that the variance of the geometric distribution is
$$V[X] = \frac{1-\pi}{\pi^2}$$
where π is the probability of success. □

Standard deviation The standard deviation of a discrete density with variance $V[X]$ is $\sqrt{V[X]}$.

The standard deviation describes a typical deviation of a value of X away from $E[X]$. The units of the standard deviation are identical to those of X.

5.6 The binomial

Recall that in a Bernoulli experiment, we have either success with probability π or failure with probability $1 - \pi$. The binomial density addresses the question of the probability of m successes in n independent repetitions of a Bernoulli experiment.

Example 5.10. Suppose that 20% of the people in a crowd at a concert liken Mozart's music to bubble gum. The experiment is picking a person at random and asking if he thinks that Mozart's music reminds him of bubble gum. Yes is a success. Assign 1 to success and 0 to failure. Let the rv X be the number of successes. What is the probability that 2 out of 4 chosen people say yes? Let us enumerate the possible outcomes.

Person:	first	second	third	fourth
	1	1	0	0
	1	0	1	0
	1	0	0	1
	0	1	1	0
	0	1	0	1
	0	0	1	1

Because $\pi = 0.2$, the first outcome has a probability of

$$P(X = 2 | n = 4 \text{ in the order } \{1,1,0,0\}) = \pi \times \pi \times (1-\pi) \times (1-\pi)$$
$$= 0.2 \times 0.2 \times 0.8 \times 0.8$$
$$= 0.0256 \ .$$

The remaining outcomes have the same probability. Because the events are independent, the probability that $X = 2$ is the sum of the probabilities of each of the rows above. Two successes and two failures in 4 repetitions can occur in 6 different ways. Therefore, we must add 0.0256 six times. Thus,

$$P(X = 2 \text{ in any order } | n = 4) = 6 \times 0.0256 = 0.1536 \ . \qquad (5.6)$$

Arrangements of 1 and 0 above is a combination of 2 in 4 slots. □

To generalize the example, denote by $n = 4$ the number of trials and by $m = 2$ the number of successes. We can write equality (5.6) thus

$$P(X = m | \pi, n) = \binom{n}{m} \pi^m (1-\pi)^{n-m}$$
$$= \binom{4}{2} \pi^2 (1-\pi)^2$$
$$= 0.1536 \ .$$

The same result (with round off error) is obtained from calling for the binomial density with two successes in four trials with probability of success $= 0.2$:

```
> round(dbinom(2, 4, 0.2), 3)
[1] 0.154
```

The so-called binomial coefficient

$$\binom{n}{m} := \frac{n!}{m!(n-m)!}$$

is the number of ways that m successes can occur in n trials (a combination). You can calculate it in R with choose(n,m). Because the events are independent, the probability of m successes is π^m. The remaining $n - m$ are failures with probability $(1-\pi)^{n-m}$. So to arrive at the probability, we must add $\pi^m \times (1-\pi)^{n-m}$ as many times as $\binom{n}{m}$. To formally define the family of binomial densities and distributions, we abandon the restriction that the number of successes is an integer. To construct the binomial, we denote the event of 0 successes in n Bernoulli trials with probability of success π by E_0, the event of one success by E_1, ..., the event of n successes by E_n. Therefore,

$$A = \{E_0, E_1, \ldots, E_n\}.$$

We map the events to the rv by assigning the index of E_i to i, $i = 0, 1, \ldots, n$ where i is the number of successes in n trials. Next, we let $P(X(E_i) = x) = \pi_i$ for $x = i$. From the construction we see that $P(X(A)) = 1$. We also see that $A \cup \overline{A} = \mathbb{R}$. Thus we define the

Binomial density Let $n \in \mathbb{Z}_{0+}$. The probability of x successes in n independent Bernoulli trials with probability of success π,

$$P(X = x|\pi, n) = \begin{cases} \binom{n}{x}\pi^x (1-\pi)^{n-x} & \text{for } x = 0, 1, \ldots, n \\ 0 & \text{otherwise} \end{cases}$$

is called the *binomial density*.

Note that n is *not* the number of repetitions of the experiment. It is the number of trials in a single experiment. We now also have

Binomial distribution The function

$$P(X \leq x|\pi, n) = \int_{-\infty}^{x} \sum_{i=0}^{x} \binom{n}{i}\pi^i (1-\pi)^{n-i} \delta(\xi - i) \mathrm{d}\xi$$

defines the two-parameter (π and n) binomial distribution.

Now that we know that the rv of discrete densities and distributions takes on any value on the real line, we can simplify the notation. Because $P(X = x|\pi) = 0$ except for $x = 0, 1, \ldots, n$, we can ignore the integral sign—as was the case in (5.4)—and sum over the values of x for which $P(X = x|\pi) > 0$. So from now on, instead of writing the binomial density or distribution as above, we will write them with respect to the rv X as

$$P(X = m|\pi, n) = \binom{n}{m}\pi^m (1-\pi)^{n-m}, \quad m = 0, 1, \ldots, n,$$

$$P(X \leq m|\pi, n) = \sum_{i=0}^{m} \binom{n}{i}\pi^i (1-\pi)^{n-i}$$

while keeping in mind that the binomial rv X can take a value x where x is a real number. Let us see what the binomial densities and distributions look like.

Example 5.11. Let $n = 10$ and $\pi = 0.3$ or $\pi = 0.7$. Then Figure 5.5 illustrates two binomial densities for $X = x$ with parameters π and n. Note the longer tail to the right or to the left for $\pi = 0.3$ or 0.7, respectively. To produce Figure 5.5, follow the script on page 138 changing line 6 from

```
d <- dgeom(x, p[i])
```

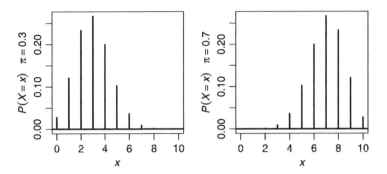

Figure 5.5   Binomial densities.

to

```
d <- dbinom(x, n, p[i])
```

Figure 5.6 illustrates the corresponding distributions. To obtain the figure, replace line 6 in the script on p. 138 with

```
d <- pbinom(x, n, p[i])
```

and the plot from type = 'h' to type = 's'. Note that R responds correctly to the following:

```
> dbinom(5.2, 10, 0.5)
[1] 0
Warning message:
non-integer x = 5.200000
> pbinom(5.2, 10, 0.5)
[1] 0.623046875
> pbinom(-1, 10, .5)
[1] 0
> dbinom(-1, 10, .5)
[1] 0
```

□

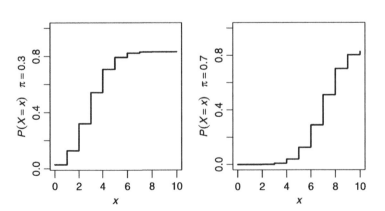

Figure 5.6   Binomial distributions of the densities illustrated in Figure 5.5.

5.6.1 Expectation and variance

We leave the proof of the following for Exercise 5.16:

Expected value of the binomial The expected value of the binomial density with n trials and a probability of success π, is

$$E[X] = n\pi .$$

Variance of the binomial The variance of the binomial n trials and a probability of success π, is
$$V[X] = n\pi(1-\pi) .$$

Example 5.12. A survey of 1 550 men in the UK revealed that 26% of them smoke (Lader and Meltzer, 2002). Pick a random sample of $n = 30$. Then

$$E[X] = 30 \times 0.26 = 7.8$$

and
$$V[X] = 30 \times 0.26 \times 0.74 = 5.772 .$$

We interpret these results as saying that if we were to take many samples of 30 men, we expect 7.8% of them to be smokers, with a variance of 5.772. □

5.6.2 Decision making with the binomial

We often use distributions to decide if an assumption is plausible or not. The binomial distribution is a good way to introduce this subject, which later mushrooms into statistical inference. We introduce the subject with an example and then discuss the example.

Example 5.13. Consider the data introduced in Example 5.12 and assume that the survey results represent the UK population of men. That is, the probability that a randomly chosen man smokes is $\pi = 0.26$. We wish to verify this result. Not having government resources at our disposal, we must use a small sample. *After deciding on how to verify the government's finding, we will select a random sample of 30 English men from the national telephone listing and ask whether they smoke or not.* □

If the sample happens to represent the population, between 7 or 8 of the respondents should say yes. Because of the sample size, it is unlikely that we get exactly 7 or 8 positive responses, even if in fact 26% of all English men smoke. So, we devise

Decision rule 1 If the number of smokers in the sample is between 6, 7, 8 or 9, then we have no grounds to doubt the government's report. Otherwise, we reject the government's report.

Because we use a single sample, we will never know for certain whether the government's finding is true. So we must state our conclusion with a certain amount of probability (belief) in our conclusion. Let M be the number of smoking men in a sample of 30 and *assume that the government finding is true*. What is the probability that we will conclude from the sample that the government's claim is true? Based on our decision rule, the probability that the number of smokers in our sample, M, will be 6, 7, 8 or 9 is

$$P(6 \leq M \leq 9 | \pi = 0.26, n = 30) = \sum_{i=6}^{9} \binom{n}{i} \times 0.26^i \times 0.74^{n-i} = 0.596 \ .$$

With R, we obtain this result in one of two ways: using the binomial density,

```
> n <- 30 ; PI <- 0.26 ; i <- 6 : 9
> round(sum(dbinom(i, n, PI)), 3)
[1] 0.596
```

or better yet, the binomial distribution

```
> round(pbinom(9, n, PI) - pbinom(5, n, PI), 3)
[1] 0.596
```

What is the probability that we will conclude that the government's claim is *not* true? That is, the number of smokers in our sample should be less than 6 and greater than 9. So

$$P(M \leq 5) + P(M \geq 10) = 1 - 0.596 = 0.404 \ .$$

Next, we collect the data and find that we have 8 smokers in our sample. We therefore conclude that we have no grounds to reject the government's finding. How much faith do we have in our decision? Not much. The probability that there will be between 6 and 9 smokers in a sample of 30—assuming that government's finding is correct—is 0.596.

Does our decision rule make sense? Not really. Why? Because we have to make a decision based on a single sample and the distinction between "right" and "wrong" is not very clear (0.596 vs. 0.404). Can we improve upon the decision rule? Let us see.

Decision rule 2 If the number of smokers in the sample is 4, or 5, ..., or 11, then we have no grounds to doubt the government's report. Otherwise, we reject the government's report.

We examine this decision rule with the assumption that the government claim is correct ($\pi = 0.26$). Now

$$P(4 \leq M \leq 11 | \pi = 0.26, n = 30) = \sum_{i=4}^{11} \binom{n}{i} \times 0.26^i \times 0.74^{n-i} = 0.905$$

and

$$P(M \leq 3 | \pi = 0.26, n = 30) + P(M \geq 12 | \pi = 0.26, n = 30) = 1 - 0.905$$
$$= 0.095 \ .$$

In R:

```
> round(pbinom(11, n, PI) - pbinom(3, n, PI), 3)
[1] 0.905
```

Next, we collect the data and find that we have 9 smokers in our sample. We therefore conclude that we have no grounds to reject the government's finding. How much faith do we have in our decision? Much. The probability that there will be between 4 and

11 smokers—assuming the government's finding is correct—is 0.905. That is, given $\pi = 0.26$, if we repeat the sample many times, then 90.5% of them will have between 4 and 11 smokers.

So which decision rule is better? On the basis of the analysis so far, Rule 2. If this is the case, then perhaps we can do even better.

Decision rule 3 If the number of smokers in the sample is between 0 and 30, then we have no grounds to doubt the government's report. Otherwise, we reject the government's report.

Now

$$P(0 \leq M \leq 30 | \pi = 0.26, n = 30) = \sum_{i=0}^{n} \binom{n}{i} \times 0.26^i \times 0.74^{n-i} = 1.0\,.$$

Something is wrong. Based on this decision rule, we will never reject the government's finding; no matter how many smokers turn up in our sample (we might as well not sample at all). What happens is that by willing to accept a wider range of smokers in the sample as a proof of the government's finding, we lose the ability to distinguish among other possibilities. The true proportion of smokers may be $\pi = 0.30$. Let us see why.

So far, we assumed that the government was right and we used the sample's data to reach a decision. Assume that the government is wrong and that the true proportion of smokers is $\pi = 0.30$. Let use now use the three decision rules:

$$P(6 \leq M \leq 9 | \pi = 0.30, n = 30) = \sum_{i=6}^{9} \binom{n}{i} \times 0.30^i \times 0.70^{n-i} = 0.512\,,$$

$$P(4 \leq M \leq 11 | \pi = 0.30, n = 30) = \sum_{i=4}^{11} \binom{n}{i} \times 0.30^i \times 0.70^{n-i} = 0.831$$

and

$$P(0 \leq M \leq 30 | \pi = 0.30, n = 30) = \sum_{i=0}^{n} \binom{n}{i} \times 0.30^i \times 0.70^{n-i} = 1.0\,.$$

Thus, if the government's claim is wrong and we offer an alternative (of $\pi = 0.30$), then the wider the range we select for accepting the government's finding, the higher the probability we will not reject the government's finding in spite of the fact that our alternative might be true. So what shall we do? The rule of thumb is to choose the narrowest range of M that makes sense. A more satisfactory answer will emerge later, when we discuss statistical inference.

5.7 The Poisson

The Poisson density models counting events, usually per unit of time (rates). It is also useful in counting intensities—events per unit of area, volume and so on. It is one of the most widely used densities. It applies in fields such as physics, engineering and biology. In astronomy, the density is used to describe the spatial density of galaxies and stars in different regions of the Universe. In engineering, it is routinely used in queuing theory. In physics, the Poisson is used to model the emission of particles. The spatial distribution of plants (see Pielou, 1977) is often described with the Poisson. Geneticists use it to model the distribution of mutations. Wildlife biologists sometimes use the Poisson to model the distribution of animals' droppings. In neuroscience, the Poisson is used to model impulses emitted by neurons.

The Poisson density depends on a single intensity parameter, denoted by λ. The mechanism that gives rise to this density involves the following assumptions:

1. Events are rare (the probability of an event in a unit of reference, such as time, is small).
2. Events are independent.
3. Events are equally likely to occur at any interval of the reference intensity unit.
4. The probability that events happen simultaneously is negligible (for all practical purposes it is zero).

The first assumption can be satisfied in any counting process by dividing the interval into many small subintervals. Small subintervals also ensure that the fourth assumption is met. The second and third assumptions must be inherent in the underlying process. This does not diminish the importance of the Poisson process—we often use the second and third assumption as a testable hypothesis about the independence of events. Furthermore, the Poisson process can be generalized to include intensity that is a function (e.g. of time) as opposed to a fixed parameter. We now skip the details of mapping events to rv and assigning them probabilities and move directly to the definition of the

Poisson density Denote the intensity of occurrence of an event by λ. Then

$$P(X = x|\lambda) = \begin{cases} \dfrac{\lambda^x}{x!}e^{-\lambda} & \text{for } x \in \mathbb{Z}_{0+} \\ 0 & \text{otherwise} \end{cases}.$$

is called the Poisson density.
Poisson distribution The function

$$P(X \leq x|\lambda) = \sum_{i=0}^{x} \frac{\lambda^i}{i!}e^{-\lambda}.$$

is called the Poisson distribution.

Often, the interval is time; so λ is in unit of counts per unit of time (a rate). In such cases, we write the Poisson density for the time interval $[0, t]$ as

$$P(X = x|\lambda, t) = \begin{cases} \dfrac{(\lambda t)^x}{x!}e^{-\lambda t} & \text{for } x \in \mathbb{Z}_{0+} \\ 0 & \text{otherwise} \end{cases}$$

where $x \in \mathbb{Z}_{0+}$. To simplify the notation, we will usually write the Poisson densities and distributions as

$$P(X = m|\lambda) = \frac{\lambda^m}{m!}e^{-\lambda},$$

$$P(X \leq m|\lambda) = \sum_{i=0}^{m} \frac{\lambda^i}{i!}e^{-\lambda}.$$

To see why the Poisson is so useful, consider the following examples:

Example 5.14. We start with a population of n individuals, all born at time 0. If death of each individual is equally likely to occur at any time interval, if the probability of death of any individual is equal to that of any other individual and if deaths are independent of each other, then the density of the number of deaths in a subinterval is Poisson. Similar considerations hold for the spatial distribution of objects. Instead of deaths, we may count defective products, or any other counting process. □

In the next section, we show that the Poisson density approximates the binomial density. Here, we introduce an example that uses this approximation to demonstrate the widespread phenomena to which Poisson densities apply.

Example 5.15. We start with a large cohort of individuals and follow their lifetimes during the interval $[0, t]$. As in Example 5.14, we assume that individuals die independently of each other and each has the same probability of dying at any time. Evidence suggests that the Poisson mortality model applies to trees and birds. Now divide $[0, t]$ into n subintervals, each of length t/n. Take n to be large enough so that the subintervals t/n are very short. Therefore, the probability of more than one death during these short subintervals is negligible and we can view the death of an individual during any subinterval t/n as a Bernoulli trial—an individual dies during the subinterval with probability π or survives with probability $1 - \pi$. Because the subintervals are short, π is small. Also, because the death of an individual during a particular subinterval is independent of the death of other individuals during other subintervals, we have the binomial density. The probability that m individuals die during $[0, t]$ is binomial; i.e.

$$P(X = m|\pi, n) = \binom{n}{m} \pi^m (1 - \pi)^{n-m} . \tag{5.7}$$

The expected number of deaths during $[0, t]$ is $E[X] = n\pi$. Denote the death rate by λ. The number of deaths during $[0, t]$ is approximately λt. Therefore, $E[X] \approx \lambda t$ or $\pi \approx \lambda t / n$. As $n \to \infty$, we can assume that $n\pi \to \lambda t$. With this in mind, we rewrite (5.7)

$$P(X = m|\lambda t, n) = \binom{n}{m} \left(\frac{\lambda t}{n}\right)^m \left(1 - \frac{\lambda t}{n}\right)^{n-m} .$$

Using the Poisson approximation to the binomial (Equation 5.8 below), we obtain

$$\lim_{n \to \infty} \binom{n}{m} \left(\frac{\lambda t}{n}\right)^m \left(1 - \frac{\lambda t}{n}\right)^{n-m} = \frac{(\lambda t)^m}{m!} e^{-\lambda t} .$$

□

The example demonstrates the fact that any phenomenon in nature in which the events are rare (π is small), n is large (many short subintervals are used), the events are independent and the probability of the event is constant, the Poisson density applies.

5.7.1 The Poisson approximation to the binomial

The Poisson approximation to the binomial holds when the expectation of the binomial, $n\pi$, is on the order of 1 and the variance, $n\pi(1 - \pi)$ is large. Both conditions imply that π is small and n is large. Under these conditions,

$$\binom{n}{m}\pi^m(1-\pi)^{n-m} \approx \frac{(n\pi)^m}{m!}e^{-n\pi}. \tag{5.8}$$

The approximation improves as $n \to \infty$ and $\pi \to 0$. This approximation was first proposed by Poisson in 1837.

Let $\lambda \in \mathbb{R}$, $n \in \mathbb{Z}_+$ and $m \in \mathbb{Z}_{0+}$. We wish to show that

$$\lim_{n \to \infty} \binom{n}{m}\left(\frac{\lambda}{n}\right)^m\left(1-\frac{\lambda}{n}\right)^{n-m} = \frac{\lambda^m}{m!}e^{-\lambda}.$$

Now λ and m are constants, so the left hand side can be written as

$$\frac{\lambda^m}{m!}\lim_{n\to\infty}\frac{n!}{(n-m)!(n-\lambda)^m}\left(1-\frac{\lambda}{n}\right)^n.$$

The identity

$$\lim_{n\to\infty}\left(1-\frac{\lambda}{n}\right)^n = e^{-\lambda}$$

is well known. Also,

$$\lim_{n\to\infty}\frac{n!}{(n-m)!(n-\lambda)^m} = \lim_{n\to\infty}\frac{n(n-1)\cdots(n-m+1)}{(n-\lambda)\cdots(n-\lambda)} = 1$$

because m and λ are fixed and $n \to \infty$. Therefore

$$\frac{\lambda^m}{m!}\lim_{n\to\infty}\frac{n!}{(n-m)!(n-\lambda)^m}\left(1-\frac{\lambda}{n}\right)^n = \frac{\lambda^m}{m!}e^{-\lambda}.$$

It can also be shown that

$$\sum_{m=0}^{\infty}\frac{\lambda^m}{m!}e^{-\lambda} = 1.$$

Therefore,

$$P(X = m|\lambda) = \frac{\lambda^m}{m!}e^{-\lambda}$$

is a density.

Example 5.16. Suppose that the density of mutations in a collection of subpopulations is Poisson with parameter λ mutations per unit of time. What is the probability that there is at least one mutant in a particular subpopulation?

$$P\{X \geq 1|\lambda\} = 1 - P\{X = 0|\lambda\} = 1 - e^{-\lambda}. \qquad \square$$

5.7.2 Expectation and variance

Next, we show that the Poisson's expected value and variance can be obtained in a closed form.

Expected value of the Poisson density The expected value of the Poisson density with parameter λ is

$$E[X] = \lambda.$$

Here is why:

$$E[X] = \sum_{m=0}^{\infty} m \frac{\lambda^m}{m!} e^{-\lambda}$$

$$= \lambda \sum_{x=1}^{\infty} \frac{\lambda^{m-1}}{(m-1)!} e^{-\lambda}.$$

Now for $n = m - 1$ we have

$$E[X] = \lambda \sum_{n=0}^{\infty} \frac{\lambda^n}{n!} e^{-\lambda}$$

$$= \lambda P(X \leq \infty | \lambda)$$

$$= \lambda.$$

5.7.3 Variance of the Poisson density

The variance of the Poisson density with parameter λ is

$$V[X] = \lambda.$$

Here is why: Let

$$P(X = m | \lambda) = \frac{\lambda^m}{m!} e^{-\lambda}.$$

Then

$$V[X] = \sum_{m=0}^{\infty} (m - \lambda)^2 P(X = m | \lambda)$$

$$= \sum_{m=0}^{\infty} m^2 P(X = m | \lambda) - 2\lambda \sum_{m=0}^{\infty} m P(X = m | \lambda)$$

$$+ \lambda^2 \sum_{m=0}^{\infty} P(X = m | \lambda)$$

$$= \sum_{m=0}^{\infty} m \times m \frac{\lambda^m}{m!} e^{-\lambda} - 2\lambda^2 + \lambda^2$$

$$= \lambda \sum_{m=1}^{\infty} m \frac{\lambda^{m-1}}{(m-1)!} e^{-\lambda} - \lambda^2.$$

For $n = m - 1$ we write

$$V[X] = \lambda \sum_{n=0}^{\infty} (n+1) \frac{\lambda^n}{n!} e^{-\lambda} - \lambda^2$$

$$= \lambda \sum_{n=0}^{\infty} n \frac{\lambda^n}{n!} e^{-\lambda} + \lambda \sum_{n=0}^{\infty} \frac{\lambda^n}{n!} e^{-\lambda} - \lambda^2$$

$$= \lambda^2 + \lambda - \lambda^2$$

$$= \lambda.$$

158 Discrete densities and distributions

Example 5.17. We return to the suicide bombings by Palestinian terrorists (Example 5.2). This time, we pool all the data instead of discussing attacks by Hamas only. In all, 278 attacks were reported in 1 102 days. To obtain the Poisson parameter λ, we divide the period into 10-day intervals. There were 111 such intervals. Therefore,

$$\lambda = \frac{278}{111} \approx 2.505 \ .$$

There were 17 10-day intervals during which no attacks happened, 25 in which one attack happened and so on (Table 5.2). These frequencies are shown as dots in Figure 5.7. The expected values were obtained by multiplying the sum of the Frequency column by

$$P(X = m|2.505) = \frac{2.505^m}{m!} e^{-2.505}$$

where m corresponds to the Count column. These are represented as the theoretical density in Figure 5.7. We shall later see that the fit thus obtained is "good."

Table 5.2 Frequency and Poisson based expected frequency of the number of suicide attacks by Palestinian terrorists on Israelis.

Count	Frequency	Expected
0	17	9
1	25	23
2	22	28
3	21	24
4	10	15
5	5	7
6	8	3
7	0	1
8	0	0
9	1	0
10	1	0
11	0	0
12	0	0
13	0	0
14	1	0

Let us see how to produce Figure 5.7 from the raw data. It is worthwhile to go through the details because they involve data manipulations. First, we load the data and examine the first few rows:

```
> load('terror.rda')
> head(terror)
  Julian       Date Killed Injured Org.1 Org.2 Org.3
1  14880  9/27/2000      1       0  None
2  14882  9/29/2000      1       0  None
3  14884  10/1/2000      1       0  None
```

```
4  14885 10/2/2000      2       0  None
5  14891 10/8/2000      1       0  None
6  14895 10/12/2000     1       0  None
```

We need to divide the events into 10-day intervals and then count the number of attacks during these intervals. It is most convenient to work with the Julian date. So we decide on the number of breaks in the 10-day intervals:

```
> (n.breaks <- ceiling((max(terror$Julian) -
+   min(terror$Julian)) / 10))
[1] 111
```

Next, we cut the dates into the 10-day intervals:

```
> head(cuts <- cut(terror$Julian, n.breaks,
+   include.lowest = TRUE), 4)
[1] [14879,14889] [14879,14889] [14879,14889] [14879,14889]
111 Levels: [14879,14889] (14889,14899] (14899,14909] ...
```

As you can see, cut() returns a vector of factors that represent the intervals. This makes tabulation and counting easy:

```
> attacks <- table(cuts)
> (a <- table(attacks))
attacks
 0  1  2  3  4  5  6  9 10 14
17 25 22 21 10  5  8  1  1  1
```

We need to use table() twice: First to count the number of occurrences of the 10-day intervals and second to count the frequency of these occurrences. So there were 17 10-day intervals in which no attacks occurred, 25 intervals in which 1 attack occurred and so on. We now have a problem. No occurrences are data. For example, there were no 10-day intervals in which 7 attacks had occurred (similarly for 8, 11, 12

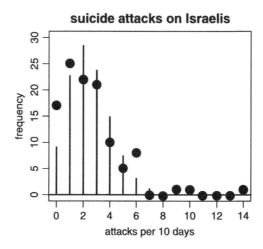

Figure 5.7   The empirical (dots) and the theoretical (fitted) Poisson (see Table 5.2).

160 Discrete densities and distributions

and 13 attacks per 10 days). We therefore need to fill in the blanks. Let us do it with array indices. Note that a is a table. In it, the number of attacks (per 10 days) are represented as the dimension names of the table. We turn them into integers with

```
> (idx <- as.numeric(names(a)) + 1)
[1]  1  2  3  4  5  6  7 10 11 15
```

(we add 1 because we are going to use idx as a vector of indices). Next, we create a sequence from zero to the maximum attack rate (for plotting later) and a vector of zero frequencies:

```
x <- 0 : (max(idx) - 1)
frequency <- rep(0, length(x))
```

Doing this

```
(frequency[idx] <- a)
```

results in this:

```
[1] 17 25 22 21 10  5  8  0  0  1  1  0  0  0  1
```

and we thus have zeros included as data. To obtain the theoretical (Poisson) density, we estimate λ from the mean attack rate:

```
> (lambda <- length(terror[, 1]) / n.breaks)
[1] 2.504505
```

So now we have

```
> z <- 0 : (length(frequency) - 1)
> expected <- round(sum(frequency) * dpois(z, lambda), 0)
> lets.see <- rbind(attacks = z, frequency = frequency,
+    expected = expected)
> d <- list()
> d[[1]] <- dimnames(lets.see)[[1]]
> d[[2]] <- rep('', 15)
> dimnames(lets.see) <- d
> lets.see
```

```
attacks    0  1  2  3  4  5 6 7 8 9 10 11 12 13 14
frequency 17 25 22 21 10  5 8 0 0 1  1  0  0  0  1
expected   9 23 28 24 15  7 3 1 0 0  0  0  0  0  0
```

for the number of attacks per 10 days (see Table 5.2). frequency is our empirical density. Finally, we compare the empirical density to the theoretical density with

```
> plot(x, frequency, pch = 19, cex = 1.5, ylim = c(0, 30),
+    ylab = 'frequency', xlab = 'attacks per 10 days',
+    main = 'suicide attacks on Israelis')
> lines(x, dpois(x, lambda) * n.breaks)
```

(points and line segments in Figure 5.7). Because the example indicates that the density of attacks is Poisson, we conclude that the attacks were equally likely at any time. Further, the attacks were independent of each other in their timing. □

5.8 Estimating parameters

We have seen that densities have parameters. For example, the probability of success, π, in the binomial and the intensity, λ, in the Poisson, are parameters. Often we have some reason to believe that data come from an underlying density. To examine this belief, we fit the density to the data. This means that we use some criterion to search for the best value of the parameters. The criterion is based on some function of the data and the parameters. Here is an example.

Example 5.18. We have n observations (counts per unit of time) that we believe are from a process that obeys the Poisson density. Each count, denoted by $x_i \in \mathbb{Z}_{0+}$, is independent of another count. If the parameter value is λ, then

$$P(X = x_i | \lambda) = \frac{\lambda^{x_i}}{x_i!} e^{-\lambda}.$$

Because the observations are independent, we write

$$\mathcal{L}(\lambda | X_1 = x_1, \ldots, X_n = x_n) := P(X_1 = x_1 | \lambda) \times \cdots \times P(X_n = x_n | \lambda) \quad (5.9)$$

$$= \prod_{i=1}^{n} P(X_i = x_i | \lambda) = \prod_{i=1}^{n} \frac{\lambda^{x_i}}{x_i!} e^{-\lambda}.$$

The notation $\prod_{i=1}^{n} x_i$ is defined thus:

$$\prod_{i=1}^{n} x_i := x_1 \times \cdots \times x_n.$$

$\mathcal{L}$ in (5.9) is called the *likelihood function*. It is a function that captures the probability of observing the data (x_i) given λ. Naturally, we are interested in the value of λ that maximizes the likelihood of the data. In (5.9), the only unknown is λ. That value of λ that maximizes $\mathcal{L}$ is called the maximum likelihood estimate of λ. We denote this maximum value by $\widehat{\lambda}$. So we established a function (the likelihood function) and the criterion (maximization) which allows us to choose the appropriate value for λ. □

Generally, we write (5.9) like this:

$$\mathcal{L}(\boldsymbol{\theta} | \boldsymbol{X} = \boldsymbol{x}) := \prod_{i=1}^{n} P(X = x_i | \boldsymbol{\theta})$$

where P is any density, $\boldsymbol{\theta} := [\theta_1, \ldots, \theta_m]$ holds the values of the m parameters of P, $\boldsymbol{X} = [X_1, \ldots, X_n]$ and $\boldsymbol{x} = [x_1, \ldots, x_n]$. Note that $\boldsymbol{x}$ represents data (known values). For an arbitrary function $f(x)$, each value of x is mapped to a single value in $\log f(x)$ and the opposite is also true. Also, $\log f(x)$ is a monotonic function of $f(x)$. This means that if, for some values of x, $f(x)$ increases, decreases or remains unchanged, so will $\log f(x)$. Therefore, if $\widehat{\boldsymbol{\theta}}$ maximizes $\mathcal{L}$, it is also maximizes $\log \mathcal{L}$. So instead of using $\mathcal{L}$ as the criterion to maximize, we use

$$L(\boldsymbol{\theta} | \boldsymbol{X} = \boldsymbol{x}) := \log \mathcal{L}(\boldsymbol{\theta} | \boldsymbol{X} = \boldsymbol{x}) \quad (5.10)$$

$$= \sum_{i=1}^{n} \log P(X_i = x_i | \boldsymbol{\theta}).$$

L is called the log maximum likelihood function. Thus we arrive at the following definition:

Maximum likelihood estimator (MLE) The value θ that maximizes $L\left(\theta | X = x\right)$ given in (5.10) is called the maximum likelihood estimator of θ. We denote this value by $\widehat{\theta}$.

Example 5.19. Continuing with Example 5.18, we wish to maximize

$$L\left(\lambda | X = x\right) = \sum_{i=1}^{n} \log\left(\frac{\lambda^{x_i}}{x_i!} e^{-\lambda}\right).$$

To simplify the notation, we write L instead of $L\left(\lambda | X = x\right)$. There are several ways to find the maximum of L. One is to take the derivative, equate it to zero and solve for λ; i.e.

$$L = \frac{\partial}{\partial \widehat{\lambda}} \left[\sum_{i=1}^{n} \log\left(\frac{\widehat{\lambda}^{x_i}}{x_i!} e^{-\widehat{\lambda}}\right)\right] = 0.$$

The notation here implies that we first take the derivative with respect to $\widehat{\lambda}$ and then solve for $\widehat{\lambda}$. Rewriting the last equation, we obtain

$$L = \frac{\partial}{\partial \widehat{\lambda}} \left[\sum_{i=1}^{n} \left(\log\left(\widehat{\lambda}^{x_i}\right) - \log\left(x_i!\right) + \log\left(e^{-\widehat{\lambda}}\right)\right)\right]$$

$$= \frac{\partial}{\partial \widehat{\lambda}} \left[\sum_{i=1}^{n} \left(x_i \log\left(\widehat{\lambda}\right) - \log\left(x_i!\right) - \widehat{\lambda}\right)\right] = 0.$$

Switching the sum and derivative and taking the derivatives we obtain

$$\sum_{i=1}^{n} \left(x_i \frac{1}{\widehat{\lambda}} - 1\right) = 0$$

which simplifies to

$$\frac{1}{\widehat{\lambda}} \sum_{i=1}^{n} x_i - n = 0$$

or

$$\widehat{\lambda} = \frac{1}{n} \sum_{i=1}^{n} x_i.$$

Thus, the MLE of λ in the Poisson is the mean of the sample, $\overline{X}$. □

Depending on the densities, obtaining MLE may not be analytically tractable. In such cases, we must rely on numerical optimization. We write the ML function and use appropriate R functions to find the values of parameters that maximize the ML function. One such function is `optim()`, a general-purpose optimization function. `fitdistr()` provides MLE for some densities and thereby you can avoid relying on `optim()` directly. There are numerous other functions that provide MLE. We will demonstrate some of these soon.

In Exercises 5.18 and 5.19 you are asked to show the following MLE:

$$\text{binomial:} \quad \widehat{\pi} = n_S/n \,, \quad \widehat{\sigma}^2 = n\widehat{\pi}(1-\widehat{\pi}) \,.$$
$$\text{Poisson:} \quad \widehat{\lambda} = \overline{X} \,, \quad \widehat{\sigma}^2 = \widehat{\lambda} \,.$$

Here $\widehat{\pi}$ is the MLE of π, $\widehat{\sigma}^2$ is the MLE of σ^2 and n_S is the number of successes in n trials. The estimates of the density parameters are based on specific data.

5.9 Some useful discrete densities

Recall that $P(X = x|\boldsymbol{\theta})$ denotes a density and $P(X \leq x|\boldsymbol{\theta})$ denotes a distribution. $\boldsymbol{\theta}$ is the set of the density's parameters. So for a named density (distribution), if we supply x (along with the necessary parameters) to an appropriate function in R, we obtain the corresponding values for the named density (distribution). Suppose that the named density is binomial. Then dbinom(x,...) provides the value of the binomial density for x. pbinom(x,...) responds with the value of the binomial distribution for x.

Often, we wish to know the value of x such that $P(X \leq x|\boldsymbol{\theta}) = p$, where p is a probability for a named density. In this case, x is called a *quantile*. For the binomial, we obtain it with qbinom(p,...). Finally, to generate a random number from a named density, we use, for example, rbinom(...). The same rule of prefixing with d, p, q or r holds for all named densities (distributions) available in R.

Next, we discuss briefly some useful discrete densities and distributions that are available in R. The list is not comprehensive.

5.9.1 Multinomial

The binomial models situations where there are n Bernoulli trials with probability of success π. The multinomial is a generalization of the binomial in the following sense: Instead of having only two outcomes (success or failure) in a single trial, we might have m outcomes (categories) in a single trial. Each possible outcome has a probability associated with it (π_i, $i = 1, \ldots, m$). We wish to know the joint probability of $X_1 = x_1, \ldots, X_m = x_m$ successes in n trials, where $n = x_1 + \cdots + x_m$.

Density

The notation in the case of multivariate densities can be complicated. However, it is worthwhile to go through it at least once so that the basic ideas are clear.

Example 5.20. A pride of lions chase a prey. The chase can end with one of the following $m = 3$ outcomes: $1 =$ the prey is killed, $2 =$ the prey escapes without injury and $3 =$ the prey escapes with injury. Suppose that the corresponding probabilities are $\pi_1 = 0.2$, $\pi_2 = 0.7$ and $\pi_3 = 0.1$. That is, $\boldsymbol{\pi} := [\pi_1, \pi_2, \pi_3] = [0.1, 0.7, 0.1]$. The outcome of each chase is independent of the outcome of another chase. Let $n = 10$ be the number of chases by the pride. Here n is the number of trials in a single experiment. We are interested in the probability that on the lth experiment with 10 trials, $x_{l1} = 2$ ended in the prey killed, $x_{l2} = 5$ ended with the prey escaping and $x_{l3} = 3$ ended with the injured prey escaping ($l \in \mathbb{Z}_+$). We write the sample space this way

$$S = \{E_1(i,j,k), \ldots, E_l(i,j,k), \ldots\} \,, \quad \{i+j+k=n\} \cap \{\mathbb{Z}_{0+} \times \mathbb{Z}_{0+} \times \mathbb{Z}_{0+}\} \,.$$

The notation indicates that although $E_l(i,j,k)$ is independent of any other event, i, j and k are dependent (they must sum to 10). We now create the following rv $\boldsymbol{Y}(E_l(i,j,k)) := (i,j,k)$. With this understanding, we simplify the notation and write the outcome of the lth experiment thus: $\boldsymbol{X}_l := \boldsymbol{Y}(E_l(i,j,k))$. We are interested in the outcome of a single experiment, so we may drop the subscript on $\boldsymbol{X}$. Let $\boldsymbol{x} := (i,j,k)$. Then from our definitions, the multivariate *density* is

$$P(\boldsymbol{X} = \boldsymbol{x}) = P(\boldsymbol{Y}(E_l(i,j,k)) = (i,j,k)) \ .$$

□

$\boldsymbol{X}$ is said to have a *multinomial density* if its joint probability function is

$$P(\boldsymbol{X} = \boldsymbol{x}|\boldsymbol{\pi},n,m) = \begin{cases} \dfrac{n!}{\prod\limits_{i=1}^{m} x_i} \prod\limits_{i=1}^{m} \pi_i^{x_i} & \text{for } x_i \in \mathbb{Z}_{0+} \\ 0 & \text{otherwise} \end{cases}.$$

The distribution of the multinomial requires multidimensional integrals, so we shall not write it formally. Interestingly, when $\boldsymbol{X}$ is multinomial with parameters n and $\boldsymbol{\pi}$, then the rv

$$Y = \sum_{i=1}^{m} \frac{(X_i - n\pi_i)^2}{n\pi_i}$$

has approximately a χ^2 (read "chi-square") density with $m-1$ degrees of freedom. Here $n\pi_i$ is the expected number of successes of the ith outcome (with m possible outcomes in each trial) and n is the total number of successes. Y then measures the relative deviation of the obtained results from the experiment X_i, compared to the expected result.

Estimating parameters

In the case of the binomial, we estimated π from the number of successes divided by the number of trials ($\pi \approx \widehat{\pi} := n_S/n$). Similarly, we estimate each $\pi_i \approx \widehat{\pi}_i := x_i/n$, where x_i are the number of successes of the ith possible outcomes.

Applications

In genetics, the multinomial arises in the Hardy-Weinberg law. The law states that for a large population at equilibrium where mating is random, the frequency of genotypes with respect to alleles A and a in a diploid population is $(1-\pi)$, $2\pi(1-\pi)$ and π^2 for genotypes AA, Aa, and aa, respectively. Other examples are the probabilities of using different habitat types by animals, colors of organisms that belong to a species and so on.

Example 5.21. The ecological and animal behavior literature are replete with the following scenario: You record the location of an animal n times. The animal's habitat is divided into m types. Geographical analysis indicates that the proportion of each habitat type in the area is $\pi_1, \ldots, \pi_m$. Based on the n locations, you estimate π_i by the x_i/n where x_i is the number of points where the animal was located in habitat i. Do the location-derived proportions differ from the expected proportions $n\pi_i$ where $i = 1, \ldots, m$?

To examine numerical results, let us create some data. We imagine 1000 locations, in six habitats:

```
> n <- 1000 ; hab <- LETTERS[1 : 6] ; m <- 6
```

Next, we imagine that the proportion of the habitats are:

```
> PI <- c(0.1, 0.2, 0.3, 0.2, 0.1, 0.1)
```

Next, we create 1 000 random multinomial deviates

```
> set.seed(100) ; nqd((d <- rmultinom(1, n, PI)))

    105
    218
    296
    186
     87
    108
```

and obtain a data frame

```
> (df <- data.frame(habitat = hab, observed = d, PI,
+       expected = round(n * PI, 3)))
  habitat observed  PI expected
1       A      105 0.1      100
2       B      218 0.2      200
3       C      296 0.3      300
4       D      186 0.2      200
5       E       87 0.1      100
6       F      108 0.1      100
```

Finally we compare the observed to the expected "data"

```
> plot(df$habitat, df$expected, ylim = c(0, 300),
+      xlab = 'habitat', ylab = 'multinomial frequencies')
> points(df$habitat, df$observed, pch = 20, cex = 3)
```

(Figure 5.8). We set the plot character to 20 with pch in points() and triple the point size with cex. As expected, the fit between the observed and expected data is is good. We will learn later how to test the goodness of fit. □

5.9.2 Negative binomial

The negative binomial is a generalization of the geometric density. It can also be regarded as a generalization of the Poisson density where the variance exceeds the mean. The conditions that give rise to this density are discussed in detail by Bliss and Fisher (1953). Let $n \in \mathbb{Z}_{0+}$. The negative binomial addresses the following question: What is the density of the number of Bernoulli trials required to achieve the nth success? Equivalently, the question is: What is the number of failures until we achieve the nth success?

166 Discrete densities and distributions

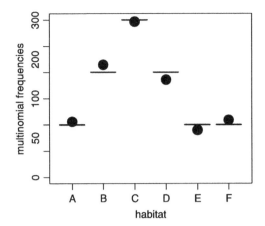

Figure 5.8   Multinomial expected frequencies (horizontal line sections) and observed frequencies (circles).

Density

Let X be the number of failures before the nth success. Then X is said to have the *negative binomial* density with parameters n and π if

$$P\{X = x|\pi, n\} = \begin{cases} \binom{x+n-1}{n-1} \pi^n (1-\pi)^x & \text{for } x \in \mathbb{Z}_{0+} \\ 0 & \text{otherwise} \end{cases}. \quad (5.11)$$

Here is why: The number of failures until the first success is $(1-\pi)^{x_1-1}\pi$, until the second success $(1-\pi)^{x_2-1}\pi$, ..., until the nth success is $(1-\pi)^{x_n-1}\pi$. Because sequences are independent, we write

$$(1-\pi)^{x_1-1}\pi \cdots (1-\pi)^{x_n-1}\pi = (1-\pi)^x \pi^n$$

where $x = x_1 + \cdots + x_n$. Now the last trial is a success. The remaining $n-1$ successes can be assigned to the remaining $x-1$ trials in $\binom{x+n-1}{n-1}$ ways. Hence we have (5.11).

The general definition of the negative binomial does not require that n be a nonnegative integer. However, because we are interested in the physical interpretation of the density, we will focus our attention on positive integers.

Parameter estimation

The expectation and variance of a negative binomial rv X are

$$E[X] = \frac{n(1-\pi)}{\pi}, \quad V[x] = \frac{n(1-\pi)}{\pi^2}.$$

To estimate the sample-based parameters we write the negative binomial in a more general way:

$$P(X = x) = \left(1 + \frac{m}{k}\right)^{-k} \frac{\Gamma(k+x)}{x!\,\Gamma(k)} \left(\frac{m}{m+k}\right)^x. \quad (5.12)$$

The function $\Gamma(\alpha)$ is called the gamma function. It is defined by

$$\Gamma(\alpha) = \int_0^\infty e^{-x} x^{\alpha-1} dx.$$

For an integer α, say $\alpha = n$,

$$\Gamma(n) = (n-1)!$$

Now use the following to estimate m and k

$$\widehat{m} = \overline{X} \qquad \widehat{k} = \frac{\overline{X}^2}{S^2 - \overline{X}} \tag{5.13}$$

where $\overline{X}$ and S^2 are the sample mean and variance (see Anscombe, 1948). The estimation of the parameters is difficult and should be avoided unless necessary. Estimation methods often fail or are unstable.

Applications

The variance of the binomial is smaller than its mean. For the Poisson the variance and the mean are equal. The variance of the negative binomial is larger than its mean: A common feature in data. The phenomenon of count data (usually modeled with the Poisson) with variance larger than the mean is called *overdispersion*. Ecologically oriented applications of the negative binomial are discussed by Krebs (1989). The negative binomial is also applicable in the following situations:

- Denote by p the probability of success in a sequence of independent Bernoulli trials. From (5.12) we conclude that $p = m/(m+k)$. If k is an integer, then the negative binomial is the density of the number of successes up to the kth failure.
- In cases where the parameter λ varies over time, instead of using the Poisson, the negative binomial may be a good candidate.
- The negative binomial may be a plausible model of the density of insect counts, when they hatch in clumps, the density of the count of plants when their distribution is clumped, the distribution of ant-hills in space (where X is the distance between hills) and so on.
- The negative binomial is applicable as a population-size model (birth/death process) when the birth and death rates per individual are constant, with a constant rate of immigration.

Example 5.22. Bliss and Fisher (1953) published one of the earliest applications of the negative binomial density. They studied the distribution of the number of ticks on a sheep. Table 5.3 shows agreement between the observed and expected frequencies. To compute the expected frequency from the observed distribution, compute the sample mean and variance and then m and k using (5.13). Then substitute these values in (5.12) and compute $N \times P(X = x)$ for $x = 0, 1, \ldots, 10$ where the total number of observations is $N = 60$ (see Exercise 5.21). □

Here is an example in which we use the negative binomial in R.

Table 5.3 Ticks sheep.

Ticks	Observed	Expected
0	7	6
1	9	10
2	8	11
3	13	10
4	8	8
5	5	5
6	4	4
7	3	2
8	0	1
9	1	1
10+	2	2

Example 5.23. Consider a Bernoulli trial with probability of success $\pi = 0.2$. We wish to conduct 100 experiments. In each experiment, we stop as soon as we get 20 successes. The density of X (the number of trials until we achieve 20 successes) is negative binomial. Let us see what the density and distribution look like. First, we set the parameters, generate random values and calculate their mean (we need the mean later):

```
n <- 100 ; k <- s <- 20 ; PI <- 0.2 ; set.seed(200)
X <- rnbinom(n, size = s, prob = PI) ; m <- mean(X)
```

Next, we plot the histogram and superimpose on it the true density and the sample based density:

```
> par(mfrow = c(1, 2))
> h(X, xlab = 'count')
> x <- 0 : 400
> lines(x, dnbinom(x, size = s, prob = PI), lwd = 3)
> lines(x, dnbinom(x, size = s, mu = m), lwd = 3,
>    lty = 2, col = 'red')
```

(the code for h() is on p. 40). The sample-based density (the broken line in Figure 5.9) is generated from the target number of successes (20) and from the mean of the sample, m. To compare the sample-based distribution to the true distribution, we plot the empirical cumulative distribution function with ecdf() and add the lines for the true distribution with pnbinom():

```
plot(ecdf(X), main = '', ylab = expression(italic(P(X<=x))))
lines(x, pnbinom(x, size = s, prob = PI), type = 's')
```

Note the use of type = s. This results in a step plot. □

In Section 17.1 we apply the negative binomial to the density of U.S. casualties in Iraq.

5.9.3 Hypergeometric

The hypergeometric is the multivariable extension of the negative binomial. The density can be describes as follows. Suppose that a population of size n consists of n_1

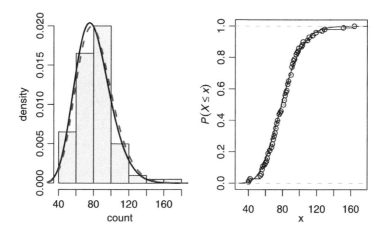

Figure 5.9   The negative binomial density (left) and distribution (right).

types A and $n_2 = n - n_1$ types a. Let the rv X denote the number of types A in a sample of size k, taken without replacement. The hypergeometric describes the density of X.

Density

The rv X is said to have hypergeometric density if given n_1, n_2 and k, its density is

$$P(X = x | n_1, n_2, k) = \begin{cases} \dfrac{\binom{n_1}{x}\binom{n_2}{k-x}}{\binom{n_1+n_2}{k}} & \text{for } x \in \mathbb{Z}_{0+} \\ 0 & \text{otherwise} \end{cases}.$$

To see that the density represents the process just described let us think about $X \in \mathbb{Z}_{0+}$ for a moment. We realize that the number of ways to select a sample of size k from a population of size n is $\binom{n}{k}$. The number of ways to select x from n_1 is $\binom{n_1}{x}$. For each of those, we can select the remaining $k - x$ from n_2 in $\binom{n_2}{k-x}$ number of ways. Thus, the number of samples having x type a is $\binom{n_1}{x}\binom{n_2}{k-x}$. To get the probabilities, we divide the last expression by $\binom{n}{k}$. These probabilities may be computed for $\max(0, k - n_2) \leq x \leq \min(k, n_1)$.

Parameter estimation

To derive the expectation and variance, let $Y \in \mathbb{Z}_{0+}$. The rv Y denotes the number of types A in a sample of size k, taken without replacement from a population with two types. Define

$$X_i = \begin{cases} 1 & \text{if type } A \text{ occurred} \\ 0 & \text{otherwise} \end{cases}.$$

Then $Y = \sum_{i=1}^{k} Y_i$. We have $E[Y_i] = n_1/n$ and $E[Y] = kn_1/n = kp$. This is also the $E[X]$ where X is a binomial rv. In other words, sampling with or without replacement have identical expectations. Now $Y_i^2 = Y_i$. Therefore, $E\left[X_i^2\right] = n_1/n$ and

$$V[Y_i] = \frac{n_1}{n} - \left(\frac{n_1}{n}\right)^2 = \frac{n_1}{n}\left(1 - \frac{n_1}{n}\right) = p(1-p).$$

The joint distribution of (Y_i, Y_j), $i \neq j$ are identical because all such pairs are equally likely to occur, regardless of the values of i and j. Therefore,

$$Cov[Y_i, Y_k] = Cov[Y_1, Y_2] = E[Y_1 Y_2] - E[Y_1]E[Y_2], \quad i \neq k.$$

Also,

$$\begin{aligned} E[Y_1 Y_2] &= 1 \cdot P\{Y_1 Y_2\} \\ &= 1 \cdot P\{Y_1 = 1 \text{ and } Y_2 = 1\} \\ &= \frac{n_1}{n}\frac{n_1 - 1}{n - 1}. \end{aligned}$$

Therefore,

$$\begin{aligned} Cov[Y_j, Y_k] &= \frac{n_1}{n}\frac{n_1 - 1}{n - 1} - \left(\frac{n_1}{n}\right)^2 \\ &= -\frac{n_1}{n}\frac{n - n_1}{n}\frac{1}{n - 1}. \end{aligned}$$

Therefore, we have

$$\begin{aligned} V[Y] &= \sum_{i=1}^{k} V[Y_i] + 2\sum_{j<l} Cov[y_j, y_l] \\ &= kp(1-p) - 2\binom{k}{2}p(1-p)\frac{1}{n-1} \\ &= kp(1-p)\frac{n-k}{n-1}. \end{aligned}$$

For sampling with replacement we have $kp(1-p)$.

Applications

The distribution is applicable in genetics to situations where we model genotypes of haploids with 2 alleles and random mating. In quality control, it is used to determine the number of items that should be tested for quality in a particular batch.

Example 5.24. A population of n objects consists of n_1 defective objects and n_2 non-defective objects. Let X be the number of defective objects in a sample of size k. If we are at least 90% certain that the population has at least 1 defective object, we will discard the population. What should be the sample size k such that we are 90% certain that at least 1 defective object is in the sample?

Let X denote the number of defective objects in the sample. We wish to choose the smallest sample size k, such that

$$P\{X \geq 1\} \geq 0.9. \qquad \square$$

Example 5.25. A state lottery requires 6 matches from a total of 53 numbers. Let Y be the number of matches. Then the probability of drawing y matches is

$$P\{Y = y\} = \frac{\binom{6}{y}\binom{47}{6-y}}{\binom{53}{6}}.$$

Therefore, the probability of winning is

```
> n1 <- 6 ; n2 <- 47 ; k <- 6 ; Y <- 6
> dhyper(Y, n1, n2, k)
[1] 4.355879e-08
```

The same result can be obtained with

```
> choose(n1, x) * choose(n2, k - x) / choose(n1 + n2, k)
[1] 4.355879e-08
```

The function choose(a, b) returns the binomial coefficient for $\binom{a}{b}$. □

5.10 Assignments

Exercise 5.1. Four patients, A, B, C and D are being readied for today's surgeries. The surgeon estimates that A will need 1000 cc of blood transfusion during the operation, B will need 2000, C will need 3000 and D will need 4000 cc. The anesthetist was late to work and only two patients will be operated upon today. To avoid the appearance of favoritism, the surgeon is going to choose the first patient randomly and operate on him. He then is going to choose the second patient randomly and operate on him. List the possible values for each of the following random variables.

1. Let X denote the total amount of blood needed for the day. List the possible values of X.
2. List the probabilities for these values.
3. After computing (1), the surgeon realizes that the blood supply is low, and he cannot afford to spend more than 5000 cc for the day. Yet, he wishes to operate on two patients. What is the probability that patient A will be chosen for surgery? Patient B? C? D?

Exercise 5.2. An experiment consists of the event of rolling a die with an even number of dots showing face up *and* flipping a coin with either H or T showing up.

1. List all possible outcomes.
2. What are the probabilities associated with each outcome?
3. Assign a value of 2 to H and 4 to T and $6 - n$ to the number of points with face up. Here n denotes the values 2, 4 or 6. Let X be the rv with values x_i = value from the outcome of flipping the coin $+6 - n$. What values does the rv X take?
4. What are the probabilities, $P(X = x_i)$, that X takes?
5. Is P in (4) a distribution function? Why?
6. Plot P in (4).

172 Discrete densities and distributions

Exercise 5.3. The director of a small local clinic needs to decide how many flu shots he should order before the flu season begins. Let X be the number of people who show up or call for the shot. The manager examines past years' data and obtains the following probability distribution for X:

	p
less than 105	0.065
105	0.033
106	0.055
107	0.081
108	0.106
109	0.126
110	0.133
111	0.126
112	0.106
113	0.081
114	0.055
115	0.033

He decides to order 110 shots. The clinic is going to be open for shots on Wednesday, from 9:00 to 10:00 am. Some people call before to reserve a shot, others just walk in.

1. What is the probability that everybody who shows up or called before gets the shot?
2. Suppose that 110 people called to reserve a shot. You walk in at 9:00 am sharp and are told that all the shots are reserved for people who called. However, some of those who call do not show up. You could wait an hour and if people do not show up, you will be the first on line to get the shot. What is the probability that you will get the shot if you wait?
3. Another person walked in at 9:10. He is the fourth one to walk in. What is the probability that he will get the shot?

Exercise 5.4. Northwestern Minnesota is prone to widespread flooding in the Spring. Suppose that 20% of all farmers are insured against flood damage. Four farmers are selected at random. Let M denote the number among the four who have flood insurance.

1. What is the probability distribution of M?
2. What is the most likely value of M?
3. What is the probability that at least two of the four selected farmers have flood insurance?

Exercise 5.5. One in 1 000 pedestrians in a busy intersection gets hit by a car. Accidents are independent.

1. Plot the density of the number of pedestrians crossing the intersection until the first accident occurs.
2. Plot the distribution of the above.
3. Let X be the number of pedestrians that crossed the intersection. How many pedestrians crossed the intersection until the first accident if $P(X \leq m) = 0.05$, 0.10, 0.90, 0.95?

Exercise 5.6. A sample of 20 students is drawn from a population where 60% of the student body is female.

1. Plot the density of the number of females in the sample.
2. Plot the distribution of the number of females in the sample.
3. Plot the density of the number of males in the sample.
4. Plot the distribution of the number of males in the sample.
5. Let X be the number of females. How many females will be in the sample if $P(X \leq m) = 0.05, 0.10, 0.90, 0.95$?
6. Let X be the number of males. How many males will be in the sample if $P(X \leq m) = 0.05, 0.10, 0.90, 0.95$?

Exercise 5.7. The U.S. National Science Foundation conducts surveys about college graduates. In 2001, 79.5% of the graduates who qualified for the survey (based on the survey's definition of college graduates) responded (see http://srsstats.sbe.nsf.gov/htdocs/applet/docs/techinfo.html. We wish to know why some did not answer. We have the list of graduates, but we do not know who did not respond. So we contact random people from the list until we find one that did not answer.

1. Define the event of interest.
2. What are the outcomes and their associated probabilities?
3. Define an appropriate rv that reflects how many college graduates we contact until we find one who did not respond to the survey.
4. Construct the distribution function of this rv.
5. Plot it for $n = 1, 2, \ldots, 10$ where n is the number of graduates we ask until we find one that did not respond.

Exercise 5.8. Let X be the ticket price for a political fund-raising dinner. Suppose that the probability distribution of X is:

x	100.00	120.0	140.00	160.00	180.00	200.00
p	0.22	0.2	0.18	0.16	0.13	0.11

1. What is the probability that a randomly selected attendee paid more than $140 for the ticket? Less than $160?
2. Compute the expected value and the standard deviation of X.

Exercise 5.9. The probability that a female wolf gives birth to a male is 0.5. Give the probability distribution of the rv variable X = the number of female puppies in a litter of size 5.

Exercise 5.10. The probability distribution of the size of a wolf litter is

Litter size (x)	1	2	3	4	5	6	7	8
$P(X = x)$	005	0.10	0.12	0.30	0.30	0.11	0.01	0.01

1. What is the expected litter size?
2. What is the probability that X is within 2 of its expected value?
3. What is the variance of the litter size?
4. What is the standard deviation of the litter size?
5. What is the probability that the number of pups is within 1 standard deviation of the expected value of litter size?

6. What is the probability that the number of pups in the litter is more than two standard deviations from its expected value?

Exercise 5.11. Show that for the binomial distributions with parameters n and π,

1. $E[X] = n \times p$
2. $V[X] = np(1-p)$

Exercise 5.12. You take a multiple-choice exam consisting of 50 questions. Each question has 5 possible responses of which only one is correct. A correct answer is worth 1 point. You have not studied for the exam and therefore decide to guess the correct answers. Let $X =$ the number of correct responses on the test.

1. What is of probability distribution of X?
2. What score do you expect?
3. What is the variance of X?

Exercise 5.13. You are given the following distribution of the rv X:

x	26.00	38.00	34.00	38.00	28.00	27.00	37.00	21.000
p	0.10	0.15	0.14	0.15	0.11	0.11	0.15	0.095

Is this possible?

Exercise 5.14. Given the following scenarios, identify the most appropriate probability distribution and draw the shape of the density and the shape of the distribution in each case.

1. Choose a deer and identify its sex.
2. The number of quarters you insert into a slot machine until you win.
3. The number of individual plants in a plot.
4. The number of females in a sample of 10 students.
5. The time until a light bulb is out.

Exercise 5.15. For the geometric distribution, prove that $E[X] = 1/\pi$.

Exercise 5.16. Show that for the binomial density with parameters π and n:

1. $E[X] = n\pi$.
2. $V[X] = n\pi(1-\pi)$.

Exercise 5.17. For the geometric distribution, prove that $V[X] = (1-\pi)/\pi^2$.

Exercise 5.18. For the binomial density, our data resulted in n_S successes in n trials. Use the MLE technique to prove the following:

1. $\widehat{\pi} = p := n_S\,/\,n$.
2. $\widehat{\sigma}^2 = np(1-p)$ where σ^2 is the variance of the density and $\widehat{\sigma}^2$ is the MLE of σ^2.

Exercise 5.19. For the Poisson density, our data resulted in x_i counts, $i = 1, \ldots n$. Use the MLE technique to prove the following:

1. $\widehat{\lambda} = \overline{X}$ where $\overline{X}$ is the mean of the sample.
2. $\widehat{\sigma}^2 = \overline{X}$.

Exercise 5.20. A working hypothesis is that plants are randomly distributed and independent of each other in an area. You establish a large number of 1×1 m^2 plots in the area and count the number of plants in each plot. The mean number of plants per plot is 2.5.

1. Plot the hypothesized density of the number of plants in a plot.
2. Plot the distribution of the above.
3. Let X be the number of plants in a plot. How many plants might be in a plot if $P(X \leq m) = 0.05, 0.10, 0.90, 0.95$?

Exercise 5.21. Use R to compute the expected values column in Table 5.3.

6

Continuous distributions and densities

In Chapter 5, we discussed densities first and then distributions. Now that we know about distributions, it is convenient to start with distributions and then move on to densities.

6.1 Distributions

Let $X \in \mathbb{R}$ and $x \in \mathbb{R}$. Then we define

Continuous probability distribution The function $P(X \leq x | \boldsymbol{\theta})$ (where $\boldsymbol{\theta}$ is a vector of parameters), is called the probability distribution function (distribution for short) of the rv X.

Example 6.1. Denote the two-hour time interval between $a = 13{:}00$ hours and $b = 15{:}00$ hours by $[a, b]$. Intelligence reports that a suicide bomber is going to explode herself in a busy intersection in Baghdad anytime within $[a, b]$. Let E_t be the event that the bombing occurred at t. Then

$$B := \{E_t : t \in [a, b]\} , \quad \overline{B} = \{E_t : t \notin [a, b]\} , \quad S = B \cup \overline{B} , \quad t \in \mathbb{R} .$$

The notation reads "B is the set of all events E_t such that the explosion occurred at t between 13:00 and 15:00." Recall that $\overline{B}$ is the complement of B and S is the event space. To obtain the distribution of these events, we must construct the rv $T(E_t)$ and assign a probability to E_t for all possible t. Because of the way E_t is defined, we simply map $T(E_t)$ to t. Instead of constructing probabilities for E_t, we construct probabilities for the compound event

$$B_t = \{E_\tau : \tau \leq t, \, t \in \mathbb{R}\} .$$

With our definition of T, we simply have $T(B_t) \leq t$. Once we assign probabilities to all B_t, our distribution is defined.

178 Continuous distributions and densities

Obviously
$$P(B_t) = 0, \quad t \leq a$$
(the explosion cannot occur before 13:00) and the probability that it will occur precisely at $t = a$ is zero (a is one among an infinite number of points). Now consider the middle point of the interval $[a, b]$ and let $t \leq a + (b-a)/2$. Imagine the line segment $(a, b]$ with the point $a + (b-a)/2$ in the middle. Drop a fine-tipped needle from such a height over the line that its tip will be equally likely to hit anywhere to the left or right of the middle point. If the tip hits outside $(a, b]$, ignore the outcome (i.e. this will be our $\varnothing$ event) because no event is defined for this case. Drop the needle many times and count the number of times its tip hits to the left or to the right of the middle of the interval. In this idealized experiment, the tip will hit half of the times in the interval $(a, a + (b-a)/2]$. Therefore, for $t = a + (b-a)/2$ we have $P(B_t) = 1/2$.

What we have just illustrated is that the relative size of the section from a to t (for t inside $[a, b]$) is in fact the probability of
$$P(T(B_t|a, b) = t) = P(T(E_t|a, b) \leq t)$$
or simply $P(T \leq t|a, b)$. Therefore,
$$P(T \leq t|a, b) = \frac{t-a}{b-a}, \quad t \in (a, b].$$
Finally, because any probability for $t > b$ is zero,
$$P(T \leq t|a, b) = 1, \quad t > b.$$
The probability the explosion may occur any time before $t = 14{:}00$ hours is
$$P(T \leq 14{:}00|a, b) = \frac{14{:}00 - 13{:}00}{15{:}00 - 13{:}00} = \frac{1}{2}.$$
(see top left inset in Figure 6.1). Here is the script for this example:

```
a <- 13 ; b <- 15 ; x <- seq( -1, b + 1, length = 2001)
par(mfrow = c(2, 2))
xlabel <- c('', '', rep(expression(italic(t)), 2))
ylabel <- c('', expression(italic(P(T < t))), '',
  expression(italic(P(T==t))))
y <- cbind(punif(x), punif(x, a, b),
  dunif(x), dunif(x, a, b))
x.limits <- rbind(c(-0.1, 1.1), c(a - 0.1, b + 0.1),
  c(-0.1, 1.1), c(a - 0.1, b + 0.1))
for(i in c(2, 1, 4, 3)){
  plot(x, y[, i], type = 'l', xlab = xlabel[i],
    xlim = x.limits[i, ], ylab = ylabel[i],
    ylim = c(0, 1.1), lwd = 2)
}
```

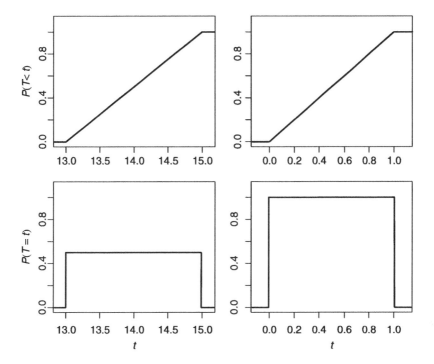

Figure 6.1 Uniform distributions and their corresponding densities. Left column: the suicide bombing example. Right column: the standard uniform.

Let us examine the script. In line 1 we set the parameters for the uniform and with seq(), we let x hold the x-values to be plotted. We have four plots to produce, so in line 2 we tell the graphics device to accept a matrix of 2×2 plots (par() and mfrow). In lines 3 to 5 we set the labels (including expression() and italic()) as vectors of length four (the number of panels we are going to draw). In lines 6 and 7 we set the y-data to be plotted. Note the use of punif() and dunif() to produce the distributions and densities respectively (lines 6 and 7). In lines 8 and 9 we set the limits for the x-axis. The plots are produced within the loop in lines 10 to 14. □

The family of distributions illustrated in Figure 6.1 defines the

Uniform distribution The rv X has a uniform distribution on the interval $[a,b]$ if

$$P(X \leq x | a, b) = \begin{cases} 0 & \text{for } x < a \\ \dfrac{x-a}{b-a} & \text{for } a \leq x \leq b \\ 1 & \text{for } b < x \end{cases}. \tag{6.1}$$

The uniform with $a = 0$ and $b = 1$ is called the *standard uniform* distribution. Another distribution that is closely related to the uniform is the exponential.

Example 6.2. Let events occur equally likely at any time interval with mean $1/\lambda$ events per hour. Every time an event occurs, reset a time interval to 0 and observe the time until the next event occurs. Denote by T the time interval between two

180 Continuous distributions and densities

consecutive events. In other words, $T = t$ means that the next event did not occur until t. Obviously, as t increases, $P(T \leq t|a,b)$ decreases. The probability that the next event does not occur at $t = 0$ is 1 (two events cannot occur at exactly the same time). It can be shown that $P(T \leq t|\lambda)$ decreases exponentially with a rate λ as time increases:

$$P(T \leq t|\lambda) = \begin{cases} 1 - e^{-\lambda t} & \text{for } t \geq 0 \\ 0 & \text{for } t < 0 \end{cases}. \tag{6.2}$$

□

From Example 6.2 we have the following definition:

Exponential distribution (6.2) defines the exponential distribution.

6.2 Densities

Once we define continuous distributions, the definition of continuous densities follows immediately:

Continuous probability density The continuous density probability function (density, for short) of a rv with distribution $P(X \leq x|\boldsymbol{\theta})$ is

$$P(X = x|\boldsymbol{\theta}) := \frac{\mathrm{d}P(X \leq x|\boldsymbol{\theta})}{\mathrm{d}x}. \tag{6.3}$$

Example 6.3. In Example 6.1 we constructed the uniform distribution. The corresponding densities for two distributions are shown in the bottom two insets of Figure 6.1. To see this, we take the derivative of $P(X \leq x|a,b)$ in (6.1):

$$P(X = x|a,b) = \frac{\mathrm{d}P(X \leq x|a,b)}{\mathrm{d}x} = \frac{1}{b-a}.$$

Therefore, the uniform density is

$$P(X = x|a,b) = \begin{cases} 0 & \text{for } x < a \\ \frac{1}{b-a} & \text{for } a \leq x \leq b \\ 0 & \text{for } x > b \end{cases}. \tag{6.4}$$

□

Because $P(X = x|\boldsymbol{\theta})$ is continuous, we cannot interpret it as the probability at $X = x$. Why? Because there are uncountably infinite values of x between a and b and the probability of choosing a specific value is therefore zero. The way around this is to interpret $P(X = x|\boldsymbol{\theta})$ as a limit (see the alternative definition of $P(X = x|\boldsymbol{\theta})$ below).

From Example 6.3 we have the following definition:

Uniform density The uniform density is given by (6.4).

Here is another example of a continuous density:

Example 6.4. Continuing with Example 6.2,

$$P(T = t|\lambda) = \frac{\mathrm{d}\left(1 - e^{-\lambda t}\right)}{\mathrm{d}t} = \lambda e^{-\lambda t}, \quad t \geq 0.$$

The interval between events cannot be negative and we have

$$P(X = x|\lambda) = \begin{cases} \lambda e^{-\lambda t} & \text{for } t \geq 0 \\ 0 & \text{for } t < 0 \end{cases}. \tag{6.5}$$

□

And so we define

Exponential density (6.5) defines the exponential density.

6.3 Properties

From the definitions of continuous distributions and densities and the discussion above, we deduce their properties. Because of our insistence on treating discrete densities and distributions as functions of $X \in \mathbb{R}$, all of the properties of continuous densities and distributions apply to discrete densities and distributions (Section 5.3).

6.3.1 Distributions

$P(X \leq -\infty|\boldsymbol{\theta}) = 0$
  This property is an immediate consequence of the definition of a rv, where we require that $X > -\infty$.

$P(X \leq \infty|\boldsymbol{\theta}) = 1$
  This property is an immediate consequence of the definition of a rv, where we require that $X < \infty$.

$P(X \leq x_1|\boldsymbol{\theta}) \leq P(X \leq x_2|\boldsymbol{\theta})$ for $x_1 \leq x_2$
  Let A be the set of events that combine to give $P(X(A) \leq x_1|\boldsymbol{\theta})$ and B the set of events that combine to give $P(X(B) \leq x_2|\boldsymbol{\theta})$. Then, $A \subset B$. A direct consequence of the definition of probability is then that $P(A|\boldsymbol{\theta}) \leq P(B|\boldsymbol{\theta})$ and therefore $P(X \leq x_1|\boldsymbol{\theta}) \leq P(X \leq x_2|\boldsymbol{\theta})$. Functions that have this property are called *nondecreasing*.

$P(x_1 < X \leq x_2|\boldsymbol{\theta}) = P(X \leq x_2|\boldsymbol{\theta}) - P(X \leq x_1|\boldsymbol{\theta})$ for $x_1 \leq x_2$
  To see this, let A be the set of events that give $P(X(A) \leq x_1|\boldsymbol{\theta})$ and B the set of events that give $P(x_1 < X \leq x_2|\boldsymbol{\theta})$. From their definition, we conclude that the events A and B are mutually exclusive. Therefore

$$P(X \leq x_2|\boldsymbol{\theta}) = P(X \leq x_1|\boldsymbol{\theta}) + P(x_1 < X \leq x_2|\boldsymbol{\theta})$$

or

$$P(x_1 < X \leq x_2|\boldsymbol{\theta}) = P(X \leq x_2|\boldsymbol{\theta}) - P(X \leq x_1|\boldsymbol{\theta}).$$

We can use the last property to derive another definition of continuous densities. From (6.3) and the definition of derivatives, we have

$$P(X = x|\boldsymbol{\theta}) = \lim_{\Delta x \to 0} \frac{P(X \leq x + \Delta x|\boldsymbol{\theta}) - P(X \leq x|\boldsymbol{\theta})}{\Delta x}.$$

From $P(x_1 < X \leq x_2|\boldsymbol{\theta}) = P(X \leq x_2|\boldsymbol{\theta}) - P(X \leq x_1|\boldsymbol{\theta})$ for $x_1 \leq x_2$, we have

Alternative definition of continuous density

$$P(X = x|\boldsymbol{\theta}) = \lim_{\Delta x \to 0} \frac{P(x \leq X \leq x + \Delta x|\boldsymbol{\theta})}{\Delta x}.$$

6.3.2 Densities

$P(X = x|\boldsymbol{\theta}) \geq 0$

Because $P(X \leq x|\boldsymbol{\theta})$ is non-decreasing, its derivative, $P(X = x|\boldsymbol{\theta}) \geq 0$.

$\int_{-\infty}^{\infty} P(X = x|\boldsymbol{\theta}) \, dx = 1$

The fundamental theorem of calculus states that

$$\int_a^b P(X = x|\boldsymbol{\theta}) \, dx = P(X \leq b|\boldsymbol{\theta}) - P(X \leq a|\boldsymbol{\theta}) . \qquad (6.6)$$

From the properties of distributions, we have

$$P(X \leq \infty|\boldsymbol{\theta}) - P(X \leq -\infty|\boldsymbol{\theta}) = 1 - 0 = 1 .$$

$P(x_1 \leq X \leq x_2|\boldsymbol{\theta}) = \int_{x_1}^{x_2} P(X = x|\boldsymbol{\theta}) \, dx$ for $x_1 < x_2$

This property is an immediate consequence of the definition of $P(X = x|\boldsymbol{\theta})$:

$$\begin{aligned} P(x_1 \leq X \leq x_2|\boldsymbol{\theta}) &= P(X \leq x_2|\boldsymbol{\theta}) - P(X \leq x_1|\boldsymbol{\theta}) \\ &= \int_{-\infty}^{x_2} P(X = x|\boldsymbol{\theta}) \, dx - \int_{-\infty}^{x_1} P(X = x|\boldsymbol{\theta}) \, dx \\ &= \int_{x_1}^{x_2} P(X = x|\boldsymbol{\theta}) \, dx . \end{aligned}$$

$P(X = x|\boldsymbol{\theta}) = 0$

This property is a direct consequence of (6.6).

The next example illustrates the properties of continuous distributions and densities with the uniform distribution.

Example 6.5. Personal observations of songbirds indicate that shortly after dawn, they spend 1–3 hours feeding. A simple assumption is that the probability that a songbird spends any amount of time feeding from 1–3 hours per morning is 1. Let X be the amount of time a songbird spends feeding in the morning. Then

$$P(X = x|0,3) = \begin{cases} 0 & \text{for } x < 1 \\ \frac{1}{2} & \text{for } 1 \leq x \leq 3 \\ 0 & \text{for } x > 3 \end{cases}, \quad P(X \leq x|0,3) = \begin{cases} 0 & \text{for } x < 1 \\ \frac{x}{2} & \text{for } 1 \leq x \leq 3 \\ 1 & \text{for } x > 3 \end{cases}.$$

From these equations we see that $P(X \leq -\infty|0,3) = 0$, $P(X \leq \infty|0,3) = 1$ and for $x_1 \leq x_2$ we have $P(X \leq x_1|0,3) \leq P(X \leq x_2|0,3)$. Obviously $P(X = x|0,3) \geq 0$ and the area under the density from $-\infty$ to ∞ is $\int_{-\infty}^{\infty} P(X = x|0,3) \, dx = 1$.

Let us use R to verify these properties with the standard uniform $P(X \leq x|0,1)$:

```
> c(punif(-Inf), punif(Inf))
[1] 0 1
> x.1 <- 0.6 ; x.2 <- 0.7 ; punif(x.1) <= punif(x.2)
[1] TRUE
```

Here `punif(x)` corresponds to $P(X \leq x|0,1)$ and `Inf` in R is ∞. □

For $P(X = x|\theta)$ satisfying the properties we just discussed, it should be clear by now that the fundamental difference between continuous and discrete densities is this: *For continuous densities, $X \in \mathbb{R}$ and $P(X = x|\theta) \geq 0$. For discrete densities, $P(X = y|\theta) > 0$ where y is a discrete subset of x.*

6.4 Expected values

In direct parallel to the expected values of discrete distributions, we have

Expected value of a continuous rv The expected value of a continuous rv X with density $P(X = x|\theta)$ is

$$E[X] = \int_{-\infty}^{\infty} x P(X = x|\theta)\,dx\ .$$

Example 6.6. For the uniform density we have

$$\begin{aligned}
E[X] &= \frac{1}{b-a}\int_a^b x\,dx \\
&= \frac{1}{b-a}\left(\frac{1}{2}b^2 - \frac{1}{2}a^2\right) = \frac{1}{2}\frac{1}{b-a}\left(b^2 - a^2\right) \\
&= \frac{1}{2}\frac{(b-a)(b+a)}{b-a} \\
&= \frac{a+b}{2}\ .
\end{aligned}$$

This is what one might expect—in the long run, the expected value will be in the middle between a and b. □

Example 6.7. For the exponential density (6.5) and for $\lambda > 0$ we obtain

$$E[T] = \int_0^\infty t\lambda e^{-\lambda t}dt = \frac{1}{\lambda}\ .$$

□

Example 6.8. Figure 5.2 shows a histogram of the number of people killed per attack by Hamas terrorists (data sources are given in Example 5.2). The smooth curve in

184 Continuous distributions and densities

the figure shows the exponential density (6.5) with $\lambda = 1/6.698 = 0.149$, where $E[X] = 1/\lambda = 6.698$ is the average (expected) number of people killed per attack. □

Example 6.9. Continuing with Example 5.2, the histogram and exponential density of time between attacks are shown in Figure 6.2. The mean number of days between attacks was 21.881. The fitted exponential density is drawn with

$$\lambda = \frac{1}{21.881} = 0.046 \ .$$

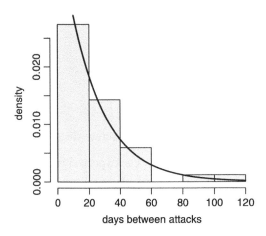

Figure 6.2   Days between attacks on Israelis by Hamas.

As we discussed, a uniform random density of events in time gives rise to an exponential density of intervals between events. Therefore, we conclude that there is evidence to support the hypothesis that the attacks occurred uniformly randomly in time. We will make the statement "evidence to support" more rigorous in due time. Figure 6.2 was produced with the following script:

```
load('terror.by.Hamas.rda')
terror <- terror.by.Hamas
lambda <- 1 / mean(terror$Killed)
j1 <- terror$Julian ; j2 <- j1
j1 <- j1[-length(j1)] ; j2 <- j2[-1]
h(j2 - j1, xlab = 'days between attacks')
x <- 0 : 120
lines(x, dexp(x, 1 / mean(j2 - j1)))
```

(the code for h() is on p. 40) which should be self-explanatory by now. □

6.5 Variance and standard deviation

The definitions of variance and standard deviation of continuous densities are similar to those of discrete densities:

Variance of a continuous rv The variance of a continuous rv X with density $P(X = x)$ is

$$V[X] = \int_{-\infty}^{\infty} (x - E[X])^2 P(X = x|\boldsymbol{\theta}) \, dx \ .$$

Standard deviation of a continuous rv The standard deviation of a continuous rv X with variance $V[X]$ is

$$S[X] = \sqrt{V[X]} \ .$$

Example 6.10. The variance of the uniform is given by

$$V[X] = \int_a^b \left(x - \frac{a+b}{2}\right)^2 \frac{1}{b-a} dx$$
$$= \frac{(a-b)^2}{12} \ .$$
□

Example 6.11. For $\lambda > 0$, the variance of the exponential is given by

$$V[X] = \int_0^\infty \left(x - \frac{1}{\lambda}\right)^2 \lambda e^{-\lambda x} dx = \frac{1}{\lambda^2} \ .$$

For the Hamas attacks, we found $\lambda = 0.046$ where $1/\lambda$ is the mean (expected) number of days between attacks. The variance of the days between attacks is $1/\lambda^2 = 472.59$. The standard deviation is 21.74 days between attacks. □

6.6 Areas under density curves

According to our definitions, $P(X = x|\boldsymbol{\theta})$ is either discrete or continuous. Therefore,

$$P(X \leq x|\boldsymbol{\theta}) = \int_{-\infty}^{x} P(X = \xi|\boldsymbol{\theta}) \, d\xi \tag{6.7}$$

for both discrete and continuous densities. From (6.7) we conclude that $P(X \leq x|\boldsymbol{\theta})$ is the area under $P(X = \xi|\boldsymbol{\theta})$ for ξ from $-\infty$ up to x. Using the rules of integration,

$$P(X > b|\boldsymbol{\theta}) = 1 - P(X \leq b|\boldsymbol{\theta})$$
$$= \int_{-\infty}^{\infty} P(X = \xi|\boldsymbol{\theta}) \, d\xi - \int_{-\infty}^{b} P(X = \xi|\boldsymbol{\theta}) \, d\xi$$
$$= \int_{b}^{\infty} P(X = \xi|\boldsymbol{\theta}) \, d\xi$$

and

$$P(a < X \leq b|\boldsymbol{\theta}) = P(X \leq b|\boldsymbol{\theta}) - P(X \leq a|\boldsymbol{\theta})$$
$$= \int_{-\infty}^{b} P(X = \xi|\boldsymbol{\theta}) \, d\xi - \int_{-\infty}^{a} P(X = \xi|\boldsymbol{\theta}) \, d\xi \tag{6.8}$$
$$= \int_{a}^{b} P(X = \xi|\boldsymbol{\theta}) \, d\xi \ .$$

186 Continuous distributions and densities

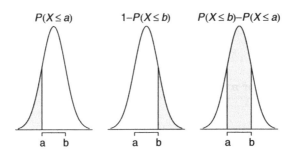

Figure 6.3   Distributions and areas under densities.

Example 6.12. The shaded areas in Figure 6.3 represent the values of $P(X \leq a|\theta)$, $1 - P(X \leq b|\theta)$ and $P(X \leq b|\theta) - P(X \leq a|\theta)$ for the so-called normal density. Let us see how to produce $P(X \leq a|\theta)$ in R (the remaining figures are produced similarly):

```
1  x <- seq(-4, 4, length = 1000) ; y <- dnorm(x)
2  plot(x, y, axes = FALSE, type = 'l', xlab = '', ylab = '',
3     main = expression(italic(P(X<=a))))
4  abline(h = 0)
5  x1 <- x[x <= -1] ; y1 <- dnorm(x1)
6  x2 <- c(-4, x1, x1[length(x1)], -4) ; y2 <- c(0, y1, 0, 0)
7  polygon(x2, y2, col = 'grey90')
8  axis(1, at = c(-1, 1), font = 8,
9     vfont = c('serif', 'italic'), labels = c('a', 'b'))
```

In line 1 we set x's range from -4 to 4 and assign the values of the normal densities at x values to y with a call to dnorm() (we will discuss the normal density in detail later). In lines 2 and 3 we plot without axes (axes = FALSE) and set the main label to italic for the expression $P(X \leq a)$. In line 4 we plot a horizontal line at $x = 0$. Next, we need to create a polygon for $x = -4$ to 4 at $y = 0$, then from $x = -1$ to $P(X = 1)$, then $P(X = x)$ for $x = -1$ to -4 and shade it. In lines 5 and 6 we set the x and y vertices of the desired polygon. In line 7 we call polygon() with the x and y coordinates of the polygon. We fill the polygon with the color grey90. Finally, in lines 8 and 9 we call axis(). Note the call to vfont(). Producing various fonts in a graph is a specialized topic (see R's documentation). □

Here is a numerical example.

Example 6.13. In Example 6.5, we constructed the uniform distribution for the feeding times of songbirds. We now calculate some probabilities of interest. For example, the probability that a bird spends between 4 and 6 hours a day feeding is the area under the curve between 4 and 6 in Figure 6.4. This area is

$$P(4 \leq X \leq 6|4, 6) = (6 - 4) \times 0.5 = 1$$

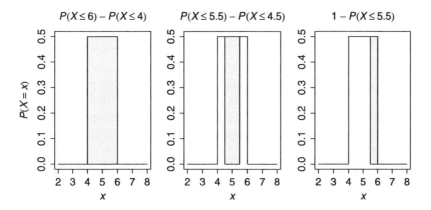

Figure 6.4   Probabilities and areas under the uniform density.

Or more formally according to (6.8)

$$P(X \leq 6|4,6) - P(X \leq 4|4,6) = \int_4^6 \frac{1}{2} dx = 1 \ .$$

Similarly, the probability that the bird spends between 4.5 and 5.5 hours a day feeding is

$$P(4.5 \leq X \leq 5.5|4,6) = (5.5 - 4.5) \times 0.5 = 0.5 \ .$$

The probability that the bird spends more than 5.5 hours a day feeding is

$$P(X > 5.5|4,6) = (6 - 5.5) \times 0.5 = 0.25$$

(Figure 6.4). We interpret each of these cases as follows: With a large number of observations, about 100% of the birds will spend between 4 and 6 hours feeding, 50% between 4.5 and 5.5 and 25% more than 5.5 hours a day feeding. The areas under the curve in these 3 cases represent the corresponding probabilities. The script for this example is almost identical to the script in Example 6.12.   □

6.7 Inverse distributions and simulations

If $P(X \leq x|\theta)$ is a given distribution, then we can calculate its value for a given x.

Example 6.14. Consider the uniform with parameters 0 and 4 and the exponential with parameter 0.2. To obtain the probabilities that $x \leq 2$ and 10, we do

```
> round(c(punif(2, 0, 4), pexp(10, 0.2)), 4)
[1] 0.5000 0.8647
```

This means that if we repeatedly draw a random value from the uniform, then 50% of these values will be ≤ 2. Similarly, about 86.5% of the values from the exponential will be ≤ 10.   □

Often, we are interested in the inverse of a distribution. That is, given a probability value p, we wish to find the *quantile*, x, such that $P(X \leq x|\theta) = p$.

Example 6.15. Continuing with Example 6.14,

```
> round(c(qunif(0.5, 0, 4), qexp(0.8647, 0.2)), 2)
[1]  2 10
```
□

Next, we want to draw a random value from a density. This means that if certain values of x are more probable under the density than others, then they should appear more frequently in a random sample from the density. For example, for the densities shown in Figure 6.3, most of the random values should be clumped around the center of the density (these have the highest probabilities of occurring). The process of generating random values from a density is called *simulation* (some call it Monte Carlo simulation). Simulations is a wide topic and details can be found in books such as Ripley (1987) and Press et al. (1992). To generate random values from a particular density, we first realize that *a priori*, there is no reason to prefer one random value (quantile) over another. However, the density itself should produce more quantiles for those values that are more probable under it. To generate a random value, x, from a known distribution $P(X \leq x|\boldsymbol{\theta}) = p$, we define the inverse of the distribution by $x = P^{-1}(p|\boldsymbol{\theta})$. Because we have no *a priori* reason to prefer one value of p over another, we use the uniform on $[0, 1]$ to generate a random value of p and then use $P^{-1}(p|\boldsymbol{\theta})$ to generate x. In other words, in the case of $P(X \leq x|\boldsymbol{\theta}) = p$, x is given and p is therefore known. In the case of $P^{-1}(p|\boldsymbol{\theta})$, p is the rv (with a uniform density with parameters 0 and 1) and so x is also a rv. In order not to confuse issues, we stray, in this case, from the convention that a rv is denoted by an upper case letter.

Example 6.16. Let us generate a random value from the exponential distribution $P(X \leq x|\lambda) = 1 - e^{-\lambda t}$. To generate x, solve $p = 1 - e^{-\lambda x}$ for x:

$$x = -\frac{\log(1-p)}{\lambda} . \tag{6.9}$$

Now generate a random deviate p from a uniform distribution on $[0, 1]$ and use it in (6.9) to compute x. We achieve this in R with:

```
> round(rexp(5, 0.1), 3)
[1]  7.291 12.883  6.723  4.265 11.154
```

Here we produce five random values from the exponential distribution with parameter $\lambda = 0.1$. Figure 6.5 illustrates the process of generating three random values from the exponential distribution. The figure was produced with

```
1  x <- seq(0, 10, length = 101) ; lambda <- 0.5
2  set.seed(10) ; u <- runif(3) ; r.x <- qexp(u, lambda)
3  plot(x, pexp(x, lambda), type = 'l', xlim = c(0, 10))
4  for(i in 1 : 3){
5    arrows(-1, u[i], r.x[i], u[i], code = 2,
6      length = 0.1, angle = 20)
7    arrows(r.x[i], u[i], r.x[i], -0.04, code = 2,
8      length = 0.1, angle = 20)
9  }
```

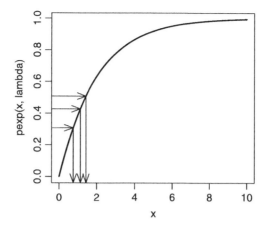

Figure 6.5 Generating random values from the exponential distribution.

In line 1 we set the values of x for which we generate values from the exponential distribution with parameter $\lambda = 0.5$. In line 2 we set.seed() to 10 (so that we can repeat the simulation), produce three random probabilities from the standard uniform and obtain their corresponding values from the exponential distribution with qexp() (here q stands for quantile). We then plot the distribution with the values of x and the corresponding values from the distribution with pexp(). Finally, we loop through the three values and draw arrows() with the arrows at the end of the line segments (code = 2). The angle and length of the arrows are set to 20 and 0.1. □

6.8 Some useful continuous densities

Here we discuss some useful continuous densities and the situations under which they arise. For further discussion about these and many other densities, consult Johnson et al. (1994).

6.8.1 Double exponential (Laplace)

The double exponential, also called the Laplace, is the density of the difference between two independent rv with identical exponential densities.

Density and distribution

The density of the double exponential is

$$P(X = x | \mu, \sigma) = \frac{1}{2\sigma} \exp\left[-\left|\frac{x - \mu}{\sigma}\right|\right].$$

Here μ and σ are the location and scale parameters. The *standard double exponential* is

$$P(X = x | 0, 1) = \frac{1}{2} \exp\left[-|x|\right]$$

and the distribution is

$$P(X \leq x|\mu, \sigma) = \frac{1}{2}\left[1 + \text{sign}(x - \mu)\left(1 - \exp\left[-\frac{|x - \mu|}{\sigma}\right]\right)\right]$$

where $\text{sign}(x)$ is $+$ if $x > 0$, $-$ if $x < 0$ and zero if $x = 0$.

Estimating parameters

For the double exponential, $E[x] = \mu$ and $V[x] = 2\sigma^2$. For a sample of size n with mean $\overline{X}$, we estimate μ and σ with

$$\widehat{\mu} = \overline{X}, \quad \widehat{\sigma} = \frac{1}{n}\sum_{i=1}^{n}|X_i - \overline{X}|.$$

Applications

Example 6.17. During the breeding season, bull elk fight with other male elk for the privilege to mate with females. Let X_1 be the giving up time of the losing bull on its first match and X_2 on the second. Assume that giving up times on the first or second fights are independent with the same mean. Then $Y = X_1 - X_2$ is a double exponential rv. □

Here are R functions for generating random double exponential, density and distribution values:

```
rdouble.exp <- function(n, mu = 0, sigma = 1){
  return(rexp(n, 1 / sigma) *
    ifelse(runif(n) <= 0.5, -1, 1))
}

ddouble.exp <- function(x, mu = 0, sigma = 1){
  return(1 / (2 * sigma) * exp(-abs((x - mu) / sigma)))
}

pdouble.exp <- function(x, mu = 0, sigma = 1){
  return(1/2 * (1 + sign(x - mu) *
    (1 - exp(-abs(x - mu) / sigma))))
}
```

Let us put these functions to the test.

Example 6.18. We wish to verify that the double exponential functions above work. So we generate random values with

```
> set.seed(5) ; y <- rdouble.exp(10000)
```

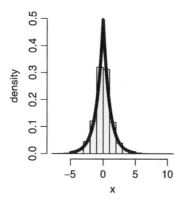 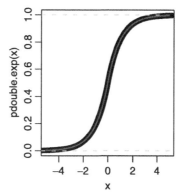

Figure 6.6 Left: density of a random sample from the double exponential (histogram) and the true and estimated densities (both curves are nearly identical). Right: the corresponding distributions.

Next, we plot the histogram of the random values and superimpose on it the theoretical and estimated values (Figure 6.6 left) with

```
> par(mfrow = c(1, 2))
> h(y, xlab = 'x', ylim = c(0, 0.5))
> lines(x, ddouble.exp(x), type = 'l')
> mu.hat <- mean(y)
> sigma.hat <- sum(abs(y - mu.hat))/ length(y)
> lines(x, ddouble.exp(x, mu.hat, sigma.hat), lwd = 3)
```

The histogram, estimated and true densities are nearly identical. Finally, we compare the true distribution, the empirical distribution and the estimated distribution (Figure 6.6 right) with

```
> plot(x, pdouble.exp(x), col = 'blue', pch = 21)
> lines(ecdf(y))
> lines(x,  pdouble.exp(x, mu.hat, sigma.hat), col = 'red')
```

Again, note the nearly perfect agreement among these three. □

6.8.2 Normal

In statistics, the normal is the most important of all densities.

Density and distribution

The rv X is said to have a *normal density* if

$$P(X = x|\mu, \sigma) = \frac{1}{\sigma\sqrt{2\pi}} \exp\left[-\frac{1}{2}\left(\frac{x-\mu}{\sigma}\right)^2\right]$$

where μ and σ are the location and scale parameters. When $\mu = 0$ and $\sigma^2 = 1$, the rv X is said to have *standard normal density*. The closed form of the distribution is not known and must be computed numerically. The standard normal rv is often denoted by Z.

Estimating parameters

It turns out that
$$E[X] = \mu, \quad V[X] = \sigma^2.$$

Define the sample variance
$$S^2 := \frac{1}{n-1} \sum_{i=1}^{n} (X_i - \overline{X})^2.$$

To estimate μ and σ^2, use
$$\widehat{\mu} = \overline{X}, \quad \widehat{\sigma}^2 = \frac{1}{n} \sum_{i=1}^{n} (X_i - \overline{X})^2 = \frac{n-1}{n} S^2,$$

where $\overline{X}$ is the sample mean and n is the sample size. Because the estimates of the mean and variance are based on a sample, they themselves are realizations of random variables. It can be shown that these two random variables—the sample mean and the sample variance—are independently distributed.

Departures from normality can be of two types. One is from symmetry, called *skewness*, and the other is reflected in differences in the proportion of the data that are in the center and tails of the distribution, called *kurtosis*. These departures are characterized by two additional parameters. We estimate skewness and kurtosis with

$$\widehat{\gamma}_1 = \frac{n \sum_{i=1}^{n} (X_i - \overline{X})^3}{(n-1)(n-2) \widehat{\sigma}^3},$$

$$\widehat{\gamma}_2 = \frac{(n+1) n \sum_{i=1}^{n} (X_i - \overline{X})^4}{(n-1)(n-2)(n-3) \widehat{\sigma}^4}.$$

The estimates of skewness and kurtosis measure departures from normality. Small values of both indicate normality. Negative $\widehat{\gamma}_1$ indicates skewness to the left, while positive indicates skewness to the right. Negative $\widehat{\gamma}_2$ indicates long tails, positive indicates short tails. Skewness and kurtosis are discussed further in Johnson et al. (1994). It is easy to show that any linear combination of independent normally distributed random variables is also normally distributed.

Applications

The normal distribution is widely used in statistics. We shall meet its applications as we proceed.

6.8.3 χ^2

If Z_i, $i = 1, \ldots, \nu$, are independent standard normal, then the distribution of the rv $X = \sum_{i=1}^{\nu} Z_i^2$ is *chi-square* with ν degrees of freedom. Heuristically, the degrees of freedom in a statistical model are defined as

$$\nu = n - m - 1$$

where n is the number of data points and m is the number of parameters to be fitted to a statistical model of the data.

Density and distribution

The χ^2 density is a special case of the gamma density with parameters $1/2$ and $1/2$ (gamma is discussed in Section 6.8.7). The χ^2_ν density is

$$P(X = x|\nu) = \frac{e^{-x/2} x^{\nu/2-1}}{2^{\nu/2} \Gamma(\nu/2)}$$

where ν denotes the degrees of freedom and Γ is the *gamma function* (not to be confused with the gamma density), defined as

$$\Gamma(\alpha) = \int_0^\infty t^{\alpha-1} e^{-t} dt \ . \tag{6.10}$$

The distribution is given by

$$P(X \leq x|\nu) = \frac{\Gamma_x(\nu/2, x/2)}{\Gamma(\nu/2)}$$

where Γ_x is the so-called *incomplete gamma* function, given by

$$\Gamma_x(\alpha) = \int_0^x t^{\alpha-1} e^{-t} dt \ .$$

If X and Y are $\chi^2_{\nu_1}$ and $\chi^2_{\nu_2}$, then $X + Y$ is $\chi^2_{\nu_1+\nu_2}$. Figure 6.7 shows the χ^2_ν for ν = 5, 10, 15 (solid, broken and dotted curves). Note that the density becomes less skewed as the number of degrees of freedom increases. The figure was produced with the following code:

```
1  x <- seq(0, 35, length = 101)
2  nu <- c(5, 10, 15)
3  ylabel <- c('dchisq(x, nu)', 'pchisq(x, nu)')
4  par(mfrow = c(1, 2))
5  plot(x, dchisq(x, nu[1]), type = 'l', ylab = ylabel[1])
6  lines(x, dchisq(x, nu[2]), lty = 2)
7  lines(x, dchisq(x, nu[3]), lty = 3)
8  text(locator(), labels = c('nu = 5', 'nu = 10', 'nu = 15'),
9      pos = 4)
10 plot(x, pchisq(x, nu[1]), type = 'l', ylab = ylabel[2])
```

```
11  lines(x, pchisq(x, nu[2]), lty = 2)
12  lines(x, pchisq(x, nu[3]), lty = 3)
13  text(locator(), labels = c('nu = 5', 'nu = 10', 'nu = 15'),
14      pos = 4)
```

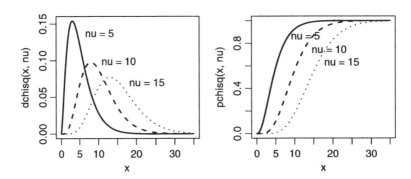

Figure 6.7   The χ^2 density and distribution for three different degrees of freedom.

The code is fairly standard by now. Note the use of lty (line type) in lines 6, 7, 11 and 12. Lines of type 2 result in broken curves and lines of type 3 result in dotted curves. Also note the use of text() and locator() in lines 8, 9, 13 and 14. text() draws text where locator() picks the coordinates from a mouse click. The text is provided in labels. So in the first click, we get 'nu = 5' on the plot where we click the mouse. pos = 4 specifies that the label should be drawn to the right of the location clicked.

Estimating parameters

The expectation and the variance are

$$E[X] = n, \quad V[X] = 2n.$$

In Section 5.8, we discussed the MLE method for estimating parameters. There, we briefly mentioned that in some cases, it is not possible to solve the MLE for a density analytically. In such cases one has to rely on numerical solutions. One of the functions that accomplish this task in R is called fitdistr(). Let us see how we might use this function to estimate the parameters from a sample we believe comes from a χ^2 density.

Example 6.19. We draw 1 000 values from a χ^2_{10} like this:

```
> n <- 1000 ; set.seed(1000) ; df <- 10 ; X <- rchisq(n, df)
```

Next, we examine the histogram of the data with

```
> h(X, ylim = c(0, 0.1), xlab = 'x')
```

(Figure 6.8). Because we know the parameter value (df = 10), we can compare the empirical density to the theoretical density with

```
> lines(x, dchisq(x, df), lwd = 2, col = 'red')
```

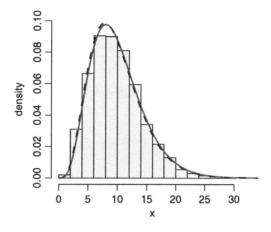

Figure 6.8 χ^2_{10}

(solid curve in Figure 6.8). Now let us use MLE to estimate the parameter df:

```
> df.hat <- fitdistr(X, densfun = 'chi-squared',
+    start = list(df = 5))
```

df.hat is a list containing the MLE of the parameter and its standard error:

```
> df.hat
       df
  9.8281250
 (0.1331570)
```

Next, we compare the empirical density $\left(\chi^2_{\widehat{df}}\right)$ to the theoretical one:

```
> lines(x, dchisq(x, df.hat[[1]]), lwd = 2,
+    lty = 2, col = 'blue')
```

(broken curve in Figure 6.8). Looks good! □

Applications

The χ^2 describes the density of the variance of a sample taken from a normal population. It is used routinely in nonparametric statistics, for example in testing association of categorical variables in contingency tables. It is also used in testing goodness-of-fit. We shall discuss all of these topics later. Testing for goodness-of-fit with χ^2 is useful because the only underlying assumption about the observations is that they are independent.

6.8.4 Student-t

The Student-t density (called after its discoverer, Student) is used to draw conclusions from small samples from normal populations. It behaves much like the normal, but it has longer tails.

Density

If Z is a standard normal rv and U is χ^2_ν, and Z and U are independent, then $Z/\sqrt{U/\nu}$ is said to come from a t *density* with ν degrees of freedom.

196 Continuous distributions and densities

The density of t with ν degrees of freedom is

$$P(X = x|\nu) = \frac{(1 + x^2/\nu)^{-\nu/2}}{B(1/2, \nu/2)\sqrt{\nu}}$$

where B is the beta function and ν is a positive integer (usually denotes the degrees of freedom). The beta function is

$$B(\alpha, \beta) = \int_0^1 t^{\alpha-1}(1-t)^{\beta-1}\, dt \; . \tag{6.11}$$

The tails of the t density are flatter than those of the normal. Here is an example that compares the standard normal to t.

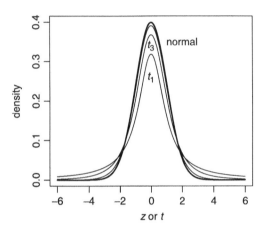

Figure 6.9 t_1, t_3 and t_{12} compared to the standard normal.

Example 6.20. Figure 6.9 illustrates the convergence of t to standard normal as the number of degrees of freedom increases. It was produced with the following script:

```
1  x <- seq(-6, 6, length = 201)
2  plot(x, dnorm(x), xlab = expression(paste(italic(z), ' or ',
3      italic(t))), ylab = 'density', type = 'l', lwd = 2)
4  df <- c(1, 3, 12) ; for (i in df) lines(x, dt(x, i))
5  labels <- c('normal', expression(italic(t[1])),
6      expression(italic(t[3])))
7  atx <- c(2, 0, 0) ; aty <- c(0.35, 0.26, 0.34)
8  text(atx, aty, labels = labels)
```

We plot the t density with a call to dt() for 1, 3 and 12 degrees of freedom in line 4. Note the production of subscripted text (t_1 and t_3) with calls to expression() in lines 5 and 6. □

Estimating parameters

For $\nu = 1$, the density has no expectation. Otherwise it is zero. The variance is

$$V[X] = \frac{\nu}{\nu - 2}.$$

Applications

The t density is used routinely in testing pairs of samples for significant differences in means. The importance of the density arises from the following fact. For the normal rv X with parameters μ and σ, the sample mean

$$\overline{X} = \frac{1}{n} \sum_{i=1}^{n} x_i$$

and the sample variance

$$S^2 = \frac{1}{(n-1)} \sum_{i=1}^{n} (X_i - \overline{X})^2$$

themselves are random variables. They have densities (named the sampling density) and it can be shown that they are independent random variables. Furthermore, they are related to each other by the rv

$$T = \frac{\overline{X} - \mu}{S^2/\sqrt{n}}$$

where T comes from t_{n-1} (the t density with $n - 1$ degrees of freedom). In R, the functions dt(), pt(), qt() and rt() provide access to the density, distribution, inverse distribution and random values from the density, respectively.

6.8.5 F

Let U and V be independent chi-square random variables with ν_1 and ν_2 degrees of freedom. Define the ratio

$$X := \frac{U/\nu_1}{V/\nu_2}.$$

Then X has the so-called F density.

Density and distribution

The density of F is

$$P(X = x | \nu_1, \nu_2) = \frac{\Gamma\left(\frac{\nu_1 + \nu_2}{2}\right) \left(\frac{\nu_1}{\nu_2}\right)^{\nu_1/2} x^{\nu_1/2 - 1}}{\Gamma(\nu_1/2)\, \Gamma(\nu_2/2) \left(1 + \nu_1 x/\nu_2\right)^{\frac{\nu_1 + \nu_2}{2}}}$$

where ν_1 and ν_2 are the degrees of freedom and the gamma function Γ is defined in (6.10). The distribution is

$$P(X \le x | \nu_1, \nu_2) = 1 - I_K(\nu_2/2, \nu_1/2)$$

198 Continuous distributions and densities

where
$$K := \frac{\nu_2}{\nu_2 + \nu_1 x}, \quad I_K(X, \alpha, \beta) := \frac{\int_0^x t^{\alpha-1}(1-t)^{\beta-1} dt}{B(\alpha, \beta)}.$$

I_K is the called the *beta regularized function*. Its numerator is the incomplete beta function and its denominator, B, is the beta function defined in (6.11). Figure 6.10 illustrates the F density and distribution for $\nu_1 = 1$, $\nu_2 = 1$ (solid curves) and $\nu_1 = 10$, $\nu_2 = 1$ (broken curves). If X comes from a t_n, then X^2 is distributed according to F with 1 and n degrees of freedom. To obtain Figure 6.10, use df() and pf() instead of dchisq() and pchisq() in the code on p. 193.

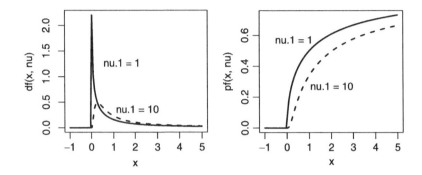

Figure 6.10 The F density and distribution.

Estimating parameters

One rarely needs to estimate the parameters of F because it is usually used for testing hypotheses. For a standard F (the location and dispersion parameters are 0 and 1),

$$E[X] = \frac{\nu_2}{\nu_2 - 2}, \quad V[X] = \frac{2\nu_2^2(\nu_1 + \nu_2 - 2)}{\nu_1(\nu_2 - 2)^2(\nu_2 - 4)}, \quad \nu_2 > 4.$$

Applications

The F density is used routinely in analysis of variance. It is also used to test for equality of two variances—the so-called F-test. We will meet applications of the F later.

6.8.6 Lognormal

Let Y come from a normal density and $X = \log Y$. Then the density of X is lognormal.

Density and distribution

The lognormal density is

$$P(X = x | \mu, \sigma) = \frac{1}{x\sigma\sqrt{2\pi}} \exp\left[-\frac{1}{2}\left(\frac{\log(x-\mu)}{\sigma}\right)^2\right], \quad \sigma > 0.$$

Here μ and σ are the location and shape parameters. When $\mu = 0$ and $\sigma = 1$, we have the so-called *standard lognormal* density

$$P(X = x|0, 1) = \frac{1}{x\sqrt{2\pi}} \exp\left[-\frac{1}{2}(\log x)^2\right].$$

The distribution of the standard lognormal is given by the standard normal $P(Z \leq z|0, 1)$ where $z = \log x$.

Figure 6.11 illustrates the densities and distributions for $P(X = x|0, 1)$ and $P(X \leq x|0, 1)$ (solid curves) and $P(X = x|0, 2)$ and $P(X \leq x|0, 2)$ (dashed curves). To obtain the figure, use dlnorm() and plnorm() instead of dchisq() and pchisq() in the code on p. 193. Note that the log of the parameter values are given, not the values themselves.

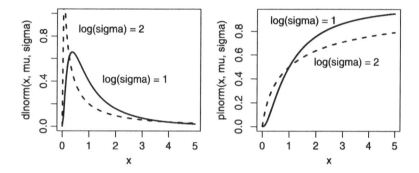

Figure 6.11   The lognormal for $\log \mu = 1, 1$ and $\log \sigma = 1, 2$.

Estimating parameters

The maximum likelihood estimates of μ and σ^2 are

$$\widehat{\mu} = \frac{1}{n}\sum_{i=1}^{n} \log X_i, \quad \widehat{\sigma}^2 = \frac{1}{n-1}\sum_{i=1}^{n} (\log X_i - \widehat{\mu})^2.$$

For $P(X = x|0, 1)$ we have

$$E[X] = \exp\left[\frac{1}{2}\sigma^2\right], \quad V[X] = \exp[\sigma^2]\left(\exp[\sigma^2] - 1\right).$$

Applications

The lognormal is most frequently used in reliability applications and survival analysis. It is also used in modeling failure times. Limpert et al. (2001) provide a detailed survey of the widespread applications of the lognormal.

6.8.7   Gamma

The gamma density arises in many applications, in particular in processes where the distribution of times between events is exponential and the counts are Poisson.

Density and distribution

The gamma density is given by

$$P(X = x|\alpha, \sigma) = \frac{1}{\sigma^2 \Gamma(\alpha)} x^{\alpha-1} \exp\left[-\frac{x}{\sigma}\right], \quad \alpha, \sigma > 0 \qquad (6.12)$$

where α and σ are the shape and scale parameters. Here Γ is the gamma function (see Equation 6.10). For $\alpha = 1$, (6.12) reduces to the exponential (see Equation 6.5).

Estimating parameters

We have

$$E[X] = \sigma\alpha, \quad V[X] = \sigma^2 \alpha.$$

There are several methods to estimate the parameters based on data. The method of moments is one approach. According to this method, the parameters are estimated with

$$\hat{\sigma} = \frac{S^2}{\overline{X}}, \quad \hat{\alpha} = \frac{\overline{X}^2}{S^2}$$

where $\overline{X}$ and S^2 are the sample mean and variance, respectively.

Applications

There is an interesting link between the gamma and exponential and Poisson densities.

Example 6.21. Suppose that the lifetime of individual i, in a population of n individuals, is a rv X_i with exponential density with rate parameter λ (see Equation 6.5). Furthermore, suppose that the lifetime of an individual is independent from that of others in the population. Then it can be shown that the density of $X = X_1 + \cdots + X_n$ is gamma with the shape parameter λ and the scale parameter n

$$P(X = x|\lambda, n) = \frac{1}{\lambda \Gamma(n)} \left(\frac{x}{\lambda}\right)^{n-1} \exp\left[-\frac{x}{\lambda}\right];$$

i.e. the sum of the lifetimes of all individuals is gamma. □

Another interesting relationship is the following. If $X_1, \ldots, X_n$ are independent rv from gamma each with parameters $\alpha_1, \ldots, \alpha_n$ and σ, then $Y = X_1 + \cdots + X_n$ is a gamma rv with parameters $\alpha = = \alpha_1 + \cdots + \alpha_n$ and σ. Also, if X_1 and X_2 are independent from gamma with parameters (α_1, σ) and (α_2, σ), then $Y = X_1/(X_1 + X_2)$ has a beta density (see Equation 6.13) with parameters (α_1, α_2).

Here is an example of applying the r, d and p versions of the gamma with R.

Example 6.22. Figure 6.12 illustrates the densities and distributions for $P(X = x|2, 1)$ and $P(X = x|2, 2)$ (solid curves) and $P(X = x|1, 2)$ and $P(X \leq x|1, 2)$ (dotted curves). The following code produced the figure:

```r
alpha <- 2 ; sigma <- 2
set.seed(10) ; r.gamma <- rgamma(1000, alpha, scale = sigma)
x <- seq(0, 20, length = 201)
par(mfrow = c(1, 2))
h(r.gamma, xlab = 'x', ylim = c(0, 0.2))
lines(x, dgamma(x, alpha, scale = sigma), type = 'l')
m <- mean(r.gamma) ; v <- var(r.gamma)
sigma.hat <- v / m ; alpha.hat <- m^2 / v
lines(x, dgamma(x, alpha.hat, scale = sigma) , lwd = 3)

plot(x, pgamma(x, alpha, scale = sigma),
   col = 'blue', pch = 21)
lines(ecdf(r.gamma))
lines(x,  pgamma(x, alpha.hat, scale = sigma.hat),
   col = 'red')
```

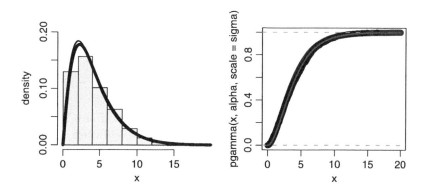

Figure 6.12   The gamma density for $\alpha = 2$ and $\sigma = 1, 2$.

The script is similar to the script on page 193 with the appropriate substitutions for the gamma.   □

6.8.8 Beta

The beta pertains to random variables that take values between zero and one. As such, it is used in modeling probabilities and proportions.

Density and distribution

The beta density is given by

$$P(X = x | \alpha, \beta) = \frac{\Gamma(\alpha + \beta)}{\Gamma(\alpha)\Gamma(\beta)} x^{\alpha-1} (1-x)^{\beta-1} , \quad \alpha, \beta > 0 . \tag{6.13}$$

Continuous distributions and densities

The beta distribution is given by

$$P(X \leq x|\alpha, \beta) = I(x|\alpha, \beta), \quad \alpha, \beta > 0$$

where

$$I(x|\alpha, \beta) = \frac{\int_0^x t^{\alpha-1}(1-t)^{\beta-1} dt}{B(\alpha, \beta)}$$

is the called the *beta regularized function*. Its numerator is the incomplete beta function and its denominator, B, is the beta function (see Equation 6.11).

Estimating parameters

The expectation and variance of the beta distribution are

$$E[X] = \frac{\alpha}{\alpha + \beta}, \quad V[X] = \frac{\alpha\beta}{(\alpha + \beta)^2(\alpha + \beta + 1)}.$$

Applications

The beta density is extremely useful in Bayes analysis (see for example Berger, 1985; Gelman et al., 1995). It is also useful in analyzing ratios. In fact, it arises naturally from the gamma distribution. Let X be a standard uniform rv. It turns out that the density of the ith highest value of X in a sample of size $i + j - 1$ is beta, $P(X = x|j, i)$. In the next example we verify this claim numerically.

Example 6.23. Suppose that the proportion of a population of n neurons that fire in any particular time interval is uniform between 0 and 1. Firing in one interval is independent of firing in another. We sample 12 neurons 10 000 times and record the proportion of these that fire. What is the density of the third highest proportion of neurons firing?

Here $i = 3$, $j = 12 - 3 + 1 = 10$. Our rv X is the third highest proportion of neurons that fire. Let us generate data by simulation. We set the data like this:

```
> j <- 10 ; n.samples <- 10000
> i <- 3 ; i.largest <- vector()
```

Next, we repeat the following 10 000 times: Take a sample of size 12 values from a standard uniform, sort the proportion and record the third highest value:

```
> for(k in 1 : n.samples){
+   x <- sort(runif(i + j - 1), decreasing = TRUE)
+   i.largest[k] <- x[i]
+ }
```

Next, we plot the histogram of the data and the corresponding beta density with parameters i and j:

```
> par(mfrow = c(1, 2))
> h(i.largest, xlab = 'x')
> xx <- seq(0, 1, length = 101)
> lines(xx, dbeta(xx, j, i))
```

Finally, we examine the beta distribution $P(X \leq x|10,3)$ and the empirical (simulation-based) distribution with:

```
> plot(ecdf(x), ylab = expression(italic(P(X<=x))), main = '',
+    xlim = c(0, 2))
> lines(xx, pbeta(xx, j, i))
```

Figure 6.13 illustrates the results. We will learn how to evaluate the goodness-of-fit between the theoretical and empirical densities later. □

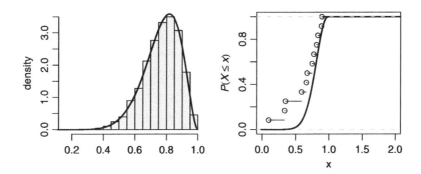

Figure 6.13 Density and distribution of third highest occurrence of a uniform rv compared to the theoretical beta.

6.9 Assignments

Exercise 6.1. A professor never dismisses his class early. Let X denote the amount of time past the hour (in minutes) that elapses before the professor dismisses class. The probability that he dismisses the class is equal for any late dismissal between 0 and 10 minutes. He never dismisses the class more than 10 minuets late.

1. What is the density of X?
2. Plot the density and the distribution of X.
3. What is the probability that at most 5 min elapse before dismissal?
4. What is the probability that between 3 and 5 min elapse before dismissal?
5. What is the expected value of the time that elapses before dismissal? Explain.
6. If X has a uniform distribution on the interval from a to b, then it can be shown that the standard deviation of X is $(b-a)/\sqrt{12}$. What is the standard deviation of elapsed time until dismissal?
7. What is the probability that the elapsed time is within 1 standard deviation of its mean value on either side of the mean?

Exercise 6.2.

1. What is the probability that an event will occur if its density is uniform between one and four?
2. What is the expected value of the event?
3. Its variance?

4. What is the probability that the value of the event will be less than four?
5. What is the probability that the value of the event will be less than or equal to 4?
6. What is the probability that the value of the event will be between two and three?

Exercise 6.3. Trees die randomly, independent of each other and a tree might die at any moment in time. The average death rate is 2 trees per month.

1. What is the probability density of the time between two consecutive deaths?
2. What is the expected number of deaths per month?
3. Its variance?
4. What is the probability that 10 months pass between two consecutive deaths?
5. What is the probability that at least 2 months pass between two consecutive deaths?
6. What is the probability that between 2 and 3 month pass between two consecutive deaths?

Exercise 6.4. The density of bill length of a bird species is normal with mean of 12 mm and standard deviation of 2 mm.

1. What is the expected bill length?
2. Its variance?
3. What is the probability that a randomly picked individual will have a bill length of 12 mm?
4. What is the probability that a randomly picked individuals will have bill length of at least 12 mm?
5. What is the probability that a randomly picked individuals will have bill length of at most 12 mm?

Exercise 6.5. Let X be the weight of deer in the winter. Mean weight is 150 kg with standard deviation 20 kg. The density of weight in the population is normal.

1. What is the value of x such that less than 5% of the population weigh less than x?
2. What is the value of x such that more than 95% of the population weigh more than x?

Exercise 6.6. Given the standard normal:

1. What is the value of z such that $P(Z \leq z) = 0.95$?
2. What is the value of z such that $P(Z \leq z) = 0.05$?
3. What is the value of z such that $P(Z \leq z) = 0.975$?
4. What is the value of z such that $P(Z \leq z) = 0.025$?

7

The normal and sampling densities

One of the central issues in statistics is how to make inferences about population parameters from a sample. For example, when we sample organisms and measure their weight, we may be interested in the following question: What is the relationship between the mean weight of the organisms in the sample and the mean weight of organisms in the population? A sample is a random collection of observations from a population of interest. Here population refers to the collection of all objects of interest. All objects in the population and all possible subsets of objects in the population constitute the sampling space. Because the sample values are rv, any function of the sample values is a rv. Such functions have well-defined densities, called the sampling densities. With knowledge of a sampling density, we can infer something about the corresponding population parameters.

We shall study the sampling densities of a sample mean (denoted by $\overline{X}$), sample proportion (denoted by p) and sample intensity (denoted by l). Here p is the proportion of objects in a sample that have some property and l is a count of the occurrences of an event of interest over a unit of reference (such as time). Both p and l are rv and are two cases where we depart from our convention that upper case letters represent rv and lower case letters values that they may take. We are interested in the relationship between the sampling densities of $\overline{X}$, p and l and values of μ, π and λ from the normal, binomial and Poisson densities. By its subject matter, this chapter has one foot in this part and one in the next part. For balance, we placed it here.

7.1 The normal density

The normal density is

$$\phi(x|\mu,\sigma) := P(X = x|\mu,\sigma) = \frac{1}{\sqrt{2\pi\sigma^2}} \exp\left[-\frac{1}{2}\left(\frac{x-\mu}{\sigma}\right)^2\right]$$

Statistics and Data with R: An applied approach through examples Y. Cohen and J.Y. Cohen
© 2008 John Wiley & Sons, Ltd.

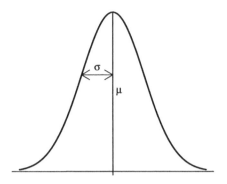

Figure 7.1 The normal density with location parameter μ and scale σ.

where μ and σ are the *location* and *scale* parameters (see Section 6.8.2 and Figure 7.1). Figure 7.1 was produced with the following script:

```
1  x <- seq(-3, 3, length = 501)
2  plot(x, dnorm(x), axes = FALSE, type = 'l', xlab = '',
3    ylab = '') ; abline(h = 0)
4  x <- 0 ; lines(c(0, 0), c(dnorm(x), -0.01))
5  x <- -1 ; lines(c(-1, 0), c(dnorm(x), dnorm(x)))
6  arrows(-1, dnorm(x), 0, dnorm(x), code = 3, length = 0.1)
7  text(0.2, 0.2, expression(italic(mu)))
8  text(-0.5, 0.26, expression(italic(sigma)))
```

In line 1 we create a vector, x, between -3 and 3 with increments that result in 501 elements. In lines 2 and 3 we plot the standard normal density with no axes and no labels. The call to dnorm(x) returns a vector of 501 values of the normal density. Each value corresponds to an element of x. In line 3 we draw a horizontal line at zero with a call to abline(). In line 4 we draw the vertical line from the top of the density (dnorm() at x = 0) to slightly below the horizontal zero line (−0.01). In line 5 we draw a horizontal line from x = −1 to 0 at y = dnorm(-1). In line 6 we add arrows() to the line. We ask for arrows at both ends (code = 3). The named argument length specifies the edges of the arrow head (in inches). In lines 7 and 8 we draw the letters μ and σ in the appropriate locations with calls to text() and expression().

The normal distribution is given by

$$\Phi(X \leq x|\mu, \sigma) := P(X \leq x|\mu, \sigma) = \frac{1}{\sqrt{2\pi\sigma^2}} \int_{-\infty}^{x} \exp\left[-\frac{1}{2}\left(\frac{\xi - \mu}{\sigma}\right)^2\right] d\xi \,.$$

There is no known closed form solution to the integral above. Tables for the standard normal distribution $P(X \leq x|0, 1)$ are published in some statistics books. If you use R, there is no reason for you to use such tables.

7.1.1 The standard normal

Because of its ubiquity, we denote the standard normal density ($\mu = 0$ and $\sigma = 1$) and distribution with $\phi(x)$ and $\Phi(x)$ and drop the dependence on parameter values so that

$$\phi(x) = \frac{1}{\sqrt{2\pi}} e^{-\frac{1}{2}x^2}.$$

Example 7.1. Figure 7.2 compares the standard normal density to another normal density with the scale and location parameters as shown. Here is the script that produced Figure 7.2:

```
1  x <- seq(-3, 5, length = 501)
2  plot(x, dnorm(x), type = 'l', ylim = c(0, 1), xlab = 'x',
3      ylab = 'density')
4  lines(x, dnorm(x, 2, .5))
5  text(0, .44, expression(paste(italic(mu[X]) == 0, ', ',
6      italic(sigma[X]) == 1)))
7  text(2, .84, expression(paste(italic(mu[Y]) == 2, ', ',
8      italic(sigma[Y]) == 0.5)))
```

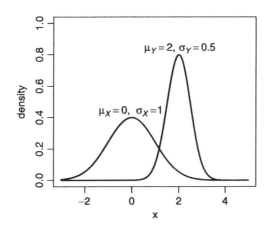

Figure 7.2   The standard normal density compared to another normal density.

Except for the calls to text(), the code is nearly identical to the code that produces Figure 7.1. In particular, the call to expression() in lines 5 to 8 produces the subscripts (see help(plotmath)). This is done with mu[Y] and sigma[Y]. The effect of the square brackets within a call to expression() is to produce the subscripts. Also, because we wish to display two mathematical expression (one for μ and one for σ), we need to paste() their expressions. □

To distinguish a standard normal rv from other rv, we write Z instead of X. It is important to learn how to interpret the areas under the standard normal curve (see also Section 6.6). Here is an example.

208 The normal and sampling densities

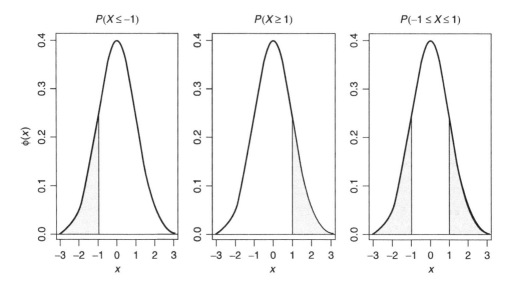

Figure 7.3   Areas under the standard normal density.

Example 7.2. Three areas under the standard normal curve are illustrated in Figure 7.3. The first shows

$$\Phi(-1) = \int_{-\infty}^{-1} \phi(x)\,dx\,.$$

This is the area under the curve from $-\infty$ to -1 standard deviation to the left of the mean. The second shows

$$1 - \Phi(1) = 1 - \int_{-\infty}^{1} \phi(x)\,dx = \int_{1}^{\infty} \phi(x)\,dx\,.$$

The third shows

$$\Phi(1) - \Phi(-1) = \int_{-\infty}^{1} \phi(x)\,dx - \int_{-\infty}^{-1} \phi(x)\,dx\,.$$

This is the area between ± 1 standard deviations away from the mean. Note that all of these areas are expressed in term of standard deviations away from the mean and in terms of the standard normal distribution $\Phi(x)$. The script that produced Figure 7.3 goes like this:

```
pg <- function(x, i){
  if (i == 1){
    x1 <- x[x <= -1] ; y1 <- dnorm(x1)
    x2 <- c(-3, x1, x1[length(x1)], -3)
  }
  if (i == 2){
    x1 <- x[x >= 1] ; y1 <- dnorm(x1)
```

```
8       x2 <- c(1, x1, x1[length(x1)], 1)
9     }
10    if(i == 3){
11      x1 <- x[x >= -1 & x <= 1] ; y1 <- dnorm(x1)
12      x2 <- c(-1, x1, 1, - 1)
13    }
14    y2 <- c(0, y1, 0, 0)
15    polygon(x2, y2, col = 'grey90')
16  }
17
18  xl <- expression(italic(x))
19  yl <- c(expression(italic(phi(x))), '', '')
20  m <- c(expression(italic(P(X <= -1))),
21     expression(italic(P(X >= 1))),
22     expression(paste(italic('P('), italic(-1 <= X),
23     italic(phantom()<= 1), ')', sep = '')))
24  par(mfrow = c(1, 3))
25  x <- seq(-3, 3, length = 501) ; y <- dnorm(x)
26  for(i in 1 : 3){
27    plot(x, y, type = 'l', xlab = xl, ylab = yl[i],
28      main = m[i]) ; abline(h = 0) ; pg(x, i)
29  }
```

Much of the code resembles that of Figures 7.1 and 7.2 and is shown here for the sake of completeness. □

To determine the values of the shaded areas in Figure 7.3 with R, keep the following in mind:

$$\phi(z) = \text{dnorm}(z) , \qquad (7.1)$$
$$\Phi(z) = \text{pnorm}(z) .$$

Example 7.3. To find the probability that Z takes on values ≤ -1.76, that is, $\Phi(-1.76) := P(Z \leq -1.76)$, we do:

```
> pnorm(-1.76)
[1] 0.039204
```

To find the probability that $P(-1.76 < Z \leq 1.76)$, we note that

$$P(-1.76 < Z \leq 1.76) = \Phi(1.76) - \Phi(-1.76)$$
$$= \text{pnorm}(1.76) - \text{pnorm}(-1.76)$$

so[1]

```
> pnorm(1.76) - pnorm(-1.76)
[1] 0.9216
```

---

[1] Because the normal is a continuous density, $\leq$ and $<$ give the same results. To be consistent with discrete densities, we will keep the distinction.

Finally, to determine

$$P(Z \le -1.76) + P(Z > 1.76) = \Phi(1.76) + (1 - \Phi(1.76))$$
$$= \text{pnorm}(1.76) + 1 - \text{pnorm}(-1.76)$$

we use

```
> 1 - pnorm(1.76) + pnorm(-1.76)
[1] 0.078408
```

Of course we chose -1.76 arbitrarily. □

In the next example, we do the reverse of what we did in Example 7.3: instead of finding the probability that Z takes on values to the left, right or between given standard deviations away from the mean, we wish to determine the values of z such that

$$\Phi(z) = p \quad \text{or} \quad \Phi^{-1}(p) = z \ .$$

To accomplish this in R, we use

$$\Phi^{-1}(p) = \text{qnorm(p)} \ . \tag{7.2}$$

Here is an example.

Example 7.4. To determine z such that $\Phi(z) = 0.67$, we do:

```
> qnorm(0.67)
[1] 0.43991
```

To determine z such that $P(Z > z) = 0.05$, we note that it is the same as $\Phi(z) = 0.95$. So

```
> qnorm(0.95)
[1] 1.6449
```

To determine z such that $P(-z < Z \le z) = 0.95$ (recall that the normal is a symmetric density), we use:

```
> c(qnorm(0.025), qnorm(0.975))
[1] -1.9600  1.9600
```

The first element of the vector is $\Phi(-z) = 0.025$ and the second is $P(Z > z) = 0.975$, so that the probability of Z between $-z$ and z is 0.95. We "concatenate" (c()) the results to obtain a vector. □

So far, we dealt with the standard normal. In the next section we explain how to use arbitrary normals; that is, normals with any value of μ and σ.

7.1.2 Arbitrary normal

To determine probabilities (areas under the density curve) for arbitrary normals, i.e. those with mean (μ) and standard deviation (σ) different from 0 and 1, we can standardize the density, find the desired values (as in Examples 7.3 and 7.4) and then reverse the standardization process. If you use R, you do not need to go through the conversion process. Instead of using (7.1) and (7.2), use

$$\phi(x|\mu,\sigma) = \text{dnorm}(x, \text{mu}, \text{sigma}),$$
$$\Phi(x|\mu,\sigma) = \text{pnorm}(x, \text{mu}, \text{sigma})$$

and

$$\Phi^{-1}(p|\mu,\sigma) = \text{qnorm}(p, \text{mu}, \text{sigma}).$$

Example 7.5. Let X be a rv distributed according to the normal with $\mu = 30$ and $\sigma = 2$. We wish to find the probability that X is in the interval $(27, 36]$. Here is how:

```
> mu <- 30 ; sigma <- 2
> pnorm(36, mu, sigma) - pnorm(27, mu, sigma)
[1] 0.9318429
```
□

Let us look at an example where these ideas are applicable.

Example 7.6. The results from the data in this example were published in Krivokapich et al. (1999). The data were obtained from the University of California, Los Angeles Department of Statistics site at http://www.stat.ucla.edu/. The researchers in this study wished to determine if a drug called Dobutamine could be used to test for a risk of having a heart attack. The reason for looking for such a drug is that the normal heart stress test (running on a treadmill) cannot be used with older patients. In all, 553 people participated in the study. One of the variables measured was the base blood pressure—the participant's blood pressure before the test. The histogram of the data with the normal density superimposed is shown in Figure 7.4. Here we refer to the sample of 533 people as the population, *not* as a sample from some population. The data are in Stata format, so we import it with

```
> library(foreign) ; cardiac <- read.dta('cardiac.dta')
```

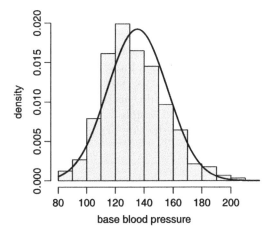

Figure 7.4 Histogram of base blood pressure. Data are from Krivokapich et al. (1999).

The mean and the standard deviation are

```
> (mu <- mean(cardiac$basebp))
[1] 135.3244
> (sigma <- sqrt(sum((cardiac$basebp - mu)^2) /
+    length(cardiac$basebp)))
[1] 20.75149
```

We select a random record from the data and wish to know the probability that the recorded base blood pressure is between 125 and 145 mmHg:

$$P(125 < X \leq 145|\mu, \sigma) = \Phi(145|\mu, \sigma) - \Phi(125|\mu, \sigma)$$
$$= \text{pnorm}(145, \text{mu}, \text{sigma}) -$$
$$\text{pnorm}(125, \text{mu}, \text{sigma})$$

which gives

```
> pnorm(145, mu, sigma) - pnorm(125, mu, sigma)
[1] 0.3692839
```

This we interpret as saying that if we sample a single record from the data many times, about 37% will be in the range between 125 and 145. Here is the script that produced Figure 7.4:

```
1  library(foreign) ; cardiac <- read.dta('cardiac.dta')
2  h(cardiac$basebp, xlab = 'base blood pressure')
3  x <- seq(80, 220, length = 201)
4  mu <- mean(cardiac$basebp)
5  sigma <- sqrt(sum((cardiac$basebp - mu)^2) /
6     length(cardiac$basebp))
7  lines(x, dnorm(x, mu, sigma))
```

The script demonstrates how to import data from Stata version 5-8 or 7/SE binary format into a data frame (Stata is a widely used commercial statistical software). First we load the package `foreign` with a call to `library()` (line 1). Next, we import the Stata binary file `cardiac.dta` to the data frame `cardiac` with a call to `read.dta()`. The remaining code should be familiar by now. Note the direct calculation of the standard deviation in lines 5 and 6. We do this because the functions `var()` and `sd()` in R compute the sample variance and sample standard deviation, not the population variance and standard deviation, i.e. the sum of squares are divided by $n - 1$, *not* by the sample size n.   □

7.1.3 Expectation and variance of the normal

Recall (Section 6.8.2) that $E[x] = \mu$ and $V[x] = \sigma^2$, i.e. the location and scale parameters of the normal density are also its expected value and variance. This may not be the case for other densities. Furthermore, the MLE (Section 5.8) of μ and σ are $\widehat{\mu} = \overline{X}$ and $\widehat{\sigma} = S[X]$, where $S[X]$ is the sample standard deviation. We demonstrate this fact with an example. The example is *not* a proof of the assertion; it is an illustration.

Example 7.7. Consider a normal density with parameters $\mu = 10$ and $\sigma = 2$. We draw a random sample of size $= 100$ from the density and plot a histogram of the sample values. Next, we plot the normal density with $\mu = 10$ and $\sigma = 2$ and with the sample mean $\overline{X}$ and sample standard deviation S (Figure 7.5). Here is the script for this example:

```
1  mu <- 10 ; sigma <- 2 ; x <- seq(0, 20, length = 101)
2  set.seed(4) ; X <- rnorm(100, mu, sigma)
3  h(X, xlab = 'x')
4  lines(x, dnorm(x, mu, sigma), lwd = 3)
5  lines(x, dnorm(x, mean(X), sd(X)))
```

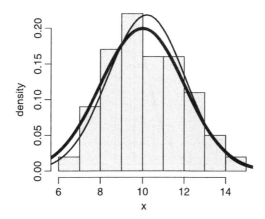

Figure 7.5 Histogram of a sample of 100 random values from the normal density with $\mu = 10$ and $\sigma = 2$. Thick curve shows the normal with μ and σ; thin curve shows the normal with the sample mean and sample standard deviation.

In line 1 we create a vector of x values. In line 2, we set the seed of the random number generator to 4 with a call to set.seed(4). This allows us to repeat the same sequence of random numbers every time we call rnorm(), which we do in line 2. In line 3 we draw a histogram of the 100 values of X, where X is our sample from a population with normal density with mean and standard deviations of 10 and 2. Because X is a sample, its mean and standard deviation are different from mu and sigma. In line 4 we draw the population normal with a call to the normal density dnorm() with the population parameters mu and sigma. To distinguish this line, we plot it thick with the line width named argument lwd = 3. Finally, in line 5 we draw the normal approximation of the sample by calling dnorm() with the sample mean and sample standard deviation with calls to mean() and sd(). □

7.2 Applications of the normal

Because of the central limit theorem (which we discuss in Section 7.6.1), the normal is widely applicable. Here, we discuss a few useful applications.

7.2.1 The normal approximation of discrete densities

Often, we have values of rv from discrete densities. We wish to investigate the normal approximate to these values. This requires that we first bin the data into a range of values, construct a histogram and then fit a normal curve to the histogram. The fit requires finding sample-based values that approximate μ and σ such that the normal curve approximates the middle height and the spread of the histogram bars best. Here is an example.

Example 7.8. The data for this example are fabricated. An observer records the number of animals visiting a water hole in Kruger National Park, South Africa. A total of 1 000 hours were recorded, where the beginning of a particular 1-hour interval was selected at random from a set of integers between 1 and 24. A histogram of the data are shown in Figure 7.6 which was produced with

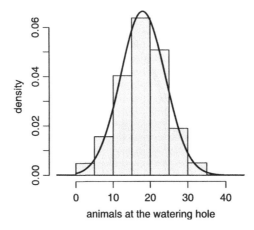

Figure 7.6 Density of the number of animals counted (per hours) in a watering hole at Kruger National Park, South Africa.

```
> set.seed(1) ; y <- rnorm(1000, 18, 6)
> h(y, xlab = 'animals at the watering hole')
> x <- seq(0, 40, length = 1001) ; lines(x, dnorm(x, 18, 6))
```

The mean and standard deviation of the data are 17.93 and 6.21. A plot of the normal density with $\mu = 17.93$ and $\sigma = 6.21$ is superimposed. It seems to fit the data well. Therefore, we accept the data as representing the true density of the number of animals in a watering hole, as opposed to a sample. Because the number of animals per hour is a discrete rv, it makes sense to calculate the probability that $X = 20$ even though the normal density is continuous and $P(X = 20) = 0$. To find this probability, we write

$$\phi(20|17.93, 6.21) = P(19.5 < X \leq 20.5|17.93, 6.21)$$
$$= \text{pnorm}(20.5, 17.93, 6.21) - \text{pnorm}(19.5, 17.93, 6.21)$$

or

```
> mu <- 17.93 ; sigma <- 6.21
> pnorm(20.5, mu, sigma) - pnorm(19.5, mu, sigma)
[1] 0.06071193
```

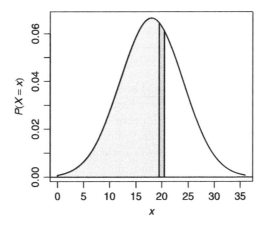

Figure 7.7   Normal approximation to a discrete rv.

The dark area in Figure 7.7 corresponds to the desired probability. Here is the script that produced the figure:

```
1   mu <- 18 ; sigma <- 6
2   boundary <- c(mu - 3 * sigma, mu + 3 * sigma)
3   x <- seq(boundary[1], boundary[2], length = 1001)
4   y <- dnorm(x, mu, sigma)
5   plot(x, y, type='l', xlab = expression(italic(x)),
6     ylab = expression(italic(P(X==x))))
7   abline(h = 0)
8   x1 <- x[x <= 20.5] ; y1 <- dnorm(x1, mu, sigma)
9   x2 <- c(boundary[1], x1, x1[length(x1)], boundary[2])
10  y2 <- c(0, y1, 0, 0)
11  polygon(x2, y2, col = 'grey80')
12  x1 <- x[x <= 19.5]
13  y1 <- dnorm(x1, mu, sigma)
14  x2 <- c(boundary[1], x1, x1[length(x1)], boundary[2])
15  y2 <- c(0, y1, 0, 0)
16  polygon(x2, y2, col = 'grey90')
```

In lines 5 and 7 we plot the density and add a horizontal line at zero. In line 8 to 11 we produce the shaded polygon with a right side at 20.5. In lines 12 to 16 we produce the shaded polygon with right side at 19.5. The polygons are shaded with different grays (`grey80` and `grey90`). □

7.2.2 Normal approximation to the binomial

Under certain conditions, the normal can be used to approximate the binomial. Before stating these conditions, let us convince ourselves that the approximation is valid with an example.

216 The normal and sampling densities

Example 7.9. We set $n = 20$ and $\pi = 0.4$ and calculate the density of the binomial,

$$P(X = x|n, \pi) = \binom{n}{x} \pi^x (1-\pi)^{n-x}$$

for $x = 0, 1, \ldots, 20$ and zero otherwise. The result is the left stick plot in Figure 7.8. Next, we set $\mu = n\pi$ and $\sigma = \sqrt{n\pi(1-\pi)}$ and plot the normal density with parameters μ and σ. The result is the smooth curve in the left panel of Figure 7.8. For the right panel, we set $n = 4$ and $\pi = 0.04$ and calculate the density of the binomial for $x = 0, 1, \ldots, 4$ and zero otherwise. The result is the right stick plot in Figure 7.8. Again, we set $\mu = n\pi$ and $\sigma = \sqrt{n\pi(1-\pi)}$ and plot the normal density with parameters μ and σ. The result is the smooth curve in the right panel of Figure 7.8. In the first case, where $\pi = 0.4$, the approximation looks good; in the second, where $\pi = 0.04$, it does not. The script for this example is:

```
1   par(mfrow = c(1, 2)) ; n <- c(20, 4) ; PI <- c(0.4, 0.04)
2   ylimits <- rbind(c(0, 0.2), c(0, 1.0))
3   ylabel <- c(expression(italic(P(X == x))), '')
4   for(i in 1 : 2){
5     xy <- list(cbind(0 : n[i], dbinom(0 : n[i], n[i], PI[i])),
6       seq(0, n[i], length = 1001))
7     plot(xy[[1]], type = 'h', lwd = 2, ylim = ylimits[i, ],
8       xlab = expression(italic(x)), ylab = ylabel[i])
9     lines(xy[[2]], dnorm(xy[[2]], n[i] * PI[i],
10      sqrt(n[i] * PI[i] * (1 - PI[i]))))
11    abline(h = 0, lwd = 2)
12  }
```

We make no apologies for the fact that it is somewhat terse. In line 1 we set the plotting device to accept 2 figures. We also set the parameters n and PI. In lines 2 and 3 we construct the vectors that hold the limits and labels for the y-axes. In lines 4 to 12 we create the left and right panels of Figure 7.8. We first create a list to hold

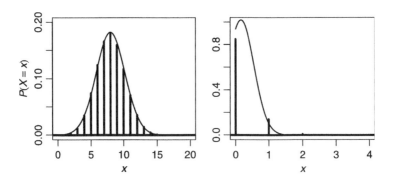

Figure 7.8   Normal approximation to two binomial densities.

the data for the panel. The list contains the following elements: the values of x and y for the plots and the values of x (1 001 of them) that are used in plotting the smooth normal approximation. Then in lines 7 and 8 we construct the sticks (type = 'h') with a line width lwd = 2 and the appropriate limits on the y-axis along with the labels. In lines 9 and 10 we plot the smooth curve of the normal with $\mu = n\pi$ and $\sigma = \sqrt{n\pi(1-\pi)}$. In line 11 we add a horizontal line at $y = 0$ to indicate that the binomial is defined for $X \in \mathbb{R}$. □

Example 7.9 illustrates a well-known theorem with the following application:

The normal approximation to the binomial Let the number of successes X be a binomial rv with parameters n and π. Also, let

$$\mu = n\pi, \quad \sigma = \sqrt{n\pi(1-\pi)}.$$

Then if

$$n\pi \geq 5, \quad n(1-\pi) \geq 5,$$

we consider $\phi(x|\mu,\sigma)$ an acceptable approximation of the binomial.

Let $m_1, m_2 = 0, 1, \ldots, n$. Then under the conditions above, we use the following normal approximation to calculate binomial probabilities of interest:

$$P(m_1 < X \leq m_2|\mu,\sigma) = \Phi(m_2 + 0.5|\mu,\sigma) - \Phi(m_1 - 0.5|\mu,\sigma)$$
$$= \text{pnorm}(\text{m.2} + 0.5, \text{mu}, \text{sigma})$$
$$- \text{pnorm}(\text{m.1} - 0.5, \text{mu}, \text{sigma}).$$

For left-tail probability, we use

$$P(X \leq m_2|\mu,\sigma) = \Phi(m_2 + 0.5|\mu,\sigma)$$
$$= \text{pnorm}(\text{m.2} + 0.5, \text{mu}, \text{sigma})$$

and for right-tail probability we use

$$P(X > m_1|\mu,\sigma) = 1 - \Phi(m_1 - 0.5|\mu,\sigma)$$
$$= 1 - \text{pnorm}(\text{m.1} - 0.5, \text{mu}, \text{sigma}).$$

Example 7.10. In Example 5.13 we discussed a survey of 1 550 men in the UK. Of these, 26% were smokers (Lader and Meltzer, 2002). We assume that the survey results represent the UK population of men and that $\pi = 0.26$ is the proportion of English men who smoke. Now we pick 250 English men randomly and assume that they smoke independently (this assumption will not be true if the sample was taken from a small area, or from within families). Thus, we have a binomial density with $n = 250$ and $\pi = 0.26$. We wish to establish the probability that between 80 and 90 people in the sample will turn out to be smokers. First, we check that $n \times \pi = 250 \times 0.26 = 65 > 5$. Also, $n \times (1-\pi) = 250 \times 0.74 = 185 > 5$. Therefore, we may use the normal approximation to the binomial. So

$$\mu = 65, \quad \sigma = \sqrt{250 \times 0.26 \times 0.74} \approx 6.94.$$

Because

```
> PI <- 0.26 ; n <- 250 ; mu <- n * PI
> sigma <- sqrt(n * PI * (1 - PI))
> (p.approx <- pnorm(90.5, mu, sigma) -
+   pnorm(79.5, mu, sigma))
[1] 0.01815858
```

we conclude that
$$P(80 \leq X \leq 90|250, 0.26) \approx 0.018 \,.$$

The exact value is obtained with

```
> (p.exact <- pbinom(90, n, PI) - pbinom(79, n, PI))
[1] 0.01968486
```

This results in a

```
> (p.approx - p.exact) / p.exact * 100
[1] -7.753617
```

% underestimate. The probability that more than 50 people in the sample are smokers is
$$P(X > 49.5|250, 0.26) = 1 - \Phi(49.5|250, 0.26)$$
$$\approx 0.99 \,;$$

i.e.

```
>  1- pnorm(49.5, 65, 6.94)
[1] 0.98724
```

The probability that less than 45 people in the sample are smokers is
$$\Phi(45.5|250, 0.26) \approx 0.002 \,;$$

i.e.

```
> pnorm(45.5, 65, 6.94)
[1] 0.0024786
```

Note the adjustments with 0.5. We use them because the rv (number of smokers) is discrete. □

7.2.3 The normal approximation to the Poisson

In Section 5.7 we saw that one way to write the Poisson density is
$$P(X = x|\lambda) = \frac{\lambda^x}{x!}e^{-\lambda}$$

for $x \in \mathbb{Z}_{0+} \cap \mathbb{R}$ and zero otherwise. We also saw that under some conditions, the Poisson approximates the binomial. Because the normal approximates the binomial, we expect that the normal also approximates the Poisson. Here is an example.

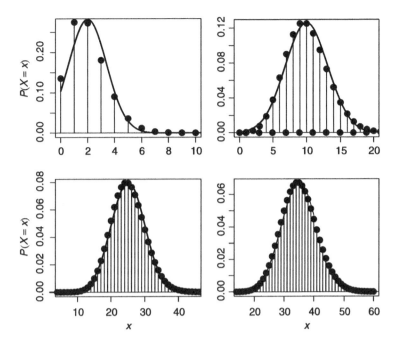

Figure 7.9 Normal approximation to the Poisson.

Example 7.11. Figure 7.9 illustrates the normal approximation to Poisson for $\lambda = 2, 10, 25, 35$. Note the increasingly better approximation. Here is the script for this example:

```
1   rm(list = ls())
2   x <- 0 : 60 ; lambda <- c(2, 10, 25, 35)
3   y <- seq(0, 40, length = 501)
4   xl <- expression(italic(x)) ; xlabel <- c('', '', xl, xl)
5   yl <- expression(italic(P(X==x)))
6   ylabel <- c(yl, '', yl, '')
7   xlimit <- rbind(c(0, 10), c(0, 20), c(5, 45), c(15, 60))
8   par(mfrow = c(2, 2))
9   for(i in 1 : 4){
10    plot(x, dpois(x, lambda[i]), type = 'h', xlab = xlabel[i],
11      ylab = ylabel[i], xlim = xlimit[i, ]) ; abline(h = 0)
12    points(x, dpois(x, lambda[i]), pch = 19, cex = 1.5)
13    lines(y, dnorm(y, lambda[i], sqrt(lambda[i])))
14  }
```

In lines 1–7 we prepare for the four panels in Figure 7.9 and in lines 8–13 we draw them. □

Example 7.11 illustrates a well-known theorem with the following application:

The normal approximation to the Poisson Let the number of counts per interval unit, X, be a Poisson rv with parameter λ. If $\lambda \geq 20$, then we consider $\phi\left(x \mid \lambda, \sqrt{\lambda}\right)$ an acceptable approximation of the Poisson.

7.2.4 Testing for normality

Some of the classical statistical methods we will discuss rely on the assumption that a sample of n observations is drawn from a normal population. Thus, we often need to test for normality. One useful, albeit informal, test is with normal scores.

Normal scores

Suppose that you chose a sample of size $n = 10$ from a population distributed according to the standard normal. Sort the 10 values from smallest to largest. Next, choose another sample of 10 and sort them from smallest to largest. Repeat the process, say, 100 times and then calculate the mean for the smallest value, the next smallest values, up to the largest value (the 10th value). If you repeat the process many times, you will come up with the following 10 means (from smallest to largest) in many samples of 10 values from standard normal:

```
> library(SuppDists)
> normOrder(10)
 [1] -1.5387755 -1.0013504 -0.6560645 -0.3757113 -0.1226703
 [6]  0.1226703  0.3757113  0.6560645  1.0013504  1.5387755
```

(SuppDists is a package of supplemental distributions and normOrder() is the density of normal order scores). Next, consider a sample of size 10 from a normal density with μ and σ. First, standardize the sample values. Suppose that the sample values represent a perfect normal. Then the standardized values represent a perfect standard normal. Therefore, your sample should have the same sequence of numbers as above. A plot of the sample against the expected values above should result in the points aligned along a 45° straight line. Because no sample is perfect, the points will not align exactly. Yet, this approach can be used to judge how well the data approximate the normal density.

Example 7.12. Personal data of 10 bill lengths (in cm) of a song bird species from the Sierra Nevada are

```
> bill.length <- c(2.50, 2.83, 2.95, 3.24, 3.32, 3.43, 3.60,
+      3.82, 4.00, 4.40)
```

A plot the data against standard normal scores, with a best-fit line, is shown in Figure 7.10. Apparently the data are from a normal density. The plot was achieved with:

```
> score <- normOrder(10)
> plot(score, bill.length) ; r = lm(bill.length ~ score)
> abline(reg = r)
```

The score vector is created in the first statement and plotted against the data in the second. To show the best fit line between the data and the scores, we create a linear

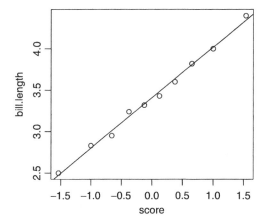

Figure 7.10   Bill lengths vs. normal scores.

model (we shall talk about linear models and regression later) with a call to `lm()` in the third statement. Here `bill.length` and `score` represent variables. They are separated by the ~ symbol. This indicates to R that the left and right hand sides relate though a formula—in our case, the linear formula `bill.length` $= a + b \times$ `score`, where a and b are to be computed by `lm()`. Finally, we add the regression line with a call to `abline()` with the named argument `reg`. Such a call extracts the regression parameters a and b from the object `r` and plots the appropriate line.   □

Q-Q plots

The quantile-quantile (Q-Q) plot is a graphical technique. It is used to determine if two data sets come from populations with a common density. We shall use it to examine normality. Q-Q plots are sometimes called probability plots, especially when data are examined against a theoretical density. To construct the plot, we use quantiles, defined as follows:

x%-quantile We say that q is the $x\%$ quantile if $x\%$ of the data values are $\leq q$.

Here is the value of the 10% quantile of the standard normal:

```
> qnorm(0.1)
[1] -1.281552
```

and here is the 50% quantile (the median) of the standard normal:

```
> qnorm(0.5)
[1] 0
```

We plot the quantiles of the normal against the quantiles of the data. If the data come from a population with a normal density, the points should fall along a straight line. This should be true for at least the points in the interquartile range (between the 25 and 75% quantiles). Interpretation of Q-Q plots requires some experience. The next series of examples will give you a feel for how to judge the density of data against the normal with Q-Q plots. We will compare normal data to normal density and centered, right and left skewed data to the normal density.

222 The normal and sampling densities

Example 7.13. Here we produce "empirical" data from a normal density and compare them to the theoretical normal density (Figure 7.11). The left panel is drawn with

```
> par(mfrow = c(1, 2))
> set.seed(1) ; x <- rnorm(101, .5, .15)
> h(x, xlab = 'x') ; y <- seq(0, 1, length = 101)
> lines(y, dnorm(y, .5, .15), type = 'l')
```

and the right with

```
> qqnorm(x, main = '') ; qqline(x)
```

From the left panel we conclude that the empirical density corresponds to the theoretical density. The right panel supports this conclusion. □

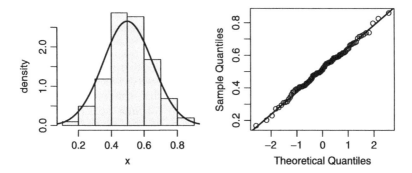

Figure 7.11 Theoretical (normal) vs. a random sample from the theoretical density (left) and the corresponding Q-Q plot.

Now let us see an example with data from a density with tails fatter than those of the normal.

Example 7.14. The data are from a centered density with no tails (we choose the beta; see Section 6.8.8) are compared to the normal. Figure 7.12 was produced with

x <- rnorm(101, .5, .15)

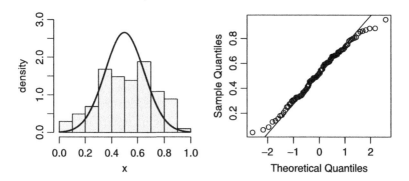

Figure 7.12 Theoretical (normal) vs. a random sample of values from a tail-less density (left) and the corresponding Q-Q plot.

in the code on page 222 replaced with

```
x <- rbeta(101, 2, 2)
```

Because the tails are not as flat as those of the normal, we see departures from the Q-Q line at both ends (right panel of Figure 7.12). □

Next, we examine the case where the data are from a density that is skewed to the left.

Example 7.15. From the right panel of Figure 7.13 we conclude that left-skewed data produce a larger departure at the lower tail than otherwise. To produce the figure, use

```
x <- rbeta(101, 1.5, 3)
```

instead of rnorm() in the code on page 222 □

Finally, let us compare right-skewed data to the normal.

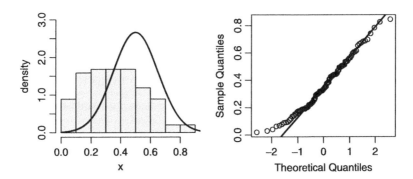

Figure 7.13   Theoretical (normal) vs. a random sample of values from a left-skewed density (left) and the corresponding Q-Q plot.

Example 7.16. From the right panel of Figure 7.14 we conclude that data from density with tails fatter than the normal on the right display the shown departure from the Q-Q-line. To produce the figure, use

```
x <- rbeta(101, 6, 3)
```

instead of rnorm() in the code on page 222 □

Other tests of normality

The tests of normality discussed thus far are semi formal. Formal tests, which we discuss only briefly here, are the one-sample Kolmogorov-Smirnov (K-S) test (Chakravarti et al., 1967), the Anderson-Darling test (Stephens, 1974) and the Shapiro-Wilk normality test (Shapiro and Wilk, 1965). The latter is used most frequently.

The Kolmogorov-Smirnov (K-S) test is used to decide if a sample comes from a population with a specific density. It is used because the distribution of the K-S test statistic does not depend on the underlying cumulative distribution function being tested and because it is an exact test. Its limitations are: The test applies to continuous

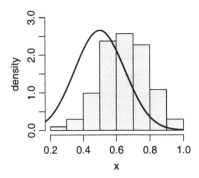 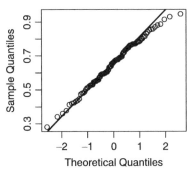

Figure 7.14 Theoretical (normal) vs. a random sample of values from a right-skewed density (left) and the corresponding Q-Q plot.

densities only; it is more sensitive near the center of the density than at the tails than other tests; the location, scale and shape parameters cannot be estimated from the data (if they are, then the test is no longer valid).

The Anderson-Darling test is used to test if a sample of data comes from a specific density. It is a modification of the K-S test and gives more weight to the tails of the density than does the K-S test. It is generally preferable to the K-S test.

Example 7.17. The results of one of the midterm tests in a statistics class were as follows:

```
> midterm
 [1]  61  69  55  47  49  58  66  57  73  56  45  67  88  61
[15]  62  85  83  76  71  82  84  73  86  81  57  74  89  71
[29]  93  59  89 108  90  87  71
```

Here,

```
> round( c(mean = mean(midterm), sd = sd(midterm)), 1)
mean   sd
72.1 15.0
```

Is the density of the test scores normal? We run the K-S test thus:

```
> ks.test(midterm, 'pnorm', mean(midterm), sd(midterm))

        One-sample Kolmogorov-Smirnov test

data:  midterm
D = 0.0958, p-value = 0.9051
alternative hypothesis: two.sided

Warning message:
cannot compute correct p-values with ties in:
ks.test(midterm, "pnorm", mean(midterm), sd(midterm))
```

We will learn to interpret the results later. For now, suffice it to say that a large p-value (larger than, say, 0.05) indicates that the sample is not different from normal with the sample's mean and standard deviation. □

To find out about the Anderson Darling test in R, we do

```
> help.search('Anderson Darling')
```

which tells us that

ad.test(nortest)
                       Anderson-Darling test for
                       normality

In other words, the ad.test() resides in the package nortest. The Shapiro-Wilk test can be applied with shapiro.test().

7.3 Data transformations

Many of the statistical tests that we use rely on the assumption that the data, or some function of them, are distributed normally. When they do not, we may transform the data to obtain normality. Once the statistical procedure is applied, we then reverse the transformation so that we may interpret the data.

Log and square root transformations are used routinely. Data with values ≤ 0 cannot be thus transformed. Both transformations reduce the variability in the data. In bivariate (or multivariate) data analysis, transformations are applied to various variables differently to arrive at appropriate linear models.

Example 7.18. Figure 7.15 illustrates the log and square root transformation. Observe the better (but not satisfactory) conformity to the assumption of linearity of the normal Q-Q plot of the log-transformed data. The square root transformation looks even better. The figure was obtained with this script:

```
1  par(mfrow = c(3, 2)) ; set.seed(5) ; x<- rexp(100, .5)
2  xx <- list(x = x, 'log(x)' = log(x), 'sqrt(x)' = sqrt(x))
3  xlabels <- c('x', 'log(x)', 'sqrt(x)')
4  for(i in 1 : 3){
5    h(xx[[i]], xlab = xlabels[i])
6    qqnorm(xx[[i]], main = '') ; qqline(xx[[i]])
7  }
```

In line 1 we set the graphics window to accept the 6 panels, set.seed() and generate 100 random values from the exponential density (see Section 6.2) with parameter $\lambda = 0.5$. In line 2 we create a list with each component corresponding to x, $\log(x)$ and $\sqrt{x}$. In lines 4–7 we draw the six panels. To test x and its transformations for normality, we use the Shapiro-Wilk test:

```
> test <- mapply(shapiro.test, xx)
> rbind(x = unlist(test[1 : 2, 1]),
+       'log(x)' = unlist(test[1 : 2, 2]),
+       'sqrt(x)' = unlist(test[1 : 2, 3]))
         statistic.W    p.value
x             0.8482443 1.021926e-08
log(x)        0.9373793 1.339048e-04
sqrt(x)       0.9786933 1.050974e-01
```

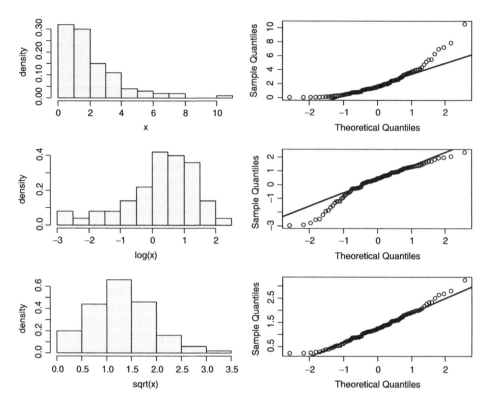

Figure 7.15 Top: histogram of 100 random points from exponential density with parameter $\lambda = 0.5$ and the corresponding normal Q-Q plot. Middle: The log-transformed data. Bottom: the square root transformed data.

We will discuss hypothesis testing and p-values later (in Section 10.5). Suffice it to say that the p-value of the square root transformed data is the largest. Therefore, the square root transformation does the best job of conforming to normality. In the code above, mapply() applies shapiro.test() to each component of the list xx and stores the results in the list test. For easier access to the components of test, we unlist() it and row bind the results with rbind(). □

See Venables and Ripley (2002) for further details about data transformations.

7.4 Random samples and sampling densities

One of the fundamental assumptions we make when we sample a population is that the sample is random. Furthermore, if a sample is random, any computed value from it is a rv, which, in turn, has a density. In talking about random samples and sampling densities, we must distinguish between a theoretical density, a density of some trait in the population and the corresponding density in a subset of the population.

7.4.1 Random samples

Let a population consist of $N \in \mathbb{Z}_+$ objects. We are interested in the population density of values, $x \in \mathbb{R}$, of a certain property of these objects. Suppose that the theoretical density of x is $P(X = x|\boldsymbol{\theta})$ where $\boldsymbol{\theta} := [\theta_1, \ldots, \theta_m]$ m. Pick a uniform random value, p_1, between zero and one and assign $x_1 = P^{-1}(p_1)$ to one object in the population. Repeat this process for all N objects. Then

$$\lim_{N \to \infty} P(X = x_i|\boldsymbol{\theta}) = P(X = x|\boldsymbol{\theta}) \ , \quad i = 1, \ldots, n$$

where on the left is the density of x in the population and on the right is the theoretical density. To bring events into the picture, we simply let E_i, $i = 1, \ldots, N$, be the event that the ith chosen object has a property value x so that $X(E_i) = x_i$. From now on, we shall write X instead of $X(E_i)$.

Example 7.19. Consider a population of 3×10^8 people (about the U.S. population at the time of writing). Let E_i be the event that the ith person in the population is h cm tall. Then it makes sense to define the rv $X(E_i) = E_i$. It is well known that the theoretical density of height is $\phi(x|\mu, \sigma)$ where μ and σ are the mean and standard deviation of height in an infinite population. Note that X is not discrete because we assume a theoretical density of a population with $N \to \infty$. □

Now let the experiment be choosing $n < N$ objects from the population, i.e.

$$E := \{E_1, \ldots, E_n\} \ .$$

We define a

Sample Let $n < N$. Then

$$\boldsymbol{X} := [X_1, \ldots, X_n]$$

is a sample.

Samples may be *dependent* or *independent*.

Independent sample If

$$P(X_1 = x_1|\boldsymbol{\theta}) = \cdots = P(X_n = x_n|\boldsymbol{\theta})$$

then we say that the sample is independent. Otherwise it is *dependent*.

Example 7.20. A sample of objects from a small population without replacement is dependent. For all practical purposes, a sample from a large population without replacement is independent. □

Example 7.20 points to the need of the following:

Rule of thumb Let $\boldsymbol{X}$ and $\boldsymbol{Y}$ be samples without and with replacement, each of size n from a population of size N. Then if $P(X_i = x_i|\boldsymbol{\theta}) \geq 0.95 \times P(Y_i = y_i|\boldsymbol{\theta})$, $i = 1, \ldots, n$, we consider $\boldsymbol{X}$ an independent sample.

The density $P(X_i = x_i|\boldsymbol{\theta}) = p_i$, $i = 1, \ldots, n$ (where n is the sample size) has the inverse $P^{-1}(p_i|\boldsymbol{\theta}) = x_i$.

Simple random sample Consider a population of N objects and draw from it a sample of n objects. If each individual in the population has the same probability of being chosen for the sample, then we say that the sample is a simple random sample, or *random sample* for short.

We will mostly deal with independent samples where each object has the same probability of being chosen. Here is an example of how to obtain a random sample.

Example 7.21. We have a population of 1 000 objects. Each object is assigned an index, i ($i = 1, \ldots, 1\,000$). To draw a sample of 10 objects from the population, do:

```
> x <- 1 : 1000
> sample(x, 10)
[1] 194 774 409 234 591 700 684 582 272  21
```

The first statement produces the "names" of the objects. Then we take a sample of 10, which in this case happens to be objects numbered 194, 774 and so on. □

From the discussion in Section 6.7, we conclude that a sample needs to be random if the density of values of a trait of interest in the sample is to represent its density in the population. This brings us to the concept of

Statistic A statistic is a computed value from a random sample.

Formally, a statistic, Y, is defined as a function, $f(\boldsymbol{X})$ that maps $\boldsymbol{X} \in \mathbb{R}^n$ to $Y \in \mathbb{R}$.

Example 7.22. The mean of a random sample

$$\overline{X}(\boldsymbol{X}) := \sum_{i=1}^n p_i X_i = \frac{1}{n} \sum_{i=1}^n X_i$$

is a function of all the sample's values and is therefore a statistic. So is the variance of a random sample

$$V(\boldsymbol{X}) = \sum_{i=1}^n p_i \left(X_i - \overline{X}\right)^2 = \frac{1}{n} \sum_{i=1}^n \left(X_i - \overline{X}\right)^2 .$$

From here on, we shall denote the mean and variance of a sample by $\overline{X}$ and V with the understanding that they are in fact functions of the sample $\boldsymbol{X}$. Note that $V(\boldsymbol{X})$ is not the same as the sample-based best estimate of the population variance, $S^2 := \widehat{\sigma}^2$, where the denominator is $1/(n-1)$, *not* $1/n$. □

The relationships between the sample, population and theoretical densities are illustrated in Figure 7.16. As $n \to N$, the density of the sample approaches that of the population. Similarly, as $N \to \infty$, the population density approaches the theoretical density. Because they reflect values of objects, the sample and population densities can be obtained only from density histograms of the property values of the population objects.

7.4.2 Sampling densities

Because a statistic is a rv, it also has a density. We thus have

Sampling density The probability density of a statistic is called the sampling density of the statistic.

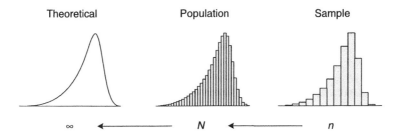

Figure 7.16   From left to right: theoretical, population and sample densities.

A *sampling density* is a property of a statistic, which itself is obtained from a sample. A *sample density* is a property of the sample values, not of a statistic of the sample values. We obtain the sample density from a single sample and the sampling density from many samples, each used to calculate a single value of the statistic. The sampling density of samples' statistic from a single population is not unique. It depends on the sampling scheme. Because we will always use simple samples, we will not be concerned with this issue.

What is the sampling density of a statistic? As we shall see, for some statistics, the sampling densities are known. When we do not know the sampling density, we can always resort to numerical methods to determine it. In the next example, we learn how to construct the sampling density of the variance for samples from an exponential density. We will do the computations inefficiently. The next section will teach you how to execute them efficiently.

Example 7.23. The left and right panels of Figure 7.17 show the sampling density of the variance for samples from the exponential density. We shall see soon why the right panel is there. To learn how to construct a sampling density, let us examine the relevant code snippet:

```
> set.seed(10)
> v <- vector()
> for(i in 1 : 50000) v[i] <- var(rexp(20))
```

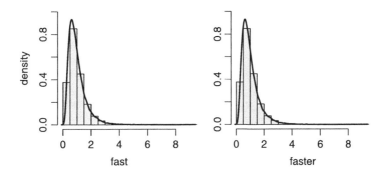

Figure 7.17   The sampling density of sample variance for samples from an exponential density.

230   The normal and sampling densities

The loop produces a vector of 50 000 variances, each computed for a random sample of 20 from the exponential (rexp(20)) with the default parameter value $\lambda = 1$. We shall talk about the remaining code that produces Figure 7.17 in the next section.   □

7.5 A detour: using R efficiently

If you run the code in Example 7.23 you will discover that its execution is slow. Let us see how we can improve on such code.

7.5.1 Avoiding loops

As we discussed,

```
> v <- vector()
> for(i in 1 : 50000) v[i] <- var(rexp(20))
```

produces a vector of variances from 50 000 repetitions of a random sample of 20 values from the exponential density with the parameter $\lambda = 1$. We can accomplish the same task in about half as much time with

```
> v <- rexp(50000 * 20) ; g.l <- gl(50000, 20)
> mapply(var, split(v, g.l))
```

First, we generate a vector, v, of random values from the exponential density. Next, we use gl() (for generate levels) to generate a factor vector (g.l) of 50 000 levels, each repeated 20 times. We split() v into a list of 50 000 components, each of length 20. Finally, we mapply() the function var() to each component of the list. This produces the same results as the loop (compare the left and right panels of Figure 7.17). We claim that avoiding the for loop cuts execution time by about half. In the next section, we prove this statement.

7.5.2 Timing execution

Here is the script that produces Figure 7.17 and times execution:

```
1  par(mfrow = c(1,2)) ; ylimits <- c(0, 1)
2  R <- 50000 ; n <- 20
3
4  s.loop <- function(R, n){
5    set.seed(10)
6    v <- vector()
7    for(i in 1 : R) v[i] <- var(rexp(n)) ; v
8  }
9  fast <- system.time((v <- s.loop(R, n)))[1 : 3]
10 h(v, xlab = 'fast', ylim = ylimits) ; lines(density(v))
11
12 s <- function(R, n){
```

```
13      set.seed(10)
14      g.l <- gl(R, n); v <- rexp(R * n)
15      mapply(var, split(v, g.l))
16    }
17    faster <- system.time((v <- s(R, n)))[1 : 3]
18    h(v, xlab = 'faster', ylab = '', ylim = ylimits)
19    lines(density(v))
20
21    s.1 <- function(R, n){
22      set.seed(10)
23      m <- matrix(rexp(R * n), nrow = n, ncol = R)
24      apply(m, 2, var)
25    }
26
27    fastest <- system.time((v <- s.1(R, n)))[1 : 3]
28    h(v, xlab = 'fastest', ylab = '', ylim = ylimits)
29    lines(density(v))
30
31
32    cpu <- rbind(faster, fastest)
33    dimnames(cpu)[[2]] <- c('user', 'system', 'elapsed')
34    print(cpu)
35
```

In line 2, we set the number of repetitions to 50 000 and the sample size to 20. In lines 4 to 8, we declare the function s.loop(). It produces the data from which we obtain the sampling density using the for loop. In line 9, we assign the data that s.loop() produces to the vector of variances, v and wrap the assignment with the function system.time(). This function returns a vector whose first three elements contain the amount of time the central processing unit (CPU) spent on user related tasks, system-related tasks and the time elapsed from beginning to end of executing its argument. The argument to system.time() is any valid R expression. In our case, the argument is the call to s.loop() and the assignment to v. The first three elements of the vector that system.time() returns are assigned to fast (still in line 9). In line 10 we plot the density of v and fit a smooth density to it with a call to density(), hence the left panel of Figure 7.17.

The same tasks accomplished in lines 4–10 are accomplished in lines 12–19, this time with mapply() (compare the left and right panels in Figure 7.17). Here are the CPU times obtained from the script:

```
         user system elapsed
fast    18.27   1.06   19.45
faster   9.21   0.00    9.22
```

The execution times are not the same for different calls to system.time() with the same expression because the CPU's background tasks differ. The changes, however, are small.

Another way to avoid the loop is with

```
> s.1 <- function(R, n){
+     set.seed(10)
+     m <- matrix(rexp(R * n), nrow = n, ncol = R)
+     apply(m, 2, var)
+ }
```

where here we put the random exponential values in a matrix and `apply()` `var()` to the matrix columns (50 000 of them) with the unnamed argument constant 2. This approach, is slightly slower than the list approach. There is a subtle point here that we shall not pursue. You are invited to explore it by switching the values of R and n in line 2.

7.6 The sampling density of the mean

The law of large numbers and the central limit theorem are often referred to as the fundamental laws of probability. As we shall see, much of statistical inference relies on the central limit theorem.

7.6.1 The central limit theorem

Let $X_1, \ldots, X_n$ be a set of n independent rv. Each of these rv has an arbitrary density with mean μ_i and finite variance σ_i^2. Then the limiting density of $X := X_1 + \cdots + X_n$ is normal with mean and standard deviation

$$\sum_{i=1}^{n} \mu_i, \quad \sqrt{\sum_{i=1}^{n} \sigma_i^2}.$$

By limiting density we mean that as $n \to \infty$, the density of X approaches the normal.

If X_i are independent and identically distributed, then all μ_i are equal (denote them by μ) and all σ_i^2 are equal (denote them by σ^2). Now the limiting density of X is normal with mean and standard deviation

$$n\mu, \quad \sigma\sqrt{n}.$$

In practice, the approach to normality is fast. When n is about 30 we may already use the assumption that X is normal. Note that we place no restrictions on the probability density of the population but one: that it has a finite variance.

7.6.2 The sampling density

Consider a finite sample of size n from a population with mean μ and standard deviation σ. Denote the mean of this sample by $\overline{X}$. Because $\overline{X}$ is a rv, it has a density. We call it the *sampling density of the sample mean*, or, for short, the *sampling density of the mean*. This density has a mean and a standard deviation.

The sampling density of the mean Let $\overline{X}$ be the mean of a sample of size n from a population with arbitrary density with mean μ and standard deviation σ. Then the sampling density of $\overline{X}$ is

$$\phi\left(\overline{X} \,\middle|\, \mu, \frac{\sigma}{\sqrt{n}}\right).$$

To see this, we note that according to the central limit theorem

$$\lim_{n \to \infty} E\left[X_1 + \cdots + X_n\right] = n\mu.$$

Therefore,

$$\lim_{n \to \infty} E\left[\frac{1}{n}(X_1 + \cdots + X_n)\right] = \frac{1}{n} \lim_{n \to \infty} E\left[(X_1 + \cdots + X_n)\right]$$
$$= \frac{1}{n} n\mu = \mu.$$

Also,

$$\lim_{n \to \infty} V\left[X_1 + \cdots + X_n\right] = n\sigma^2.$$

Therefore,

$$\lim_{n \to \infty} V\left[\frac{1}{n}(X_1 + \cdots + X_n)\right] = \frac{1}{n^2} \lim_{n \to \infty} V\left[X_1 + \cdots + X_n\right]$$
$$= \frac{1}{n^2} n\sigma^2 = \frac{\sigma^2}{n}.$$

The standard deviation of the sampling density of the mean plays an important role in statistics. It is thus goes by the name

The standard error is defined as $\sigma/\sqrt{n}$ where n is the sample size and σ is the population standard deviation.

The following example demonstrates that the sampling density of a sample mean from an arbitrary population probability density is normal with mean μ and standard error $\sigma/\sqrt{n}$.

Example 7.24. We introduced the data about the U.S. Department of Defense confirmed reports of U.S. military casualties in Iraq in Example 2.16 (see http://icasualties.org). Let us examine the density of the days between consecutive reported casualties. One of the columns of the data frame includes the Julian day of the reported casualty. So here is what we do:

```
> par(mfrow = c(1, 2))
> load('casualties.rda')
> h(d <- diff(casualties$Julian), xlab = 't')
```

After loading the data, we use `diff()` to difference consecutive Julian dates, store the new vector in d and draw the density (see left panel of Figure 7.18). The density is reminiscent of the exponential, but it has holes (zeros) in it and drops more precipitously than the exponential does. In short, it looks like none of the theoretical

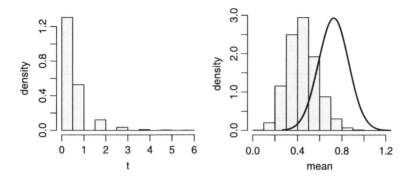

Figure 7.18 The density of the "population" of days between consecutively reported U.S. military casualties in Iraq.

densities we discussed thus far. Next, we draw 10 000 samples from the casualties "population" (with replacement of course),[2] each of size 30:

```
> set.seed(10)
> m <- matrix(sample(d, 30 * 10000, replace = TRUE),
+     nrow = 30, ncol = 10000)
> h(apply(m, 2, mean), xlab = 'mean')
```

From these, we construct the sampling density of the mean (right panel of Figure 7.18). To verify that this density is approximately normal with

```
> c(mu = mean(d), sigma = sd(d))
      mu     sigma
0.4608985 0.7457504
```

we superimpose $\phi\left(\overline{X}\,|\,\mu, \sigma/\sqrt{n}\right)$:

```
> x <- seq(0, 1.2, length = 201)
> lines(x, dnorm(x, mean(d), sd(d) / sqrt(n)))
```

This is not a proof that the sampling density is normal, it is an experimental illustration of a well known result from mathematical statistics. □

7.6.3 Consequences of the central limit theorem

The discussion so far focused on large samples. We know that the sampling density of the mean for samples from a population with mean μ and standard deviation σ is normal with μ and $\sigma/\sqrt{n}$. The sampling density of means for small samples (say $n < 30$) from a normal population with μ and σ is also normal with μ and $\sigma/\sqrt{n}$. We can now summarize the consequences of the central limit theorem for the sampling density of the mean.

Denote by $\mu_{\overline{X}}$ the mean of the sampling density of a sample mean and by $\sigma_{\overline{X}}$ its standard deviation. The corresponding population parameters are μ and σ. Then:

1. $\mu_{\overline{X}} = \mu$: the sampling density of $\overline{X}$ is centered at μ.
2. $\sigma_{\overline{X}} = \sigma/\sqrt{n}$: as the sample size increases, the standard deviation of the sampling density (the standard error) of $\overline{X}$ decreases.

---

[2] At the time of writing, thank goodness, the number of casualties was far less than 10 000.

3. When n is large, the sampling density of $\overline{X}$ approximates the normal, regardless of the probability density of the population.
4. When the population density is normal, so is the sampling density of $\overline{X}$ for any sample size n.
5. When the population density is not normal and n is small, we cannot assume that the sampling density of the mean is normal.

In practice, we have

A rule of thumb For a sample size of $n > 30$, we consider the sampling density of the mean from any population density with mean μ and standard deviation σ to be normal with mean μ and standard deviation $\sigma/\sqrt{n}$.

With this in mind, we can compute probabilities for ranges of values.

Example 7.25. The results of the final examination in a statistics course were $\mu = 52.5$, $\sigma = 12.1$ and there were 35 students in the class. Consider the class as the population. We wish to determine the probability that a student's score in a randomly picked sample of 10 scores will be between 45 and 55.
Based on the sampling density of the mean, we have

$$P\left(45 < X \leq 55 \,\middle|\, 52.5, 12.1/\sqrt{10}\right) = \Phi\left(55 \,\middle|\, 52.5, 12.1/\sqrt{10}\right) \\ - \Phi\left(45 \,\middle|\, 52.5, 12.1/\sqrt{10}\right)$$

and the answer is

```
> mu <- 52.5 ; se <- 12.1 / sqrt(10)
> pnorm(55, mu, se) - pnorm(45, mu, se)
[1] 0.71825
```

As usual, we interpret this result to mean that if we take many many samples of 10, then about 72% of the students in the samples will have scores between 45 and 55%. □

When n is small and the population density is not normal, all we can say is that

1. $\mu_{\overline{X}} = \mu$: the mean of the sampling density of the mean equals the mean of the population.
2. $\sigma_{\overline{X}} = \sigma/\sqrt{n}$: the standard deviation of the sampling density of the mean equals the standard error.

In other words, we do not know the sampling density of $\overline{X}$. One way around this is to proceed with the bootstrap (see, for example, Efron and Tibshirani, 1993). We shall return to the bootstrap method later.

7.7 The sampling density of proportion

In statistical analyses, we are often interested in the proportions of individuals in a population that exhibit a certain trait. For example, we may be interested in the proportion of young fish in the population, the proportion of the population that voted, the proportion of the sample that is recaptured after marking, the proportion of birds' tag return and so on. In such cases, we label the object of concern (a person, a bird, a trap, a patient, an answer) that exhibits the trait as S (success) and the one that does not as F (failure). Here S and F refer to events, *not* to statistics. We denote the proportion of S in the population by π and in the sample by p where by definition, p is a rv. Here is one case where we stray from our usual convention that upper case letters denote rv for p is a sample-based function and is therefore a rv.

7.7.1 The sampling density

We wish to study the sampling density of p where

$$p := \frac{\text{number of successes in the sample}}{\text{sample size}} = \frac{n_S}{n}$$

(n_S is a rv). It turns out that, with some constraints, we can approximate the binomial with the normal (see Section 7.2.2). With these constraints, the central limit theorem applies to the sampling density of proportions.

Recall that for $n\pi \geq 5$ and $n(1-\pi) \geq 5$, we may approximate the rv n_S (the number of successes) with the normal with

$$E[n_S] = n\pi = \mu, \quad V[n_S] = n\pi(1-\pi) = \sigma^2.$$

Here n is the number of trials and π is the population probability of success. Therefore, the sampling density of np (where the rv p is the proportion of successes in n trials) is

$$\phi(np\,|\,\mu,\sigma), \quad \mu = n\pi, \quad \sigma = \sqrt{n\pi(1-\pi)}.$$

From the properties of expectation and variance that

$$E[ax] = aE[x], \quad V[ax] = a^2 V[x]$$

(where a is a constant) we obtain

$$E[p] = E\left[\frac{n\pi}{n}\right] = \pi, \quad V[p] = \frac{1}{n^2}n\pi(1-\pi) = \frac{\pi(1-\pi)}{n}.$$

To summarize,

The sampling density of the sample proportion (for short, the *sampling density of proportions*) Let the rv p be the proportion of successes in a binomial experiment with n trials and with probability of success π. Then the sampling density of p is

$$\phi\left(p\,\Big|\,\pi, \sqrt{\frac{\pi(1-\pi)}{n}}\right).$$

In the following example, we examine the effect of increasing the number of trials on the normal approximation to the sampling density of proportions.

Example 7.26. The data for this example are from International Program Center (2003). We find that in 2000, the sex ratio of males/females of all ages in the West Bank was 1.0 342. The ratio of males in the population is therefore 0.50 842. Imagine sampling the population of Palestinians in the West Bank. With S denoting a male, the density of males in the sample is binomial, with parameters n (sample size) and $\pi \approx 0.508$ (ratio of males in the population). We choose $n = 7$ and $n = 40$. The top

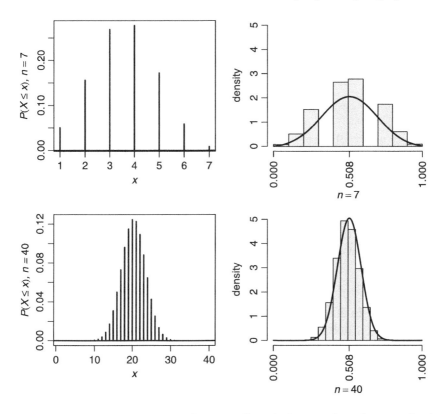

Figure 7.19 The binomial density (left panel) and the sampling density of p (right panel). Histograms show the simulated sampling density (for 5 000 repetitions) and curves the theoretical normal sampling density. Top panel is for $n = 7$ trials and bottom for 40 trials.

panel of Figure 7.19 shows the binomial with parameters 0.508 and 7 (left) and the normal approximation (curve) to the sampling density of p for 5 000 repetitions.

The bottom panel of Figure 7.19 illustrates the results for $n = 40$. Note that both the binomial density and the simulated sampling density converge to their corresponding normal density. To produce Figure 7.19, we first set the necessary constants, parameters and axis labels:

```
par(mfrow = c(2, 2)) ; PI <- 0.508 ; n <- c(7, 40)
sigma <- sqrt(PI * (1 - PI) / n)
R <- 5000 ; x <- seq(0, 1, length = 201)
xlabel.1 <- expression(italic(x))
xlabel.2 <- c(expression(italic(paste(n == 7))),
   expression(italic(paste(n == 40))))
ylabel <- c(expression(italic(paste(P(X<=x), ', n = ', 7))),
   expression(italic(paste(P(X<=x), ', n = ', 40))))
```

238 The normal and sampling densities

The constants that we use are self-explanatory. The setting of labels is not so, but we did discuss paste(), expression() and italic() before (e.g. Examples 3.8 and 5.2). The simulation and drawing are produced with this chunk of script:

```
set.seed(7)
for (i in 1 : 2){
  plot(1 : n[i], dbinom(1 : n[i], n[i], PI), type = 'h',
    lwd = 2, xlab = xlabel.1, ylab = ylabel[i])
  abline(h = 0, lwd = 2)
  p <- rbinom(R, n[i], PI) / n[i]
  h(p, xlim = c(0, 1), ylim = c(0, 5), axes = FALSE,
    xlab = xlabel.2[i])
  axis(1, at = c(0, PI, 1), las = 2) ; axis(2)
  lines(x, dnorm(x, PI, sigma[i]))
}
```

Let us elaborate. In lines 2–4 we plot the binomial density with parameters $n = 7$ (or 40) and $\pi = 0.508$. In line 5 we produce 5 000 random number of success for the given number of trials. To obtain $p = \widehat{\pi}$, we divide each value by the appropriate number of trials. So in lines 6 and 7, we obtain the simulated sampling density of p. Fitting the axis tick marks in a small plot is tricky. So we show the histogram without axes by setting axes = FALSE. Then in line 8 we add the axes; first the x and then the y (by setting the argument to axis to 1 or 2, respectively). When we draw the x-axis, we ask to put the tick marks at = zero, π and 1. To prevent the tick labels from colliding, we ask to draw them perpendicular to the axis (hence las = 2). Finally, in line 9 we draw the theoretical sampling density. □

7.7.2 Consequence of the central limit theorem

In Example 7.26, π is nearly 0.5. If π is closer to zero or one, we need ever larger number of trials to use the normal approximation to the sampling density of p. Denote by μ_p the proportion of successes for the sampling density of the probability of success. The population density is binomial with parameters n and π. From the central limit theorem and with the observation we made about Figure 7.19 we conclude that:

1. $\mu_p = \pi$: the sampling density of p is centered at π.
2. $\sigma_p = \sqrt{\pi(1-\pi)/n}$: as the number of trials increases, the standard deviation of the sampling density of p decreases.
3. The sampling density of p approaches normal as n increases.
4. The farther π is from 0.5, the larger the value of n that is needed for the normal approximation to be accurate.

To address the last point, we have

A rule of thumb If $n\pi \geq 5$ and $n(1-\pi) \geq 5$, then the central limit theorem may be used for the sampling density of p.

7.8 The sampling density of intensity

Recall that intensity refers to counts per unit of something. For example, rates refer to counts per unit of time. Examples are the arrival rate of patients to an emergency room, birth rate, cancer rate, the number of plants per m², the number of organisms per unit of volume and so on. As we saw in Section 5.7, the Poisson density is appropriate for modeling such phenomena. Also, in Section 7.2.3, we saw that for the Poisson (intensity) parameter $\lambda > 20$, we can use the normal approximation with $\mu = \lambda$ and $\sigma^2 = \lambda$.

7.8.1 The sampling density

Let n_C be the number of counts in n units of intervals from a population with intensity parameter λ (counts per unit interval). Then

$$E\left[n_C\right] = n\lambda\,, \quad V\left[n_C\right] = n\lambda\,.$$

Therefore, for the sample based intensity (the rv $l = n_C / n$) we have

$$E\left[l\right] = E\left[\frac{n_C}{n}\right] = \frac{1}{n}E\left[n_C\right] = \frac{n\lambda}{n} = \lambda\,,$$
$$V\left[l\right] = V\left[\frac{n_C}{n}\right] = \frac{1}{n^2}V\left[n_C\right] = \frac{n\lambda}{n^2} = \frac{\lambda}{n}\,.$$

So from the central limit theorem we obtain

The sampling density of the sample intensities (or the *sampling density of intensities*) Let the rv l be the number of counts per n unit intervals from a population with Poisson density with parameter λ. Then the sampling density of l is

$$\phi\left(l \bigg| \lambda, \sqrt{\frac{\lambda}{n}}\right)\,.$$

In the next example, we illustrate the properties of the sampling density of l.

Example 7.27. The R library UsingR includes data about murder rates in 30 Southern US cities (see documentation about the data). We examine the histogram with

```
> library(UsingR) ; data(south) ; n <- c(5, 10, 100)
> par(mfrow = c(2, 2))
> h(south, xlab = 'murder rate', ylim = c(0, 0.1))
```

loading the package and attaching the data south. Then we prepare the window to accept four drawings and finally draw the histogram (Figure 7.20). The empirical density is perhaps Poisson. We consider the data to be the population. So we calculate λ = the mean of the data. We draw the theoretical density with:

```
> lambda <- mean(south)
> x <- 0 : 30
> lines(x, dpois(x, lambda), type = 'h', lwd = 2)
> abline(h = 0, lwd = 2)
```

It looks "good" and we move on:

```
> set.seed(100)
> ylab = c('', 'density', '')
> for(i in 1 : 3){
+   m <- matrix(rpois(n[i] * 500, lambda), nrow = n[i],
+       ncol = 500)
+     l <- apply(m, 2, mean)
+   h(l, ylim = c(0, 1.1), xlim = c(10, 18),
+       xlab = bquote(italic(list(l,~~ n==.(n[i])))),
+       ylab = ylab[i])
+   x <- seq(0, 20, length = 201)
+   lines(x, dnorm(x, lambda, sqrt(lambda/n[i])))
+ }
```

The for loop illustrates what happens to the sampling density as the sample increases from 5 to 10 to 100. We need the loop (over the index i) three times for the three sample sizes. For each of these, we matrix() the number of Poisson random deviates, rpois(), into m. Then, using apply() with its second argument set to 2, we mean() the matrix columns into l (the length of these columns represent the sample size and the number of columns represent the repetitions) for the population murder rate lambda. The 500 values of the rv l are then used to compare the simulated sampling densities (the next three histograms in Figure 7.20) to the theoretical normal sampling

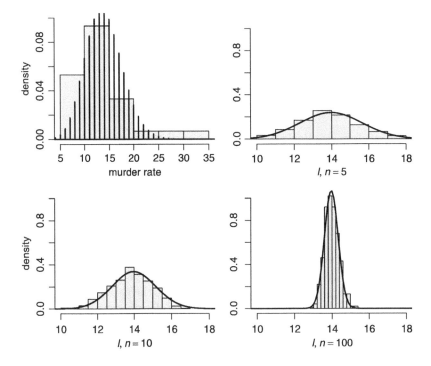

Figure 7.20 Sampling density of the intensity of murders in some Southern U.S. cities.

density with mean λ and standard error $\sqrt{\lambda/n}$. As n increases, the standard error decreases and all densities are centered around the population rate λ. □

7.8.2 Consequences of the central limit theorem

Denote by μ_l the mean intensity for a sample of size n from a Poisson population with parameter λ. From the central limit theorem and from the observation we made about Figure 7.20 we conclude that:

1. $\mu_l = \lambda$: the sampling density of l is centered at λ.
2. $\sigma_l = \sqrt{\lambda/n}$: as the sample size increases, the standard deviation of the sampling density of l decreases.
3. The sampling density of l approaches normal as n increases.

7.9 The sampling density of variance

To discuss the sampling density of the variance of any density, we need the idea of a central moments of the density $P(X = x|\boldsymbol{\theta})$. The mth *central moment* is defined as

$$\psi_m := \int_{-\infty}^{\infty} P(X = x|\boldsymbol{\theta}) (x - \psi)^m \, dx$$

(where $\psi := \psi_1$). The corresponding central moments of a sample of size n are defined thus:

$$C_n := \frac{1}{n} \sum_{i=1}^{n} (X_i - C)^n$$

where $\overline{X} = C$ is the sample mean. Let S^2 be the sample variance. Then

$$E\left[S^2\right] = E\left[C_2\right] = \frac{n-1}{n} \psi_2 \tag{7.3}$$

is the expected value (the mean) of the sample variance and

$$E\left[V\left[S^2\right]\right] = E\left[V\left[C_2\right]\right] = \frac{(n-1)^2}{n^3} \psi_4 - \frac{(n-1)(n-3)}{n^3} \psi_2^2 \tag{7.4}$$

is the expected variance of the sample variance. In words, the mean of the sampling density of the sample variance is given by (7.3) and its variance by (7.4). To proceed, we must determine ψ_2 and ψ_4. This may be accomplished if the density $P(X = x|\boldsymbol{\theta})$ is specified.

Example 7.28. The sampling density of the variance of a sample from a normal population is known analytically. For the normal, $\psi_2 = \sigma^2$ and $\psi_4 = 3\sigma^4$. Therefore, for the normal, the first and second central moments (mean and variance) of the sampling density of the sample-variance are

$$C_1[V] = \frac{n-1}{n} \sigma^2 \,, \quad C_2[V] = \frac{2(n-1)}{n^2} \sigma^4 \,.$$

In our usual vernacular, this means that for the sampling density of the variance we have

$$\mu_V = \frac{n-1}{n} \sigma^2 \,, \quad \sigma_V = \frac{1}{n} \sigma^2 \sqrt{2(n-1)} \,.$$

In other words, the sampling density of the variance is centered (asymptotically) around the population variance σ^2. The standard error, $\sqrt{2(n-1)}\sigma^2/n$, goes to zero as $n \to \infty$. We can say even more. The sampling density of the sample variance is known as Pearson type III; it is given by

$$P(V = v) = \frac{\left(\frac{n}{2\sigma^2}\right)^{(n-1)/2}}{\Gamma\left(\frac{n-1}{2}\right)} v^{(n-3)/2} \exp\left[-\frac{n}{2\sigma^2}v\right].$$

You can use parpe3(), quape3() and cdfpoe3() (in the package lmomco) to compute the moments, quantiles and distribution of the Pearson type III. □

7.10 Bootstrap: arbitrary parameters of arbitrary densities

In this section we discuss the bootstrap method. Because the bootstrap we use provides confidence intervals for the estimated mean, we also discuss the exact methods for estimating confidence intervals. We introduce the bootstrap method with an example. To compare our results to those that appear in the literature routinely, we shall use a common data set.

Example 7.29. The data are about the percentage of the Swiss population in 1888 with years of education (Tukey, 1977):

```
> edu <- c(12, 9, 5, 7, 15, 7, 7, 8, 7, 13, 6, 12, 7, 12,
+       5, 2, 8, 28, 20, 9, 10, 3, 12, 6, 1, 8, 3, 10, 19,
+       8, 2, 6, 2, 6, 3, 9, 3, 13, 12, 11, 13, 32, 7, 7, 53,
+       29, 29)
> length(edu)
[1] 47
```

We wish to estimate the population variance. There are 47 observations. We take a sample of 47 with replacement from edu and compute its variance. Next, we take another sample with replacement and compute its variance. We repeat the process 1 000 times and thus get 1 000 variances. We can now compute the sampling density of the variance. It can be shown that if the sample represents the population, then as the number of repetitions increases, the mean of the sampling density of the variance approaches the population variance. To obtain the variance statistic, we first create a bootstrap object with this call:

```
> library(boot)
> bs <- boot(edu, var, seed = 1)
```

(see help(boot) for details). Now we can examine the density of the variance with

```
> plot(bs, main = '', ylab = 'density', col = 'gray90')
```

which produces Figure 7.21. To obtain the confidence interval, we call

```
> summary(bs)
Call:
boot(data = edu, statistic = var, seed = 1)

Replications: 1000
```

```
Statistics:
     Observed   Bias   Mean    SE
var    92.46   -3.573  88.88  37.50
```

Empirical percentiles:
```
     2.5%    5%    95%  97.5%
var 28.07  36.26 160.2  173.3
```

bca confidence limits:
```
     2.5%    5%    95%  97.5%
var 43.19  48.78 194.6  219.2
```

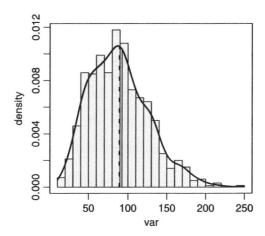

Figure 7.21 Density of the bootstrap variance of the Swiss education data. Solid vertical line indicates the bootstrap variance, broken line the variance computed from the data.

Here bca stands for bootstrap bias-correct, adjusted confidence limits. From the summary, our point estimate of the variance is 88.88 while the observed variance is 92.46. This results in a bias of −3.57. Our 95% confidence interval is between 43.19 and 219.2. The density of the bootstrap variance is not normal (Figure 7.22). This confirms the need for bootstrap to obtain the confidence interval. □

Note that in the example we do not assume a density; we construct it. Also, we constructed a confidence interval for the variance. With the bootstrap, we can obtain point estimate and confidence intervals on any statistic we desire. We shall discuss point estimates and confidence intervals later.

7.11 Assignments

Exercise 7.1. Determine the following standard normal curve areas:

1. Area to the left of 1.76
2. Area to the left of −.65
3. Area to the right of 1.40
4. Area to the right of −2.52

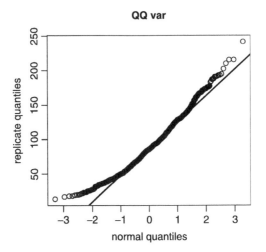

Figure 7.22 Normal Q-Q plot of the Swiss education data.

5. Area between -2.22 and 0.63
6. Area between -1 and 1
7. Area between -3.5 and 3.5

Exercise 7.2. Z is a rv with a standard normal density. Determine

1. $P(Z < 2.4)$
2. $P(Z \leq 2.4)$
3. $P(Z < -1.25)$
4. $P(1.15 < Z < 3.4)$
5. $P(-0.75 < Z < -0.65)$
6. $P(-2.80 < Z < 1.35)$
7. $P(1.9 < Z)$
8. $P(-3.35 \geq Z)$
9. $P(Z < 4.9)$

Exercise 7.3. Determine the value of z that satisfies the following (Z is standard normal):

1. $P(Z < z) = 0.6$
2. $P(Z < z) = 0.5$
3. $P(Z < z) = 0.004$
4. $P(-z < Z < z) = 0.8$
5. $P(Z < z) = 0.90$
6. $P(Z < z) = 0.95$
7. $P(-z < Z < z) = 0.99$
8. $P(z > Z) = 0.05$
9. $P(z > Z) = 0.025$
10. $P(z < Z) = 0.01$

Exercise 7.4. The following table is from Table 10, Sample et al. (1997). It gives weights of brown bats (in g).

```
brown.bat
       state    sex n average   sd
1 New Mexico   male  5    8.47 0.81
2 New Mexico female  3    6.96 0.27
3     Indiana   male  6    6.03   NA
4     Indiana female 40    6.99   NA
```

Suppose that the distributions of weights in the New Mexico and Indiana populations are normal. For Indiana, use the standard derivations of the New Mexico population.

1. Complete the last two columns of the following table:

```
       state    sex n average   sd P(X<8) P(6.5<X<7.5)
1 New Mexico   male  5    8.47 0.81
2 New Mexico female  3    6.96 0.27
3     Indiana   male  6    6.03 0.81
4     Indiana female 40    6.99 0.27
```

2. You are told that a bat from one of these two states was caught. It weighs 8.5 g. Where was it most likely to be collected from? What was its most likely sex?

Exercise 7.5. The mean weight of a bird in a population is 31 g and the standard deviation is 0.2 g. The distribution of weights in the population is normal. Determine:

1. The probability that the weight for a randomly selected bird exceeds 30.5 g.
2. The probability that the weight for a randomly selected bird is between 30.5 and 31.5 g. Between 30 and 32 g.
3. Suppose that a bird is classified as underweight if its weight is less than 30.4 g. What is the probability that at least one bird in a sample is underweight?

Exercise 7.6. The rv X denotes the IQ score of a randomly selected individual. Suppose that the density of X is approximately normal with mean 100 and standard deviation 15. Furthermore, X must be an integer. Calculate the following:

1. The probability that a randomly selected person has an IQ of 100.
2. The probability that a randomly selected person has an IQ of 100 or less. 110 or less.
3. What percentage of the population will have an IQ between 75 and 125?
4. What percentage of the population will have an IQ larger than 120?

Exercise 7.7. Suppose that the distribution of the weight of young seals in a certain population is approximately normal with mean 150 kg and standard deviation 10 kg. The scale is accurate to the nearest kg.

1. What is the probability that a randomly selected young of the year seal weighs 120 kg at most?
2. What is the probability that a randomly selected young of the year seal will weigh at least 125 kg?
3. What is the probability that it will weigh between 135 and 160 kg?

Exercise 7.8. From http://www.cdc.gov/nchs/data/hus/tables/2003/03hus061 .pdf we learn that in 1990–92, the percentage of U.S. adults that smoked (adult population is defined as 18 years or older, age adjusted) was 27.9. In 1999–2001, that percentage dropped to 21.1. Consider these percentages as reflecting population values.

1. What is the probability that 50 individuals smoke in a sample of 200 individuals from the 1990–92 population? From the 1999–2001 population?
2. What is the probability that 100 individuals smoke in a sample of 200 individuals from the 1990–92 population? From the 1999–2001 population?
3. What is the probability that between 50 and 100 individuals smoke in a sample of 200 individuals from the 1990–92 population? From the 1999–2001 population?
4. What is the probability that between 51 and 99 individuals smoke in a sample of 200 individuals from the 1990–92 population? From the 1999–2001 population?

Exercise 7.9. Suppose that 25% of the anthrax scares are false alarms. Let X denote the number of false alarms in a random sample of 100 alarms. What are the approximate values of the following probabilities?

1. $P(20 \leq X \leq 30)$
2. $P(20 < X < 30)$
3. $P(35 \leq X)$
4. The probability that X is more than 2 standard deviations from its mean.

Exercise 7.10. Live traps manufactured by a certain company are sometimes defective.

1. If 5% of such traps are defective, could the techniques introduced thus far be used to approximate the probability that at least five of the traps in a random sample of size 50 are defective? If so, calculate this probability; if not, explain why not.
2. Compute the probability that at least 20 traps in a random sample of 500 traps are defective.

Exercise 7.11. The following is a sample of 20 independent measurements of the concentration of a pollutant in ppm, along with the expected standard normal score for a sample of 20.

```
    ppm  score
1    25 -1.867
2    27 -1.408
3    31 -1.131
4    36 -0.921
5    36 -0.745
6    37 -0.590
7    38 -0.448
8    41 -0.315
9    41 -0.187
10   42 -0.062
11   43  0.062
12   43  0.187
13   53  0.315
14   55  0.448
15   57  0.590
```

16	62	0.745
17	76	0.921
18	78	1.131
19	89	1.408
20	103	1.867

Are the data approximately normal?

Exercise 7.12. The amount of time that an individual animal spends near a water hole at Kruger National Park in South Africa is a normal rv with mean 60 min and standard deviation of 10 min.

1. What is the probability that the next observed animal will spend more than 45 min at the water hole?
2. What amount of time is exceeded by only 10% of the animals?

Exercise 7.13. We need to make sure that the following data are approximately distributed according to the normal. Would you use a transformation and if yes, which one?

0.7552 1.1816 0.1457 0.1398 0.4361 2.8950 1.2296 0.5397
0.9566 0.1470 1.3907 0.7620 1.2376 4.4239 1.0545 1.0352
1.8760 0.6547 0.3369 0.5885 2.3645 0.6419 0.2941 0.5659
0.1061 0.0594 0.5787 3.9589 1.1733 0.9968 1.4353 0.0373
0.3240 1.3205 0.2035 1.0227 0.3017 0.7252 0.7515 0.2350

Exercise 7.14.

1. Explain the difference between a population characteristic and a statistic.
2. Does a statistic have a density? Does a population parameter have a density?

Exercise 7.15. Describe how you would select a random sample from each of the following:

1. students enrolled at a university;
2. books in a bookstore;
3. registered voters in your state;
4. subscribers to the local daily newspaper.

Exercise 7.16. We have a population with measurements $X = 1, 2, 3, 4$.

1. A random sample of 2 is selected without replacement and with order important. There are 12 possible samples. Compute the sample mean for each of the 12 possible samples and show the sampling distribution in a table.
2. A random sample of 2 is selected with replacement. Show the sampling distribution in a table.

Exercise 7.17. We have a population with measurements $X = 5, 3, 3, 4, 4$. Here $\mu = 3.8$. Suppose the researcher does not know this value, but wishes to estimate it from samples. The statistics available are: the sample mean, the sample median and the average of the largest and smallest values in the sample. The researcher decides to use a random sample of 3. Order is not important. Therefore, there are 10 possible samples. For each of the 10 samples, compute the 3 statistics. Construct the sampling distribution for each of these statistics. Which statistic would you recommend for estimating μ? Explain.

Exercise 7.18. A population consists of 5 values: 8, 14, 16, 10, 11.

1. Compute the population mean.
2. Select a random sample of 2 (write 1, ..., 5 on 5 slips of paper and draw 2 of them at random). Compute the mean of the sample.
3. Repeat the procedure for 25 samples of 2, calculating the mean for each pair.
4. Draw a histogram of the mean of the 25 samples. Are most of the sample means near the population mean? Do the values of sample means differ a lot from sample to sample, or do they tend to be similar?

Exercise 7.19. What are the sampling distributions of the following?

1. The means of samples of size 34 from a normal population.
2. The means of samples of size 22 from a normal population.
3. The proportions of males of a sample of 30 from a population with approximately equal number of males and females.

Exercise 7.20.

1. Create a vector of 101 values of x, where x is between -3 and 3 where the vector values are from $F(X \leq x)$ where F is the standard normal density.
2. Plot the results.
3. For the same values of x, create a vector of $f(x)$ where f is the standard normal.
4. Plot the results.

Exercise 7.21. For the following exercises, use `set.seed(2)` in the appropriate places.

1. Create a random sample of 100 values from the standard normal.
2. Report the variance and standard deviation of this sample.
3. Create a random sample of 1 000 values from the standard normal.
4. Report the variance and the standard deviation.
5. Which of the means and standard derivations were closer to the theoretical density (sample size 100 or 1 000)? Why?

Part III

Statistics

8

Exploratory data analysis

Deduction and induction are two major approaches to science. In deduction, one reaches a conclusion from known facts. In induction, the known facts are believed to support the conclusions with high probability. In a nutshell, Exploratory Data Analysis (EDA) takes more of an inductive approach than say, formal hypothesis testing, where the approach is mostly deductive. Consequently, EDA is *not* a collection of unique statistical techniques. It is an approach which emphasizes using data to generate hypotheses. As such, it lets the data "speak" for themselves and is particularly appropriate for massive amounts of data.

EDA uses primarily data summaries and graphical techniques to examine the data (for outliers for example), explore potential cause and effect or trends in data and to model relations among related variables in data. EDA was originated by Tukey (1977), followed by works such as Velleman and Hoaglin (1981) and Chambers et al. (1983). Among the useful EDA graphical methods are histograms, Q-Q plots and box plots. Among the frequently used numerical methods are measures of the center and spread of data, Chebyshev's rule, the empirical rule and correlation.

The difference between EDA and the classical (hypothesis testing) approach is illustrated in Figure 8.1. The illustrated differences are idealized. In reality, we move between the stages freely, but always with a major approach in mind. Most statistical analyses use random sample data and assume some underlying population density. Consequently, we make two basic:

Assumptions

1. Random samples (all objects from a population have the same probability of membership in a sample).
2. The population density did not change while a sample was obtained and its variance is finite.

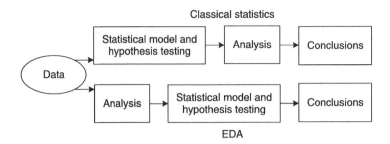

Figure 8.1   The classical statistics vs. EDA approaches.

8.1 Graphical methods

Simple graphical methods can be used to examine the two assumptions above. Some, we have discussed: histograms in Section 3.3; scatter plots and paired scatter plots in Section 3.5; lattice plots in Section 3.6. To complete the picture, we discuss run-sequence plots.

Let $\boldsymbol{X} := X_1, \ldots, X_n$ be a sample. The assumptions imply that the order of the values of X_i, $i = 1, \ldots, n$ is not important. This leads to the statistical model

$$X_i = a + \varepsilon_i$$

where a is some constant and $\varepsilon_i = X_i - a$. For densities with location and scale, we usually assume that $a = \mu$, mean $\varepsilon = 0$ and $V[\varepsilon] = \sigma^2$ where μ and σ are the location and scale parameters of the density (mean and standard deviation in the case of the normal). If the observations are independent, then ε_i must be independent and we say that ε_i are independent and identically distributed (iid) random variables. To examine the assumptions, we use run-sequence plots. They consist of the vector index, i, plotted on the x-axis and X_i on the y-axis. In time series, i often maps to dates.

Example 8.1. In Example 2.16 we introduced the time series of the U.S. military casualties in Iraq, as reported by the U.S. Department of Defense. Let us assume that the casualties count is a Poisson process. Then the model is

$$X_i = \lambda + \varepsilon_i$$

where the location (mean) of $\varepsilon_i = 0$ and the standard deviation is $\sqrt{\lambda}$. So as in the code in Example 2.16, we produce the casualty counts per 10-day intervals. These counts are the mean and variance of the presumed density. The top panel of Figure 8.2 shows the standard deviation and the bottom the counts, crossed by the mean counts $\widehat{\lambda}$. The data show no trend in the mean, but there are cycles in the variance (which means that ε_i are not iid). Note the two exceptionally bloody 10-day periods that began on April 3, 2004 and on November 8, 2004. This inspection of the data leads one to conclude that the variance of the data is larger than its mean. Thus, the data are overdispersed and instead of using the Poisson to model it, we might use the negative binomial (see Section 5.9.2).   □

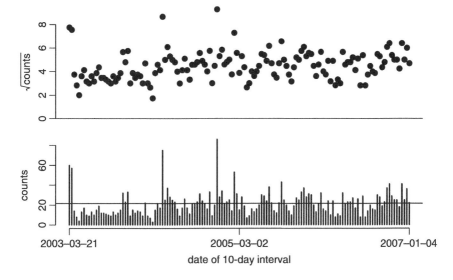

Figure 8.2   Run sequence plot of the U.S. Department of Defense reported U.S. military casualties in 10-day intervals in Iraq.

8.2 Numerical summaries

Numerical methods in EDA allow us to examine data summaries with regard to central tendency of the data and their dispersion.

8.2.1 Measures of the center of the data

A computed value that reflects the center of the data is often considered a typical data value. There are different ways to assess a central value of data. These depend on how values are distributed across the population or the sample.

Mean

For data that are clumped around some central value, the mean of the sample reflects typical population values. The sample mean is the expected value of the sample, where each value has a probability of $1/n$ (where n is the sample size). If the random values are not equally probable, the mean must be weighted by the values' respective probabilities. To wit, we define the weighted mean as

Weighted sample mean Let $\boldsymbol{X} := [X_1, \ldots, X_n]$ be a random sample of objects with $\boldsymbol{p} = [p_1, \ldots, p_n]$, the probability of selecting each object from the population. Then the weighted sample mean is the rv

$$\overline{X} := \sum_{i=1}^{n} p_i X_i \,. \tag{8.1}$$

In matrix notation, $\overline{X} := \boldsymbol{X} \cdot \boldsymbol{p}$, where the dot product is defined in (8.1).

Sample mean We say that $\overline{X}$ is the sample mean if $p_i = 1/n$ for the weighted sample mean, i.e.

$$\overline{X} := \frac{1}{n} \sum_{i=1}^{n} X_i .$$

We distinguish between sample mean and

Population mean For a population of size N and with $x_1, \ldots, x_N$, the population mean is

$$\mu := \frac{1}{N} \sum_{i=1}^{N} x_i . \tag{8.2}$$

We are also interested in the sample-based estimate of the population mean.

MLE estimate of the population mean The MLE estimate of the population mean, denoted by $\widehat{\mu}$, is the sample mean $\overline{X}$.

If we analyze a random sample from a population, we have to assume a (empirical or theoretical) probability density for the population. So we have four mean-related values: the sample mean, $\overline{X}$, the population mean, μ, the expected value of probability density of the population, $E[x]$ and the estimated population mean, $\widehat{\mu}$. If we can consider the population infinite, then $\mu = E[x]$. Otherwise, μ is given in (8.2). Let us compare these quantities by example.

Example 8.2. The density of height in humans is thought to be normal. Consider a population of $N = 501$ on a small island in Fiji. Let the theoretical density of height (in cm) be $\phi(x\,|\,180, 10)$. For all practical purposes, we consider INFINITY $= 10\,000$ to be infinitely large. So we plot the theoretical density with

```
> x <- seq(120, 240, length = INFINITY)
> y <- dnorm(x, 180, 10)
> plot(x, y, type = 'l',
+    xlab = 'x', ylab = 'density')
```

(Figure 8.3). We draw $E[x]$ thus:[1]

```
> abline(v = 180, lwd = 2, col = 'blue')
```

Next, we take a random sample from the theoretical density of 501 (the population). From it, we obtain a mean height and draw it: This is our μ:

```
> set.seed(79) ; population <- rnorm(N, 180, 10)
> abline(v = mean(population),
+    lwd = 4, col = 'red', lty = 2)
```

Finally, we draw a sample of $n = 30$ from the population, calculate its mean ($\widehat{\mu} = \overline{X}$) and draw it:

```
> X <- sample(population, n)
> abline(v = mean(X), lty = 3)
```

---

[1] We use x instead of X to emphasize that x is *not* a rv.

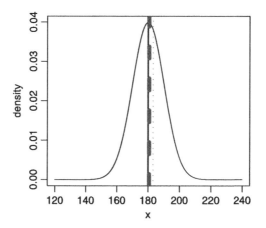

Figure 8.3   The curve is the theoretical density $\phi(x\,|\,E[x],S[x])$. The solid line is $E[X] = 180$, the dashed line is the population mean μ and the dotted line is the sample-based estimate of the theoretical mean, namely $\widehat{\mu} = \overline{X}$.

Table 8.1   Differences between theoretical density, population density and sample-derived values for the mean and standard deviation.

Density	Population	Sample
$E[X] = 180$	$\mu = 180.57$	$\widehat{\mu} = \overline{X} = 183.14$
$S[X] = 10$	$\sigma = 10.23$	$\widehat{\sigma} = S[X] = 8.83$
$X \in \mathbb{R}$	$N = 501$	$n = 30$

The arguments col and lty and lwd specify the color, line type (solid, dashed or dotted) and line width. The remaining results are shown in Table 8.1. The upshot is this: If the population from which we draw the sample is large enough, then $\mu \approx E[X]$. Otherwise, we have to distinguish between the theoretical density and the population density (which in this case becomes an empirical density). We shall always distinguish between the sample density and the population density. □

Here is an example with a small population.

Example 8.3. For this example, we use estimates of the height (in ft) of the 18 faculty members in a Department at a University. First, we create the name and height vectors:

```
> name <- c('Ira A', 'David A', 'Todd A', 'Robert B',
+    'Yosef C', 'James C', 'Francesca A', 'David F',
+    'Rocky G', 'Peter J', 'Anne K', 'Kristen N', 'Ray N',
+    'James P', 'Peter S', 'George S', 'Ellen S', 'Bruce V')
> height <- c(5 + 4 / 12, 6 + 11 / 12, 5 + 11 / 12,
+    5 + 11 / 12, 6, 5 + 10 / 12, 5 + 10 / 12, 5 + 11 / 12,
+    5 + 3 / 12, 5 + 10 / 12, 5 + 8 / 12, 5 + 7 / 12,
+    5 + 10 / 12, 5 + 9 / 12, 5 + 10.5 / 12, 5 + 10.5 / 12,
+    5 + 10 / 12, 6)
```

Next, we combine them into a data frame

```
> faculty <- data.frame(name, height)
```

Because we will use these data again, we save the data frame to a file

```
> save(faculty, file = 'faculty.rda')
```

Recall that by convention, we save R data with the object name, appended with the .rda extension. Thus, after loading faculty.rda, we obtain the data frame object named faculty. You may use your own convention. Consider the faculty as our population. First, we wish to examine a run sequence plot for the sorted (by height) population and identify() some interesting values. So we load the data and sort them

```
> load('faculty.rda')
> head(faculty, 3)
    name   height
1  Ira A 5.333333
2 David A 6.916667
3  Todd A 5.916667
> idx <- sort(faculty$height, decreasing = TRUE,
+    index.return = TRUE)
```

To sort the data frame on a particular column, we need to establish the indices of the sorted column. Thus, we use sort() on faculty$height; set the sort order to decreasing and ask sort() to return the sorted index of faculty$height by specifying index.return = TRUE. We store the returned list in idx:

```
> idx
$x
 [1] 6.92 6.00 6.00 5.92 5.92 5.92 5.88 5.88 5.83 5.83 5.83
[12] 5.83 5.83 5.75 5.67 5.58 5.33 5.25

$ix
 [1]  2  5 18  3  4  8 15 16  6  7 10 13 17 14 11 12  1  9
```

The ix element of idx contains the sorted indices of faculty$height. Now to sort the data frame, we do

```
> f <- faculty[idx$ix, ]
```

plot f and identify() points of interest:

```
> plot(f$height, xlab = 'sorted faculty index',
+    ylab = 'height (ft)')
> identify(f$height, label = f$name)
[1]  1 17 18
```

(Figure 8.4). Next, we compare the population mean to the sample mean for $n = 5$:

```
> round(mean(faculty$height), 2)
[1] 5.84
```

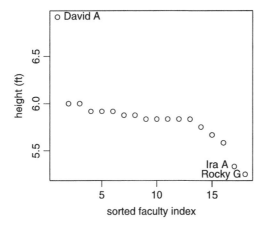

Figure 8.4   Sorted faculty heights.

or $\mu \approx 5.84$. Here we round the value of the mean height to two decimal digits because the values are estimates. It makes no sense to report them with high accuracy. Next we compute the height of a random sample of the faculty. So we wrote the numbers 1 through 18 on 18 small pieces of paper, put them in a box, mixed them and picked 5 of these pieces of paper. The numbers 5, 7, 10, 14 and 3 turned up. Therefore, the heights in the random sample data were Yosef C = 6.00, Francesca A = 5.83 and so on. The mean of the sample was $\overline{X} = 5.87$. In R, we do

```
> idx
Error: Object "idx" not found
> idx <- c(5, 7, 10, 14, 3)
> round(mean(faculty$height[idx]), 2)
[1] 5.87
```

or more directly

```
> set.seed(1) ; mean(sample(faculty$height, 5))
[1] 5.866667
```

Note the error message. Because we want to use an index vector named `idx`, we first verify that no such object or function exist in R. `sample()` takes a random sample from its first argument. `set.seed()` ensures that we repeat the same sequence of random numbers. □

The mean of a sample is sensitive to extreme values. Extreme values may arise in a sample either by chance or because the population density is skewed. In the former case, we call these values outliers. In the latter, extreme values in the sample represent extreme values in the population; they are therefore *not* outliers. If we happen to have outliers in a sample from a symmetric population density, or an asymmetric population density, then the sample mean may not represent typical values in the population.

Example 8.4. Consider the data in Example 8.3. As Figure 8.5 illustrates, David A is unusually tall and Rocky G and Ira A are short. They therefore influence the mean unduly in the sense that it no longer represents a typical observation.

258 Exploratory data analysis

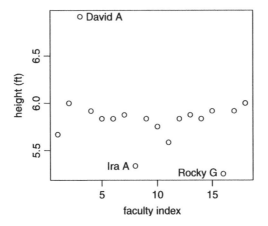

Figure 8.5   Faculty height in a department.

Figure 8.5 was produced with the following script:

```
1  load('faculty.rda')
2  plot(faculty$height, xlab = 'faculty index',
3      ylab = 'height (ft)')
4  identify(faculty, label = faculty$name)
```

Here we annotate a plot with the low level plotting function identify(). In line 1 we load the faculty data frame. In line 2 we plot the data with the appropriate labels. In line 4 we label whichever points we wish—the extreme points in this case. The function identify() is applied to the active graphics window. It identifies the data point nearest the point that you click. Once a point is identified, identify() labels this point by the value of the corresponding faculty$name, based on the value of the faculty$height. This can be done because for each mouse click near a point, identify() returns the index of the point in the data. Because we identify a vector of labels with the named argument label, identify() draws this label. Once we identify all the points of interest (3 points in our case), we hit the escape key (or click the right button and then click on stop). This terminates the process of labeling points on the plot. □

Median

To get around the problem of outliers or to represent a typical value in a population with skewed density, we define the

Median the middle value of the data.

To compute the sample median, first sort the data and then find the value of the middle observation. If the number of data points is odd, then the middle observation is chosen such that there is equal number of observations with values smaller and larger than the identified observation. If the number of observations is even, locate the middle two observations and compute their mean. This will then be the median. The R function `median()` does what its name implies. Here is an example.

Example 8.5. Two homeless persons—with zero annual income—are sitting in a bar complaining about how poor they are. Bill Gates[2] enters the bar and one person says to the other: "On the average, we are extremely wealthy; on the median, we are exactly as poor as we were before." □

Here is a more substantial (and not a funny) example.

Example 8.6. Figure 8.6 shows the trend in U.S. mean and median income (see Table A-1 in DeNavas-Walt et al., 2003, p. 17). The disparity in income between rich and poor becomes obvious when the median is compared to the mean. This is so because the density of income in the U.S. population is highly skewed. Figure 8.6 was obtained as follows: First we `load()` the data:

```
> load('us.income.rda')
```

Here are the first few rows of the data:

```
> head(us.income, 3)
  year median  mean
1 2002  42409 57852
2 2001  42900 59134
3 2000  43848 59664
```

Next, we plot mean/1 000 vs. year with the following arguments: `type = 'l'`, meaning the plot type is lines between the points; `ylim = c(30, 60)` means the limits of the y-axis are between 30 and 60. The label of the y-axis is set with the named argument `ylab`:

```
> plot(us.income$year, us.income$mean / 1000,
+      type = 'l', ylim = c(30, 60), ylab = 'income in $1000')
```

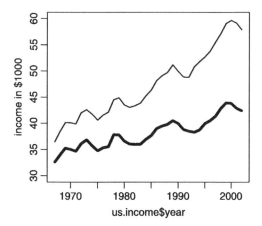

Figure 8.6 Mean (thin) and median (thick) income per household in the U.S., in $1 000.

---

[2] who at the time of writing is one of the richest people on Earth.

Finally, we do

```
> lines(us.income$year, us.income$median / 1000, lwd = 3)
```

This adds `lines()` between the points of the median / 1 000 and sets the line's width to 3 with `lwd = 3`. Income distribution is usually measured with the Gini Coefficient. We show the income distribution to illustrate the difference between mean and median in skewed densities. □

If the population density is symmetric, then the population mean and median coincide.

Mode

Like the median, the mode is a location measure that is insensitive to extreme data values:

Mode The most frequently occurring value in a sample.

For symmetrically distributed data, the mode is an indicator of the population center. For skewed densities, the mode indicates the bulk of the observation values. When ties occur, there are as many modes as there are ties.

Example 8.7. Consider the weight of 10 Nashville warblers (in g) from the Sierra Nevada (personal data):

12 14 17 10 8 12 9 16 13 10

The modes are 12 g and 10 g. □

Trimmed mean

If you wish to report a mean, but minimize the influence of outliers on it, then use the

x% trimmed mean The mean with $(x/2)\%$ of the largest values and $(x/2)\%$ of the smallest values removed from the data.

To compute a trimmed mean, sort the data and remove equal percentage of the data from the top and bottom. When reporting values of trimmed means, be sure to report the percent of the data trimmed. With R, you trim means with the function `mean()` and the named argument `trim`. Specify `trim` as proportion of the data to be trimmed from one side; this proportion will be trimmed from the other side automatically. Before you trim a mean, be sure to justify why. After all, if the density is likely to produce extreme values, they should not be trimmed because they do represent the underlying density. This is the case, for example, with the exponential density.

Example 8.8. Consider the faculty height data in Example 8.3. There are 18 observations. If we trim $1/18 \times 100 \approx 5.5\%$ of the data from the top and 5.5% of the data from the bottom, we remove David D and Rocky G from the data. The untrimmed mean is

```
> load('faculty.rda')
> mean(faculty$height)
[1] 5.842593
```

and the trimmed is

```
> mean(faculty$height, trim = 1 / 18)
[1] 5.8125
```

Not much of a change. □

8.2.2 Measures of the spread of data

The spread of the data values reflects how probable data values are. A sample from a density with little spread will have predictable values.

Range

One way to express the spread of data in a sample is using the

Range The value of the difference between the smallest and largest values in the data.

The range is not very informative about the spread of the data. Here is an example.

Example 8.9. Consider x and y and their ranges:

```
> (xy <- list(x = c(0, rep(100, 4)),
+              y = c(100, 140, 180, 200, 100)))
$x
[1]   0 100 100 100 100

$y
[1] 100 140 180 200 100

> mapply(range,xy)
       x   y
[1,]   0 100
[2,] 100 200
```

Both x and y have a range of 100. However, y is more variable than x:

```
> mapply(var, xy)
   x    y
2000 2080
```

We sometimes use the sample's range to estimate its variance. □

Variance, standard deviation and coefficient of variation

We define

Sample variance We also have

$$S^2 := \frac{1}{n-1} \sum_{i=1}^{n} (X_i - \overline{X})^2 \qquad (8.3)$$

where n is the sample size.

Population variance The variance of a population of size N is

$$\sigma^2 := \frac{1}{N} \sum_{i=1}^{N} (x_i - \mu)^2$$

where μ is the population mean.

To compute the sample variance, we divide by $n-1$, *not* by n. This is so because it turns out that by dividing the sample sum of squares by $n-1$ we get a sample variance that is not biased. As was the case for the mean, we have the sample variance, S^2, the population variance, σ^2, a sample-based estimate of the population variance, $\hat{\sigma}^2 = S^2$ and the population underlying density variance $V[X]$. Both $V[X]$ and σ^2 are *not* rv. If we can assume that for all practical purposes the population is infinite, then $\sigma^2 \approx V[X]$. Corresponding to the sample and population and density variances we have their standard deviations, S, σ and $S[X] := \sqrt{V[X]}$. The units of the sample variance are the square of the units of measurement. The units of the standard deviation are the same as the unit of the measurement. The standard deviation is interpreted as the magnitude of a typical deviation from the mean.

Example 8.10. The data for this example were obtained from the WHO (see Example 2.7 for data source and description). Here we wish to compare the mortality rate of children under the age of 5 (per 1 000) in Western Africa and Northern Europe. Table 8.2 shows the data along with means, variances and standard deviations. Note the following:

- There are missing values. These must be handled properly. We exclude them from the computations.
- The variance and the standard deviation are calculated by dividing the sum of squares by $n-1$, not by n, because we treat these data as samples.
- n refers to the numbers of observations for which there are no missing data; not to the sample size.

Table 8.2 Western Africa and Northern Europe children (under 5) mortality per 1000 children under 5.

Western Africa		Northern Europe	
Country	Mortality	Country	Mortality
Benin	155.73	Channel Islands	6.53
Burkina Faso	160.20	Denmark	6.49
Cape Verde	35.87	Estonia	11.44
Cte d'Ivoire	173.08	Faeroe Islands	4.79
Gambia	134.07	Finland	4.17
Ghana	93.40	Iceland	7.03
Guinea	175.78	Ireland	17.65
Guinea-Bissau	209.81	Isle of Man	11.25
Liberia	229.33	Latvia	5.89
Mali	180.96	Lithuania	4.34
Mauritania	156.35	Norway	6.52
Niger	209.94	Sweden	4.34
Nigeria	132.69	United Kingdom	6.52
Saint Helena			
Senegal	112.08		
Sierra Leone	307.31		
Togo	136.39		
Mean	162.69		7.83
Variance	3 758.19		16.56
Standard deviation	61.30		4.07

The example requires new and useful function calls in R. So we isolate the R implementation for this example in the next example. □

In the next example we demonstrate how to use the database access capabilities of R for Example 8.10.

Example 8.11. First we present the script and then we analyze it.

```
1   western.africa <- 1 ; northern.europe <- 2
2   region.name <- c('Western Africa', 'Northern Europe')
3   file.name <- c('WesternAfrica.tex','NorthernEurope.tex')
4
5   # 1. import the data
6   library(RODBC) ; c <- odbcConnect('who')
7   sqlTables(c) ; who <- sqlFetch(c, 'MyFormat')
8   odbcClose(c) ; save(who, file = 'who.fertility.mortality.rda')
9
10  # comment/uncomment for desired table
11  # region = western.africa
12    region = northern.europe
13
14  # 2. make data frame
15  ifelse(region == western.africa, rows <- c(50 : 66),
16      rows <- c(135 : 147))
17  mort <- who$'under 5 mort'[rows]
18  stats <- c(mean(mort, na.rm = TRUE),
19      var(mort, na.rm = TRUE),
20      sd(mort, na.rm = TRUE))
21  mort <- data.frame(c(mort, stats))
22  rnames <- c(as.character(who$country[rows]),
23      '\\hline Mean', 'Variance',
24      'Standard deviation')
25  dimnames(mort) <- list(rnames, c('Mortality'))
26
27  # 3. table
28  library(Hmisc)
29  cap1 <- paste(region.name[region],
30      ', children (under 5) mortality')
31  cap2 <- 'per 1000 children under 5.'
32  latex(mort, file = file.name[region],
33      caption = paste(cap1, cap2),
34      label = paste('table:',region.name[region],'mortality'),
35      cdec = 2,
36      rowlabel = 'Country',
37      na.blank = TRUE,
38      where = '!htbp',
39      ctable = TRUE)
```

The code here demonstrates several useful features of R. In particular, it demonstrates:

1. how to import data from a database directly into R (lines 6 to 8);
2. how to create a data frame, subset data and deal with missing values (lines 15 to 25);
3. how to create a LaTeX table such as Table 8.2 (lines 28 to 39).

In lines 1 through 3 we create some objects that we need to use later to modify the script output if desired. We now go through the first two topics (the third is presented for the sake of completion). The topics are independent. If you are not interested in a particular one, skip it. The code is not simple. Study it carefully and you will obtain useful skills.

1. How to import data from a database directly into R (lines 6 to 8)

This task is accomplished with the package RODBC. See Example 2.10 for further details.

2. How to create a data frame, subset data and deal with missing values (lines 15–25).

In the who data frame, rows 50–66 contain data about countries in Western Africa. Rows 135–147 contain data about countries in Northern Europe. Based on the choice of region, we assign the appropriate rows to index vectors in lines 15 and 16. The function ifelse() takes three arguments: The first is the condition—region == western.africa. The second is executed if the condition is true and the third is executed if the condition is false.

In line 17 we store the mortality data for the chosen region in mort. In lines 18–20, we store the mean, variance and standard deviation of the mortality with calls to mean(), var() and sd(). There are missing data, so each of these functions is called with na.rm = TRUE—otherwise, the requested values will be returned as NA by the respective calls. We add the stats to mort in line 21. The concatenated vector is then casted into a data.frame named mort. In lines 22–24 we assign names to the rows of mort. The names are the country names from the who data. Here we must coerce the factor column country into character strings with a call to as.character(). Without this, the return value will be the integer value of the factors, not the string value. To the row names vector rnames we also add the names of the stats. The \\hline in line 23 is a LaTeX command, so we shall leave it at that. In line 25, we assign the row names and the column name Mortality to the mort data.frame with dimnames(). Here are the first few rows of mort:

```
> mort
                 Mortality
Channel Islands   6.532000
Denmark           6.494000
Estonia          11.444000
Faeroe Islands         NA
Finland           4.792000
```

The data frame is now ready for creating a LaTeX table. We will not discuss the rest of the code as it produces Table 8.2. The code is presented for the sake of completeness. However, note the use of the library(Hmisc), a rather useful library by Frank Harrell. □

All of the measures we discussed thus far have units. In Example 8.10, the units of measurement are deaths of children under 5 per 1000 per year. The units of the mean are the units of measurement. To derive the units of the variance, consider a sample of X_i, $i = 1, \ldots, n$ where X_i are measured in calories (denoted by cal). We write the formula for the variance and below it the formula in units:

$$S^2 = \frac{1}{n-1} \sum_{i=1}^{n} (X_i - \overline{X})^2 ,$$

$$\text{cal}^2 = \frac{1}{\text{no units}} \sum_{i=1}^{n} (\text{cal} - \text{cal})^2 .$$

In the units formula, $n - 1$ is a count and therefore has no units. In the sum, we subtract cal from cal. The result is therefore expressed in cal. Then we square the result. This gives cal². Then we sum cal². Summing units preserves the unit, so we end with a sum expressed in cal². Dividing this sum by a unitless quantity, we end with the units of the variance, cal². Obviously, the units of the standard deviation are those of the units of measurement.

Sometimes, we wish to make comparisons among observations that we measure in different units. In such cases we use the

Coefficient of variation (CV)

$$\text{CV} := 100 \times \frac{S}{\overline{X}} .$$

The CV is rarely used to compare values for populations, so we will not invent notation for the population CV. Because S and $\overline{X}$ have the same units as the measurement, the CV carries no units. We multiply the ratio $S/\overline{X}$ by 100 to express CV as a percentage.

Example 8.12. From Table 8.2 we find

```
> sqrt(3758.19) / 162.69
[1] 0.3768153
> sqrt(16.56) / 4.07
[1] 0.999852
```

for the CV of Western Africa and Northern Europe. Thus, relatively speaking, under-five mortality in Western Africa is (roughly) uniformly high across countries; not so in Northern Europe. □

As was the case for the mean, the measures of data spread we discussed thus far are sensitive to outliers. The next measure we discuss is not.

Interquartile range

Exactly as was the case with the median, we sort the data and find the

Lower quartile The value of the observation for which 25% of the values are smaller.
Upper quartile The value of the observation for which 25% of the values are larger.
Interquartile range (IQR) The difference between the upper and lower quartiles,

$$\text{IQR} = \text{upper quartile} - \text{lower quartile} .$$

As for the median, when the number of observations is even, the value of the quartile is the mean of the two observations at the respective location. Because IQR is used mostly with samples, as opposed to with populations, we will not invent a population notation for it. Here is an example that illustrates the sensitivity of variance to outliers and the insensitivity of IQR.

Example 8.13. Returning to Example 8.3, we compare the variances with and without the tallest faculty member, David A:

```
> round(c(with = var(faculty$height),
+   without = var(faculty$height[-2])), 2)
   with without
   0.11    0.04
```

And the IQR

```
> rbind(with = summary(faculty$height),
+   without = summary(faculty$height[-2]))
         Min. 1st Qu. Median  Mean 3rd Qu.   Max.
with     5.25   5.771  5.833 5.843   5.917  6.917
without  5.25   5.750  5.833 5.779   5.917  6.000
```

summary() gives the summary statistics and -2 excludes the second element from the faculty$height vector. □

To obtain any quantile (not only quartiles), use quantile(). This function presents one method (another is ecdf()) to build an empirical density of the data, which may be compared to a presumed density.

Example 8.14. In Example 7.6, we compared the empirical density of base blood pressure (hist() with freq = FALSE) to the normal (Figure 7.4). Here are the same results, this time comparing the distributions of base blood pressure. After loading the data, we compute sample-based estimates of the normal parameters:

```
> (mu.hat <- mean(cardiac$basebp))
[1] 135.3244
> (sigma.hat <- sd(cardiac$basebp))
[1] 20.77011
```

Next, we plot the normal distribution with the estimated parameters:

```
> par(mfrow = c(1, 2))
> x <- seq(80, 220, length = 201)
> plot(x, pnorm(x, mu.hat, sigma.hat), type = 'l')
```

(smooth curve in left panel, Figure 8.7). To compute the empirical distribution, we create a vector of probabilities (for which the data-based quantiles will be computed), obtain the quantiles and add points to the left panel of Figure 8.7:

```
> p <- seq(0, 1, length = 21)
> q <- quantile(cardiac$basebp, probs = p)
> points(q, p)
```

For visual comparison, we draw the empirical distribution with

```
> plot(ecdf(cardiac$basebp))
```

As we can see, quantile() is the inverse of ecdf(). □

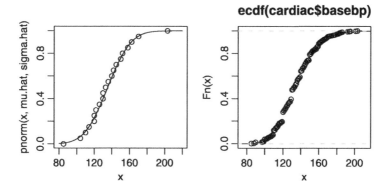

Figure 8.7 Left: the normal distribution with sample-based mean and standard deviation (smooth curve) and the empirical distribution derived from `quantile()` (points). Right: the empirical distribution from `ecdf()`. x is the base blood pressure.

8.2.3 The Chebyshev and empirical rules

This topic is not considered traditionally in EDA. So far we discussed data summaries that provide a numerical value that we can use for comparisons. Often, we are interested in making probability statements that relate to the data in general. For example, we may wish to know the probability that a random value of the data will be within a certain range of values. This generalizes the idea of the value of a typical observation. We may also be interested in graphical methods that aid in visualizing data properties such as mean, IQR and so on. We discuss such methods next.

Chebyshev's rule

This rule gives a lower limit on the number of observations that fall within specified standard deviations of the mean. All we need to know is the mean and standard deviation of the sample. We do not need to know anything about the density of the data.

Chebyshev's rule For k greater than 1, at least $1 - 1/k^2$ proportion of the observations are within k standard deviations of the mean.

With this rule, we construct Table 8.3. The last column of the table shows that 75% of the data fall within 2 standard deviation of the mean and 95% of the data fall within 4.5 standard deviations of the mean.

Example 8.15. The data for this example were obtained from United Nations (2003). We are interested in the percent population growth per year for countries around the world. The data summarize values for 1995–2000. Because the data are based on surveys, X, the percent growth per year, is a rv with:

```
> options(stringsAsFactors = TRUE)
> UN <- read.table('who-population-data-2002.txt',
+       header = TRUE, sep = '\t')
> names(UN)[c(6, 7, 8, 9, 11)] <- c('% urban', '% growth',
```

268   Exploratory data analysis

```
+     'birth rate', 'death rate', 'under 5 mortality')
> save(UN, file = 'UN.rda')
>
> (X.bar <- mean(UN$'% growth', na.rm = TRUE))
[1] 1.355263
> (S <- sd(UN$'% growth', na.rm = TRUE))
[1] 1.150491
```

Table 8.3   Chebyshev's rule.

Standard deviation	$1 - 1/k^2$	Proportion
2	1−1/4	0.75
3	1−1/9	0.89
4	1−1/16	0.94
4.472	1−1/20	0.95
5	1−1/25	0.96
10	1−1/100	0.99

options() directs functions such as data.frame() and read.table() to convert input strings to factors. If you set this option to FALSE, then strings will remain so in the newly created data frame. (Note how we rename UN's columns.) In at least 95% of the countries around the world the growth rate is between

```
> Chebyshev <- c(low = X.bar - 4.472 * S,
+    high = X.bar + 4.472 * S)
> round(Chebyshev, 2)
  low high
-3.79 6.50
```

per year. We will return to this conclusion in a moment.   □

Empirical rule

If the data are close to normal, we can obtain a narrower estimate of the proportion of the population within a certain range of values than we do with Chebyshev's rule. Thus, we have

Empirical rule If the histogram of the data is approximately normal, then roughly:
  68% of the observations are within one standard deviation of the mean.
  95% are within two standard deviations of the mean.
  99.7% are within three standard deviations of the mean.

Example 8.16. Continuing with Example 8.15, we examine the empirical density with

```
> par(mfrow = c(1, 2))
> h(UN$'% growth', xlab = '% growth per year')
> x <- seq(-2, 5, length = 201)
> lines(x, dnorm(x, X.bar, S))
```

(left panel, Figure 8.8). The superimposed $\phi(x\,|1.36, 1.15)$ indicates that the empirical density is approximately normal. Therefore, in at least 95% of Earth's nations, the growth rate is between

```
> empirical <- c(low = X.bar - 2 * S,
+    high = X.bar + 2 * S)
> round(rbind(Chebyshev, empirical), 2)
            low high
Chebyshev -3.79 6.50
empirical -0.95 3.66
```

Note the narrower range of the empirical estimate, compared to the Chebyshev estimate. For good measure, we also draw the empirical and theoretical, $\Phi(x\,|\hat{\mu},\hat{\sigma})$, densities

```
> mu.hat <- X.bar ; sigma.hat <- S
> x <- seq(min(UN$'% growth'), max(UN$'% growth'), length = 201)
> plot(x, pnorm(x, mu.hat, sigma.hat), type = 'l',
+    xlab = 'quantile (% growth / year)',
+    ylab = expression(italic(Phi(x))))
> p <- seq(0, 1, length = 51)
> q <- quantile(UN$'% growth', probs = p)
> points(q, p)
```

(right panel, Figure 8.8). See Example 8.14 for explanation of the code. □

8.2.4 Measures of association between variables

The *correlation coefficient* measures how strongly two variables are related. Various types of relations may be observed in the following example and we are looking for ways to quantify them.

Example 8.17. Figure 8.9 shows various associations between X and Y: no apparent association (top left):

```
> X <- rnorm(20, 0, .25) ; Y <- rnorm(20, 0, .25)
> plot(X, Y, axes = FALSE, xlim = c(-1, 1), ylim = c(-2, 2),
+    xlab = '', ylab = '') ; abline(h = 0) ; abline(v = 0)
```

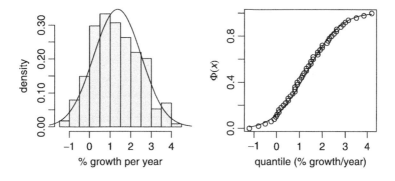

Figure 8.8   Density and distribution of annual % growth rate for 228 nations.

270 Exploratory data analysis

positive association (top right):

```
> X <- seq( -1, 1, length = 20) ; Y <- X + rnorm(X, 0, .25)
> plot(X, Y, axes = FALSE, xlim = c(-1, 1), ylim = c(- 2, 2),
+    xlab = '', ylab = '') ; abline(h = 0) ; abline(v = 0)
```

negative association (bottom left):

```
> Y <-  -X + rnorm(X, 0, .25)
> plot(X, Y, axes = FALSE, xlim = c(-1, 1), ylim = c(-2, 2),
+    xlab = '', ylab = '') ; abline(h = 0) ; abline(v = 0)
```

and quadratic association:

```
> Y <-  -.5 + X * X + rnorm(X, 0, .25)
> plot(X, Y, axes = FALSE, xlim = c(-1, 1), ylim = c(-2, 2),
+    xlab = '', ylab = '') ; abline(h = 0) ; abline(v = 0)
```

In all associations, we add `rnorm()` "noise" to the otherwise deterministic equation. In a deterministic equation, x predicts y with certainty. □

Next, we develop ways to quantify the relationship between two variables.

Covariance and Pearson's correlation coefficient

Consider a population of N objects. We are interested in the relation between the values of two population traits, (x_i, y_i), for object i. To proceed, we first need to construct the mathematical space in which we make our observations. So we define a

Euclidean product Let (x_i, y_i) be a pair of trait values for an object i from a population of size N and define

$$\boldsymbol{x} := [x_1, \ldots, x_N] \ , \ \boldsymbol{y} := [y_1, \ldots, y_N] \ , \ x_i, y_i \in \mathbb{R} \ .$$

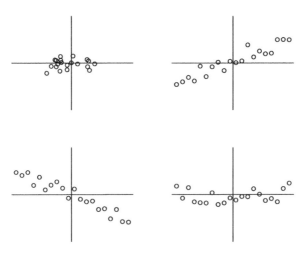

Figure 8.9   Various associations between X and Y.

Then
$$\boldsymbol{x} \times \boldsymbol{y} := \begin{bmatrix} (x_1, y_1) & \cdots & (x_1, y_N) \\ \vdots & \ddots & \vdots \\ (x_N, y_1) & \cdots & (x_N, y_N) \end{bmatrix}$$

is said to be the Euclidean product of $\boldsymbol{x}$ and $\boldsymbol{y}$, where $\boldsymbol{x} \times \boldsymbol{y} \in \mathbb{R} \times \mathbb{R}$ (also written as $\mathbb{R}^2$).

Here $\mathbb{R}^2$ defines the familiar Euclidean plane. Each point in the plane is defined by the pair (x_i, y_i). Thus, (X_i, Y_i), $i = 1, \ldots, n$, is a random sample of size n from the population where $(X_i, Y_i) \in \mathbb{R}^2$. To quantify the association (positive, negative or none) between the pairs of rv values, we use the

Sample covariance For a sample of size n,

$$S_{XY} := \frac{1}{n-1} \sum_{i=1}^{n} (X_i - \overline{X})(Y_i - \overline{Y}) \tag{8.4}$$

(we divide by $n-1$ and not by n for the same reason we did in equation 8.3) is the sample covariance.

Parallel to the sample covariance, we have the

Population covariance For a population of size N,

$$\sigma_{xy} := \frac{1}{N} \sum_{i=1}^{N} (x_i - \mu_x)(y_i - \mu_y) \tag{8.5}$$

(where μ_x and μ_y are the population means of x and y) is the population covariance.

Densities also have covariances, but we shall not discuss them here. From both (8.4) and (8.5) we observe the following qualitative relations:

- If X_i, Y_i have a positive relationship; i.e. as X_i gets larger (smaller) so does Y_i, then $S_{XY} > 0$.
- If X_i, Y_i have a negative relationship; i.e. as X_i gets larger (smaller), Y_i gets smaller (larger)), then $S_{XY} < 0$.
- If X_i, Y_i have no relationship; i.e. as X_i gets larger (smaller) Y_i gets either larger or smaller), then $S_{XY} \approx 0$.

The magnitude of the covariance depends on the units of measurement. To get around this problem, we note that in the sum in (8.4), we multiply the units of X by the units of Y. Therefore, to standardize the measure of association between X and Y, we divide by a quantity that is the product of these units. Thus, we divide the average of the relation between X and Y in equation (8.4) by $S_X \times S_Y$ and define

Pearson's sample correlation coefficient (R_{XY}) For a sample of size n from a population with paired traits x_i, $y_i \in \mathbb{R}^2$,

$$R_{XY} := \frac{1}{(n-1) S_X S_Y} \sum_{i=1}^{n} (X_i - \overline{X})(Y_i - \overline{Y}) . \tag{8.6}$$

272 Exploratory data analysis

Pearson's population correlation coefficient (ρ_{XY}) For a population of size N with paired traits x_i, y_i,

$$\rho_{xy} := \frac{1}{N\sigma_x\sigma_y} \sum_{i=1}^{N} (x_i - \mu_x)(y_i - \mu_y) \ . \tag{8.7}$$

Using the definition of S_{XY} in (8.4), we have

$$R_{XY} = \frac{S_{XY}}{S_X S_Y} \ . \tag{8.8}$$

In words, Pearson's sample correlation coefficient is the covariance between X and Y, scaled by the product of the standard deviation of X and the standard deviation of Y. When no ambiguity arises, we drop the subscripts on R and ρ. Note that S_{XY}, R and ρ map values in $\mathbb{R}^2$ into $\mathbb{R}$.

Example 8.18. The following data are from Focazio et al. (2001). The title of the document, "Occurrence of Selected Radionucleotides in Ground Water Used for Drinking Water in the United States: A Reconnaissance Survey, 1998," says it all. Let us do some EDA. First, the data description:

```
> load('wells.info.rda')
> wells.info
         name                          explanation
1     USGS.SN                    USGS serial number
2     reading  reading in pCi/L (pico-Curie per liter)
3          sd                    standard deviation
4         mdc      minimum detectable concentration
5  nucleotide            Ra-224, Ra226 or Ra228
```

Next, some data massaging:

```
> load('wells.nucleotides.rda')
> s <- split(wells.nucleotides, wells.nucleotides$nucleotide)
> nuc <- data.frame(s[[1]]$reading, s[[2]]$reading,
+   s[[3]]$reading)
> names(nuc) <- names(s)
```

We split the data frame by the `nucleotide` factor (with levels Ra224, Ra226 and Ra228). Then we put together a new data frame, nuc, with the factor levels as columns. Finally, we name the columns accordingly. Now

```
> pairs(nuc)
```

produces Figure 8.10. Obviously, there are positive relations between pairs. From the increasing scatter of the points with increasing concentration, we conclude that a log-log transformation might accentuate the paired relations. So we do

```
> round(cor(nuc, use = 'pairwise.complete.obs'), 2)
      Ra224 Ra226 Ra228
Ra224  1.00  0.54  0.55
Ra226  0.54  1.00  0.52
```

```
Ra228  0.55  0.52  1.00
> round(cor(log(nuc), use = 'pairwise.complete.obs'), 2)
      Ra224 Ra226 Ra228
Ra224  1.00  0.61  0.63
Ra226  0.61  1.00  0.70
Ra228  0.63  0.70  1.00
```

cor() computes the Pearson R by default. We tell it what to do with missing values by assigning appropriate value to the named argument use. From the correlation matrix we find that increase of the concentration of any of the nucleotides is associated with increase of the other two. Based on this finding, we might recommend that in future research, only one nucleotide should be measured, but in more wells.

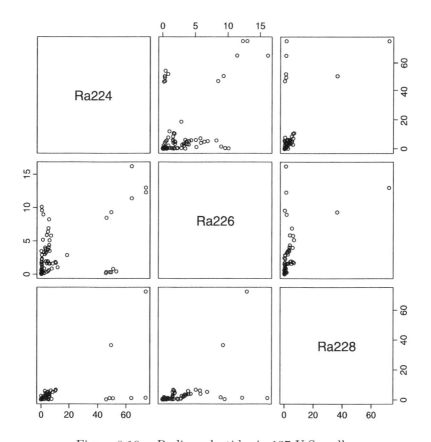

Figure 8.10   Radionucleotides in 137 U.S. wells.   □

Properties of the correlation coefficient

- In (8.6), the units in the numerator are the units of X times the units of Y. In the denominator we have a count (i.e. $n-1$) which has no units and then the standard deviations of X and Y, which carry the units of X and the units of Y. Therefore, R has no units. This allows us to compare R values across samples of different entities and different units of measurement.

274 Exploratory data analysis

- We can exchange the role of the variables and assign Y to X and X to Y. This will not affect the value of R.
- The values of R range from -1 to 1. When R is close to zero (e.g. between -0.2 and 0.2) we conclude that there is no relationship between X and Y. When R is close to -1 (e.g. -1.0 to -0.8), we conclude that there is negative relationship between X and Y. Finally, when R is between 0.8 and 1.0, we conclude that there is positive relationship between X and Y.
- $R = 1$ or $R = -1$ only when the points that represent the data fall exactly on a straight line.

Note that large values of R do not necessarily imply a simple linear relationship between Y and X. They imply trends.

Example 8.19. Let

```
> x <- c(1, 2, 3, 4, 5, 6, 7, 8, 9, 10, 20)
> set.seed(111) ; Y <- x^2 + rnorm(length(x), 0, 10)
> plot(x, Y) ; cor.test(x, Y)[[4]]
      cor
0.9442931
```

(as in Figure 8.11). Here $R \approx 0.94$, yet the relation between X and Y is not simple linear ($Y = a + bX$). □

Like the mean and variance, Pearson's correlation coefficient is unduly affected by outliers; thus, our next topic.

Spearman's rank correlation coefficient

The sample Spearman's rank correlation coefficient, denoted by R_S, overcomes the effect of outliers on R by considering the rank of the data, not their magnitude. The computation of R_S is simple. In the data, replace the smallest value of X by 1, the next smallest by 2 and so on. Do the same for Y (do not change the paired order

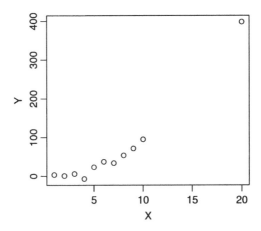

Figure 8.11   A realization of $Y = x^2 + \varepsilon$ where the density of ε_i is $\phi(x_i | 0, 10)$, $i = 1, \ldots, 12$.

of the observations). Next, compute Pearson's correlation coefficient. Because values were transformed to ranks, the computation can be simplified:

The sample Spearman's rank correlation coefficient For a sample of size n,

$$R_S = \frac{12}{n(n-1)(n+1)} \sum_{i=1}^{n} \left(\text{rank}(X_i) - \frac{n+1}{2} \right) \left(\text{rank}(Y_i) - \frac{n+1}{2} \right).$$

R_S has the same properties as R. $R[x, y]$ is the population Spearman's rank correlation coefficient.

Example 8.20. For the data in Example 8.19, Spearman's rank correlation is more appropriate than Pearson's:

```
> Pearson <- cor.test(x, Y)[[4]]
> Spearman <- cor.test(x, Y, method = 'spearman')[[4]]
> round(c(Pearson = Pearson, Spearman = Spearman), 2)
 Pearson.cor Spearman.rho
        0.94         0.93
```

Because we explore rank relation, the outlier influences R_S like any other point. ☐

8.3 Visual summaries

In addition to graphical methods for EDA (Section 8.1), there are techniques to examine data summaries visually. Prominent among them are box plots, lag plots and dot charts. We discussed the latter in Example 3.6, so here we discuss box plots and lag plots.

8.3.1 Box plots

Box plots are used with categorized numerical data. For example, in an experiment you may measure plant growth under control and treatment conditions. Then the factor has two levels and the numerical data are the plants' dry weight. Box plots are (roughly) standard. In the next example, we display and interpret box plots.

Example 8.21. We continue with the UN data (Example 8.15). Let us summarize the % growth rate by continent:

```
> par(mar = c(14, 4, 4, 2) + 0.1)
> boxplot(UN$'% growth' ~ UN$continent, las = 2,
+     main = '% growth rate by continent')
> identify(UN$'% growth' ~ UN$continent, labels = UN$country)
```

First, we recognize that some continent names are long and will not fit on the category axis (the x-axis). So we specify the margins parameter, `mar` (which sets the plot margins in line units), for the bottom, left, top and right of the plot. Then we use the formula that plots the growth as a function of continent with `UN$'% growth' ~ UN$continent`. Because `continent` is a factor, `boxplot()` knows how to group the data and display them (Figure 8.12). Finally, we recognize two countries in Europe: one with exceptionally high and one with exceptionally low growth

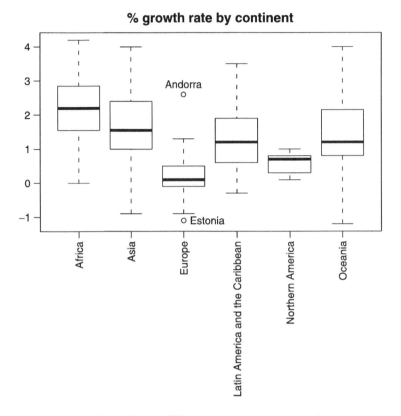

Figure 8.12 Growth rate (% per year) for countries by continent.

rate. To identify them, we use `identify()`. Consequently, the plot is waiting for mouse clicks. We tell `identify()` how to find the data for the plot exactly as we specify them for `boxplot()`. The labels for the points come from UN$country.

Let us interpret the box plot. Take, for example, Africa. The lowest horizontal line section is the lower *whisker*. The lower boundary of the rectangle is the lower *hinge*, which is placed at the first quartile. Quartiles are produced with `quantile()`. The first quartile is produces with `quantile(x, 1/4)`. The thick line is the *median*. The upper boundary of the rectangle is the upper hinge, the third quartile, which is produced by `quantile(x, 3/4)`. The upper horizontal line section is the upper whisker. Whiskers are 1.5 quartiles away from the mean. Beyond hinges, points are plotted individually. They may be suspected outliers. If the median is not in the middle between the whiskers, then the density of the data is not symmetric. The unnecessary proliferation of names for the box plot can be replaced by "boundary of the box" and the "outer vertical line sections." □

8.3.2 Lag plots

Lag plots relate to time series data. Because the rv values are time dependent, we can no longer talk about simple random samples. Thus, the classical statistical methods we discuss do not apply (see the classic text by Box and Jenkins, 1976). The dependency

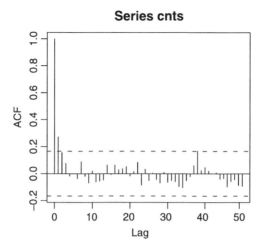

Figure 8.13 Autocorrelation function for the 10-day intervals of the U.S. casualties in Iraq. The lags span 500 days. The broken horizontal lines are 95% confidence on the lagged correlations.

structure of the time series might be of interest because if there is none, we can proceed with the usual statistical methods. Let X_t be a discrete time series data ($t \in \mathbb{Z}_{0+}$). Then in lag plots, we explore the relationship

$$X_t \quad \text{vs.} \quad X_{t-1}, \ldots, X_{t-m}$$

where m is the maximum lag. For example, consider a series of daily maximum temperatures for say 101 days. The pairs (X_t, X_{t-1}), $t = 1, \ldots, 100$, is a (not necessarily simple) sample of $n = 100$ consecutive daily maximum temperatures. We can thus use this sample to compute R. When many lags are combined, we obtain the *autocorrelation function*. The analysis is not limited to a single time series.

Example 8.22. We go back to the U.S. military casualties in Iraq (the data were introduced in Example 2.16). What can we say about the correlation between the number of casualties during a particular 10-day interval and the count during a 10-day interval 200 days earlier? 190 days earlier? We do

```
> load('Iraq.cnts.rda')
> acf(cnts, lag.max = 50)
```

and thus obtain Figure 8.13. Apparently, the number of casualties during a time interval depends on the number of casualties one interval earlier and none other. This indicates that deaths among the time intervals are independent (the broken horizontal strip shows the boundaries of the 95% confidence level—we will talk about these later).   □

8.4 Assignments

Exercise 8.1. The following data (e-Digest of Environmental Statistics, 2003a) show the concentrations of mercury in Cod and Plaice (from the Southern Blight, the North

278 Exploratory data analysis

Sea) and Whiting (from Liverpool Bay, Irish Sea) for 1983 through 1996 (mg/kg wet weight).

Year	Cod	Plaice	Whiting1	Whiting2
1983	0.09	0.08	0.17	0.14
1984	0.08	0.06	0.12	0.11
1985	0.08	0.05	0.13	0.09
1986	0.08	0.04	0.13	0.11
1987	0.08	0.05	0.12	0.11
1988	0.08	0.06	0.12	0.12
1989	0.10	0.05	0.13	0.10
1990	0.09	0.05	0.13	NA
1991	0.06	0.05	0.11	NA
1992	NA	NA	NA	0.10
1993	0.07	0.05	0.13	0.09
1994	0.07	0.05	NA	NA
1995	NA	NA	NA	0.09
1996	0.07	NA	NA	NA

The data are in uk.metals.in.fish.txt.
Compute:

1. The mean for each species.
2. The median for each species.
3. The mode for each species.
4. Compare the values and interpret.

Exercise 8.2. In Exercise 8.1, we computed the means and the medians of tissue mercury for 3 species. For Whiting1:

1. Identify potential outlier(s).
2. Compare the means with and without the outlier.
3. What percentage would you use for the computation of trimmed mean for whiting? Why?

Exercise 8.3. Over 65% of residents in Minneapolis - St. Paul earn less than the average. Flow can this be? Explain.

Exercise 8.4. Use the data in Exercise 8.1.

1. Compute the range for each of the species.
2. Can you identify any relationship between the range and the mean for the 4 species?
3. Compute the interquartile range for each species.
4. Compute the variance for each species.
5. Which of the two measures—interquartile range or range—would you use to estimate the variance if the latter is not available?

Exercise 8.5. About mean and variance:

1. Give two sets of five numbers that have the same mean but different standard deviations.

2. Give two sets of five numbers that have the same standard deviation but different means.

Exercise 8.6. The data are shown in Exercise 8.1. Consider Whiting1 and Whiting2 separately.

1. What are the variances of tissue mercury for each species?
2. Standard deviations?
3. What are the units of the mean, variance and standard deviation?
4. Interpret the data.
5. Compute the CV of the data and explain the results.

Exercise 8.7. For this exercise, combine the data for Whiting1 and Whiting2 in Exercise 8.1. Recall that both Whiting data sets were collected from Liverpool Bay, Irish Sea. The E.U. and the Oslo/Paris Commissions (OSPARCOM) Environmental Quality Standard for mercury is 0.30 mg mercury per kg of wet flesh from a representative sample of commercial fish species.

1. What are the mean and standard deviation of mercury in Whiting (mg/kg wet weight)?
2. Suppose that the density of mercury concentrations in Whiting is not known. What is the range of mercury concentrations (mg/kg wet weight) in which you would expect 75% of the fish to be?
3. Suppose that the density of mercury concentrations in Whiting is known to be normal. What is the range of mercury concentrations (mg/kg wet weight) in which you would expect 95% of the fish?
4. Draw a histogram of the data. Based on it, which rule would you use to draw conclusions about the density of the data: Chebyshev's or Empirical?
5. Based on the results above, would you eat a Whiting from Liverpool Bay? Justify.

Exercise 8.8. The following are data of the number of 100 admissions to a typical U.S. emergency room per hour:

```
 3.77591   5.90821   0.72853   0.69898   2.18034  14.47484
 6.14781   2.69841   4.78284   0.73523   6.95368   3.81015
 6.18802  22.11967   5.27272   5.17622   9.38018   3.27373
 1.68467   2.94240  11.82258   3.20946   1.47060   2.82933
 0.53036   0.29720   2.89356  19.79466   5.86656   4.98406
 7.17643   0.18634   1.62005   6.60234   1.01755   5.11363
 1.50870   3.62607   3.75771   1.17514   5.39941   5.14123
 6.46131   6.26553   2.77321   1.50641   6.46562   4.97278
 2.57087  10.03916   2.11121  10.89386  16.08895   2.78915
 2.97309   4.88698   1.04933   1.54724   5.52968   3.87094
 0.44837   5.54088   1.23632   7.85993  24.16406   2.15566
13.65195   5.68416   4.06684   4.18503   8.92383  11.56236
14.54944   1.42795   1.94393   0.26028   1.75935   7.82621
 4.07268  13.79622   1.93097   5.04128   4.09257   0.29631
11.41927   4.02085   7.91848   6.16896   6.72822  10.50189
 5.17549   2.23726   5.21804   1.30414   3.40615   1.31869
 2.23303   1.05303   0.66286   1.74444
```

The data are in ERAdmissionsRate.txt.

1. Show the histogram of the data.
2. Compute the mean and the standard deviation of the number of arrivals per hour.
3. What proportion of the data are within 0 to 14.5182 arrivals per hour?

Exercise 8.9. For this exercise, use the UK fish contaminants data (e-Digest of Environmental Statistics, 2003a). The data are in uk.metals.in.fish.txt.

For Cod only:

1. Compute the upper and lower quartile for each contaminant.
2. Compute the interquartile range for each contaminant.
3. How large or small should an observation be to be considered an outlier for each contaminant?
4. How large or small should an observation be to be considered an extreme outlier for each contaminant?
5. Are there any mild or extreme outliers in the data for each contaminant? If yes, which are they?
6. Construct a box plot for each contaminant and draw conclusions from it.

Exercise 8.10. The average and standard deviation for the midterm were 50 and 20. For the final they were 55 and 10. Your test score on the midterm was 75 and on the final 70. The instructor is going to "curve" the results. On which test did you do better?

Exercise 8.11. Regarding Pearson's correlation coefficient (r) and Spearman's rank correlation coefficient (r_S) :

1. Give an example of data of interest to you where Pearson's r is more appropriate than Spearman's rank r. Show the data and compute r.
2. Give an example of data of interest to you where Spearman's rank r is more appropriate than Pearson's r. Show the data and compute r_S.

In both cases, explain why you prefer one over the other.

Exercise 8.12. For this exercise, use the faculty height data introduced in Example 8.3.

1. Reproduce Figure 8.5 with Ira A. identified on the plot.
2. Reproduce Figure 8.5 with height sorted from tallest to shortest. You need to use sort() with the arguments decreasing and index.return set to TRUE. sort() returns a list that contains a vector of the sorted indices of faculty height. Access the vector of indices (a component of the returned list) to sort both columns of the faculty data frame.
3. On the plot produced in (2), show the data points and connect them with a broken line.

Exercise 8.13. For this exercise, consult the code for Example 8.10.

1. Copy the WHO excel file to a convenient location on your system.
2. Create a named ODBC connection to it (from your system's control panel).
3. Import the mortality data to R.

4. Produce results as shown in Table 8.2 (no need to produce a LaTeX table) for Northern, Eastern, Western and Southern Africa.
5. Comment on the results.

Exercise 8.14. The following data (e-Digest of Environmental Statistics, 2003 a) show the atmospheric inputs of metals from UK sources to the North Sea from 1987 to 2000.

	Arsenic	Cadmium	Chromium	Copper	Nickel	Lead	Titanium	Zinc
1987	NA	98	280	970	360	3500	770	5400
1988	NA	100	320	930	340	3500	970	5300
1989	210	77	260	1100	440	3000	1100	6100
1990	200	67	150	590	340	1400	500	4900
1991	140	59	130	1400	450	2400	800	5800
1992	57	31	180	350	250	880	630	2800
1993	57	34	48	330	150	760	240	2700
1994	69	29	59	400	220	980	390	5700
1995	57	27	140	480	290	940	320	3100
1996	56	23	130	380	250	880	310	4100
1997	70	28	71	450	67	810	480	3300
1998	34	12	99	240	79	430	280	2300
1999	38	13	67	280	72	400	180	1400
2000	48	19	55	330	83	392	83	3500

The data are in UKAtmosphericInput.txt.

1. Use pairs() to display the relationship between pairs of variables.
2. Use cor.test() and a for loop twice to create a correlation matrix that shows correlations between pairs of heavy metals' dumping into the North Sea by the U.K.
3. Which metals seem to be most correlated? Why?
4. Which metals seem to be least correlated? Why?

Exercise 8.15. The following data (e-Digest of Environmental Statistics, 2003 b) show sources of pollution by enumeration area in 2001. Data source: Advisory Committee on Protection of the Sea (ACOPS), e-Digest of Environmental Statistics, Published August 2003, Department for Environment, Food and Rural Affairs http://www.defra.gov.uk/environment/statistics/index.htm

	tanker	fishing	support	coastal.tanker	cargo	
1	2	3	0	4	6	
2	0	2	1	0	4	
3	0	1	0	2	4	
4	0	1	0	3	3	
5	0	10	0	1	2	
6	0	12	0	4	4	
7	1	4	1	1	0	
8	0	26	0	0	4	
9	0	3	0	1	0	

282 Exploratory data analysis

```
10      0       3       6               2       2
     pleasure.craft  wreck  other
1             0       0       3
2             0       1       0
3             1       0       2
4             7       0       8
5             2       0       7
6             0       0       2
7             1       0       1
8             1       1       4
9             0       0       0
10            1       0       0
```

The data are in UK.pollution.by.enumeration.txt).
Because the data represent enumeration, it is appropriate to use rank correlations.

1. Create a rank correlation matrix for the data above.
2. When you run the rank correlation, you get a warning message. Explain it.
3. Which types of vessels seem to be correlated?
4. Which do not?
5. Speculate about the results.

Exercise 8.16. The following data (e-Digest of Environmental Statistics, 2003 a) describe metal contaminants (mg/kg wet weight) analyzed in fish muscle. Pesticides and PCBs were analyzed in fish liver. Total DDT = ppDDE + ppTDE + ppDDT, Total HCH = a HCH + g HCH. PCBs were measured on a formulation basis (as Arcolor 1254). For 1993 data, only larger fish were available.

The data are saved in a list file named uk.metals.in.fish. Load the list. Find the data in the list and then:

1. Draw a box plot of the concentrations of metals in fish tissue.
2. What conclusions can you draw from the plots about:
   (a) mean concentration in the 4 species?
   (b) variance of concentrations of metals in the 4 species?
   (c) the symmetry of the density of concentration of metals in the tissue of these 4 species?
   (d) How are these related to the life-history of the species?

9
Point and interval estimation

In this chapter, we put to work our understanding of sampling densities. Our goal is to estimate the value of a population parameter (e.g. mean, proportion, variance, rate) from a sample. Once a single value is estimated from the sample, we wish to say something about the corresponding population value. Because the estimates are sample-based, they are rv. Thus, their relation to the population values are uncertain. We quantify this uncertainty with interval estimates. We state the probability that a computed interval contains the population value.

For the most part, the generic approach is to study the sampling density of a sample-based estimate of a population a parameter. We will learn how to make probability statements about the population parameter value based on a sample estimate of the parameter and the latter's sampling density. This is where the central limit theorem plays a crucial role. Table 9.1 lists the notation we follow in this chapter. Because we do not wish to limit our point and interval estimates to the normal density only, we must start with some general considerations and follow them with density-specific expositions.

Table 9.1 Notation

Parameter	Population	Estimate	Sample	Units
Mean	μ	$\widehat{\mu}$	$\overline{X}$	Measurement
Variance	σ^2	$\widehat{\sigma}^2$	S^2	(Measurement)^2
Standard deviation	σ	$\widehat{\sigma}^2$	S	Measurement
Ratio or proportion	π	$\widehat{\pi}$	p	Unit-free
Intensity	λ	$\widehat{\lambda}$	l	Count per unit of measurement

Statistics and Data with R: An applied approach through examples Y. Cohen and J.Y. Cohen
© 2008 John Wiley & Sons, Ltd.

9.1 Point estimation

Consider a population described by the density family $P(X = x | \boldsymbol{\theta})$ where $\boldsymbol{\theta} := [\theta_1, \ldots, \theta_m]$ is the set of the density's parameters. Density family refers to a single density with different possible values for $\boldsymbol{\theta}$. A single set of values for $\boldsymbol{\theta}$ specifies the density exactly. By our definitions, $P(X = x | \boldsymbol{\theta})$ is a continuous or discrete density and $X \in \mathbb{R}$. We address the case where $\boldsymbol{\theta} \in \mathbb{R}^m$. We wish to estimate $\boldsymbol{\theta}$ based on a sample from the population. Among the available techniques to estimate $\boldsymbol{\theta}$ are the method of moments, maximum likelihood estimators (MLE) and Bayes estimators. We discuss MLE only.

We should mention, however, the R package lmomco (for L-moment and L-comoments). L-moments are useful in the estimation of density parameters. Parameter estimates based on L-moments are generally better than standard moment-based estimates. The estimators are robust with respect to outliers and their small sample bias tends to be small. L-moment estimators can often be used when MLE are not available, or are difficult to compute. R functions that use numerical approaches to obtain MLE often require a guess of starting values. You may use the results from L-moments estimation as starting values.

9.1.1 Maximum likelihood estimators

We quickly summarize the ideas of MLE first introduced in Section 5.8. Let $\boldsymbol{X} := [X_1, \ldots, X_n]$ be a sample from a population with density $P(X = x | \boldsymbol{\theta})$. One realization of the sample (one sample data) is $\boldsymbol{x} := [x_1, \ldots, x_n]$. Given the data, we write the likelihood function as

$$\mathcal{L}(\boldsymbol{\theta} | \boldsymbol{x}) := \prod_{i=1}^{n} P(X = x_i | \boldsymbol{\theta})$$

and its log as

$$L(\boldsymbol{\theta} | \boldsymbol{x}) := \log \mathcal{L}(\boldsymbol{\theta} | \boldsymbol{x}) = \sum_{i=1}^{n} \log P(X = x_i | \boldsymbol{\theta}). \quad (9.1)$$

We then define

Maximum likelihood estimator (MLE) The value $\boldsymbol{\theta}$, denoted by $\widehat{\boldsymbol{\theta}}$, that maximizes (9.1) is called the maximum likelihood estimator of $\boldsymbol{\theta}$.

Another definition, which has important consequences in our work is

Statistic If $\widehat{\boldsymbol{\theta}}$ is free of unknown parameters, then $\widehat{\boldsymbol{\theta}}$ is said to be a statistic.

The definition implies that the estimation rule (e.g. MLE) that defines $\widehat{\boldsymbol{\theta}}$ is free of unknown parameters. It does not imply that the density of $\widehat{\boldsymbol{\theta}}$ is free of unknown parameters. We shall always assume that our estimators are statistics and therefore use *statistic* or *MLE* interchangeably. If L is differentiable, then we may determine $\widehat{\boldsymbol{\theta}}$ by solving

$$\frac{\partial}{\partial \theta_i} L(\boldsymbol{\theta} | \boldsymbol{x}) = 0, \quad i = 1, \ldots, m$$

(see Examples 5.18 and 5.19). If it is not, we need to determine $\widehat{\boldsymbol{\theta}}$ numerically.

Example 9.1. Our sample is $X = [X_1, \ldots, X_n]$. We wish to estimate $\theta = [\mu, \sigma^2]$ for the normal. The likelihood function is

$$\mathcal{L}(\theta|x) = \frac{1}{\left(\sqrt{2\pi\sigma^2}\right)^n} \exp\left[-\frac{1}{2}\sum_{i=1}^{n}\left(\frac{x_i - \mu}{\sigma}\right)^2\right]$$

and the log likelihood is

$$L(\theta|x) = -\frac{n}{2}\log\left(2\pi\sigma^2\right) - \frac{1}{2}\sum_{i=1}^{n}\left(\frac{x_i - \mu}{\sigma}\right)^2. \qquad (9.2)$$

So we need to simultaneously solve

$$\frac{\partial}{\partial \mu} L(\theta|x) = \frac{1}{\sigma^2}\sum_{i=1}^{n}(x_i - \mu) = 0$$

and

$$\frac{\partial}{\partial \sigma^2} L(\theta|x) = -\frac{n}{2\sigma^2} + \frac{1}{2\sigma^4}\sum_{i=1}^{n}(x_i - \mu)^2 = 0.$$

This gives

$$\hat{\mu} = \overline{X}, \quad \hat{\sigma} = \frac{n-1}{n} S^2$$

where S^2 is the sample variance. □

The example illustrates that sometimes there is a difference between the MLE of a population parameter (e.g. $\hat{\sigma}^2$) and the sample-based estimate (e.g. S^2).

9.1.2 Desired properties of point estimators

To proceed, we need the definition of

Mean squared error (MSE) Let $E\left[\hat{\theta} - \theta\right]$ be the expected difference between a parameter and its estimate. Then

$$\left(E\left[\hat{\theta} - \theta\right]\right)^2 = V\left[\hat{\theta}\right] + \left(E\left[\hat{\theta}\right] - \theta\right)^2 \qquad (9.3)$$

(where V denotes variance) is called the mean squared error.

With MSE in mind, we have the following desired properties of point estimators:

Unbiased We define the bias of an estimator to be $E\left[\hat{\theta}\right] - \theta$. If $E\left[\hat{\theta}\right] - \theta = 0$, we say that $\hat{\theta}$ is an *unbiased estimator* of θ. From (9.3) we conclude that for an unbiased estimator, MSE $= V\left[\hat{\theta}\right]$.

Precision An estimator $\hat{\theta}_1$ is said to be more precise than estimator $\hat{\theta}_2$ if $V\left[\hat{\theta}_1\right] < V\left[\hat{\theta}_2\right]$.

Consistency We say that $\widehat{\theta}$ is a consistent estimator of θ if $\widehat{\theta}$ is unbiased and $V[\widehat{\theta}] \to 0$ as the sample size $n \to \infty$.

Efficiency An estimator $\widehat{\theta}_1$ is said to be more efficient than estimator $\widehat{\theta}_2$ if $V\left[\widehat{\theta}_1\right] / V\left[\widehat{\theta}_2\right] < 1$.

Best estimator The variance of the most efficient amongst all estimators is the smallest. Such an estimator is called the *best estimator*.

Example 9.2. Let $X_1, \ldots, X_n$ be independent and identically distributed rv from a population with mean μ and variance σ^2. Then $E\left[\overline{X}\right] = \mu$ is an unbiased estimator and
$$\left(E\left[\overline{X} - \mu\right]\right)^2 = V\left[\overline{X}\right] = \frac{\sigma^2}{n}.$$
Recall that we defined the standard error (the standard deviation of the sampling density of $\overline{X}$) to be $\sigma/\sqrt{n}$. □

It is difficult to establish these desired properties for small samples. Thus, we often rely on asymptotic (or large sample) properties. By asymptotic properties we mean that as $n \to \infty$, the estimator converges to a limit (either the estimator itself or a probability).

Asymptotic desired properties of point estimators

1. If $\lim_{n \to \infty} \left(E\left[\widehat{\theta}\right] - \theta\right) = 0$ then $\widehat{\theta}$ is said to be an asymptotically unbiased estimator of θ.
2. If $\lim_{n \to \infty} V\left[\widehat{\theta}\right] = \min_{\{\theta\}} V\left[\{\widehat{\theta}\}\right]$ where $\{\widehat{\theta}\}$ is the set of all estimators, then $\widehat{\theta}$ is said to be an asymptotically efficient estimator of θ.
3. If $\widehat{\theta}$ is both asymptotically unbiased and asymptotically efficient, then $\widehat{\theta}$ is said to be an asymptotically best estimator of θ.

Example 9.3. We wish to estimate μ. Possible estimators are:

1. the sample mean, $\overline{X}$;
2. a random value of the sample, X_i;
3. the smallest value in the sample, $X_{\min} := \min \mathbf{X}$;
4. $\overline{X} + 1/n$.

Regarding bias, (1) and (2) are unbiased because $E\left[\overline{X}\right] = E\left[X_i\right] = \mu$. Because $E\left[X_{\min}\right] < E\left[\overline{X}\right] = \mu$, (3) is biased and so is (4).

Regarding precision, because $\sigma^2/n < \sigma^2$, (1) is more precise than (2). The efficiency of (2) relative to (1) is $1/n$. Both (3) and (4) are biased. Therefore, their efficiency cannot be measured with a variance ratio.

Regarding asymptotic properties, (1), (2) and (4) are unbiased; (3) is. The variance of (1) and (4) are asymptotically zero. Therefore, (1) and (4) are consistent. Overall, (1) is the best estimator. □

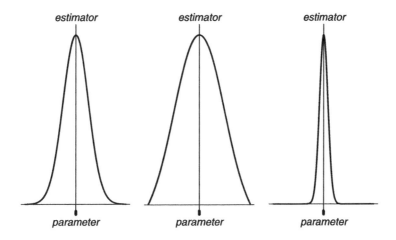

Figure 9.1 Population parameters (short line segments), estimators (long line segments) and sampling densities of the estimators. The densities are all centered around the estimator (statistic).

Figure 9.1 illustrates these ideas visually:

(a) The estimator is biased.
(b) The estimator is unbiased, but it is not as efficient as the estimator in (c).
(c) This is the best estimator.

The left panel in Figure 9.1 was produced with the following script:

```
par(mfrow = c(1,3), mar = c(2, 1, 2, 1))
x <- seq(-4, 4, length = 1000)
plot(x, dnorm(x), axes = FALSE,
  xlab = expression(plain('(a)')),
  ylab = '', type = 'l')
abline(h = 0) ; abline(v = 0)
axis(1, at = -1,
  labels = expression(italic('parameter')),
  lwd = 3)
axis(3, at = 0, labels = expression(italic('estimator')))
```

mar in line 1 sets the distance (in lines) of each graphics from the bottom, left, top and right. Note in line 3 that we plot without the axes by setting the named argument axes to FALSE. Then, in line 7 we plot the x-axis only with a call to axis() with the first unnamed argument set to 1. The named argument at tells axis() where to draw the axis line in relation to the value on the y-axis. In line 10 we plot the y-axis

288 Point and interval estimation

with the value of the first unnamed argument set to 3. The y-axis crosses the x-axis at 0. Similar code was used to draw the right two panels of Figure 9.1.

We are now ready to discuss point estimates of parameters of some specific densities.

9.1.3 Point estimates for useful densities

For some densities, parameter estimates that have the desired properties (the best estimates) are well known (Table 9.1): $\widehat{\lambda} = \overline{X}$ and $\widehat{\sigma}^2 = \widehat{\lambda}$ for the Poisson (Example 5.19 and Exercise 5.19), $\widehat{\pi} = p$ and $\widehat{\sigma}^2 = n\widehat{\pi}(1-\widehat{\pi})$ for the binomial (Exercise 5.18). Estimates for other densities were discussed in Sections 5.9 and 6.8. Here, we apply these estimates to populations with common densities and in the context of EDA.

Normal

For a population with normal density, $\overline{X}$ is the best estimate of μ. If the population density is not normal, but is symmetric with heavy tails, then a trimmed mean is a better statistic than $\overline{X}$ for estimating μ. In the next example we do some EDA and examine this issue.

Example 9.4. The data for the following example are from the United States Department of Justice (1995). They include survey results of crime on 680 U.S. college campuses for 1994. Figure 9.2 shows the density of the (log) ratio of enrollment to full-time faculty. To obtain the figure, we need to prepare the data for analysis. The data frame contains 382 columns. We extract those we need with:

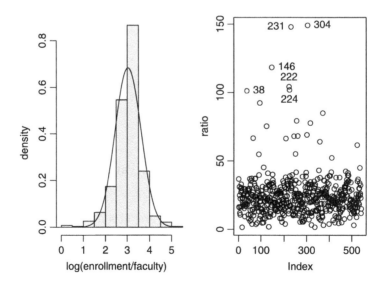

Figure 9.2   Left: empirical and theoretical densities of the log of the ratio of enrollment to full-time faculty in 541 U.S. universities and colleges for 1994. Right: run-sequence of the data.

```
> load('college.crime.rda')
> c.c <- college.crime[, c('school', 'city', 'state',
+   'enrollment', 'full-time faculty')]
```

We want to analyze a subset of the data for a particular range of enrollment and full-time faculty. Why? Because alas, in the original data, exceedingly large or small numbers designate missing values (NA in our vernacular). So we want to use

```
> condition <- (c.c[, 4] >= 2520 & c.c[, 4] <= 56348) &
+   (c.c[, 5] >= 1 & c.c[, 5] <= 10378)
```

to subset the data (columns 4 and 5 contain the enrollment and full-time faculty data). We subset the data with this:

```
> c.c <- c.c[condition, ]
```

and compute the ratio and its log

```
> r <- c.c[, 'enrollment'] / c.c[, 'full-time faculty']
> log.r <- log(r)
```

The left panel of Figure 9.2, produced with

```
> h(log.r, xlab = 'log(enrollment / faculty)')
> x <- seq(-1, 5, length = 201)
> lines(x, dnorm(x, mean(log.r), sd(log.r)))
```

reveals heavy tails (narrow center). This is confirmed with

```
> qqnorm(log.r, main = 'log(enrollment / faculty)')
> qqline(log.r)
```

(Figure 9.3). Thus, a trimmed mean would represent the center of the data (typical values) better than a mean. A run-sequence plot (right panel, Figure 9.2) also illustrates the heavy tails. We draw it with

```
> plot(r, ylab = 'ratio')
```

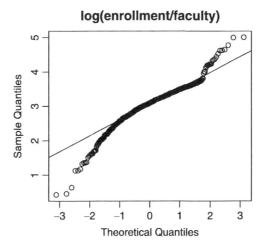

Figure 9.3 The Q-Q plot reveals heavy tails (compare to the density in Figure 9.2).

While at it, we wish to identify some schools with exceptionally high ratio of enrollment to full-time faculty. So we do:

```
> idx <- identify(r)
```

and click on those points we are interested in. We store the index of these ratios in idx and retrieve the corresponding school anmes from c.c with

```
> bad <- cbind(c.c[idx, 1 : 4], ratio = round(r[idx], 1))
```

(Table 9.2). Therefore, the best estimator of the ratio for all U.S. universities and colleges is a trimmed mean:

```
> round(c( '0%' = mean(r),
+    '10%' = mean(r, trim = 0.1),
+    '20%' = mean(r, trim = 0.2)), 1)
  0%   10%   20%
24.3  22.5  22.3
```

Table 9.2 In 1994, student enrollment to full-time faculty ratios in six out of 542 universities and colleges in the U.S. exceeded 100.

School	City	State	Ratio
Golden Gate University	San Francisco	CA	101.3
Purdue University, North Central Campus	Westville	IN	118.4
Baker College of Flint	Flint	MI	104.3
Davenport College	Grand Rapids	MI	101.9
Ferris State University	Big Rapids	MI	147.9
Fairleigh Dickinson University-Madison	Madison	NJ	149.0

Because the difference between 10% and 20% trimmed means is small, we conclude that reporting a 10% trimmed mean is adequate. □

Binomial

For the binomial density, the ratio of the number of successes to the number of trials, p, is the best estimate of the probability of success, π. The reason is that because of the central limit theorem, the sampling density of p approaches normal with $\mu_p = \pi$ and standard error $\sqrt{\pi(1-\pi)/n}$ (see Section 7.7.1).

Example 9.5. The following data and information are from e-Digest of Environmental Statistics (2003a). Radon-222 (^{222}Rn) is a radioactive decay product of naturally occurring uranium-238. It is a gas with a half-life of 3.8 days. It is known to cause lung damage if further radioactive decay occurs while it is in the lung. It is measured in units of Becquerel (Bq). In the UK, 200 Bq per m^3 is deemed an action level. Larger concentrations in a dwelling are thought to contribute significantly to the risk of lung cancer. Therefore, some remedial action to decrease radon concentrations is called for.

Radon levels in 67,800 dwellings were measured in the Cornwall area. In 15,800 of them, the concentrations of radon were above the action level. In the Greater London area, radon concentrations were measured in 450 dwellings. None of the measurements was above the action level. We wish to determine the true proportion (π) of dwellings with radon concentrations above the action level in each of the locations. To proceed, we define a Bernoulli experiment where success is a reading above the action level. The rv is X, the number of successes. The number of trials is the number of readings. The number of dwellings with readings above the action level is the number of successes. Then it makes sense to use

$$\widehat{\pi}_1 = p_1 = \frac{15\,800}{67\,800} = 0.23 \; , \; \widehat{\pi}_2 = p_2 = \frac{0}{450} = 0$$

to estimate the true ratio in Cornwall and London, π_1 and π_2. Here we tacitly assume that sample units are independent, which may not be true. □

Poisson

For the Poisson density, counts per unit of measurement, (e.g. n/T, where T denotes time, n/A, where A denotes area, n/V, where V denotes volume) are the best estimates of the intensity parameter λ. In the context of population studies, we often talk about crude and age-adjusted rates.

Crude rate For n events, counted from a population of N individuals, during period T, the crude rate is

$$d := \frac{n}{N \times T} \; .$$

Age-adjusted rate Denote by s_i the proportion of individuals of age group i in the *standard population*[1] (e.g. 100,000). Let n_i and N_i denote the number of events in age group i and the population of age group i. Then the age-adjusted rate, D, is

$$D := \sum_{i=1}^{m} s_i \frac{n_i}{N_i}$$

where m is the number of age groups.

For the age-adjusted rate, n_i/N_i is the weight of age group i. Therefore, D is interpreted as the weighted sum of m Poisson rv. These definitions can be generalized. A rate is not necessarily time-related. Counts per unit area, for example, are also rates.

Example 9.6. Here are some examples where the Poisson and "rates" are applicable:

1. The number of nerve impulses emitted per second.
2. The number of accidents per 100 cars in an intersection, per month.
3. The number of individual plants of a species per plot.
4. The number of eggs (clutch-size) laid by a female bird. □

To assume Poisson density for rates, we must invoke the usual assumptions about the process: First, that the probability of events (that are counted) is small; second, that

---

[1] The standard, or reference population, is a population for which the data are adjusted. For example, the age distribution of U.S. population in 1960 may be defined as the standard population.

the events are independent; and third, that the distribution (in time or space) of the events is uniform. In other words, they are equally likely to occur at any interval.

Example 9.7. In this example we compute the crude incidence rate of all cancers reported in the U.S. (data are from National Cancer Institute, 2004). Figure 9.4 shows the crude cancer rate in the U.S. from 1973 to 2001. Also shown are the 95% confidence intervals, to be discussed later. The figure was produced with

```
1  load('cancerCrudeRate.rda') ; attach(cancerCrudeRate)
2  plot(year, lambda, type = 's',
3      ylab = 'crude rate')
4  lines(year, lower, type = 's')
5  lines(year, upper, type = 's')
```

To obtain a step plot, we use `type = 's'` in the call to `plot()`. □

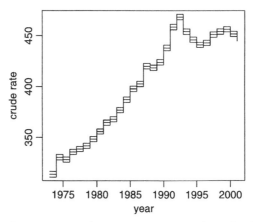

Figure 9.4   U.S. crude cancer rate (per 100 000 per year) on the date of first diagnosis for 1973 through 2001 with 95% confidence intervals. Rates are considered constant during a year; hence the step plot.

Exponential

For the exponential density, $l := 1/\overline{X}$ is the best estimate of the decay parameter, λ.

Example 9.8. Example 5.2 illustrates the fit of the exponential density of the time until the next attack by Hamas (see also Figure 5.2). □

9.1.4 Point estimate of population variance

We discussed the sampling density of a sample variance in Section 7.9. From (7.3) we conclude that the unbiased estimator of the population variance (for any continuous density), σ^2, is

$$\widehat{\sigma}^2 = \frac{n-1}{n} S^2$$

where n is the sample size and S^2 is the sample variance. Because $\hat{\sigma}^2$ is an unbiased estimator of σ^2, you might think that $\hat{\sigma}$ is an unbiased estimator of σ. Not so: The sample standard deviation consistently underestimates the population standard deviation. For convenience, we will use S to estimate σ.

9.1.5 Finding MLE numerically

R includes numerous functions to obtain numerical estimates of the MLE of population parameters from a sample. The idea is to supply the negative of the log-likelihood function along with initial guesses of the parameter values and R will try to minimize it in the parameter space. One of the most difficult issues that you will need to deal with is reasonable initial guesses for the parameters. This means that you will have to supply initial values of the parameter that are close enough to their optimal values (those values that will minimize the negative of the log-likelihood function). Because you do not know these values, any auxiliary information helps. To obtain a reasonable initial guess, try to use functions from the package lmomco.

There is no guarantee that the optimizing function (which minimizes the negative of the log-likelihood function) will find the global minimum. To address this issue, look for MLE from a variety of permutations of initial guesses. If they all converge on a single point in the parameter space, then you are lucky. There are some optimization algorithms that presumably find the global minimum (simulated annealing is one of them). However, their use is beyond our scope.

Example 9.9. Let us create a sample from a normal with known μ and σ^2:

```
> mu <- 20 ; sigma.2 <- 4 ; set.seed(33)
> X <- rnorm(100, mu, sqrt(sigma.2))
```

Next, we define $-L$ according to (9.2):

```
> log.L <- function(mu.hat = 15, sigma.2.hat = 6){
+     n <- length(X)
+     n / 2 * log(2 * pi * sigma.2.hat) +
+         1/2 * sum((X - mu.hat)^2 / sigma.2.hat)
+ }
```

In the function, we set the initial guesses for μ and σ^2 to 15 and 6. We now use mle() from the stats4 package:

```
> library(stats4)
> (fit <- mle(log.L))

Call:
mle(minuslogl = log.L)

Coefficients:
    mu.hat sigma.2.hat
  20.118984    4.022548
Warning message:
NaNs produced in: log(x)
```

`mle()` returns an object of class `mle-class` that we shall use later to analyze the parameter estimates. The warning message arises perhaps because of negative values during search for the minimum. □

9.2 Interval estimation

Let $X_1, \ldots, X_n$ be a random sample with distribution $P(X \leq x \,|\, \boldsymbol{\theta})$. We wish to obtain an interval estimate $I(X \,|\, \theta_i)$ of $\theta_i \in \boldsymbol{\theta}$. The estimate is provided in terms of the so-called *coverage probability* that $\theta_i \in I(X \,|\, \theta_i)$. Because θ_i is not known, we cannot obtain the coverage probability. We can, however, obtain a sample-based interval, e.g. $I\left(X \,\middle|\, \widehat{\theta}_i\right)$, that has a known probability of including θ_i. We refer to this probability as the *confidence coefficient* and to the interval as the *confidence interval* associated with a particular confidence coefficient.

Example 9.10. If $X_1, \ldots, X_n$ are independent and identically distributed rv (with large enough n) with mean μ and variance σ^2, then from the central limit theorem, the sampling density of $\overline{X}$ is normal with $\mu_{\overline{X}} = \mu$ and standard deviation $\sigma/\sqrt{n}$. Therefore, we let $\boldsymbol{\theta} := [\mu, \sigma/\sqrt{n}]$. The covering probability of the interval for μ, for example, is

$$I_{1-\alpha}(X \,|\, \mu) = \left[\Phi^{-1}\left(\alpha/2 \,\middle|\, \mu, \frac{\sigma}{\sqrt{n}}\right), \Phi^{-1}\left(1 - \alpha/2 \,\middle|\, \mu, \frac{\sigma}{\sqrt{n}}\right)\right] \quad (9.4)$$

is $1 - \alpha$, $\alpha \in [0, 1]$ (see Figure 7.3). Because μ and $\sigma/\sqrt{n}$ are not known, we do not know the interval. Suppose that $\sigma/\sqrt{n}$ is known, but μ is not. From the sample, we obtain $\widehat{\mu}$. Now the confidence interval (of $\widehat{\mu}$) for the confidence coefficient $1 - \alpha$ is

$$I_{1-\alpha}(X \,|\, \widehat{\mu}) = \left[\Phi^{-1}\left(\alpha/2 \,\middle|\, \widehat{\mu}, \frac{\sigma}{\sqrt{n}}\right), \Phi^{-1}\left(1 - \alpha/2 \,\middle|\, \widehat{\mu}, \frac{\sigma}{\sqrt{n}}\right)\right]. \quad (9.5)$$

Since $\widehat{\mu}$ is a single value from a single sample with the said sampling density (centered on μ), all we can say is that for repeated samples, $(1 - \alpha) \times 100\%$ of the $\widehat{\mu}$ are within the interval (9.4). Therefore, the probability that (9.5) includes μ is $1 - \alpha$. This fact is often stated as: "We are $(1 - \alpha) \times 100\%$ confident that (9.5) contains μ." □

The argument in Example 9.4 works only if μ and σ^2 are independent, or the variance is known. If they are not, then, in most cases,

$$I_{1-\alpha}\left(X \,|\, \mu, \sigma^2\right) \neq I_{1-\alpha}\left(X \,|\, \widehat{\mu}, \widehat{\sigma}^2\right).$$

Shifting the center of the interval from μ to $\widehat{\mu}$ results in change in the coverage probability because the variance changes. Next, we examine Example 9.10 through R's lens.

Example 9.11. We set the parameters

```
> alpha <- 0.05 ; mu <- 10 ; sigma <- 2 ; n <- 35
> set.seed(222) ; X <- rnorm(n, mu, sigma)
> mu.hat <- mean(X) ; S <- sd(X)
```

and estimate the "unknown" interval

```
> I.mu <- c(low = qnorm(alpha / 2, mu, sigma / sqrt(n)),
+       high = qnorm(1 - alpha / 2, mu, sigma / sqrt(n)))
```

Next, we estimate the confidence interval with σ presumed known

```
> I.mu.hat <- c(qnorm(alpha / 2, mu.hat, sigma / sqrt(n)),
+       qnorm(1 - alpha / 2, mu.hat, sigma / sqrt(n)))
```

and with σ estimated from the sample:

```
> I.mu.sigma.hat <- c(
+       qnorm(alpha / 2, mu.hat, S / sqrt(n)),
+       qnorm(1 - alpha / 2, mu.hat, S / sqrt(n)))
```

Here are the results:

```
> round(rbind(
+       'true interval' = I.mu,
+       'estimated interval, sigma known' = I.mu.hat,
+       'estimated interval, sigma unknown' =
+       I.mu.sigma.hat), 2)
                                   low  high
true interval                      9.34 10.66
estimated interval, sigma known    9.17 10.50
estimated interval, sigma unknown  9.18 10.49
```

We were lucky enough with our sample—the interval with S is narrower than the interval with σ. This does not usually happen. □

We are now ready to discuss interval estimation for some specific densities.

9.2.1 Large sample confidence intervals

Because of the central limit theorem, we can use the fact that the sampling density of the mean is normal with $\mu_{\overline{X}} = \mu$ and standard deviation $\sigma/\sqrt{n}$ where n is the sample size. This is true regardless of the population probability density; i.e. regardless of the density from which the random sample is drawn. In Sections 7.7 and 7.8 we learned how to obtain the normal sampling density for proportions and rates. With this knowledge, we also learn how to construct confidence intervals for proportions and rates.

Mean

We assume that σ^2 is known. This may seem unrealistic. After all, if we know σ^2, then we probably know μ. For large samples, we use this assumption routinely because the sample variance, S^2, is likely to be close to σ^2. We adopt the following

Rule of thumb about sample size For a sample size larger than 30 we set $\sigma^2 = S^2$.

296 Point and interval estimation

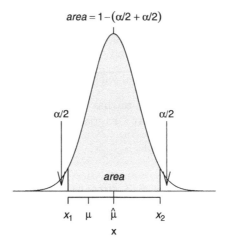

Figure 9.5   Areas excluded and included in computing confidence interval for $1-\alpha$ confidence coefficient.

To obtain the confidence interval, we first need to choose the confidence coefficient, $1-\alpha$. As Figure 9.5 illustrates, the choice of α dictates the area that the interval must cover (or equivalently, the areas under the normal that must be excluded). To determine the values of X_1 and X_2 that result in the desired confidence coefficient, let us rewrite (9.5) in R terms:

$$\begin{aligned}I_{1-\alpha}\left(X\,|\widehat{\mu}\right) &= \left[\Phi^{-1}\left(\alpha/2\,\Big|\widehat{\mu},\frac{S}{\sqrt{n}}\right),\ \Phi^{-1}\left(1-\alpha/2\,\Big|\widehat{\mu},\frac{S}{\sqrt{n}}\right)\right]\\ &= \mathtt{c(qnorm(alpha/2, mu.hat, S/sqrt(n))},\\ &\quad \mathtt{qnorm(1-alpha/2, mu.hat, S/sqrt(n)))}\end{aligned} \quad (9.6)$$

where n is the sample size, mu.hat is the sample mean and S is the sample standard deviation.

Example 9.12. We can sharpen our understanding of confidence intervals by analyzing the code that produces Figure 9.5. First, we draw a sample from a known normal:

```
> mu <- 10 ; sigma <- 2 ; n <- 35 ; alpha <- 0.05
> set.seed(5) ; X <- rnorm(n, mu, sigma)
> mu.hat <- mean(X) ; S <- sd(X)
```

This will allow us to determine if the confidence intervals capture the true mean. By assigning $\alpha = 0.05$, we are setting the confidence coefficient to 0.95. Next, we create a vector to draw the densities and shaded polygon:

```
> x <- seq( mu.hat - 4 * S / sqrt(n),
+     mu.hat + 4 * S / sqrt(n), length = 201)
```

We now plot the sampling density of mean(X):

```
> plot(x, dnorm(x, mu.hat, S / sqrt(n)), axes = FALSE,
+     type = 'l', xlab = expression(x), ylab = '',
+     cex.main = 1,
```

```
+   main = expression(italic(area) == 1 -
+   (alpha / 2 + alpha / 2)))
> abline(h = 0) ; abline(v = mu.hat)
```

We do not want axes just yet, but we want a title, so we set cex.main to 1 and then we draw the main title using expression(). Finally, we draw a horizontal line at zero and a vertical line at $\hat{\mu} = \bar{X}$. We obtain the boundaries (quantiles) of the interval according to (9.6):

```
> X.1 <- qnorm(alpha / 2, mu.hat, S / sqrt(n))
> X.2 <- qnorm(1 - alpha / 2, mu.hat, S / sqrt(n))
```

We prepare the polygon and draw it shaded with:

```
> x.CI <- seq(x.1, x.2, length = 201)
> x.poly <- c( x.1, x.CI, x.2,   x.2)
> y.poly <- c(0, dnorm(x.CI, mu.hat, S / sqrt(n)), 0, 0)
> polygon(x.poly, y.poly, col = 'grey90')
```

Here is how we prepare the annotation:

```
> l <- c(
+   expression(italic(X[1])), expression(italic(mu)),
+   expression(hat(mu)), expression(italic(X[2])))
```

For example, X[1] enclosed in expression draws X_1 and hat(mu) draws $\hat{\mu}$. We draw the x-axis (by specifying the unnamed argument 1)

```
> axis(1, at = c( X.1, mu, mu.hat, X.2), labels = l)
```

and tell axis() to draw the tick marks with at. We add the text in the shaded polygon with

```
> text(mu.hat, .1, expression(italic(area)))
```

The arrows() are drawn with

```
> lx <- X.1 - 0.1 ; hx <- X.2 + 0.1
> ly <- 0.04 ; hy <- 0.5
> arrows(lx, ly, lx, hy, code = 1, angle = 15, length = .15)
> arrows(hx, ly, hx, hy, code = 1, angle = 15, length = .15)
```

including control of the arrows' angle and length. Finally, we annotate the arrows with

```
> text(lx, hy + 0.05, expression(alpha / 2))
> text(hx, hy + 0.05, expression(alpha / 2))
```

The confidence interval is

```
> round(c(low = X.1, high = X.2), 2)
  low   high
 9.69 11.079
```

We conclude that we are 95% certain that the confidence interval captures the true mean, which we happen to know is $\mu = 10$. □

So far, we examined simulated data. In the next example, we use real data.

Example 9.13. The data for this example are from United States Department of Justice (2003). Between 1973 and 2000, there were 7 658 cases of capital punishment in the U.S. The mean (μ) and standard deviation (σ) of age at the time of sentencing were

```
> load('capital.punishment.rda')
> cp <- capital.punishment
> age <- (cp[, 26 ] * 12 + cp[, 25] -
+    (cp[, 12] * 12 + cp[, 11])) / 12
> mu <-  mean(age, na.rm = TRUE)
> sigma <- sd(age, na.rm = TRUE)
> round(c(mu = mu, sigma = sigma),2)
   mu sigma
30.31  8.91
```

Note how we calculate the age of sentencing. A few records from the data should clarify the assignment to age above:

```
> head(cp[, c(26, 25, 12, 11)], 3)
  SentenceYear SentenceMonth DOBYear DOBMonth
1         1971            11    1927       10
2         1971             3    1946        8
3         1971             6    1950        3
```

(DOB stands for date of birth). Here is a typical scenario. The U.S. Department of Justice reports that the mean age of convicts at sentencing to death was 30.3 years. We have no access to (or resources to analyze) the whole data set. We manage to get hold of 30 random records from the data or from press reports:

```
> set.seed(3) ; n <- 30 ; X <- sample(age, n)
> summary(X)
   Min. 1st Qu.  Median    Mean 3rd Qu.    Max.
  17.00   22.02   27.29   28.16   33.08   50.08
```

To obtain the 95% confidence interval, we use (9.6):

```
> mu.hat <- mean(X) ; S <- sd(X) ; alpha <- 0.05
> round(c(low = qnorm(alpha / 2, mu.hat, S / sqrt(n)),
+     high = qnorm(1 - alpha / 2, mu.hat, S / sqrt(n))), 2)
  low  high
25.31 31.01
```

We conclude that we have no grounds to reject the U.S. Department of Justice claim that the average age at sentencing to death was 30.31 years. □

Proportions

As we have seen, estimating ratios is one way to analyze presence/absence data (Example 9.5). Other examples are estimating sex ratios in animal populations, habitat selection by animals and plants, nesting success ratio, proportions of people answering yes or no to a question in a survey and the proportion of patients that die in spite of a treatment.

Here we develop confidence intervals for proportions. Let us quickly review what we know about proportions that is relevant to our discussion here. As usual, we identify a Bernoulli experiment with success or failure. We obtain a rv by assigning 1 to success and 0 to failure. Let N be the population size and n the sample size. We denote the number of objects in the population that exhibit a property that we define as success by N_S, similarly for n_S in the sample. Accordingly,

$$\pi := \frac{N_S}{N}, \quad \widehat{\pi} = p = \frac{n_S}{n}.$$

The density of the number of successes in a sample, X, is binomial with mean and variance $\overline{X} = n \times p$ and $S^2 = n \times p \times (1-p)$. In Section 7.2.2, we established the normal approximation to the binomial. The approximation holds when

$$n \times p \geq 5 \quad \text{and} \quad n \times (1-p) \geq 5. \tag{9.7}$$

In Section 7.7.2 we also stated the consequences of the central limit theorem for sample proportions:

1. $\mu_p = \pi$: the sampling density of p is centered at π.
2. $\sigma_p = \sqrt{\pi(1-\pi)/n}$: as the number of trials increases, the standard deviation of the sampling density of p decreases.
3. The sampling density of p approaches normal as n increases.

And in Exercise 5.18 we established that the MLE of π is $\widehat{\pi} = p$. Therefore, if (9.7) is satisfied, then according to (9.6), the confidence interval for confidence coefficient $1 - \alpha$ is

$$\begin{aligned} I_{1-\alpha}\left(p \mid \widehat{\pi}\right) &= \left[\Phi^{-1}\left(\alpha/2 \,\middle|\, \widehat{\pi}, \sqrt{\frac{\widehat{\pi}(1-\widehat{\pi})}{n}}\right), \Phi^{-1}\left(1 - \alpha/2 \,\middle|\, \widehat{\pi}, \sqrt{\frac{\widehat{\pi}(1-\widehat{\pi})}{n}}\right)\right] \\ &= \texttt{c(qnorm(alpha/2, pi.hat, sqrt(pi.hat} * \texttt{(1 - pi.hat)/n)),} \\ &\quad \texttt{qnorm(1 - alpha/2, pi.hat, sqrt(pi.hat} * \texttt{(1 - pi.hat)/n)))} \end{aligned} \tag{9.8}$$

where pi.hat is the sample proportion of successes and n is the sample size.

Example 9.14. In 2006, there were 45 students in a Statistics class at the University of Minnesota, 15 of them blond. Assuming that the students in the class were a random sample from the population of students at the University of Minnesota (with regard to hair color), let us estimate the number of blonds at the University with 95% confidence. Using (9.8), we write

```
> rm(list = ls())
> n <- 45 ; n.S <- 15 ; pi.hat <- n.S / n ; alpha <- 0.05
> round(c(low = qnorm(alpha / 2, pi.hat,
+     sqrt(pi.hat * (1 - pi.hat) / n)),
+   high = qnorm(1 - alpha / 2, pi.hat,
+     sqrt(pi.hat * (1 - pi.hat) / n))), 2)
 low high
0.20 0.47
```

Now let us compare this (asymptotic in the sense of the normal approximation) confidence interval to some others:

```
> library(Hmisc)
> round(binconf(n.S, n, method = 'all'), 2)
            PointEst Lower Upper
Exact           0.33  0.20  0.49
Wilson          0.33  0.21  0.48
Asymptotic      0.33  0.20  0.47
```

The Wilson and asymptotic intervals are equally wide, but centered around different locations. The exact method gives the widest interval. It is obtained directly from the binomial density. We shall talk about the exact and Wilson methods later. □

Intensities

In Section 7.2.3, we saw that the normal approximation for the Poisson is $\mu = \lambda$ and $\sigma^2 = \lambda$. In Example 5.19 we showed that the MLE of λ is $\widehat{\lambda} = l$ where l is the events count per unit interval (time, area, etc.). For $\lambda \geq 20$, we use the normal approximation to the Poisson, $\phi\left(x \mid \lambda, \sqrt{\lambda}\right)$. Therefore,

$$I_{1-\alpha}\left(l \mid \widehat{\theta}\right) = \left[\Phi^{-1}\left(\alpha/2 \mid \widehat{\lambda}, \sqrt{\widehat{\lambda}/n}\right), \Phi^{-1}\left(1 - \alpha/2 \mid \widehat{\lambda}, \sqrt{\widehat{\lambda}/n}\right)\right] \quad (9.9)$$
$$= \texttt{c(qnorm(alpha/2, lambda.hat, sqrt(lambda.hat/n)),}$$
$$\texttt{qnorm(1 - alpha/2, lambda.hat, sqrt(lambda.hat/n)))}.$$

Example 9.15. We wish to verify a claim that the average count of a plant species is 25 individuals per $1\,\mathrm{m}^2$ plot. So we count 25 plants in 100 plots. We count 24 plants in a different set of 100 plots and set the confidence coefficient to 0.95. Let R generate the data:

```
> lambda <- 25 ; n <- 2 ; alpha <- 0.05
> set.seed(10) ; counts <- sum(rpois(n, lambda))
```

We estimate λ with

```
> lambda.hat <- counts / n
```

and use (9.9) to obtain the confidence interval:

```
> round(c(lambda.hat = lambda.hat,
+    low = qnorm(alpha / 2, lambda.hat,
+        sqrt(lambda.hat / n)),
+    high = qnorm(1 - alpha / 2, lambda.hat,
+        sqrt(lambda.hat / n))), 2)
lambda.hat        low       high
     24.50      17.64      31.36
```

We conclude that we have no grounds to reject the claim. You can obtain the same results with

```
> library(epitools)
> round(pois.approx(counts, pt = 2), 2)
  x pt rate lower upper conf.level
1 49  2 24.5 17.64 31.36       0.95
```

where approximate refers to the asymptotic confidence interval as given in (9.9). The package `epitools` provides three other estimates of confidence intervals for the Poisson parameters named `pois.exact()`, `pois.daly()` and `pois.byar()`. □

9.2.2 Small sample confidence intervals

Because of the central limit theorem, large sample confidence intervals can be developed regardless of the density of the data. This is so because the sampling density of the mean is normal. What do we do when the sample size is small? Assume a specific density for the population and use that assumption to develop the sampling density of the parameter of interest. Then use knowledge of the sampling density to develop confidence intervals. In the following sections we discuss the case of small sample sizes (< 30). We address the following situations: the population density is normal, binomial or Poisson. Finally, we discuss the case where no assumption about the population density is made and we wish to estimate the confidence interval of whatever function of the sample values we please.

Normal population

Let X be a rv from a normal with mean μ and standard deviation σ. Then

$$Z = \frac{X - \mu}{\sigma}$$

is a rv with $\mu = 0$ and $\sigma = 1$ (the standard normal). Now for $n > 30$, the *sampling density* of $\overline{Z}$ is (asymptotically) normal with $\mu_{\overline{Z}} = 0$ and standard error $\sigma_{\overline{Z}} = 1/\sqrt{n}$. When $n < 30$, the sampling density of $\overline{Z}$ is t with $n - 1$ degrees of freedom. Denote by $P(Z \leq z | n-1)$ the t density with $n - 1$ degrees of freedom. Then in our framework, the confidence interval of $\widehat{\mu} = \overline{Z}$ is given by

$$I'_{1-\alpha}(Z|\widehat{\mu}) = \left[P^{-1}(\alpha/2 | n-1),\ P^{-1}(1 - \alpha/2 | n-1)\right]$$
$$= \mathtt{c(qt(alpha/2, n-1), qt(1-alpha/2, n-1))}\,.$$

To translate $I'_{1-\alpha}(Z|\widehat{\mu})$ back to the scale of X, we center the confidence interval around $\overline{X}$ and scale it by $S/\sqrt{n}$. Therefore, the modified confidence interval is now

$$I_{1-\alpha}(X|\widehat{\mu}) = \overline{X} \pm \frac{S}{\sqrt{n}} I'_{1-\alpha}(Z|\widehat{\mu})$$
$$= \mathtt{(mu.hat + S/sqrt(n)\ *} \quad (9.10)$$
$$\mathtt{c(qt(alpha/2, n-1), qt(1-alpha/2, n-1))}\,.$$

Example 9.16. The data for this example are from Patten and Unitt (2002). The authors reported wing-chord measurements for three subspecies of sage sparrow

302 Point and interval estimation

Table 9.3 Chord lengths of 3 sage sparrow subspecies.

Subspecies	Chord (mm)	sd	n
A. b. Cinera (male)	65.4	3.10	13
A. b. Canescens (male)	70.9	2.88	45
A. b. Nevadensis (male)	78.7	2.79	38
A. b. Cinera (female)	63.0	2.77	12
A. b. Canescens (female)	67.2	2.77	42
A. b. Nevadensis (female)	73.4	2.30	30

(Table 9.3). To estimate the 95% confidence interval for the mean of *A. b. Cinera* males, we implement (9.10)

```
> S <- 3.10 ; n <- 13 ; mu.hat <- 65.4 ; alpha <- 0.05
> t.ci <- c(low = qt(alpha / 2, n - 1),
+     high = qt(1 - alpha / 2, n - 1))
> round(mu.hat + S / sqrt(n) * t.ci, 2)
  low  high
63.53 67.27
```

We do not have access to the original data. If you have data, you can achieve the same results with t.test(). □

Binomial experiments

If we know that the density of a rv of interest is binomial, then we can use the approach suggested by Vollset (1993). Agresti and Coull (1998) showed that this method, called the Wilson method, works better than exact computation of confidence intervals. The sampling density of p according to the exact method is F. Let

$$\nu_1 = 2(n - n_S + 1), \quad \nu_2 = 2n_S$$

be the degrees of freedom. Then the lower value of the confidence interval is

$$p_L = \frac{n_S}{n_S + F^{-1}(1 - \alpha/2 \,|\, \nu_1, \nu_2)(n - n_S + 1)}. \quad (9.11)$$

For the upper value of the confidence interval, we first get the degrees of freedom:

$$\nu_1' = 2n_S + 2, \quad \nu_2' = 2(n - n_S)$$

and then

$$p_H = \frac{(n_S + 1)F^{-1}\left(1 - \alpha/2 \,\middle|\, \nu_1', \nu_2'\right)}{n - n_S + (n_S + 1)F^{-1}\left(1 - \alpha/2 \,\middle|\, \nu_1', \nu_2'\right)}. \quad (9.12)$$

For the Wilson method, let

$$z_C = -\Phi^{-1}\left(\frac{\alpha}{2}\right), \quad p = \frac{n_S}{n}.$$

Define

$$A := p + \frac{z_C^2/2}{n}, \quad B = z_C \sqrt{\frac{p(1-p) + \frac{z_C^2/4}{n}}{n}}.$$

Then
$$p'_L = \frac{A-B}{1+z_C^2/n}, \quad p'_H = \frac{A+B}{1+z_C^2/n}. \qquad (9.13)$$

The computations are available through binconf() in the library Hmisc. Here is an example.

Example 9.17. We record as success the event that one or more individuals of a species are identified in a one m^2 plot. We examine 30 such plots and wish to obtain the confidence interval for $\widehat{\pi}$. To compare the results to π, we simulate the data and compute the confidence interval:

```
> library(Hmisc)
> n <- 30 ; PI <- 0.4 ; set.seed(1)
> (n.S <- rbinom(1, n, PI))
[1] 10
> round(binconf(n.S, n, method = 'all'), 3)
           PointEst Lower Upper
Exact         0.333 0.173 0.528
Wilson        0.333 0.192 0.512
Asymptotic    0.333 0.165 0.502
```

The asymptotic result should *not* be used here because the sample is small (we show it for the sake of comparison). As expected, the Wilson method gives the narrowest confidence interval for 95% confidence coefficient. □

Poisson counts

Recall from Section 5.7 that the Poisson density may be written as

$$P(X = x) = \frac{\lambda^x}{x!} e^{-\lambda}$$

where λ is the intensity parameter (e.g. events per unit of time, number of individuals of a plant species in a plot). We also found that

$$E[X] = \lambda, \quad \sigma^2 = \lambda, \quad \widehat{\lambda} = l$$

(Section 5.8) where l is the sample count rate. Data that represent counts are common and we are often interested in the confidence interval estimate of the intensity parameter λ. When the sample is large, we can invoke the central limit theorem and use the large sample confidence interval procedure outlined in Section 9.2.1. When the sample is small (or large) and the sample comes from a population with Poisson density, then the sampling density of l is approximately χ^2 (see Section 6.8.3). Therefore, we can compute confidence intervals.

Let the intensity l be the count of events per unit of measurement (e.g. time, area) and $P(X \le x | \nu)$ the χ^2 distribution with ν degrees of freedom. Then (see Ulm, 1990; Dobson et al., 1991)

$$\begin{aligned} I_{1-\alpha}\left(x \,\middle|\, \widehat{\lambda}\right) &= \left[\frac{P^{-1}(\alpha/2 \,|\, 2l)}{2}, \frac{P^{-1}(1-\alpha/2 \,|\, 2(l+1))}{2}\right] \\ &= \texttt{c(qchisq(alpha/2, 2 * l)/2,} \\ &\qquad \texttt{qchisq(1 - alpha/2, 2 * (l + 1))/2)} \,. \end{aligned} \qquad (9.14)$$

Example 9.18. Consider a count of 11 deaths per day per 1 000 individuals. To estimate the 95% confidence interval, we refer to (9.14):

```
> l <- 11 ; alpha <- 0.05
> round(c(low = qchisq(alpha / 2, 2 * l) / 2,
+       high = qchisq(1 - alpha / 2, 2 * (l + 1)) / 2), 1)
 low high
 5.5 19.7
```

Compare this to the following:

```
> library(epitools)
> round(pois.exact(l), 2)
   x pt rate lower upper conf.level
1 11  1   11  5.49 19.68       0.95
```

The package `epitools` contains functions that are used routinely in epidemiological research. □

9.3 Point and interval estimation for arbitrary densities

The results in this section apply to both small and large samples. They also apply to arbitrary population parameters. Next to the worst case scenario is when we have a small sample and no idea about the sampling density of a statistic.[2] In such cases, we can use the bootstrap method to estimate confidence intervals for a statistic. You may find detailed discussion of the bootstrap method in (among others) Efron (1987), Efron and Tibshirani (1993), DiCiccio and Efron (1996) and Davison and Hinkley (1997).

In a nutshell, the bootstrap procedure repeats sampling (with replacement) and pretends that each new sample is independent and is taken from the population (as opposed to from the sample).[3] The fundamental assumption is that the sample represents the population. Implementation of the bootstrap in R is quite easy: use `sample()` with size = n (sample size) and with replace = TRUE. Calculate the statistic of interest and accumulate it in a vector. After many repetitions (in the thousands), you will have enough data to obtain the empirical sampling density and analyze it (with respect to point and interval estimates). R includes several functions that bootstrap (see also Section 7.10).

Example 9.19. One of the most celebrated power laws in biology is the relationship between metabolic rate at rest, called basal metabolic rate (BMR) (ml O_2 per hr) and body mass (g) within mammals. The data we use are from White and Seymour (2003). Consider small animals (between 10 and 20 g) from two mammalian orders, *Insectivora* (insect-eating) and *Rodentia* (rodents). The density for known species data of body mass and BMR are shown in the top panel of Figure 9.6. ($n = 13$ and 24). We are interested in the following questions: What is the sampling density of the mean

---

[2] In the worst case scenario we have no data.
[3] This is why it is called bootstrap—it is as if we are lifting ourselves by pulling on our bootstraps.

Point and interval estimation for arbitrary densities   305

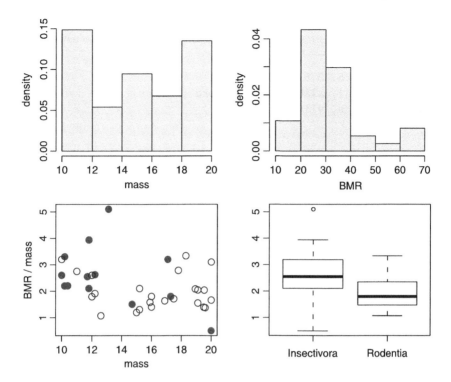

Figure 9.6 Results for known species with body mass between 10 and 20 g for two mammalian orders. Top: density of body mass and BMR. Bottom: scatter of the data for *Insectivora* (filled circles) and *Rodentia* (open circles).

of the ratio of BMR / mass for these two orders? What are the corresponding means and confidence intervals for 95% confidence coefficient?

To focus attention on the issue at hand, we shall not go over the data `bmr.rda` manipulations that allow us to answer these questions. Suffice it to say that we end with two data frames, one for *Insectivora* and the other for *Rodentia*. The relationship between mass and BMR / mass are shown in the bottom panel of Figure 9.6. Let us move on to the bootstrap. First, we set the number of bootstrap repetitions, prepare a matrix with two columns that will hold the BMR / mass means and remember the sample sizes (the number of species from each order for which data are available):

```
> B <- 10000
> n <- c(length(Insectivora), length(Rodentia))
```

Next, we compute the B means for each order using bootstrap:

```
> set.seed(3)
> m.i <- matrix(sample(Insectivora, n[1] * B,
+    replace = TRUE), ncol = B, nrow = n[1])
> m.r <- matrix(sample(Rodentia, n[2] * B,
+    replace = TRUE), ncol = B, nrow = n[2])
> mean.BMR.M <- cbind(apply(m.i, 2, mean),
+    apply(m.r, 2, mean))
```

306 Point and interval estimation

We now have enough information to examine the bootstrapped sampling densities of the mean BMR / mass for the two mammalian orders. We start with *Insectivora* (Figure 9.7 left):

```
> x.limits <- c(1.5, 3.5)
> h(mean.BMR.M[, 1], xlab = 'Insectivora',
+    xlim = x.limits, ylim = c(0, 1.25))
```

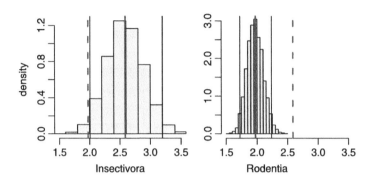

Figure 9.7 Sampling densities for the mean of BMR / mass for known species with body mass between 10 and 20 g for two mammalian orders. Left: *Insectivora*. Shown are the 95% confidence interval, mean for *Insectivora* and mean for *Rodentia* (broken line). Right: Same, this time for *Rodentia*.

We add the 95% confidence limits with

```
> ci <- matrix(ncol = 2, nrow = 2)
> ci[1, ] <- quantile(mean.BMR.M[, 1],
+    prob = c(0.025, 0.975))
> abline(v = ci[1, 1]) ; abline(v = ci[1, 2])
```

draw the mean of the ratio for *Insectivora*

```
> abline(v = mean(mean.BMR.M[, 1]), col = 'red', lwd = 2)
```

and add a broken line for the mean of *Rodentia*

```
> abline(v = mean(mean.BMR.M[, 2]),
+    col = 'red', lwd = 2, lty = 2)
```

Note that the latter's mean ratio is outside the confidence interval for the mean of the former. We repeat the same steps for *Rodentia* (Figure 9.7 right):

```
> h(mean.BMR.M[, 2], xlab = 'Rodentia', ylab = '',
+    xlim = x.limits, ylim = c(0, 3))
> ci[2, ] <- quantile(mean.BMR.M[, 2],
+ prob = c(0.025, 0.975))
> abline(v = ci[2, 1]) ; abline(v = ci[2, 2])
> abline(v = mean(mean.BMR.M[, 2]), col = 'red', lwd = 2)
> abline(v = mean(mean.BMR.M[, 1]), col = 'red',
+    lwd = 2, lty = 2)
```

In both cases, we observe that the other Order's mean ratio of BMR / mass is outside the 95% confidence interval. This raises questions about the common practice of lumping all orders of mammals into a single power law of body mass vs. BMR. □

9.4 Assignments

Exercise 9.1. Figure 9.8 shows 3 sampling distributions of 3 different statistics along with the true value of the population characteristic. Which of the statistics would you choose? Why?

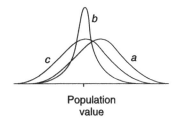

Figure 9.8   Three sampling distributions of three statistics.

Exercise 9.2. One of the criteria for choosing the best estimate of a statistic is that its sampling distribution has the smallest variance. Why is this important?

Exercise 9.3. Fill in the appropriate estimates in Table 9.4

Table 9.4 $E[X]$ and $V[X]$ denote the expected value and variance of the rv X.

Distribution	Statistic	Best sample-based estimate
normal	$E[X]$	...
	$V[X]$	...
binomial	$E[X]$	...
	$V[X]$	...
Poisson	$E[X]$	...
	$V[X]$	...

Exercise 9.4. What is the confidence level that corresponds to significance value of α?

Exercise 9.5.

1. Why does the standard deviation of the sampling distribution of $\overline{X}$ decrease as the sample size increases?
2. What other name is this standard deviation is known by?

3. Let
$$Z = \frac{\bar{X} - \mu}{\sigma/\sqrt{n}}.$$

What are the expected value and variance of Z?

Exercise 9.6. In a sample of 1 000 randomly selected people in the U.S., 320 said that they oppose abortion. Let π denote the proportion of the U.S. population that opposes abortion. Give a point estimate of π.

Exercise 9.7. The U.S. Environmental Protection Agency (EPA) publishes rules about safe values of radionucleotides (isotopes of elements that emit radiation) in water. These radionuclides are carcinogens. The EPA's rules for drinking water safety are called Maximum Contaminant Levels (MCL) in drinking water. Levels higher than MCL are considered unsafe. As of the year 2000, the MCL for Ra-226, Ra-228 and gross alpha-particle activity in community water systems are: (a) Combined Ra-226 and Ra-228: 5 pCi/L; (b) Gross alpha-particle activity (including Ra-226 but excluding radon and uranium): 15 pCi/L. The data file is named `nucleotides-usgs.txt`. The data include the following columns:

```
    name                         explanation
1   USGS.SN                      USGS serial number
2     result  reading in pCi/L (pico-Curie per liter
3         sd                     standard deviation
4        mdc minimum detectable concentration (1 SD)
5 nucleotide      of radium (Ra-224, Ra226 or Ra228)
```

(see Focazio et al., 2001). Download the data and import them into R.

1. Are the data about the variable named `result` normal? If not, what transformation would you use on this variable?
2. Give an estimate of μ. (If you use a transformation, compute μ on the transformed variable first and then reverse the transformation to provide μ.)
3. Give an estimate of σ^2 of the data (or its transform if you decide to transform it). Exclude the negative values from the estimate.
4. Estimate the probability that a randomly chosen well will exceed the MCL for the combined Ra-226 and Ra-228.
5. Create a data frame for the data and save the data frame in an R file named `nucleotides.rda`.

Exercise 9.8. Discuss how each of the following factors affects the width of the large sample confidence interval for μ when σ is known.

1. Confidence level
2. Sample size
3. Population standard deviation

Exercise 9.9. The formula used to compute a confidence interval for μ when n is large and σ is known is $\bar{X} \pm z_{1-\alpha/2} \times \sigma/\sqrt{n}$. What is the value of $z_{1-\alpha/2}$ for each of the following confidence levels?

1. 95%
2. 90%

3. 99%
4. 80%
5. 85%

Exercise 9.10. Suppose that 50 random samples of deer urine in snow are analyzed for the concentration of uric acid. Denote by μ the average concentration of urea uric acid in the population. Suppose that the sample resulted in a 95% confidence interval of $[9, 11]$.

1. Would a 90% confidence interval be narrower or wider than the given interval? Explain.
2. Consider the statement: There is a 95% chance that μ is between 9 and 11. Is the statement correct? Explain.
3. Consider the statement: If the process of selecting a sample of size 50 and then computing the corresponding 95% confidence interval is repeated 100 times, then 95 of the resulting intervals will include μ. Is the statement correct? Explain.

Exercise 9.11. The following data summarize the nucleotide data (see Exercise 9.7):

```
        mean        sd      n
Ra224  7.690137  16.939224  90
Ra226  2.088269   3.148321  90
Ra228  2.383604   8.176315  90
```

Compute

1. The 90% confidence intervals for the mean pCi/L for Ra224, Ra226 and Ra228.
2. Are the intervals similar? Do you think that there is a difference in mean concentrations of radium isotopes for Ra224, Ra226 and Ra228? Explain.

Exercise 9.12. For the following data summaries, compute the 95% confidence intervals for the mean (here SD refers to the sample standard deviation and σ refers to the population standard deviation):

1. $n = 40$, $\overline{X} = 50$, $SD = 4$, $\sigma = 5$, population distribution is normal.
2. $n = 30$, $\overline{X} = 50$, $SD = 4$, $\sigma = 5$, population distribution is normal.
3. $n = 40$, $\overline{X} = 50$, $SD = 4$, $\sigma = 5$, population distribution is unknown.
4. 10 successes in 40 independent Bernoulli trials with $\pi = 0.4$.
5. 10 successes in 40 independent Bernoulli trials.
6. 4 success in 54 independent Bernoulli trials.
7. 30 plants per 50 plots.
8. 10 plants per 50 plots.
9. $n = 20$, $\overline{X} = 10$, $SD = 2$, population distribution is normal.

Exercise 9.13. Use the package Hmisc to compute the binomial confidence intervals for confidence level = 0.95 for the following:

```
nS   n
10   40
 4   10
 3   20
10  100
```

Exercise 9.14.

1. Write two expressions that calculate the low and high values for the confidence interval of Poisson counts and for a 0.95 confidence interval.
2. Compute the 95% confidence intervals for the following:
   (a) 10 plants per 50 plots;
   (b) average of 1 plant per plot;
   (c) 5 plane landings per hour in a nearby airport.

Exercise 9.15. For this exercise, use the functions bootstrap() and summary(), which you may find in the script file named bootstrap.R.

If you write a script for this exercise, to use the functions in bootstrap.R, start your script with the statement source('bootstrap.R'). If you use the workspace in R for this exercise, then run the statement source('bootstrap.R')

1. Create a new data frame from nucleotides that looks like this:

   ```
       Ra224  Ra226  Ra228
   1  -0.200 0.1300 0.0512
   2      NA 0.1310     NA
   3   0.484 0.1090     NA
   4      NA 0.1590 0.1750
   5   0.020 0.0492 0.3370
   6      NA 0.0860 0.4350
   ```

   where the values correspond to the variable result in nucleotides.
2. Run pairs() on the new data frame. From the pairs, it seems that there are relationships among the various isotopes of radium. Use an appropriate transformation on the data to obtain linear relationships. Show the outcome of the transformation in a pairs() plot.
3. Use the function bootstrap() to create the densities of the means of the 3 isotopes. Show the densities.
4. Do the same for the variance.
5. Use summary() on the objects produced in the previous two items to provide confidence intervals on the means and variances of the 3 isotopes. What are they?
6. Draw and report conclusions about the relationships among the isotopes in wells across the U.S.

Exercise 9.16. Explain the meaning of $t_{15, 0.95}$.

Exercise 9.17. For the following, use the appropriate R functions.

1. The upper 5th percentile of a t distribution with 21 df.
2. The lower 5th percentile of a t distribution with 18 df.
3. $P(X < 2)$ where P is the t distribution with 15 df.
4. $P(X > 1.8)$ where P is the t distribution with 17 df.
5. $P(X < 0)$ where P is the t distribution with 22 df.
6. $P(X > 0)$ where P is the t distribution with 10 df.

Exercise 9.18.

1. A sample of 10 fishermen revealed that their weight (in kg) before going out on a fishing trip was

   93.7 101.8   91.6 116.0 103.3   91.8 104.9 107.4 105.8   96.9

   What is the 95% CI around the mean sample weight?
2. During the trip, 1 fisherman disappeared. Upon their return, the remaining 9 fishermen were weighed. Now the data were

   104 112 102 126 113 102 115 117 116

   What is the 95% CI now?

Exercise 9.19. Write the equation you would use to construct confidence intervals around the mean under the given conditions.

1. Sample of 35 chord lengths from birds where chord length has a uniform distribution in the population. Variance of chord length in the population is known.
2. Sample of 32 weights of birds where the weights have a normal distribution in the population. Variance in the population not known.
3. Sex ratio in a sample of 33 birds, under the assumption that the sex ratio is 1 : 1.
4. Twenty five boats arriving to a dock every day.
5. Chord length from a sample of 22 birds, where chord length has a normal distribution in the population.
6. Unknown population distribution, small sample (10).

Exercise 9.20. Identity three functions in R that compute confidence intervals for a binomial experiment with 10 trials and 5 successes. Run the functions and explain the results.

Exercise 9.21. You are interested in estimating the true value of the number of ticks per oak tree in you neighborhood. What will be the 95% interval within which you are are likely to capture the true value of the number of ticks per oak tree if you counted 30 ticks per tree?

10
Single sample hypotheses testing

Here we meet inferential statistics for the first time. So far we dealt with the problem of how to estimate a population parameter such as mean (μ), proportion (π), intensity (λ) and a range of values within which it may reside. We can use samples to test the plausibility of claims. For example, a biologist tells us that the average litter size of a sample taken from a population of wolves is 8. Based on the analysis we develop here and with a sample of our own, we will be able to make a statement about how plausible this claim is. The plausibility of a claim is stated in terms of probability. If the probability is high, then we deem our claim correct. Otherwise, we reject it as being incorrect.

For the most part, we follow the sequence in Chapter 9—we will discuss large sample hypothesis testing and then small sample. For large samples, we deal with hypothesis testing for means, ratios and intensities. We follow a similar sequence for small samples, where to make progress, we may need to assume a density. So we discuss small sample hypotheses testing for a sample from a normal, binomial or a Poisson population. Finally, we discuss the general case, where we wish to test hypotheses for an arbitrary parameter from a population with arbitrary density. Here, as in Chapter 9, we rely on the bootstrap method.

10.1 Null and alternative hypotheses

A *hypothesis* is an assumption about the value or values of a population parameter or parameters.

For example, we may assume that the average density of grasses (plants per m^2) in an area is $\lambda = 10$. We may assume that the proportion of females in the population is $\pi > 0.5$. Recall that Greek letters represent population parameters. As before, we shall use a sample statistic and its density to estimate the plausibility of an assumption.

314   Single sample hypotheses testing

We always formulate two mutually exclusive and, if possible, exhaustive hypotheses. Exhaustive hypotheses are such that their union covers all possible outcomes of an experiment, i.e. their union is the sampling space. By formulating contradictory hypotheses, we are forced to choose between them.

Example 10.1. In set theory terms, a hypothesis might be that $\theta \in A$. Then the alternative hypothesis is $\theta \in \overline{A}$ (the complement of A) or equivalently, $\theta \notin A$.   □

10.1.1   Formulating hypotheses

The standard procedure in hypothesis testing is to assume that one hypothesis is true. We then reject this hypothesis in favor of an alternative hypothesis if the sample evidence is incompatible with the original hypothesis. As a rule, the hypothesis we choose as true (before running the analysis) should be such that the burden of proof is on us. Analogous to a court case, the original hypothesis is that the accused is innocent unless proved otherwise. The alternative hypothesis is guilty and the burden of proof falls on the prosecutor. Thus, we have the following definitions.

Null hypothesis (H_0) The claim that is initially assumed true.

Alternative hypothesis (H_A) The claim that is initially assumed not true.

We assume that H_0 is true and reject it in *favor* of H_A if the sample evidence strongly suggests that H_0 is false. This is the idea of "beyond a reasonable doubt" in a court analogy. In general, we formulate H_0 such that if we do not reject it, then we take no action.

Example 10.2. Let λ be the average number of daily visits to a hospital emergency room. Assume that the number of daily visits are independent on different days and visits are independent of each other. The hospital management is interested in estimating λ, the true "population" daily visits. If $\lambda > 25$ patients per day, then management will invest $10 million in emergency room renovations. Otherwise, no investment will be considered. There are two sets of null and alternative hypotheses:

$$H_0 : \lambda \leq 25 \quad \text{vs.} \quad H_A : \lambda > 25 \quad \text{or}$$
$$H_0 : \lambda \geq 25 \quad \text{vs.} \quad H_A : \lambda < 25 \,.$$

Which one should we choose? If management chooses the first alternative, then unless there is strong evidence that $\lambda > 25$, we will not reject the hypothesis that $\lambda \leq 25$ and no investment will be made. If management chooses the second alternative, then unless there is strong evidence that $\lambda < 25$, we will not reject the hypothesis that $\lambda \geq 25$ and the investment will be made. So administrators concerned about patients' well-being should choose the second pair. Administrators concerned about profit might choose the first pair. In the first alternative, the burden of the proof is on $\lambda > 25$. In the second, it is on $\lambda < 25$.   □

The situation is a little different in the next example.

Example 10.3. Field biologists often use darts filled with anesthetics to capture animals for data collection, translocation and so on. It is important to know precisely the concentration of the active ingredient in the solution; otherwise, the animal might die or endanger its handler. A pharmaceutical company claims that the concentration of the anesthetizing agent in the solution is $\mu = 1$ ml per 10 cc saline. We wish to test the manufacturer's claim. Let μ be the mean concentration of the active ingredient in the imaginary population of all solutions that the company manufactures. Because the consequences of both cases ($\mu > 1$ or $\mu < 1$) are unacceptable, we set

$$H_0 : \mu = 1 \quad \text{vs.} \quad H_A : \mu \neq 1 .$$

In other words, we believe the manufacturer. The burden of proof that $\mu \neq 1$ is then on us. □

Suppose that the hospital administrators in Example 10.2 choose $H_0 : \lambda \leq 25$. That is, they'd rather not invest the money unless they have to. Were they to choose $H_0 : \lambda = 25$ and the evidence was in fact $\lambda < 25$, then they would have taken no action. In other words, they would have erred on the side of caution (according to their perception of caution). So instead of using $H_0 : \lambda \leq 25$ they could use

$$H_0 : \lambda = 25 \quad \text{vs.} \quad H_A : \lambda < 25 .$$

Similarly, suppose the administrators choose $H_0 : \lambda \geq 25$. That is, they'd rather invest the money. Were they to choose $H_0 : \lambda = 25$ and the evidence was in fact $\lambda > 25$, then they would have invested the money. In other words, they would have erred on the side of caution (according to their now different perception of caution). So instead of using $H_0 : \lambda \geq 25$ they could use

$$H_0 : \lambda = 25 \quad \text{vs.} \quad H_A : \lambda > 25 .$$

Let θ be the parameter of interest and let θ_0 be a specific value that θ might take. Then for the null hypothesis, we shall always choose

$$H_0 : \theta = \theta_0 \quad \text{vs. one of} \quad H_A : \theta \begin{cases} < \theta_0 \\ \neq \theta_0 \\ > \theta_0 \end{cases} . \tag{10.1}$$

The choice of H_0 vs. H_A for a specific problem is not unique. It often depends on the purpose of the study and the bias of the investigator. This is why we should always prefer to err on the side of caution; that is, we should choose the null hypothesis that is the opposite of what we would have liked.

Example 10.4. Continuing with Example 10.3, the manufacturer claims that $\mu = 1$. If the animal's safety is of more concern than the fact that we may fail to anesthetize it, then we choose

$$H_0 : \mu = 1 \quad \text{vs.} \quad H_A : \mu > 1 .$$

So if we reject H_0, we prefer the alternative hypothesis that the concentration is actually greater than 1. This will lead us to reject a solution because it contains too much of the anesthetizing ingredient more often than if we were to choose $H_A : \mu < 1$. □

The upshot? Choose H_A such that if you reject H_0, then H_A represents the more cautious action. Similar considerations apply when dealing with proportions and intensities.

Example 10.5. Suppose that a new AIDS drug is to be tested on humans. The drug has terrible side effects and is intended for patients that usually die within a month. We decide that if at least 10% of the patients to whom the drug is administered survive for two months or more, then the drug should be administered.

In this case we set

$$H_0 : \pi = 0.1 \quad \text{vs.} \quad H_A : \pi > 0.1 \, .$$

So if we reject H_0 (and we err in rejecting it), then we would administer the drug anyway. We do this because the drug is designed for gravely ill patients. We would rather give them the drug than not. □

In any case, we always have to guard against drawing the wrong conclusion. We may decide to reject H_0 while it is true. We may also decide not to reject H_0 while it is false. We deal with these kinds of errors next.

10.1.2 Types of errors in hypothesis testing

As we have seen, a statistic such as $\overline{X}$ is a rv. Therefore, its value is subject to uncertainty. This uncertainty may lead to errors in reaching conclusions about rejecting H_0 based on hypothesis testing. Because of the way H_0 and H_A are chosen, there are two fundamental types of errors.

Type I error H_0 is correct, but we reject it.

Type II error H_0 is incorrect, but we fail to reject it.

Example 10.6. Consider a court of law analogy. Our H_0 is that the accused is innocent. Under *this* hypothesis, the two possibilities are:

Type I error The accused is in fact innocent, but is found guilty.
Type II error The accused is in fact guilty, but is found innocent.

These errors arise because of uncertainty and because of the way H_0 and H_A are set—mutually exclusive and exhaustive. □

Table 10.1 Type I and type II errors.

State of nature	Action	
	H_0 not rejected	H_0 rejected
H_0 is true	No error	Type I error
H_0 is false	Type II error	No error

Table 10.1 summarizes these errors. With regard to H_0, nature can be in one of two states: true or false. We have two possible actions: declare the state of nature true or false. If H_0 is true and we reject it, then we commit a type I error. If H_0 is false and

we do not reject it, we commit a type II error. Our goal is to minimize the possibility that we commit any of these two errors based on the sample data. It is important to keep in mind that type I and II errors refer to H_0, *not* to H_A. As we shall see, minimizing the possibility of type I error conflicts with the goal of minimizing the possibility of type II error.

Example 10.7. In North America (and elsewhere), waterfowl population studies often include tagging birds. The tags are small, light-weight sleeves that go on the bird's leg. They usually include a return address and an offer of reward for those who find and return them. Tag returns are then used to analyze and draw conclusions about the populations. Suppose that in a normal year, 25% of all tags are returned. To increase this percentage, a researcher designs a one-year experiment: Increase the reward for a returned tag in the hope of increasing the proportion of tag returns. If the experiment succeeds, then the increased reward will continue. To set the hypothesis, the researcher defines π to be the true proportion of returned tags. From next year's return, the researcher wishes to test:

$$H_0 : \pi = 0.25 \quad \text{vs.} \quad H_A : \pi > 0.25 .$$

One possible outcome of the experiment is that the increased reward had no effect on the proportion of returned tags (i.e. $\pi = 0.25$). Yet, the results of the experiment lead the researcher to conclude that it did. The researcher then rejects H_0 in favor of H_A (i.e. he concludes that $\pi > 0.25$). This is a type I error.

Another possible outcome of the experiment is that the increased reward had an effect (i.e. $\pi > 0.25$). Yet, the results of the experiment lead us to conclude that it did not. The researcher then does not reject H_0 in favor of H_A (i.e. he concludes that $\pi = 0.25$). This is a type II error.

Because the conclusion is based on a sample and because a sample is subject to random variations, there is no guarantee that the researcher did not commit one of the two errors. However, the researcher's goal is to determine the likelihood of committing one of the two errors. The consequences of type I and type II errors differ. Type I error results in continuing the higher reward while the investment is not justified. Type II error results in discontinuing the higher reward while the investment is justified. □

10.1.3 Choosing a significance level

Each error type has a probability associated with it: The probability of type I error, denoted by α and the probability of type II error, denoted by β. A test of a hypothesis with $\alpha = 0.05$ is said to have *significance level* of 0.05. For one reason or another, traditional values for α are set to 0.1, 0.05 or 0.01.

When we choose $\alpha = 0.05$, we in effect say that if we were to repeat sampling the population and testing the same hypotheses many times, then we will reject a true H_0 in about 5% of the repetitions. A smaller α means that we elect to minimize the risk of rejecting a true null hypothesis. Why then not choose $\alpha = 0.01$, or better yet, $\alpha = 0.001$? The problem is that the smaller α is (the smaller the likelihood of type I error), the larger β is (the larger the likelihood of type II error). Therefore, we have

Rule of thumb Choose the largest value of α that is tolerable (traditional values for α are 0.1, 0.05 and 0.01).

Example 10.8. The U.S. president is concerned about his reelection. He therefore decides to go to war based on results from a public opinion survey. If, he says, at least 75% of the people support going to war, then we will go to war; otherwise, we will not. So he sets
$$H_0 : \pi = 0.75 \quad \text{vs.} \quad H_A : \pi < 0.75 .$$
Type I error is rejecting the null hypothesis in favor of the alternative hypothesis when the null hypothesis is true. That is, type I error means that the president might choose *not* to go to war when in fact he should have. Now suppose the president says: If at least 25% of the people object to the war, we will not go to war. So he wishes to test
$$H_0 : \pi = 0.25 \quad \text{vs.} \quad H_A : \pi < 0.25 .$$
Here type I error is rejecting $\pi = 0.25$ in favor of $\pi < 0.25$. So type I error means that the president might decide to go to war in spite of the fact that he should not.

In the first case, type I error does not have grave consequences—under the assumption that erring on the side of not going to war is better than erring on the side of going to war. So we advice the president to choose $\alpha = 0.1$ or perhaps even larger. In the second case, type I error has serious consequences and we advise the president to choose $\alpha = 0.01$ or even smaller. The better alternative in the second case is to set H_A: $\pi > 0.25$. Anyway, a politician perhaps should not decide to go to war based on public opinion. □

10.2 Large sample hypothesis testing

In the next few sections, we examine testing hypotheses for large samples where the parameters of interest are mean, proportion or intensity. To keep the notation consistent with things to come, we will, from now on, write $\mathrm{SE} := S/\sqrt{n}$ for the standard error.

10.2.1 Means

Recall from Section 9.2.1 that when the sample size $n > 30$, we may use the sample standard deviation, S, to approximate the population standard deviation, σ. Furthermore, the sampling density of the sample mean, $\phi(x\,|\widehat{\mu}, \mathrm{SE})$, is centered on $\mu_{\widehat{X}} = \mu$ (the population mean). We assume that $\mu = \mu_0$, where μ_0 is given and wish to test one of the pairs in (10.1)—with μ replacing θ—with a significance level α. We have a sample of size n (> 30) with $\overline{X}$ and standard deviation S. Because we assume that $\mu = \mu_0$ is true, the sampling density of the rv $\overline{X}$ is $\phi(x\,|\mu_0, \mathrm{SE})$. With this in mind, let us construct the test for each pair of H_0, H_A in (10.1), one at a time.

Lower-tailed test

Our pair of hypotheses is
$$H_0 : \mu = \mu_0 \quad \text{vs.} \quad H_A : \mu < \mu_0 .$$
Based on what we know (that the sampling density of $\overline{X}$ is $\phi(x\,|\mu_0, \mathrm{SE})$ and that the significance level is α), we compute
$$x_L = \Phi^{-1}(\alpha\,|\mu_0, \mathrm{SE}) \qquad (10.2)$$
$$= \mathtt{qnorm(alpha, mu.0, SE)}$$

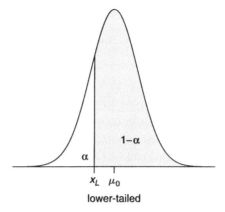

Figure 10.1 The sampling density of $\overline{X}$ under the null hypothesis is $\phi(x \mid \mu_0, \text{SE})$. x_L is the critical value in the sense that the probability of any $\overline{X}$ to the left of it is small ($= \alpha$) and therefore if we do get $\overline{X} < x_L$, then we reject H_0.

(Figure 10.1) where x_L is a *critical value* in the following sense. The probability that $\overline{X} \leq x_L$ is α, or by our notation,

$$\Phi(x_L \mid \mu_0, \text{SE}) := P\left(\overline{X} \leq x_L\right) = \alpha \ .$$

If we do get $\overline{X} \leq x_L$ from the sample, then we must reject H_0 because under it, such $\overline{X}$ is very unlikely. In fact, if $\overline{X} \leq x_L$, we conclude that $\mu < \mu_0$, which is our alternative hypothesis. Incidentally, to obtain Figure 10.1, we recycle the code we used to obtain Figure 9.5 (see Example 9.12). To wit, for lower-tailed hypothesis testing, use (10.2) to obtain x_L. If the sample-based $\overline{X} \leq x_L$, reject H_0 in favor of H_A.

Example 10.9. In Example 9.13 we constructed the confidence interval of the age at sentencing for a sample from the population of death penalty convicts in the U.S. We concluded that the true population mean was captured by the interval. Let us examine the following situation: The government says, "The mean age at sentencing to death is 30.31 years." You say, "I do not believe you. I think it is smaller. Let me see the data." They say, "We cannot give you the data, but we are willing to provide you with a random sample of 30 cases." We use the same sample as in Example 9.13 to test the hypothesis that

$$H_0 : \mu = 30.31 \quad \text{vs.} \quad H_A : \mu < 30.31 \ .$$

with $\alpha = 0.05$. So we obtain the sample, $\boldsymbol{X}$, exactly as we did in Example 9.13 and specify the data thus:

```
> mu.0 <- mean(age, na.rm = TRUE)
> n <- 30 ; alpha <- 0.05
> X.bar <- mean(X) ; S <- sd(X) ; SE <- S / sqrt(n)
```

We are now ready to implement (10.2):

```
> c(mu.0 = mu.0, x.L = qnorm(alpha, mu.0, SE), X.bar = X.bar,
+   alpha = alpha)
  mu.0   x.L X.bar alpha
 30.31 27.92 28.16  0.05
```

Because $\overline{X} > x_L$, we do not reject H_0 and consequently believe the government's claim. Similar to the confidence interval we interpret the result thus: If we were to take many samples of size n from the population of death penalty convicts, then in the limit, 95% of the samples' means will be larger than $x_L = 27.92$. Therefore, based on our rejection probability (α), we believe that our sample is typical and therefore do not reject H_0. □

Upper-tailed test

Our pair of hypotheses is

$$H_0 : \mu = \mu_0 \quad \text{vs.} \quad H_A : \mu > \mu_0 \,.$$

From the information we have ($\phi(x\,|\mu_0, \text{SE})$ and α), we compute

$$x_H = \Phi^{-1}\left(1 - \alpha\,|\mu_0, \text{SE}\right) \tag{10.3}$$
$$= \texttt{qnorm}(1 - \texttt{alpha}, \texttt{mu.0}, \texttt{S}/\texttt{sqrt(n)})$$

(Figure 10.2) where x_H is a *critical value* in the following sense. The probability that $\overline{X} > x_H$ is α, or by our notation,

$$1 - \Phi\left(x_H\,|\mu_0, \text{SE}\right) := 1 - P\left(\overline{X} \le x_H\right) = \alpha$$

is small ($= \alpha$). If we do get $\overline{X} > x_H$ from the sample, then we must reject H_0 because under it, such $\overline{X}$ is very unlikely. In fact, if $\overline{X} > x_H$, we conclude that $\mu > \mu_0$, which is our alternative hypothesis. To wit, for upper-tailed hypothesis testing, use (10.3) to obtain x_H. If the sample-based $\overline{X} > x_H$, reject H_0 in favor of H_A.

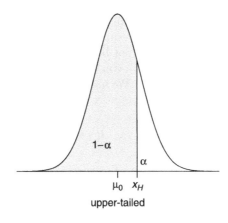

Figure 10.2 The sampling density of $\overline{X}$ under the null hypothesis is $\phi(x\,|\mu_0, \text{SE})$. x_H is the critical value in the sense that the probability of any $\overline{X}$ to the right of it is small ($= \alpha$).

Example 10.10. We continue with analysis of capital punishment as it relates to race (the data were introduced in Example 9.13). The mean age at sentencing for blacks is 28.38 years. We take a sample from the population of inmates. We assume that $\mu_0 = 28.38$ and ask: Would the sample mean make us reject the hypothesis that the mean age of the population of convicts at the time of sentencing to death is $\mu = \mu_0$? Our hypotheses are

$$H_0 : \mu = 28.38 \quad \text{vs.} \quad H_A : \mu > 28.38$$

and we choose $\alpha = 0.05$ and sample size $n = 40$. So we specify the data

```
> set.seed(5) ; n <- 40 ; alpha <- 0.05
> mu.0 <- mean(age[cp$Race == 'Black'], na.rm = TRUE)
> X <- sample(age, n)
> X.bar <- mean(X) ; S <- sd(X) ; SE <- S / sqrt(n)
```

and use (10.3)

```
> round(c(mu.0 = mu.0, x.H = qnorm(1 - alpha, mu.0, SE),
+   X.bar = X.bar, alpha = alpha), 2)
 mu.0   x.H X.bar alpha
28.38 30.86 31.09  0.05
```

Because $\overline{X} > x_H$, we reject H_0 and conclude that $\mu > 28.38$ with 95% certainty. This means that if we repeat the samples, under H_0, 95% of the means will be $\leq x_H$. Because under the null hypothesis model we obtained a rare result (5% of the means under H_0 will be $> x_H$), we reject H_0. However, we may be wrong because our sample might be one of those 5%. So we admit that we may erroneously reject a true H_0 (that $\mu = \mu_0$). Our type I error has a chance of 0.05 to occur. □

The last two examples bring up two important points. First, because we have data for the population, we do not really need to do any statistical tests for the mean. We *know* that the mean age at sentencing for the black population of convicts who were sentenced to death is 28.38 years and we *know* that it is 30.31 for the whole population. The question whether these differences are significant is no longer statistical. It is a matter of opinion whether a difference of $30.31 - 28.38 = 1.93$ is large enough to reflect social issues (such as discrimination). Second, we can go even further. If we have a very large sample, say $n = 100\,000$, with a small S, say 1, then the standard error is $\text{SE} = 0.003$. Such a small standard error reflects the fact that $\overline{X}$ is very close to μ. In other words, for all practical purposes, our sample *is* the population. Again, statistics are hardly needed in such cases. We do use statistics in Example 10.10 for heuristic reasons and to make a point: Sometimes you have to sample the population.

Two-tailed test

Our pair of hypotheses is

$$H_0 : \mu = \mu_0 \quad \text{vs.} \quad H_A : \mu \neq \mu_0 .$$

Because we are testing on both sides of μ_0, we need to specify α on the left and right extremes. Traditionally, we use $\alpha/2$ on each tail of the test. However, in decision making and risk analysis, it may make sense to choose different values for α_L on the left tail and α_H on the right tail. For now, we shall be content with equal rejection regions on both tails of the density. In Chapter 11, we will examine other possibilities. From the information we have ($\phi(x|\mu_0, \mathrm{SE})$ and α), we compute

$$x_L = \Phi^{-1}(\alpha/2 | \mu_0, \mathrm{SE}) \qquad (10.4)$$
$$= \mathtt{qnorm(alpha/2, mu.0, S/sqrt(n))}$$

and

$$x_H = \Phi^{-1}(1 - \alpha/2 | \mu_0, \mathrm{SE}) \qquad (10.5)$$
$$= \mathtt{qnorm(1 - alpha/2, mu.0, S/sqrt(n))}$$

(Figure 10.3)

where x_L and x_H are *critical values* in the usual sense. The probability that $\overline{X} \leq x_L$ or $\overline{X} > x_H$ is α, or by our notation,

$$\Phi(x_L | \mu_0, \mathrm{SE}) := P(\overline{X} \leq x_L) = \alpha/2 \text{ or}$$
$$1 - \Phi(x_H | \mu_0, \mathrm{SE}) := 1 - P(\overline{X} \leq x_H) = \alpha/2$$

is small ($= \alpha$). If we do get $\overline{X} \leq x_L$ or $\overline{X} > x_H$ from the sample, then we must reject H_0 because under it, such $\overline{X}$ is very unlikely. In fact, if $\overline{X} \notin [x_L, x_H]$, we conclude that $\mu \neq \mu_0$, which is our alternative hypothesis. To wit, for two-tailed hypothesis testing, use (10.4) to obtain x_L and (10.5) to obtain x_H. If $\overline{X} \notin [x_L, x_H]$, reject H_0 in favor of H_A.

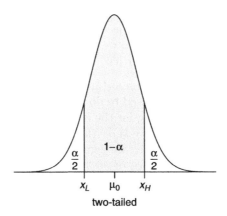

Figure 10.3   The sampling density of $\overline{X}$ under the null hypothesis is $\phi(x|\mu_0, \mathrm{SE})$. x_L and x_H are the critical values in the sense that the probability of any $\overline{X}$ to the left of x_L or to the right of x_H is small ($= \alpha$).

Example 10.11. Continuing with the capital punishment data, there have been a total of 138 women convicted to death. Let us take a sample of $n = 40$ of them and ask: Assume that for the whole population, $\mu = \mu_0 = 30.31$. Does the sample of women's age at sentencing to death confirm this assumption? Our hypotheses are

$$H_0 : \mu = 30.31 \quad \text{vs.} \quad H_A : \mu \neq 30.31 \; .$$

and we choose $\alpha = 0.05$. So we specify the data

```
> set.seed(33) ; n <- 40 ; alpha <- 0.05
> X <- sample(age[cp$Sex == 'F'], n)
> mu.0 <- mean(age, na.rm = TRUE)
> X.bar <- mean(X) ; S <- sd(X) ; SE <- S / sqrt(n)
```
and use (10.4) and (10.5)
```
> round(c(mu.0 = mu.0, x.L = qnorm(alpha/2, mu.0, SE),
+   x.H = qnorm(1 - alpha/2, mu.0, SE), X.bar = X.bar,
+   alpha = alpha / 2), 3)
  mu.0    x.L    x.H  X.bar  alpha
30.307 26.640 33.974 34.490  0.025
```

Because $\overline{X} = 34.49 \notin [x_L, x_H] = [26.64, 33.97]$, we reject H_0 and conclude that $\mu \neq 30.31$ with 95% certainty. This means that if we repeat the samples from the population of female convicts, under H_0, 95% of the means will be within the interval $[x_L, x_H]$. Because we obtained a rare result (5% of the means under H_0 will *not* be in the interval), we reject H_0. However, we may be wrong because our sample might be one of those 5%. So we admit that we may erroneously reject a true H_0 (that $\mu = \mu_0$). Our error has a chance of 0.05 to occur (this is type I error). □

Note the duality between confidence intervals (Example 9.13) and two-tailed hypotheses testing (Example 10.11). In the former, we estimate μ and construct a random interval centered on $\overline{X}$. In the latter, we assume that $\mu = \mu_0$ and construct a true interval centered on μ_0. In the former, our confidence coefficient is $1 - \alpha$, in the latter, our significance is α.

Because the normal is symmetric, we could use absolute values to make the notation (in the case of two-tailed hypotheses testing) more compact. However, we wish to keep the notation general. Because *not* all sampling densities are necessarily symmetric, x_L and x_H are not necessarily equidistant from θ_0.

10.2.2 Proportions

Recall that an unbiased estimator of a population proportion with regard to some property, π, is

$$\widehat{\pi} = p = \frac{n_S}{n}$$

where n is the sample size and n_S is the number of observations that posses the property. As we have seen in Section 7.7.1, for $n\pi \geq 5$ and $n(1 - \pi) \geq 5$, we treat the sampling density of p as approximately normal, or in the case of hypothesis testing, as

$$\phi\left(p \,\middle|\, \pi_0, \sqrt{\pi_0 (1 - \pi_0)/n}\right) .$$

Therefore, to test hypotheses, we replace θ in the pairs in (10.1) with π and proceed exactly as in Section 10.2.1 replacing μ with π and SE with $\sqrt{\pi_0 (1 - \pi_0)/n}$. There is potentially one additional step: Because the binomial is discrete and the normal is continuous, if $1/(2n) \geq |p - \pi_0|$, we need to adjust π_0 to $\pi_0 - 1/(2n)$. We will consider the normal approximation when samples are large. Therefore, we shall not use this so-called continuity correction (its use is controversial anyway).

Example 10.12. The following was analyzed by Kaye (1982). The plaintiff in *Swain v. Alabama* (1965) alleged discrimination against blacks in grand jury selection. At the time, 25% of those eligible to serve on a grand jury were blacks. Of the 1 050 individuals called to potentially serve on a grand jury, 177 were blacks. Do the data support the assertion of discrimination? If the proportion of blacks in the "sample" (those who were called for jury duty) is significantly smaller then their proportion in the population, then we yell "discrimination."

We use $\pi_0 = 0.25$. Our hypotheses are

$$H_0 : \pi = \pi_0 \quad \text{vs.} \quad H_A : \pi < \pi_0 .$$

for $\alpha = 0.01$. We have $n = 1\,050$, $n_S = 177$, $\pi_0 = 0.25$, $p = 177/1\,050 \approx 0.169$. Here $n \times \pi = 1\,050 \times 0.25 \geq 5$ and $n \times (1 - 0.25) \geq 5$. Therefore, we may proceed with the large sample test for proportions. Modifying the code in Example 10.9, we obtain

```
> n <- 1050 ; n.S <- 177 ; PI.0 <- 0.25 ; alpha <- 0.01
> SE <- sqrt(PI.0 * (1 - PI.0) / n)
> round(c(PI.0 = PI.0, x.L = qnorm(alpha, PI.0, SE),
+     p = n.S / n, alpha = alpha), 2)
 PI.0   x.L      p alpha
 0.25  0.22   0.17  0.01
```

Because $0.17 < 0.22$, we reject H_0 and conclude that in fact there is evidence of discrimination. Note that we use π_0 in the standard error term (SE) because by assuming $\pi = \pi_0$ we obtain the standard deviation of the binomial. In other words, in the case of the binomial, by virtue of assuming π, we are specifying σ. Interestingly, the court concluded that there was *no* evidence for discrimination. The judge claimed that the difference between the ratio of blacks in the population and their ratio among potential jurors (-0.081) was small! □

10.2.3 Intensities

In Section 7.2.3, we saw that the normal approximation for the Poisson is $\mu = \lambda$ and $\sigma^2 = \lambda$. We also saw that a sample-based intensity, $l = \overline{X}$, is the best estimate of λ. From Section 7.8.1 we conclude that to test hypotheses, we use

$$\phi\left(l \middle| \lambda_0, \sqrt{\frac{\lambda_0}{n}}\right) .$$

Therefore, we replace θ in the pairs in (10.1) with λ and proceed exactly as in Section 10.2.1 replacing μ with λ and SE with $\sqrt{\lambda_0/n}$.

Example 10.13. One year of data about the daily number of arrivals of visitors to Park Lomumba revealed that the number of visitors per day is Poisson with parameter $\lambda_0 = 25$. The average number of arrivals to Park Kasabubu is $l = 30$. Could this average come from a density with $\lambda_0 = 25$ with the possibility of 5% error if the answer is no?

Based on the question, we need to test

$$H_0 : \lambda = 25 \quad \text{vs.} \quad H_A : \lambda > 25$$

at $\alpha = 0.05$. Modifying the code in Example 10.10, we obtain

```
> n <- 1 ; lambda.0 <- 25 ; l <- 30 ; alpha <- 0.05
> SE <- sqrt(lambda.0 / n)
> round(c(lambda.0 = lambda.0,
+     x.H = qnorm(1 - alpha, lambda.0, SE),
+     l = l, alpha = alpha), 2)
lambda.0      x.H        l     alpha
   25.00    33.22    30.00      0.05
```

Because $30.00 < 33.22$ we conclude with high certainty that $\lambda = 30.00$ in Kasabubu. □

10.2.4 Common sense significance

At this point, we have enough understanding to discuss an important issue—the difference between statistical and common sense significance. Sometimes we may reject H_0 based on the significance test. However, the difference between values we test for may be so small, that it no longer makes sense to distinguish between them. Here is an imaginary example that clarifies the issue.

Example 10.14. A bigot claims that the IQ of his "race" is higher than the average IQ in the population, which equals 100. He sets the hypotheses to

$$H_0 : \mu = 100 \quad \text{vs.} \quad H_A : \mu > 100$$

at $\alpha = 0.01$. He reports the following about the IQ of children of his race:

$$n = 10\,000 \,, \quad \overline{X} = 100.5 \,, \quad S = 20 \,.$$

From (10.3), we obtain

```
> n <- 10000 ; alpha <- 0.01 ; mu.0 <- 100 ;
> X.bar <- 100.5 ; S <- 20 ; SE <- S / sqrt(n)
> round(c(mu.0 = mu.0, x.H = qnorm(1 - alpha, mu.0, SE),
+     X.bar = X.bar, alpha = alpha), 2)
  mu.0      x.H    X.bar    alpha
100.00   100.47   100.50     0.01
```

He therefore concludes that "to a high degree of statistical significance, children of my race are more intelligent than an average child."

The bigot's implied conclusion—that children of his race are smarter than average—is practically nonsense. Why? Because with $n = 10\,000$, the point estimate $\overline{X} = 100.5$ is very close to $\mu = 100$. In fact, he established a new population average. Furthermore, a child with an IQ of 100.5 is unlikely to do consistently better in anything that is related to IQ than a child with an IQ of 100 (a difference of 0.5 points in a population with a standard deviation of 20 is meaningless). In other words, in reality, a 0.5-point difference in IQ is meaningless—particularly in light of the inaccuracy of measuring IQ in the first place. Finally, IQ test results most likely fluctuate, even for the same person at different times, by more that 0.5, which then requires an Analysis of Variance (see Chapter 15). □

Example 10.14 illustrates the fact that with a large enough sample we can always establish significance. However, at some point, the large sample represents the population, not a sample from the population. To get around the problem of significance because of large sample sizes we should either determine the necessary sample size to detect a specified difference with a particular significance or admit that no statistics are necessary. In the latter case, whatever differences we find are hardly random and we need to address how meaningful these differences are. We address this issue in Chapter 11.

10.3 Small sample hypotheses testing

When samples are small, we can no longer invoke the central limit theorem. If the population is normal, then we can use the t density to construct hypothesis tests. Otherwise, we have to assume a density (e.g. binomial, Poisson). If we cannot assume a density and we still wish to test hypotheses about the value of an arbitrary parameter from an arbitrary density, we can use the bootstrap method. These, then, are the subjects of this section and the next.

10.3.1 Means

Here, we deal with small samples ($n < 30$) from normal populations. As was the case for confidence intervals, we use the t density for inference. Recall from Section 9.2.2 that for small samples ($n < 30$), the sampling density of $\overline{X}$ is $t(z\,|\,n-1)$, where $n-1$ are the so-called degrees of freedom and

$$Z = \frac{\overline{X} - \mu}{\sigma/\sqrt{n}} \quad \text{or} \quad \overline{X} = \mu + Z\frac{\sigma}{\sqrt{n}}.$$

We estimate σ with S. Therefore, for lower-tailed hypotheses tests with $\mathrm{SE} := S/\sqrt{n}$, we use

$$\begin{aligned} x_L &= \mu_0 - t^{-1}\left(\alpha\,|\,n-1\right)\mathrm{SE} \\ &= \mathtt{mu.0 - qt(alpha, n-1) * SE} \end{aligned} \quad (10.6)$$

and reject H_0 if $\overline{X} < x_L$. For upper-tailed tests we use

$$\begin{aligned} x_H &= \mu_0 + t^{-1}\left(1-\alpha\,|\,n-1\right)\mathrm{SE} \\ &= \mathtt{mu.0 + qt(1 - alpha, n-1) * SE} \end{aligned} \quad (10.7)$$

and reject H_0 if $\overline{X} > x_H$. For two-tailed tests we use

$$\begin{aligned} [x_L, x_H] &= \left[\mu_0 - t^{-1}\left(\frac{\alpha}{2}\,|\,n-1\right)\mathrm{SE},\, \mu_0 + t^{-1}\left(1-\frac{\alpha}{2}\,|\,n-1\right)\mathrm{SE}\right] \\ &= \mathtt{c(mu.0 - qt(alpha/2, n-1) * SE,} \\ &\quad\ \mathtt{mu.0 + qt(1 - alpha/2, n-1) * SE)} \end{aligned} \quad (10.8)$$

and reject H_0 if $\overline{X} < x_L$ or $\overline{X} > x_H$.

Example 10.15. Based on a large number of nests, the clutch size of yellow-headed blackbirds in Iowa was reported to be 3.1 eggs per nest (Orians, 1980, p. 269). A sample of 26 nests in Minneapolis showed a mean clutch size of 4.0 eggs per nest with $S^2 = 1.6$. Can we claim that the mean clutch size of the sample from Minnesota represents the mean clutch size of the Iowa population with 0.05 significance?

Here $\mu_0 = 3.1$, $\overline{X} = 4.0$, $\alpha = 0.05$, $\sigma \approx S = \sqrt{1.6}$, and $n = 26$. We wish to test

$$H_0 : \mu = 3.1 \quad \text{vs.} \quad H_A : \mu > 3.1 .$$

Based on (10.7), we obtain

```
> mu.0 <- 3.1 ; X.bar <- 4.0 ; n <- 26 ; S <- sqrt(1.6)
> SE <- S / sqrt(n) ; alpha <- 0.05
> x.H <- mu.0 + qt(1 - alpha, mu.0, SE) * SE
> round(c(mu.0 = mu.0, x.H = x.H,
+   X.bar = X.bar, alpha = alpha), 2)
 mu.0   x.H X.bar alpha
 3.10  3.79  4.00  0.05
```
and thus reject H_0. □

10.3.2 Proportions

In Section 9.2.2, we introduced two common methods to compute confidence intervals for small samples from a population with binomial density. These referred to the exact method (9.11) and (9.12) and the Wilson method (9.13). The function `binconf()` in the package `Hmisc` calculates confidence intervals for the binomial (Example 9.17). We can use it to obtain p_L and p_H for hypothesis testing. Here is how.

Example 10.16. Sex ratio at birth in sexually reproducing population is generally considered to be $1 : 1$. A sample of 40 individuals revealed a ratio of 0.45 females in the population. Is the proportion of females smaller than 0.5?

The hypotheses are

$$H_0 : \pi = 0.5 \quad \text{vs.} \quad H_A : \pi < 0.5 .$$

and we use $\alpha = 0.05$. Then

```
> library(Hmisc) ; n <- 40 ; n.S <- 20 ; p <- 0.45
> PI.0 <- n.S / n ; a <- 0.1
> x <- binconf(n.S, n, method = 'wilson', alpha = a)
> round(c(PI.0 = PI.0, p.L = x[2], p = p, alpha = a/2), 2)
 PI.0   p.L     p alpha
 0.50  0.37  0.45  0.05
```

and we cannot reject H_0. Let us explain what is going on. According to H_0, $\pi_0 = 20/40$. Now `binconf()` computes confidence intervals, so whatever α you pass on to it, it will calculate $\alpha/2$ for the tails on each side of π_0. Therefore, we set a $= 2\alpha = 0.1$. Because the sample is small, we use the Wilson method, hence `method = 'wilson'` (it is the default method anyway). In this case, `binconf()` returns a vector. We store it in x and retrieve p_L from `x[2]`. Were we to test for $H_A: \pi > \pi_0$, then we retrieve p_H from `x[3]`. To test $H_A: \pi \neq \pi_0$, pass to `binconf()` your original α, not 2α and retrieve p_L and p_H from `x[2]` and `x[3]`. □

10.3.3 Intensities

Recall from Section 9.2.2 that the Poisson density with parameter λ is

$$P(X = x) = \frac{\lambda^x}{x!}e^{-\lambda} \quad x = 1, 2, \ldots.$$

We use $\hat{\lambda} = l$ (counts per unit of measurement) to estimate λ. We compute confidence intervals for a particular confidence coefficient according to (9.14). Let $P(X \leq x | n)$ be the χ^2 distribution with n degrees of freedom. Then for lower-tailed hypothesis testing with $H_0: \lambda = \lambda_0$ and significance level α, we use

$$l_L = P^{-1}(\alpha | 2\lambda_0)/2 \qquad (10.9)$$
$$= \text{qchisq(alpha, 2 * lambda.0)}/2$$

and reject H_0 if $l < l_L$. For upper-tailed testing we use

$$l_H = P^{-1}(1 - \alpha | 2(\lambda_0 + 1))/2 \qquad (10.10)$$
$$= \text{qchisq(1 - alpha, 2 * (lambda.0 + 1))}/2$$

and reject H_0 if $l > l_H$. For two-tailed testing we use

$$[l_L, l_H] = \left[P^{-1}\left(\frac{\alpha}{2}\bigg| 2\lambda_0\right)/2, P^{-1}\left(1 - \frac{\alpha}{2}\bigg| 2(\lambda_0 + 1)\right)/2\right]$$
$$= [c(\text{qchisq(alpha/2, 2 * lambda.0)}/2, \qquad (10.11)$$
$$\text{qchisq(1 - alpha, 2 * (lambda.0 + 1))}/2]$$

and reject H_0 if $l < l_L$ or $l > l_H$.

Example 10.17. People who fish in Lake of the Woods, in Northern Minnesota, claim that on the average, they catch (and release some) 6.5 fish per day. Suppose that the number of fish caught per day is Poisson and because you are gullible, you set $\lambda = \lambda_0 = 6.5$. You went fishing on Lake of the Woods recently and encountered 3 fish per day. Given your experience, would you believe the claim with $\alpha = 0.05$?
We wish to test

$$H_0 : \lambda = 6 \quad \text{vs.} \quad H_A : \lambda < 6.$$

Using (10.9) we obtain

```
> lambda.0 <- 6.5 ; l <- 3 ; alpha <- 0.05
> round(c(lambda.0 = lambda.0,
+    l.L = qchisq(alpha, 2 * lambda.0) / 2, l = l,
+    alpha = alpha), 2)
lambda.0     l.L       l     alpha
    6.50    2.95    3.00      0.05
```

and we conclude that people's claims are more than fish stories (that is, we do *not* reject H_0). □

10.4 Arbitrary statistics of arbitrary densities

As was the case in Section 9.3, we can use the bootstrap method to obtain the sampling density of any statistic from arbitrary density of interest. From the density and for a given α, we compute critical values. In Example 9.19 we used bootstrap to build confidence intervals. In the next example, we examine the density of x_L and x_H (which can be used for confidence intervals at half their α values).

Example 10.18. Consider an experiment in which the growth rate of $n = 35$ tumors are recorded:

```
> set.seed(1) ; n <- 20
> X <- runif(n, 0, 10)
```

We calculate the confidence intervals for the confidence coefficients $1 - \alpha = 0.90$ and 0.95 with

```
> p <- c(0.025, 0.05, 0.95, 0.975)
> qnorm(p, mean(X), sd(X) / sqrt(n))
```

These confidence intervals are rv. Now put the above statements in a loop with R repetitions and collect the quantiles (associated with p and the sampling density of $\overline{X}$). These quantiles are then the sampling density of the lower and upper boundaries of the confidence intervals. Here is the chunk that does the bootstrap:

```
1   set.seed(1) ; n <- 35 ; X <- runif(n, 0, 10)
2   R <- 100000 ; p <- c(0.025, 0.05, 0.95, 0.975)
3
4   m <- matrix(sample(X, size = n * R, replace = TRUE),
5     ncol = R, nrow = n)
6   X.bar <- apply(m, 2, mean) ; SE <- apply(m, 2, sd) / sqrt(n)
7   xp <- cbind(matrix(p, nrow = R, ncol = length(p),
8     byrow = TRUE),   X.bar, SE)
9
10  x <- qnorm(xp[, 1 : 4], xp[, 5], xp[, 6])
```

Let us examine the top left panel of Figure 10.4. The histogram is the sampling density of the lower boundary of the confidence interval of $\overline{X}$ for $\alpha = 0.025$. The left-most vertical line shows the lower boundary of the sampling density of the lower confidence interval of $\overline{X}$ for the confidence coefficient $= 0.05$. The rightmost vertical line is the corresponding upper boundary. The inner vertical lines delineate the same as the outer vertical lines, but for confidence coefficient $= 0.1$. The graph to the right of the sampling density shows the normal approximation of the sampling density of the upper boundary of the confidence interval of $\overline{X}$ (the sampling density itself is shown in the lower right corner of the figure). The vertical lines were obtained this way:

```
> for(i in 1 : 4){
+   q <- quantile(x[, j], prob = p)
+   critical.x[j, ] <- q
+ }
```

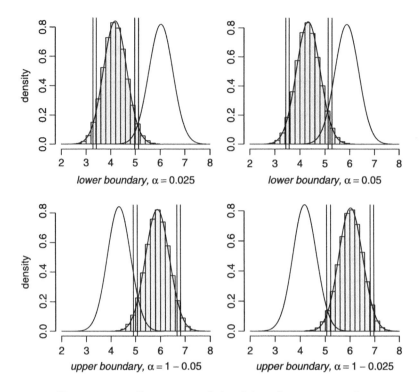

Figure 10.4   Bootstrapped densities of tumor growth.

Here are their values:

```
                    significance | lower-tailed   upper-tailed
                          alpha  |  0.025 0.050    0.050 0.025
        confidence coefficient  |  0.050 0.100    0.900 0.950
--------------------------------+------------    ------------
sampling densities:             |
lower boundary, alpha = 0.025   |  3.280 3.418    4.979 5.135
lower boundary, alpha = 0.050   |  3.426 3.566    5.125 5.280
upper boundary, alpha = 0.950   |  4.916 5.075    6.671 6.815
upper boundary, alpha = 0.975   |  5.055 5.217    6.824 6.965
```

The results can be used to test hypotheses or for confidence intervals. The remaining panels in Figure 10.4 were obtained similar to the top left. The confidence coefficients and α in the results above illustrate the duality between confidence interval and hypothesis testing. □

10.5 p-values

So far, we compared the value of our statistic to the values of x_L or x_H that we obtained from the statistic's sampling density. But if we assume a sampling density, then instead of comparing values, we can compare probabilities.

Example 10.19. Suppose we wish to use a lower-tailed test for a mean with α and μ_0 given for a large sample. We take a sample and obtain $\overline{X}$. Because under H_0 we know the sampling distribution; it is

$$\Phi(x_L \mid \mu_0, \text{SE}) = \alpha, \quad \Phi(\overline{X} \mid \mu_0, \text{SE}) = p\text{-value}.$$

Now to obtain the p-value, we do

```
> p.value <- pnorm(X.bar, mu.0, S / sqrt(n))
```

So if p-value $< \alpha$, we know that $\overline{X}$ is to the left of the lower critical value, x_L and we reject H_0. With the p-value given, we do not need to calculate x_L. □

Example 10.19 is specific. Let the sampling distribution of a statistic under H_0 be $P(X \leq x_0 \mid \theta)$ where the rv X is the statistics and x is its value we obtain from a sample. Then, in general, we define the p-values thus:

lower-tailed: p-value $= P(X \leq x \mid \theta)$,
upper-tailed: p-value $= P(X > x \mid \theta)$,
two-tailed: p-value] $= \min[P(X \leq x \mid \theta), P(X > x \mid \theta)]$.

So for the one-tailed tests, if p-value $< \alpha$ we reject H_0. For two-tailed tests, if p-value $< \alpha/2$ we reject H_0.

There is a subtle, but important point here. Just like α (which is associated with x_L and x_H), the p-value is *not* associated with a rv. It is associated with a given quantity (e.g. the computed sample mean, $\bar{x}$). To clarify this point, suppose that we obtain x_0 and y_0 for the statistic from two different samples with identical sampling density. Now suppose that both p-values indicate significance (i.e. rejection of H_0) with the p-value associated with $x_0 <$ than the p-value associated with y_0. Because x_0 and y_0 are not rv, we cannot say that x_0 is more significant than y_0. It is useful to know (and consider) that x_0 is more extreme on the sampling density than y_0, but it has nothing to do with more or less significant. Next, we repeat some of our examples with p-values this time.

Example 10.20. In Example 10.9 (lower-tailed test) we calculated

```
> round(c(mu.0 = mu.0, x.L = qnorm(alpha, mu.0, SE),
+   X.bar = X.bar, alpha = alpha), 2)
 mu.0   x.L X.bar alpha
30.31 27.92 28.16  0.05
```

and thus did not reject H_0. Instead we calculate

```
> round(c(mu.0 = mu.0, X.bar = X.bar, alpha = alpha,
+   p.value = pnorm(X.bar, mu.0, SE)), 2)
   mu.0   X.bar   alpha p.value
  30.31   28.16    0.05    0.07
```

Because $0.07 > 0.05$, we do not reject H_0. In Example 10.10 (upper-tailed test), we calculated

```
> round(c(mu.0 = mu.0, x.H = qnorm(1 - alpha, mu.0, SE),
+   X.bar = X.bar, alpha = alpha), 2)
 mu.0   x.H X.bar alpha
28.38 30.86 31.09  0.05
```

and therefore rejected H_0. Here we calculate

```
> round(c(mu.0 = mu.0, X.bar = X.bar, alpha = alpha,
+   p.value = 1 - pnorm(X.bar, mu.0, SE)), 2)
   mu.0   X.bar  alpha p.value
  28.38   31.09   0.05    0.04
```

In Example 10.11 (two-tailed), we obtained

```
   mu.0    x.L    x.H  X.bar  alpha
 30.307 26.640 33.974 34.490  0.025
```

For the p-value we have

```
> round(c(mu.0 = mu.0, X.bar = X.bar, alpha = alpha / 2,
+   p.value = min(pnorm(X.bar, mu.0, SE),
+   1 - pnorm(X.bar, mu.0, SE))), 3)
   mu.0   X.bar  alpha p.value
 30.307  34.490  0.025   0.013
```

□

In Example 10.20 we use large samples and consequently the normal sampling density. Both the normal and t are symmetric and the notation can be simplified. However, to keep the discussion general, we should keep the upper- and lower-tail calculations intact in case we use asymmetric sampling densities such as χ^2, binomial and Poisson. At any rate, as the next example illustrates, one must use these ideas with caution.

Example 10.21. In Example 10.17, we discussed lower-tailed hypothesis testing with a small Poisson sample. Let's use a two-tailed test with a somewhat different story. Recall that people claim that on the average, they catch (and release some) 6.5 fish per day in Lake of the Woods. We supposed that the number of fish caught per day is Poisson. You wish to test for $\lambda = \lambda_0 = 6.5$. Because you have no idea whether people are exaggerating or underestimating their catch rate, you decide on a two-tailed test with $\alpha = 0.1$. You went fishing and caught 5 fish in two days. So

```
> lambda.0 <- 6.5 ; l <- 2.5 ; alpha <- 0.1
> round(c(lambda.0 = lambda.0,
+   x.L = qchisq(alpha / 2, 2 * lambda.0) / 2,
+   x.H = qchisq(1 - alpha / 2, 2 * (lambda.0 + 1)) / 2, l = l,
+   alpha = alpha / 2), 3)
 lambda.0     x.L     x.H       l   alpha
    6.500   2.946  12.498   2.500   0.050
```

Because $l = 2.5 < x_L = 2.946$, you reject people's claim. To obtain p-values, we use

```
> (lower.p <- pchisq(2 * l, 2 * lambda.0))
[1] 0.02480687
> round(c(lambda.0 = lambda.0, l = l, alpha = alpha / 2,
+   p.value = lower.p), 3)
 lambda.0       l   alpha p.value
    6.500   2.500   0.050   0.025
```

Note the duality
```
> (p.value <- pchisq(l * 2, 2 * lambda.0))
[1] 0.02480687
> qchisq(p.value, 2 * lambda.0) / 2
[1] 2.5
```
Figure 10.5 illustrates the idea of the p-value. The left panel shows the sampling density for the lower-tailed density (solid curve) and the upper-tailed density (broken curve) and the regions of rejection for two-tailed test (light gray). The right panel magnifies the lower region and shows the relationship between the rejection probability (light gray) and the p-value (black) and their corresponding x_L and l_0. In this case, we reject the null hypothesis because $l_0 < x_L$ and consequently, p-value $< P(X \leq x_L | 2\lambda_0)/2$ where P is the χ^2 distribution with $2\lambda_0$ degrees of freedom. □

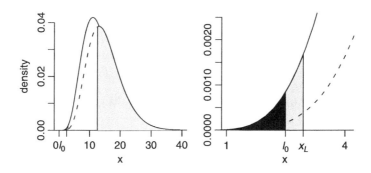

Figure 10.5 The relationship between l, x_L and their corresponding areas under the sampling density.

10.6 Assignments

Unless otherwise stated, use $\alpha = 0.05$ where necessary.

Exercise 10.1. Is $\overline{X}$ a legitimate value about which we may test hypotheses? Why?

Exercise 10.2. Which of the following statements does not follow the rules of setting up hypotheses? Why?

1. $H_0 : \mu = 15$ vs. $H_A : \mu = 15$
2. $H_0 : \pi = 0.4$ vs. $H_A : \pi > 0.6$
3. $H_0 : \mu = 123$ vs. $H_A : \mu < 123$
4. $H_0 : \pi = 0.1$ vs. $H_A : \pi \neq 0.1$

Exercise 10.3. Suppose that in protecting airplanes from colliding with birds, the local Airport Commission sets a rule that the number of nesting birds around the airport should not exceed 200. An inspection team—whose main concern is airport safety—decides to test

$$H_0 : \mu = 200 \quad \text{vs.} \quad H_A : \mu > 200$$

where μ is the average number of nesting birds around the airport during the breeding season. Would that be preferable to testing $H_A : \mu < 200$? Explain.

Exercise 10.4. Use R with the data in `teen-birth-rate-2002.txt` to produce the data in Table 10.2.

Table 10.2 Teen birth rates under the assumption of normal distribution.

	$\overline{X}$	S	n^a	Z	p^b
Black	76.14	15.60	44	4.79	0
Hispanic	88.96	23.66	48	7.05	0
White	32.51	11.71	51	−19.74	0

[a] Based on states data.
[b] p is the area under the normal that gives the probability of getting the vlaue Z or larger.

Exercise 10.5. The Faculty Senate at the University of Minnesota decided to change the grading system from straight letter grades to letter grades with + or −. Imagine that before changing the grading system the University administration decided to implement the change if more than 60% of the faculty favor the change. A random sample of 20 faculty was selected for a survey. Denote by π the proportion of all faculty that are in favor of adding + or − to the letter grade. Which pair of hypothesis tests would you recommend to the administration:

$$H_0 : \pi = 0.6 \quad \text{vs.} \quad H_A : \pi < 0.6$$

or

$$H_0 : \pi = 0.6 \quad \text{vs.} \quad H_A : \pi > 0.6 .$$

Explain.

Exercise 10.6. The U.S. Environmental Protection Agency (EPA) decides that a concentration of arsenic in drinking water that exceeds π_0, where π_0 is some constant, is unsafe. A farmer uses these guidelines to sample her well for testing for arsenic concentration.

1. Use the EPA website to find the level that the EPA chose for π_0.
2. Recommend specific hypotheses (H_0 vs. H_A) for the farmer to test.

Exercise 10.7. Imagine that a spokesman for the Russian nuclear power industry says: "From the available data, there is no convincing evidence of increased risk of death from cancer due to living near nuclear facilities. Yet, no study can prove the absence of increased risk."

1. Denote by π_0 the proportion of the population in areas near a nuclear power plants who die of cancer during a given year. Consider the hypotheses

    $$H_0 : \pi = \pi_0 \quad \text{vs.} \quad H_A : \pi > \pi_0 .$$

    Based on the quote above, did the spokesman reject H_0 or fail to reject H_0?
2. If the spokesman was incorrect in his conclusion, would he be making a type I or a type II error? Explain.
3. Do you agree with the spokesman's statement that no study can prove the absence of increased risk? Explain.

Exercise 10.8. Prairie Island is a nuclear power plant next to an Indian reservation in Minnesota. The plant releases its cooling water into the Mississippi river. Assume that by law, the plant is not allowed to discharge water warmer than 125°C into the river because warmer water causes damage to downstream ecosystems. Suppose that the Minnesota Pollution Control Agency (PCA) wishes to investigate whether the plant is in compliance. PCA employees take 50 temperature readings at random times during a month from the river water from near the plant. The hypotheses to be tested are
$$H_0 : \mu = 125 \quad \text{vs.} \quad H_A : \mu > 125 .$$
Describe the consequences of type I and type II errors. Which error would you consider to be more serious? Explain.

Exercise 10.9. Suppose that the U.S. Park Service wishes to increase the entry fee to National Parks by 20%. Before increasing the entry fee, officials decide to conduct a market survey to determine how well such an increase will be accepted by the public. The survey includes the following question: "Would a 20% increase in entry fee to a National Park reduce the number of visits to National Parks you plan next year? Denote by π_0 the fraction of the population who would answer yes. Set the hypotheses to be tested to
$$H_0 : \pi = \pi_0 \quad \text{vs.} \quad H_A : \pi > \pi_0 .$$
Describe the consequences of type I and type II errors.

Exercise 10.10. Suppose that the EPA decides that if mercury concentration of fish in a lake exceeds 5 ppm, then fishing in the lake must be banned.

1. Which pair of hypothesis would you prefer?
$$H_0 : \mu = 5 \quad \text{vs.} \quad H_A : \mu > 5$$
or
$$H_0 : \mu = 5 \quad \text{vs.} \quad H_A : \mu < 5 .$$
Explain.
2. What significance level would you prefer for the test? 0.1 or 0.01? Explain.

Exercise 10.11. An imaginary study of oil spills attempts to answer the following question: What area of the ocean surface would a 1 gal oil spill cover? Denote by μ the average area covered by 1 gal of oil spill. The researcher pours 1 gal in open water and measures the area. The experiment is repeated 50 times. The researcher wishes to test the hypotheses
$$H_0 : \mu = \mu_0 \quad \text{vs.} \quad \mu > \mu_0$$
where μ_0 is some area constant, measured in m². What is the appropriate test statistic and the rejection values for the following significance values:

1. $\alpha = 0.01$
2. $\alpha = 0.05$
3. $\alpha = 0.10$
4. $\alpha = 0.13$

Single sample hypotheses testing

Exercise 10.12. Imagine that a high-ranking politician has an affair with an intern. The politician's image-makers decide to carry a public opinion survey in which they ask registered voters to rate how much they like the politician on a scale of -5 (hate) to $+5$ (love). Denote by μ the population average of how much registered voters like the politician. The image makers are going to use a large sample z-statistic to test

$$H_0 : \mu = 0 \quad \text{vs.} \quad H_A : \mu < 0.$$

What is the appropriate critical-z for each of the following significance values?

1. 0.001
2. 0.01
3. 0.05
4. 0.1

Exercise 10.13. For each pair of p-value and α, state whether the observed p-value will lead to rejection of H_0 at the given α:

1. p-value $= 0.07$, $\alpha = 0.10$
2. p-value $= 0.2$, $\alpha = 0.10$
3. p-value $= 0.07$, $\alpha = 0.05$
4. p-value $= 0.5$, $\alpha = 0.05$
5. p-value $= 0.03$, $\alpha = 0.01$
6. p-value $= 0.003$, $\alpha = 0.001$

Exercise 10.14. Find the p-value associated with each given z-statistic for testing

$$H_0 : \mu = \mu_0 \quad \text{vs.} \quad \mu \neq \mu_0$$

where μ_0 is given:

1. $z_? = 2.2$
2. $z_? = -1.4$
3. $z_? = -0.5$
4. $z_? = 1.25$
5. $z_? = -5.2$

Exercise 10.15. A sample of 36 birds is caught in mist nets. The mean carpal length is 12.1 cm and $S = 0.2$ cm. We wish to test the hypothesis that the mean carpal length is

$$H_0 : \mu = 12 \quad \text{vs.} \quad \mu > 12.$$

1. What is the value of the Z statistic?
2. What is the p-value associated with the value of this Z?
3. Let $\alpha = 0.05$; should H_0 be rejected? State your conclusions.

Exercise 10.16. Zebras spend on the average 75 min at a watering hole. To test their sensitivity to the presence of predators in the area, we play a tape of lion roars. A sample of 100 observations with the tape played reveals that the zebras spend an average of 68.5 min at the watering hole. The standard deviation is 9.4 min. Let $\alpha = 0.01$. Set up the hypotheses, write the test statistic and determine the p-value of the sample mean. Does the experiment indicate that the tape playing shortens the time the zebras spend at the watering hole?

Exercise 10.17. Bui et al. (2001) conducted a cross-sectional survey in three districts of Quang Ninh province, Viet Nam, to find out what proportion of the people who lived there engaged in behavior that put them at risk of becoming infected with HIV and to measure their knowledge about HIV infection and AIDS. The survey was conducted in a rural district, Yen Hung; a mountainous district inhabited primarily by ethnic minority groups, Binh Lieu; and an urban district, Ha Long. Here is a sample of the data:

```
           age.at    district gender mean  sd  n
5       interview     Ha Long    men   31 9.2 210
7        marriage   Binh Leiu    men   21 3.4 210
10       marriage    Yen Hung  women   22 3.0 210
14 first intercourse Binh Leiu women   20 3.0 210
3       interview    Yen Hung    men   30 8.8 210
```

(the data may be found in `marriage-Viet-Nam.txt`. Assume that the urban district represents the population. Formulate the appropriate hypotheses and answer the following questions for a significance level of $\alpha = 0.05$.

1. Do women marry at a younger age in the rural district compared to women in the urban district?
2. Do women marry at a younger age in the mountainous district compared to women in the urban district?
3. Do men in the rural district engage in first sexual intercourse at a younger age than men in the urban district?
4. Do men in the mountainous district engage in first sexual intercourse at a younger age than men in the urban district?
5. Discuss your findings based on the results and the p-values for each of the questions above.

Exercise 10.18. Write a function that returns a hypothesis test for a small sample from a normal population. Use the following function as a guideline

```
zp <- function(alpha = 0.05 , n , nS , pi0 ,
        HA = c('greater' , 'smaller' , 'neq')){

        p <- nS / n ; se <- sqrt(p * (1 - p) / n)
        correction <- 1/(2 * n)
        ifelse(correction < abs(p - PI) , correction , 0)
        Z <- (p - PI - correction) / se
        critical.z <- qnorm(1 - alpha)
        H0 <- ifelse(Z > critical.z , 'reject' ,
                'do not reject')
        if (HA == 'smaller') {
                critical.z <- qnorm(alpha)
                H0 <- ifelse(Z < critical.z , 'reject' ,
                        'do not reject')
        }
        if (HA == 'neq') {
```

```
17              critical.z <- qnorm(1 - alpha / 2)
18              H0 <- ifelse(abs(Z) > critical.z , 'reject' ,
19                           'do not reject')
20          }
21          results <- list(Z , critical.z , alpha , H0)
22          names(results) <- c('Z' , 'critical.z' , 'alpha' , 'H0')
23          class(results) <- 'table'
24          print(results)
25      }
```

Exercise 10.19. Write a function similar to zp() (see Exercise 10.18 and code in Example 10.12) that returns a hypothesis test for a small sample from a binomial population. (Hint: use library(Hmisc) and then type binconf in R before writing the function.)

Exercise 10.20. Write a function similar to zp() (see Exercise 10.18 and code in Example 10.12) that returns a hypothesis test for a small sample from a Poisson population.

Exercise 10.21. For this exercise, use the data in capital.punishment.rda (see United States Department of Justice, 2003).

1. What is the proportion of blacks in the population of inmates for the whole data?
2. Use the Internet to find the proportion of blacks in the U.S. population at large for the year 2000 (be sure to cite the data source properly). Denote this proportion by π_0.
3. Does the proportion of black inmates on death row in 2000 represent a sample from the population at large?
4. Take a random sample of 10 cases from the data. Does the proportion of blacks in the sample represent the proportion of black inmates on death row?

Exercise 10.22. Data from pollution sensors in a busy intersection indicate that for 1 000 consecutive days, particulates in the air were above the minimum allowed once every 10 days. The data (available in an R file, named accidents.rda indicate that the probability of exceeding the minimum at each day is as likely as any other day. The local pollution control authority forces a nearby coal electricity generating plant to add extra scrubbers at a considerable cost. Let λ_A be the number of incidents of exceeding the allowed minimum per day after the scrubbers were put into operation. What would the value of λ_A need to be for you to believe that the scrubbers are effective in reducing the average number of days when pollution exceeds the allowed maximum?

Exercise 10.23. Endangered species are difficult to study because any sampling must be non-destructive and obtaining a sample of sufficient size in the first place is very difficult. An intensive study of a population of 10 female Siberian tigers in the wild resulted in the following data about litter size:
    2, 4, 3, 1, 2, 3, 5, 1, 3, 1 .
A standard deviation of litter size that is above 3 can lead to extinction of the population. Based on the data, is this value likely to occur?

Exercise 10.24. The notation $X \sim N(\mu, \sigma)$ means that X has a normal density with mean μ and standard deviation σ. $X \sim$ Binomial(n, p) means that X has a binomial density with parameters n (the number of trials) and p (the probability of success). Determine the p-value of the following and state whether they are significant or not:

1. $X = 1.96$, $X \sim N(0, 1)$, upper-tailed test.
2. $X = 1.96$, $X \sim N(0, 1)$, lower-tailed test.
3. $X = 1.96$, $X \sim N(0, 1)$, two-tailed test.
4. $X = 1.7$, $X \sim N(0, 1)$, upper-tailed test.
5. $X = 1.7$, $X \sim N(0, 1)$, lower-tailed test.
6. $X = 1.7$, $X \sim N(0, 1)$, two-tailed test.
7. $X = 18$, $X \sim$ Binomial$(50, 0.5)$, upper-tailed test.
8. $X = 18$, $X \sim$ Binomial$(50, 0.5)$, lower-tailed test.
9. $X = 18$, $X \sim$ Binomial$(50, 0.5)$, two-tailed test.
10. $X = 4$, $X \sim$ Poisson(8), upper-tailed test.
11. $X = 4$, $X \sim$ Poisson(8), lower-tailed test.
12. $X = 4$, $X \sim$ Poisson(8), two-tailed test.

11

Power and sample size for single samples

In this chapter, we are interested in the question of power and sample size. Roughly speaking, power refers to our ability to distinguish between two alternative models; i.e. between H_0: $\theta = \theta_0$ and H_A: $\theta = \theta_A$ where θ_0 and θ_A are given. Recall that type II error is the probability of *not* rejecting H_0 while it is false. Related to this error is the probability of *rejecting* the null hypothesis given that the alternative hypothesis is true. Sample size refers to the problem of deciding on a sample size based on desired power, specified detectable difference (between θ_0 and θ_A, for example) and significance level. We deal with these issues in this chapter.

Associated with the decision to reject H_0 or not are the following potential consequences:

1. Do not reject H_0 when H_0 is true (no error)
2. Reject H_0 when H_0 is false (no error)
3. Reject H_0 when H_0 is true (type I error)
4. Do not reject H_0 when H_0 is false (type II error)

Formally, we define a

Power of a test The probability of correctly rejecting H_0 for a given θ_A; i.e.

$$\text{power} := 1 - P(\text{type II error given } \theta_A)$$
$$= 1 - \beta \text{ given } \theta_A .$$

Our plan is to discuss power and sample size for means, proportions and intensities for large and then small samples.

11.1 Large sample

In this section, we show how to compute the power to distinguish between a statistical model according to H_0 and according to H_A for large sample sizes. We shall also discuss the sample size one needs to obtain given significance, power and the so-called detectable difference between the statistic under H_0 vs. under H_A. We address means (μ), proportions (π) and intensities (λ).

11.1.1 Means

Here we discuss the case where we wish to compute the power of our ability to distinguish between μ_0 and a given alternative μ_A. We also discuss sample size.

Power

To formalize the idea of power, we summarize what we have learned thus far about large sample hypotheses testing. We assume that each observed value in the sample of size n is independent of other values in the sample. We also assume that these values come from a population with mean μ and standard deviation σ. Because n is large, we invoke the central limit theorem. We take $\sigma \approx S$, the sample standard deviation. Then in the limit, the sampling density of $\overline{X}$ is normal with mean μ and standard deviation (= standard error) $S/\sqrt{n}$. With these in mind, we can decide whether we accept or reject H_0. Let us introduce the idea of power with an example.

Example 11.1. Prostate cancer often leads to prostatectomy. This in turn causes erectile dysfunction (ED). Sildenafil citrate, better known as Viagra, is then used to treat ED. To determine the efficacy of the treatment, urologists use a short form (5 questions) of the so-called International Index of Erectile Function (IIEF). Raina et al. (2003) treated patients that underwent prostatectomy for 3 years with 50 or 100 mg of Sildenafil. At the end of the first year of treatment, they reported an average score on the IIEF of 18.52 ± 1.23 ($n = 48$) on the IIEF test.

Suppose that the average test score for a population of males who underwent prostatectomy and were *not* treated for a year with Sildenafil citrate is $\mu_0 = 18$. We wish to test the hypothesis that the treatment was not effective. Namely,

$$H_0 : \mu = \mu_0 \quad \text{vs.} \quad H_A : \mu > \mu_0$$

where μ is the average score for the population of males who were treated for a year with Sildenafil citrate. We use $\alpha = 0.05$. The sample is large enough to justify the use of the central limit theorem. Therefore, based on (10.3) and similar to Example 10.10, we obtain

```
> mu.0 <- 18 ; S <- 1.23 ; n <- 48 ; SE <- S / sqrt(n)
> X.bar <- 18.52 ; alpha <- 0.05
> round(c(mu.0 = mu.0, x.H = qnorm(1 - alpha, mu.0, SE),
+   X.bar = X.bar, alpha = alpha,
+   p.value = 1 - pnorm(X.bar, mu.0, SE)), 3)
   mu.0     x.H   X.bar   alpha p.value
 18.000  18.292  18.520   0.050   0.002
```

and we conclude that the treatment was effective. The rejection region for the sampling density under H_0 is shown in Figure 11.1 (the code to produce the figure was hijacked from Example 9.12). The gray area is the type I error—rejecting the null hypothesis when it is true.

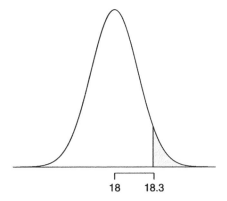

Figure 11.1 Rejection region (gray) for the Viagra treatment.

Next, suppose that in reality, the mean score on the IIEF test for the untreated population is 18.1. Denote this mean by μ_A. Under H_0, any $\overline{X}$ that falls to the left of approximately 18.3 will prompt us not to reject H_0. But here we commit an error. The probability of this error is β. It is shown as the dark region to the left of critical value of 18.3 under the null model in Figure 11.2A. To compute β in Figure 11.2A, we write

$$\beta = \Phi\left(x_H \mid \mu_A, \mathrm{SE}\right) \tag{11.1}$$
$$= \mathrm{pnorm}(\mathrm{x.H}, \mathrm{mu.A}, \mathrm{SE})$$

(where $x_H :=$ the critical value x.H) which gives

```
> mu.A <- 18.1 ; round(pnorm(x.H, mu.A, SE), 2)
[1] 0.86
```

Next, suppose that $\mu_A = 18.2$. Now β is

```
> mu.A <- 18.2 ; round(pnorm(x.H, mu.A, SE), 2)
[1] 0.7
```

(see Figure 11.2B). In Figure 11.2C and D, $\mu_A = 18.3$ and 18.4. Here is a summary of what we have done thus far:

```
> mu.A <- c(18.1, 18.2, 18.3, 18.4)
> x.H <- qnorm(1 - alpha, mu.0, SE)
> beta <- matrix(nrow = 2, ncol = length(mu.A))
> beta[1, ] <- mu.A
> beta[2, ] <- round(pnorm(x.H, mu.A, SE), 2)
> dimnames(beta) <- list(c('mu.A', 'beta'),
+   rep('',length(mu.A))) ; beta

mu.A 18.10 18.2 18.30 18.40
beta  0.86  0.7  0.48  0.27
```

From these results we conclude that if we reject $\mu_0 = 18$ in favor of $\mu_A = 18.3$, for example, then there is 48% chance that we will not reject 18 (in favor of 18.3) while $\mu_0 = 18$ is false. Figure 11.2 was produced with a script similar to that for Figure 11.1. □

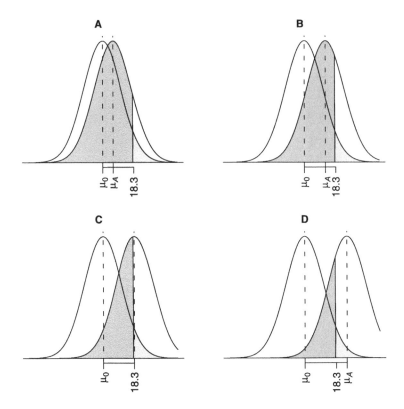

Figure 11.2 The dark gray regions indicate type II error for 4 alternative models of $\mu_A = 18.1, 18.2, 18.3$ and 18.4 in A, B, C and D. Note the decrease in the size of the type II error as μ_A moves away from $\mu_0 = 18$.

As the example illustrates, the larger the difference between μ_0 and the supposed alternative state of nature μ_A, the smaller the β. What we have just illustrated is the fact that the

Magnitude of type II error (β) decreases as the difference between a null and alternative models increases.

This motivates the definition of power of hypothesis testing with H_0: $\theta = \theta_0$ as

Power $= 1 - \beta$ The probability of correctly rejecting H_0 for a given θ_A.

Usually, as a

Rule of thumb about power A power of $1 - \beta = 0.8$ is considered acceptable.

Instead of choosing a small set of values for μ_A, we can choose a range of values to obtain the *power profile*.

Example 11.2. Continuing with Example 11.1, let us compute β and power, for $\alpha = 0.05$ and $n = 48$ for $\mu_A = 18$ to 19. Figure 11.3 illustrates the duality between the decrease in the type II error and increase in the power (the probability of rejecting

H_0 when H_A is true) with increase in $|\mu_0 - \mu_A|$. To produce Figure 11.3 we first specify the data:

```
> mu.0 <- 18 ; S <- 1.23 ; n <- 48 ; SE <- S / sqrt(n)
> mu.A <- seq(18, 19, length = 201) ; alpha = 0.05
> x.H <- qnorm(1 - alpha, mu.0, SE)
```

Note that x_H is the critical value for the upper-tailed test under H_0 (i.e. $\mu = \mu_0$). x_H is the right boundary for which we calculate β. Next, we use (11.1) to calculate β:

```
> beta <- pnorm(x.H, mu.A, SE)
```

The plotting is done with

```
> par(mfrow=c(1, 2))
> plot(mu.A, beta, type = 'l',
+    xlab = expression(italic(mu[A])),
+    ylab = expression(beta))
> plot(mu.A, 1 - beta, type = 'l',
+    xlab = expression(italic(mu[A])),
+    ylab = 'power')
```

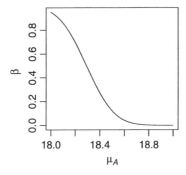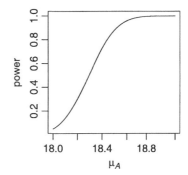

Figure 11.3   The duality between type II error (β) and power profiles for the Viagra experiment.

and given our work thus far, hardly needs an explanation.   □

Let us interpret our findings in Examples 11.1 and 11.2.

Example 11.3. The higher the test score on the IIEF, the more successful the treatment with Viagra. If the score is high enough, continued treatment is justified. The data indicate that the score was 18.52 ± 1.23 for $n = 48$. We assume that the mean score for the untreated population is $\mu = 18$ with $S = 1.23$. Therefore, under H_0, the sampling density of $\overline{X}$ is $\phi\left(x|18, 1.23/\sqrt{48}\right)$. From this, we find that at $\alpha = 0.05$, $x_H \approx 18.3$ and we reject the null hypothesis that the treatment is not effective if the sample $\overline{X} > 18.3$. The researchers found $\overline{X} = 18.52$ and we conclude that the treatment is effective with rejecting H_0 in only 5% of repeated samples (of size n) when H_0 is correct. Suppose that being conservative, the researchers decide that the mean score of the treated population is $\mu_A = 18.4$ (compared to their finding of $\overline{X} = 18.52$). Then

```
> round(1 - pnorm(x.H, 18.4, SE), 2)
[1] 0.73
```

In other words, the probability of rejecting the mean test score of the untreated population as a representative value of the mean of the treated population is 0.73. In repeated sampling, we will reach the correct conclusion (that the treatment is effective) in 73% of the samples. □

Next, we generalize the example-based results concerning power to include lower-, upper- and two-tailed hypothesis testing. Consider the rv X from a population with arbitrary density with mean μ and standard deviation σ. Take a sample of size n (> 30) from the population. Then the sampling density of $\overline{X}$ is approximately $\phi(x\,|\mu, \text{SE})$. We have three versions of hypothesis testing and therefore, three versions of power calculations. In all tests, the significance value is α.

Lower-tailed power

Because our test is
$$H_0 : \mu = \mu_0 \quad \text{vs.} \quad H_A : \mu < \mu_0 ,$$
it makes sense to consider $\mu_A < \mu_0$ only. To obtain the power, we first compute
$$x_L = \Phi^{-1}(\alpha\,|\mu_0, \text{SE}) \tag{11.2}$$
$$= \text{qnorm}(\text{alpha}, \text{mu.0}, \text{SE})$$
and then the power
$$1 - \beta = \Phi(x_L\,|\mu_A, \text{SE}) \tag{11.3}$$
$$= \text{pnorm}(\text{x.L}, \text{mu.A}, \text{SE})$$
where $\text{x.L} := x_L$.

Upper-tailed power

Our test is
$$H_0 : \mu = \mu_0 \quad \text{vs.} \quad H_A : \mu > \mu_0 .$$
Therefore, we consider only $\mu_A > \mu_0$. Recall that
$$x_H = \Phi^{-1}(1 - \alpha\,|\mu_0, \text{SE}) \tag{11.4}$$
$$= \text{qnorm}(1 - \text{alpha}, \text{mu.0}, \text{SE}) .$$
The power is then
$$1 - \beta = 1 - \Phi(x_H\,|\mu_A, \text{SE}) \tag{11.5}$$
$$= 1 - \text{pnorm}(\text{x.H}, \text{mu.A}, \text{SE}) .$$

Two-tailed power

Because our test is
$$H_0 : \mu = \mu_0 \quad \text{vs.} \quad H_A : \mu \neq \mu_0 ,$$
we separate the power calculation into the alternatives $\mu_L < \mu_0$ and $\mu_H > \mu_0$. We also wish to stay away from symmetric α (for reasons we explain in Example 11.5 below). So we specify α_L and α_H. Without further information, we may choose

some $\Delta > 0$ and specify $\mu_L = \mu_0 - \Delta$ for the power of H_0 vs. H_A: $\mu < \mu_0$ and $\mu_H = \mu_0 + \Delta$ for the power of H_0 vs. H_A: $\mu > \mu_0$. With some auxiliary information (such as when the risk of type II error when μ_0 is underestimated is higher than when μ_0 is overestimated), you may choose different distances for μ_L and μ_H from μ_0.

First, we compute the lower-tailed part of the power:

$$x_L = \Phi^{-1}(\alpha_L | \mu_0, \text{SE}) \qquad (11.6)$$
$$= \texttt{qnorm(alpha.L, mu.0, SE)}$$

and

$$1 - \beta_L = \Phi(x_L | \mu_L, \text{SE})$$
$$= \texttt{pnorm(x.L, mu.L, SE)}.$$

Next, we compute the power for the upper-tailed part:

$$x_H = \Phi^{-1}(1 - \alpha_H | \mu_0, \text{SE})$$
$$= \texttt{qnorm(1 - alpha.H, mu.0, SE)}$$

and

$$1 - \beta_H = 1 - \Phi(x_H | \mu_H, \text{SE}) \qquad (11.7)$$
$$= 1 - \texttt{pnorm(x.H, mu.H, SE)}.$$

The power is then

$$1 - \beta = (1 - \beta_L) + (1 - \beta_H). \qquad (11.8)$$

Example 11.4. The U.S. Environmental Protection Agency (EPA) publishes data about maximum allowed consumption of mercury-contaminated fish (Table 11.1).

For sharks caught off the coast of Florida, we have the data in Table 11.2 (adapted from Adams and McMichael, 1999). Suppose that we combine all of the data in Table 11.2. How much can we distinguish among the means of mercury concentrations for the various shark species and how should that affect the consumption recommendations in Table 11.1?

Table 11.1 Fish meals per month based on consumer adult body weight of 154 lbs (≈ 70 kg) and average meal size of 8 oz (≈ 225 g)

Meals/month	Mercury (ppm)
16	0.03-0.06
12	0.06-0.08
8	0.08-0.12
4	0.12-0.24
3	0.24-0.32
2	0.32-0.48
1	0.48-0.97
0.5	0.97-1.90
None	> 1.9

Table 11.2 Total mercury in sharks caught off the coast of Florida.

Species	Common name	n	Mercury (ppm) Mean	SD
Carcharhinus leucas	Bull shark	53	0.77	0.32
Carcharhinus limbatus	Blacktip shark	21	0.77	0.71
Rhizoprionodon	Atlantic sharpnose	81	1.06	0.63
Sphyrna tiburo	Bonnethead shark	95	0.50	0.36

First, we combine the data for all the shark species into a single model. This requires a weighted average of the means and the square root of the weighted averages of the variances (mu.0 and S below):

```
> X.bars <- c(0.77, 0.77, 1.06, 0.5)
> ns <- c(53, 21, 81, 95) ; n <- sum(ns)
> Ss <- c(0.32, 0.71, 0.63, 0.36)
> mu.0 <- sum(X.bars * ns)/ n
> S <- sqrt(sum(Ss * ns) / n) ; SE <- S / sqrt(n)
> round(c(mu.0 = mu.0, n = n, S = S, SE = SE), 2)
   mu.0      n      S     SE
   0.76 250.00   0.68   0.04
```

Our model is $\phi(x\,|\,0.76, 0.04)$, so we are in the category of one fish meal per month. What is our power to distinguish between 0.76 and 0.97 (the low boundary on 0.5 fish meals per month) at $\alpha = 0.05$? We set $\mu_A = 0.96$ and from (11.4) and (11.5) we obtain

```
> x.H <- qnorm(1 - alpha, mu.0, SE) ; mu.A <- 0.97
> (power <- 1 - pnorm(x.H, mu.A, SE))
[1] 0.9992515
```

This is reassuring, but not really what we want if we are concerned with consumers health. Why? Because we want to make sure that we have power to distinguish the low boundary and our current mean value. Otherwise, we might err and actually recommend two meals as opposed to one. So now we choose $\mu_A = 0.48$ and obtain the power to distinguish between $\phi(x\,|\,0.76, 0.04)$ and $\phi(x\,|\,0.48, 0.04)$, at $\alpha = 0.05$. Using (11.2) and (11.3) we obtain

```
> x.L <- qnorm(alpha, mu.0, SE) ; mu.A <- 0.48
> (power <- pnorm(x.L, mu.A, SE))
[1] 0.9999994
```

and we are reassured again that chances are that we will not reject a true null (that $\mu = 0.76$) in favor of a lower value that might lead us to recommend no more than two fish meals a month, as opposed to no more than one. In Exercise 11.2 you are asked to explore some potentially wrong recommendations for the number of fish meals per month.   □

Sample size

As we saw in the previous section, the sample's standard error enters into all of the power calculations. The standard error is based on the estimate of the population σ from the sample's standard deviation (S) and on the sample size n. Therefore, if we specify a power, before taking a sample, then the only unknowns in the power equations are the standard error and n. So if we can find a way to estimate σ, the only remaining unknown is n. Thus, we can determine the sample size that is needed to obtain a significant difference between μ_0 and μ_A for a given power and given α. For practical reasons, we adopt the following:

Rule of thumb for estimating σ If the population density is approximately symmetric, then
$$\sigma \approx \frac{\text{data range}}{4}.$$

We wish to calculate the sample size (under a large sample and normal sampling distribution) such that we obtain a significance level of α with a given power level $1 - \beta$ and a detectable difference between μ_0 and μ_A.

Lower-tailed sample size

For the power, we have
$$x_L = \Phi^{-1}\left(\alpha \,|\, \mu_0, \sigma/\sqrt{n}\right)$$
and
$$1 - \beta = \Phi\left(x_L \,|\, \mu_A, \sigma/\sqrt{n}\right).$$
Therefore,
$$x_L = \Phi^{-1}\left(1 - \beta \,|\, \mu_A, \sigma/\sqrt{n}\right).$$
Standardizing x_L both ways, we obtain
$$\mu_0 - \Phi^{-1}(\alpha \,|\, 0, 1)\frac{\sigma}{\sqrt{n}} = \mu_A + \Phi^{-1}(1 - \beta \,|\, 0, 1)\frac{\sigma}{\sqrt{n}}.$$

Solving for n and recalling that, by convention, we drop the parameters from the notation for the standard normal, we obtain

$$\begin{aligned}
n &= \frac{\sigma^2}{(\mu_0 - \mu_A)^2}\left[\left(\Phi^{-1}(\alpha) + \Phi^{-1}(1 - \beta)\right)\right]^2 \\
&= \text{sigma}\char`\^2/(\text{mu.0} - \text{mu.A})\char`\^2 * \\
&\quad (\text{qnorm(alpha)} + \text{qnorm}(1 - \text{beta}))\char`\^2.
\end{aligned} \qquad (11.9)$$

Upper-tailed sample size

For the power, we have
$$x_H = \Phi^{-1}\left(1 - \alpha \,|\, \mu_0, \sigma/\sqrt{n}\right)$$
and
$$1 - \beta = 1 - \Phi\left(x_H \,|\, \mu_A, \sigma/\sqrt{n}\right)$$
or
$$\beta = \Phi\left(x_H \,|\, \mu_A, \sigma/\sqrt{n}\right).$$

Therefore,
$$x_H = \Phi^{-1}\left(\beta\,|\mu_A, \sigma/\sqrt{n}\right)$$
or
$$\mu_0 + \Phi^{-1}(1-\alpha)\frac{\sigma}{\sqrt{n}} = \mu_A - \Phi^{-1}(\beta)\frac{\sigma}{\sqrt{n}}\,.$$

Therefore,
$$\begin{aligned}n &= \frac{\sigma^2}{(\mu_0-\mu_A)^2}\left[\left(\Phi^{-1}(1-\alpha)+\Phi^{-1}(\beta)\right)\right]^2 \\ &= \text{sigma}\hat{\,}2/(\text{mu.0}-\text{mu.A})\hat{\,}2 * \\ &\quad(\text{qnorm}(1-\text{alpha})+\text{qnorm}(\text{beta}))\hat{\,}2\,. \end{aligned} \qquad (11.10)$$

Two-tailed sample size

In the most general case, to obtain the sample size, we need to specify α_L, α_H, $\mu_L < \mu_0$, $\mu_H > \mu_0$, $1-\beta_L$ and $1-\beta_H$. Then we calculate the sample size for the lower tail, n_L, by modifying (11.9):

$$\begin{aligned} n_L &= \frac{\sigma^2}{(\mu_0-\mu_L)^2}\left[\left(\Phi^{-1}(\alpha_L)+\Phi^{-1}(1-\beta_L)\right)\right]^2 \\ &= \text{sigma}\hat{\,}2/(\text{mu.0}-\text{mu.L})\hat{\,}2 * \\ &\quad(\text{qnorm}(\text{alpha.L})+\text{qnorm}(1-\text{beta.L}))\hat{\,}2\,. \end{aligned} \qquad (11.11)$$

We obtain the sample size for the upper tail, n_H, by modifying (11.10):

$$\begin{aligned} n_H &= \frac{\sigma^2}{(\mu_0-\mu_H)^2}\left[\left(\Phi^{-1}(1-\alpha_H)+\Phi^{-1}(\beta_H)\right)\right]^2 \\ &= \text{sigma}\hat{\,}2/(\text{mu.0}-\text{mu.H})\hat{\,}2 * \\ &\quad(\text{qnorm}(1-\text{alpha.H})+\text{qnorm}(\text{beta.H}))\hat{\,}2\,. \end{aligned} \qquad (11.12)$$

Summing both we obtain
$$n = n_L + n_H\,.$$
For all cases, always adjust n upward to the nearest integer.

Example 11.5. In this example, we deal with Atlantic sharpnose (Table 11.2). The data indicate that the mean and standard deviation of mercury concentrations were 1.06 and 0.63 ppm, respectively, with a sample size of 81. Suppose that for all practical purposes, a difference of 0.1 ppm is negligible. That is, from the perspective of a consumer, there is as much risk if one eats fish with $\mu_L = 0.96$, $\mu_0 = 1.06$ or $\mu_H = 1.16$ ppm of tissue mercury. Underestimating mercury concentration may result in too high a limit on consumption (bad idea). Overestimating may result in too low a limit on consumption (not such a bad idea). Underestimating μ_0 happens when we do a lower-tailed test and reject a true H_0 in favor of $H_A: \mu < \mu_0$. This is type I error and we want to minimize it. So we choose $\alpha_L = 0.01$ and $\alpha_H = 0.025$.

What about power? Because we wish to be cautious, we do not mind much about rejecting H_0 in favor of $H_A: \mu < \mu_0$ when $\mu_L < \mu_0$ is true. So we set $1-\beta_L = 0.7$ and $1-\beta_H = 0.8$. What should the sample size be for these specifications? First, we specify some of the data and plot the power profile for the left and right sides according to (11.6) and (11.7):

```
> sigma <- 0.63 ; alpha.L <- 0.01 ; alpha.H <- 0.025
> mu.0 <- 1.06 ; n <- 81 ; SE <- sigma /sqrt(n)
> x.L <- qnorm(alpha.L, mu.0, SE)
> mu.a <- seq(0.7, mu.0, length = 201)
> power.L <- pnorm(x.L, mu.a, SE)
> mu.A <- seq(mu.0, 1.4, length = 201)
> x.H <- qnorm(1 - alpha.H, mu.0, SE)
> power.H <- 1 - pnorm(x.H, mu.A, SE)
> plot(mu.a - mu.0, power.L, type = 'l',
+    xlim = c(-0.3, 0.3),
+    xlab = expression(italic(Delta*mu)),
+    ylab = expression(italic(1-beta)))
> lines(mu.A - mu.0, power.H)
```

(left panel of Figure 11.4). To obtain an estimate of the sample size we compute n_L with (11.11) and n_H with (11.12)

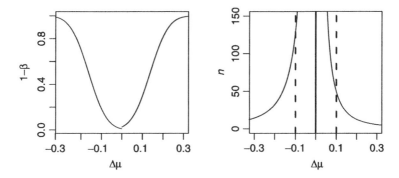

Figure 11.4   Left - Power profiles for two-tailed hypothesis testing with $\alpha_L = 0.01$ and $\alpha_H = 0.025$. Right - Sample size profiles for two-tailed hypothesis testing and asymmetric decision rules about α and $1 - \beta$.

```
> mu.0 <- 1.06 ; mu.a <- 0.96 ; mu.A <- 1.16
> beta.L <- 0.3 ; beta.H <- 0.2
> n.L <- ceiling(sigma^2 / (mu.0 - mu.a)^2 *
+    (qnorm(alpha.L) + qnorm(1-beta.L))^2)
> n.H <- ceiling(sigma^2 / (mu.0 - mu.A)^2 *
+    (qnorm(1 - alpha.H) + qnorm(beta.H))^2)
```

The necessary sample size is

```
> c(n.L = n.L, n.H = n.H, n = n.L + n.H)
n.L n.H   n
129  50 179
```

The bulk of n is eaten up by the requirements on the left side ($n_L = 129$). The reported sample size of 81 does not meet our specifications. It comes with inadequate power to distinguish among desired alternatives. To observe the sample size profiles, we do

```
> n.l <- n.L ; n.h <- n.H
> mu.a <- seq(0.7, mu.0, length = 201)
> n.L <- sigma^2 / (mu.0 - mu.a)^2 *
+     (qnorm(alpha.L) + qnorm(1-beta.L))^2
> mu.A <- seq(mu.0, 1.4, length = 201)
> n.H <- sigma^2 / (mu.0 - mu.A)^2 *
+     (qnorm(1 - alpha.H) + qnorm(beta.H))^2
> plot(mu.a - mu.0, n.L, type = 'l', ylim = c(0, 150),
+     xlim = c(-0.3, 0.3),
+     xlab = expression(italic(Delta*mu)),
+     ylab = expression(italic(n)))
> lines(mu.A - mu.0, n.H)
> abline(v = c(0.96 - mu.0, 1.06 - mu.0, 1.16 - mu.0),
+     lty = c(2, 1, 2), lwd = 2)
```

Here we calculate a sequence of n_L on a sequence of μ_L and similarly for μ_H. We then plot the lower sample size profile for the difference $\mu_L - \mu_0$ and add lines() for the upper size profile for $\mu_H - \mu_0$ (right panel of Figure 11.4). The vertical broken abline()s show the values we used for μ_L and μ_H to obtain n_L and n_H (adjusted for the difference from μ_0). The solid vertical line is μ_0 adjusted for itself. □

Because the normal is symmetric, the equations for power and sample size can be simplified. For example, the required sample size for two-sided hypothesis testing is often written as

$$n = \frac{2\sigma^2}{(\mu_0 - \mu_A)^2} \left(z_{1-\alpha/2} + z_\beta\right)^2 .$$

Albeit more cumbersome, we prefer to stick with our notation because it is closer to how one eventually address these issues with R. Furthermore, keeping the notation with the distributions explicit (e.g., $\Phi(\alpha)$ instead of z_α) allows for the flexibility in decision making that we used in Example 11.5.

11.1.2 Proportions

Recall from Section 7.7.1 that the sampling density of large sample proportions is $\phi\left(p \mid \pi, \sqrt{\pi(1-\pi)/n}\right)$. Because we deal with large samples, we shall ignore the continuity correction. If you do wish to implement this correction, you might as well use small sample or exact methods to obtain power for proportions. The methods for large sample means (Section 11.1) do not apply directly because π and $\sqrt{\pi(1-\pi)/n}$ are dependent, so the standard errors of the sampling densities are different for $\pi_0 \neq \pi_A$. At any rate, our rv is the sample proportion $p = n_S / n$ where n is the total number of trials and n_S is the number of successes in a binomial experiment.

Power

To obtain the power for lower- upper- and two-tailed hypothesis testing, replace the pairs μ_0, SE everywhere in Section 11.1.1 with π_0, $\sqrt{\pi_0(1-\pi_0)/n}$. Similarly, replace the pairs μ_A, SE everywhere in Section 11.1.1 with π_A, $\sqrt{\pi_A(1-\pi_A)/n}$. For example, from (11.2),

$$p_L = \Phi^{-1}\left(\alpha \,\big|\, \pi_0, \sqrt{\pi_0(1-\pi_0)/n}\right)$$
$$= \text{pnorm}(\text{alpha}, \text{pi.0}, \text{sqrt}(\text{pi.0} * (1-\text{pi.0})/\text{n}))$$

and from (11.3), the power for lower-tailed hypothesis testing with significance α is

$$1 - \beta = \Phi\left(p_L \,\big|\, \pi_A, \sqrt{\pi_A(1-\pi_A)/n}\right)$$
$$= \text{pnorm}(\text{p.L}, \text{pi.A}, \text{sqrt}(\text{pi.A} * (1-\text{pi.A})/\text{n}))$$

where $\text{pi.L} := \pi_L$.

Example 11.6. The data for this example were published by the European Commission, which is the executive arm of the European Union (Gallup Europe, 2003). It pertains to a European public opinion survey of attitudes toward peace in the World. In all, 7515 people were interviewed. Among the questions was the following:

"Tell me if in your opinion, it presents or not a threat to peace in the world ...?"

followed by a list of selected countries (arranged randomly). Here is the order of the data (% responding) in the report:

```
> load('eu.rda')
> eu
         country yes no undecided
1             EU   8 89         3
2         Israel  59 37         5
3           Iran  53 41         5
4    North Korea  53 40         7
5  United States  53 44         5
6           Iraq  52 44         4
7    Afghanistan  50 45         6
8       Pakistan  48 46         6
9          Syria  37 56         7
10         Libya  36 58         7
11  Saudi Arabia  36 68         7
12         China  30 65         5
13         India  22 74         5
14        Russia  21 76         4
15       Somalia  16 75        10
```

What is the power of distinguishing between $\pi_0 = 0.59$ and π_A between 0 and 1 in the case of Israel? The hypotheses we test are $H_0: \pi = \pi_0 = 0.59$ vs. $H_A: \pi \neq \pi_0$. We shall treat the power associated with the lower- and upper-tailed hypotheses as equally important.

To compute the power, we implement the necessary modifications to (11.6) through (11.8). First, the data:

```
> alpha <- 0.05 ; n <- 7515
> pi.0 <- 0.59 ; pi.A <- seq(0.55, 0.64, length = 201)
```

Next, we use power for two-tailed hypothesis testing and the necessary substitutions in (11.2) through (11.8):

```
> p.L <- qnorm(alpha / 2, pi.0,
+    sqrt(pi.0 * (1 - pi.0) / n))
> power.L <- pnorm(p.L, pi.A,
+    sqrt(pi.A * (1 - pi.A) / n))
> p.H <- qnorm(1 - alpha / 2, pi.0,
+    sqrt(pi.0 * (1 - pi.0) / n))
> power.H <- 1 - pnorm(p.H, pi.A,
+    sqrt(pi.A * (1 - pi.A) / n))
> power <- power.L + power.H
```

To see what we got, we do

```
> plot(pi.A, power,type = 'l',
+    xlab = expression(italic(pi[A])),
+    ylab = 'power')
> polygon(c(0.58, 0.58, 0.60, 0.60, 0.58),
+    c(-1,1.1,1.1,-1,-1), col = 'gray90')
> lines(pi.A, power)
```

(Figure 11.5). From the figure, the possibility of rejecting a correct π_0 is rather small.

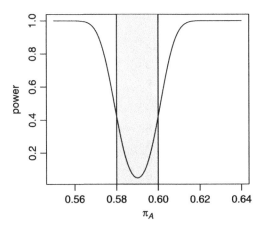

Figure 11.5 Power of detecting alternative models for EU public opinion results about Israel's threat to world peace.

It grows around a narrow band between about $\pi_A = 0.58$ and 0.59. You can use binom.power() in the package binom to obtain results identical to those in Figure 11.5, e.g.

```
> library(binom)
> y <- binom.power(pi.A, n = n, p = pi.0, alpha = alpha,
+    alternative='two.sided', method = 'asymp')
> lines(pi.A, y)
```

The curve we obtain from our computations and those of binom.power() are indistinguishable. □

Sample size

Consider a lower-tailed hypothesis test. For the power, we have

$$p_L = \Phi^{-1}\left(\alpha \,\middle|\, \pi_0, \sqrt{\frac{\pi_0(1-\pi_0)}{n}}\right)$$

and

$$1 - \beta = \Phi\left(p_L \,\middle|\, \pi_A, \sqrt{\frac{\pi_A(1-\pi_A)}{n}}\right).$$

Therefore,

$$p_L = \Phi^{-1}\left(1 - \beta \,\middle|\, \pi_A, \sqrt{\frac{\pi_A(1-\pi_A)}{n}}\right).$$

Standardizing p_L both ways, we obtain

$$\pi_0 - \Phi^{-1}(\alpha)\sqrt{\frac{\pi_0(1-\pi_0)}{n}} = \pi_A + \Phi^{-1}(1-\beta)\sqrt{\frac{\pi_A(1-\pi_A)}{n}}.$$

So

$$n = \left[\frac{\Phi^{-1}(\alpha)\sqrt{\pi_0(1-\pi_0)} + \Phi^{-1}(1-\beta)\sqrt{\pi_A(1-\pi_A)}}{\pi_0 - \pi_A}\right]^2$$

$$= (\text{qnorm(alpha)} * A + \text{qnorm}(1 - \text{beta}) * B) / \qquad (11.13)$$
$$(\text{pi.0} - \text{pi.A}))^2$$

where $A = \text{sqrt(pi.0} * (1 - \text{pi.0}))$ and $B = \text{sqrt(pi.A} * (1 - \text{pi.A}))$. With the same approach, we obtain the sample size for upper-tailed hypothesis testing as

$$n = \left[\frac{\Phi^{-1}(1-\alpha)\sqrt{\pi_0(1-\pi_0)} + \Phi^{-1}(\beta)\sqrt{\pi_A(1-\pi_A)}}{\pi_0 - \pi_A}\right]^2. \qquad (11.14)$$

For two-tailed tests, we use (11.13) and (11.14) with (potentially) different values for α and $1 - \beta$ for the lower- and upper-tailed testing:

$$n_L = \left[\frac{\Phi^{-1}(\alpha_L)\sqrt{\pi_0(1-\pi_0)} + \Phi^{-1}(1-\beta)\sqrt{\pi_L(1-\pi_L)}}{\pi_0 - \pi_L}\right]^2, \qquad (11.15)$$

$$n_H = \left[\frac{\Phi^{-1}(1-\alpha_H)\sqrt{\pi_0(1-\pi_0)} + \Phi^{-1}(\beta)\sqrt{\pi_H(1-\pi_H)}}{\pi_0 - \pi_H}\right]^2. \qquad (11.16)$$

Here the alternatives are specified as $\pi_L < \pi_0$ and $\pi_H > \pi_0$. The sample size is then

$$n = n_L + n_H. \qquad (11.17)$$

In all cases of obtaining sample sizes, if you do not know what π_0 is, be conservative. Choose $\pi_0 = 0.5$. This results in the largest value of $\sqrt{\pi_0(1-\pi_0)}$ and consequently in the lowest upper bound (infimum) on n.

Power and sample size for single samples

Example 11.7. Years after meltdown of a nuclear power plant, area residents claim that the incidence of cancer among them is higher than the incidence in the population of the country. To investigate their claim, we wish to estimate the proportion of exposed population that suffers from cancer. We want to be able to distinguish between the true proportion and alternative proportions that are 0.005 larger or smaller than the true proportion. We desire a sample size for $\alpha = 0.01$ and power $= 0.9$. We will not distinguish between the lower and upper tails. First, the data:

```
> delta <- 0.25 ; pi.0 <- 0.5 # max sample size
> pi.A <- seq(pi.0 - delta, pi.0 + delta, length = 401)
> alpha <- 0.01 ; beta <- 0.1
```

Next, n_L and n_H according to (11.15) and (11.16):

```
> n.L <- (qnorm(alpha / 2) * sqrt(pi.0 * (1 - pi.0))
+     + qnorm(1 - beta) * sqrt(pi.A[pi.A < 0.5] *
+     (1 - pi.A[pi.A < 0.5])) / (pi.0 - pi.A[pi.A < 0.5]))^2
> n.H <- (qnorm(1 - alpha / 2) * sqrt(pi.0 * (1 - pi.0))
+     + qnorm(beta) * sqrt(pi.A[pi.A > 0.5] *
+     (1 - pi.A[pi.A > 0.5])) / (pi.0 - pi.A[pi.A > 0.5]))^2
```

Let us put the results in a matrix according to (11.17)

```
> d <- cbind(
+     delta = c(pi.A[pi.A < 0.5], pi.A[pi.A > 0.5]) - 0.5,
+     n = c(n.L + n.H, n.L + n.H))
```

and plot

```
> plot(d, type = 'l', ylog = TRUE,
+     ylim = c(0, 50000), xlim = c(-0.02, 0.02),
+     xlab = expression(italic(pi[0]-pi[A])),
+     ylab = expression(italic(n)))
> abline(v = c(-0.005, 0.005), lty = 2) ; abline(h = 16500)
```

(Figure 11.6; note the use of the named argument ylog—it plots the y-axis on a log scale). The vertical broken lines show the detectable distance and the horizontal line the sample size ($\approx 16\,500$). □

11.1.3 Intensities

In Section 7.8.1, we established that the sampling density of the Poisson parameter is $\phi\left(l\,\big|\,\lambda, \sqrt{\lambda/n}\right)$. As was the case with the binomial, we cannot translate the equations in Section 11.1 directly to obtain power and sample size because the mean and the variance of the Poisson are dependent.

Power

For lower-tailed power we consider only $\lambda_A < \lambda_0$. To obtain the power, we first compute

$$l_L = \Phi^{-1}\left(\alpha\,\big|\,\lambda_0, \sqrt{\lambda_0/n}\right) \qquad (11.18)$$
$$= \mathtt{qnorm(alpha, lambda.0, sqrt(lambda.0/n))}$$

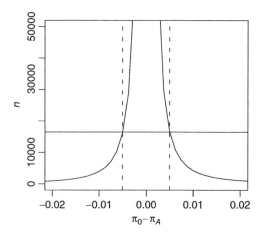

Figure 11.6 Sample size for $\alpha = 0.01$, $1 - \beta = 0.9$, $\pi_0 = 0.5$ and π_A between 0.498 and 0.502.

where α is the significance level. The power is then

$$1 - \beta = \Phi\left(l_L \,\middle|\, \lambda_A, \sqrt{\lambda_A/n}\right) \tag{11.19}$$
$$= \text{pnorm}(\text{l.L}, \text{lambda.A}, \text{sqrt}(\text{lambda.A}/\text{n}))$$

where $\text{l.L} := l_L$. For upper-tailed power we consider only $\lambda_A > \lambda_0$. We first obtain

$$l_H = \Phi^{-1}\left(1 - \alpha \,\middle|\, \lambda_0, \sqrt{\lambda_0/n}\right) \tag{11.20}$$
$$= \text{qnorm}(1 - \text{alpha}, \text{lambda.0}, \text{sqrt}(\text{lambda.0}/\text{n}))$$

and then the power

$$1 - \beta = 1 - \Phi\left(l_H \,\middle|\, \lambda_A, \sqrt{\lambda_A/n}\right) \tag{11.21}$$
$$= 1 - \text{pnorm}(\text{l.H}, \text{lambda.A}, \text{sqrt}(\text{lambda.A}/\text{n})).$$

To obtain the power for a two-tailed test, we consider $\lambda_L < \lambda_0$ and $\lambda_H > \lambda_0$. We first compute l_L as in (11.18) with α_L instead of α and then $(1 - \beta_L)$ as in (11.19). Next, we compute l_H as in (11.20) with α_H instead of α and then $(1 - \beta_H)$ as in (11.21). The power is then

$$1 - \beta = (1 - \beta_L) + (1 - \beta_H) \ .$$

Example 11.8. In fisheries, catch per unit effort (CPUE) is defined as the number of fish caught per unit effort. A unit effort is measured (or should be measured) as the amount of time a net (for example) is in the water and the volume that the net covers (this requires knowledge about the swimming behavior of the species caught). So the units might be fish per m^3-hr. Suppose that we submerge nets for $100 \, m^3$-hr and catch 35 fish. The fish population is large enough to assume sampling with replacement and fish are supposed to be caught independent of each other. We set

$\lambda_0 = 0.35$ and wish to test the power to distinguish between λ_0 and $\lambda_A = 0.4$. Then from (11.20) and (11.21):

```
> lambda.0 <- 0.35 ; lambda.A <- seq(0.35, 0.85, length = 501)
> alpha <- 0.05 ; n <- 100
> l.H <- qnorm(1 - alpha, lambda.0, sqrt(lambda.0 / n))
> power <- 1 - pnorm(l.H, lambda.A, sqrt(lambda.A/n))
```

The power profile and specific values are obtained with

```
> plot(lambda.A, power, type = 'l',
+    xlab = expression(italic(lambda[A])))
> round(c(lambda.0 = lambda.0, n = n,
+    lambda.A = lambda.A[51], alpha = alpha,
+    power = power[51]), 3)
lambda.0         n  lambda.A     alpha     power
   0.350   100.000     0.400     0.050     0.227
```

(Figure 11.7). So if there is some management decision to be made based on 0.35 vs. 0.4, it is not particularly powerful. □

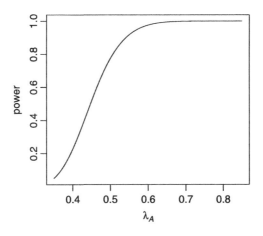

Figure 11.7   Poisson power profile.

Sample size

For lower-tailed tests, the power is obtained from

$$l_L = \Phi^{-1}\left(\alpha \,\big|\, \lambda_0, \sqrt{\lambda_0/n}\right)$$

and

$$1 - \beta = \Phi\left(l_L \,\big|\, \lambda_A, \sqrt{\lambda_A/n}\right) .$$

Therefore,

$$l_L = \Phi^{-1}\left(1 - \beta \,\big|\, \lambda_A, \sqrt{\lambda_A/n}\right) .$$

Standardizing l_L both ways, we obtain

$$\lambda_0 - \Phi^{-1}(\alpha)\sqrt{\frac{\lambda_0}{n}} = \lambda_A + \Phi^{-1}(1-\beta)\sqrt{\frac{\lambda_A}{n}}.$$

Solving for n, we get

$$n = \left[\frac{\Phi^{-1}(\alpha)\sqrt{\lambda_0} + \Phi^{-1}(1-\beta)\sqrt{\lambda_A}}{\lambda_0 - \lambda_A}\right]^2. \tag{11.22}$$

Sample sizes for upper- and two-tailed hypothesis tests are obtained in much the same way as for the binomial (with the appropriate substitutions of λ for π).

Example 11.9. Continuing with Example 11.8, we desire the sample size that will allow us to distinguish between said λ_0 and λ_A with $1 - \beta = 0.8$. Using the parallel of (11.22) for an upper-tailed test, we obtain:

```
> round(c(lambda.0 = lambda.0, n = ceiling(n[51]),
+    lambda.A = lambda.A[51], alpha = alpha,
+    power = 1 - beta), 3)
lambda.0        n lambda.A    alpha    power
    0.35  1145.00     0.40     0.05     0.80
```

So we need a sample from $1\,145$ m^3-hr. □

11.2 Small samples

In this section we address the issue of power for small samples. Here we can no longer use the normal approximation. Calculating sample sizes for small samples requires that you first calculate the sample size for large samples. Then if n turns out to be small (< 30), recalculate n as discussed below.

11.2.1 Means

In Sections 9.2.2 and 10.3.1, we established that the sampling density of the mean of a small sample ($n < 30$) from a normal population is t. In the former, we used the sampling density to construct confidence intervals and in the latter to test hypotheses. To establish the power for hypothesis tests in this case, we replace, where necessary, Φ with the t distribution $P(Z \leq z|n-1)$, where n is the sample size and $n-1$ are the degrees of freedom.

Power

For a lower-tailed test, to obtain x_L, we use

$$x_L = \mu_0 - P^{-1}(\alpha|n-1)\frac{S}{\sqrt{n}} \tag{11.23}$$

$$= \text{mu.0} - \text{qt(alpha, n} - 1) * \text{SE}$$

where S is the sample's standard deviation, n is the sample size and P^{-1} is the inverse of the t distribution for α and $n-1$ given. To determine the power for lower-tailed hypothesis tests, we need to find the area to the left of x_L under H_A, where the density of the standardized rv under H_A is t. So we define

$$z_A := \frac{x_L - \mu_A}{SE}$$
$$= (\text{x.L} - \text{mu.A})/(\text{SE})$$

and the power is

$$1 - \beta = P(Z \leq z_A \,|\, n-1) \qquad (11.24)$$
$$= \text{pt}(\text{z.A}, \text{n}-1)$$

where $\text{z.A} := z_A$.

For an upper-tailed test, we use

$$x_H = \mu_0 + P^{-1}(1-\alpha \,|\, n-1)\frac{S}{\sqrt{n}}$$
$$= \text{mu.0} + \text{qt}(1-\text{alpha}, \text{n}-1) * \text{SE},$$

$$z_A := \frac{x_H - \mu_A}{SE}$$
$$= (\text{x.H} - \text{mu.A})/(\text{SE})$$

and

$$1 - \beta = 1 - P(Z \leq z_A \,|\, n-1)$$
$$= 1 - \text{pt}(\text{z.A}, \text{n}-1).$$

For two-tailed power, specify (if you so desire) α_L, α_H, $\mu_L < \mu_0$ and $\mu_H > \mu_0$. Then use 11.23 with α_L to obtain x_L and the corresponding z_L. According to (11.24),

$$1 - \beta_L = P(Z \leq z_L \,|\, n-1)$$
$$= \text{pt}(\text{z.L}, \text{n}-1).$$

Similarly,

$$1 - \beta_H = 1 - P(Z \leq z_H \,|\, n-1)$$
$$= \text{pt}(\text{z.H}, \text{n}-1).$$

The power is then

$$1 - \beta = (1 - \beta_L) + (1 - \beta_H).$$

Sample size

To obtain the sample size, first use the appropriate equations in Section 11.1.1. Then if $n < 30$, recalculate n using the following. For a lower-tailed sample size and from (11.23),

$$x_L = \mu_0 - P^{-1}(\alpha \,|\, n-1)\frac{S}{\sqrt{n}}.$$

Also,
$$x_L = \mu_A + P^{-1}(1-\beta\,|n-1)\frac{S}{\sqrt{n}}.$$

Equating both right hand sides and solving for n we get

$$\begin{aligned}n &= \frac{S^2}{(\mu_0 - \mu_A)^2}\left[P^{-1}(\alpha\,|n-1) + P^{-1}(1-\beta\,|n-1)\right]^2 \\ &= \text{S}^2/(\text{mu.0} - \text{mu.A})^2 * \\ &\quad (\text{qt}(1 - \text{alpha}, \text{n} - 1) + \text{qt}(1 - \text{beta}, \text{n} - 1))^2.\end{aligned} \quad (11.25)$$

You can easily verify that for an upper-tailed test

$$\begin{aligned}n &= \frac{S^2}{(\mu_0 - \mu_A)^2}\left[P^{-1}(1-\alpha\,|n-1) + P^{-1}(\beta\,|n-1)\right]^2 \\ &= \text{S}^2/(\text{mu.0} - \text{mu.A})^2 * \\ &\quad (\text{qt}(1 - \text{alpha}, \text{n} - 1) + \text{qt}(\text{beta}, \text{n} - 1))^2.\end{aligned} \quad (11.26)$$

For two-tailed tests, use the appropriate substitutions in (11.25) and (11.26) to obtain n_L and n_H and then

$$n = n_L + n_H.$$

You can use power.t.test() to compute power.

11.2.2 Proportions

In Section 9.2.1 we discussed the asymptotic method to compute large sample confidence intervals for proportions. In Section 9.2.2, we discussed the exact and Wilson methods to compute confidence intervals for small samples from populations with binomial density. These methods (and others) can also be used to obtain power and sample size. When we say asymptotic method, we mean the large sample approximation with the normal. This should not be confused with methods to determine asymptotic power (see the package asypow).

Power

Take for example the exact method. The sampling density of p is F. For lower-tailed power, we first use $F^{-1}(\alpha|\nu_1, \nu_2)$ to obtain p_L under π_0 and then $F(p_L|\nu_1, \nu_2)$ under p_A to obtain the power. For π between about 0.2 and 0.8, the various methods to compute the power give approximately the same results. The package binom contains several functions that compute power, confidence intervals and so on for the binomial density. We are interested in binom.power().

Example 11.10. On the average, a pride of lions is successful in catching prey in $\pi_0 = 0.4$ of 25 attempts. What is our ability to distinguish between π_0 and $\pi_A = 0.5$ with type I error at $\alpha = 0.05$? Let us specify the data:

```
> pi.0 <- 0.4 ; pi.A <- 0.5 ; alpha <- 0.05 ; n <- 25
```

and compute the power with each available method:

```
> library(binom)
> methods <- c("cloglog", "logit", "probit", "asymp",
+   "lrt", "exact")
> results <- matrix(ncol = 1, nrow = length(methods))
> for(i in 1 : length(methods)){
+   results[i, 1] <- binom.power(pi.A, n = n, p = pi.0,
+     alpha = 0.05, alternative = 'greater',
+     method = methods[i])
+ }
> dimnames(results) <- list(methods, 'Power')
> round(results, 3)
         Power
cloglog  0.289
logit    0.253
probit   0.257
asymp    0.270
lrt      0.327
exact    0.259
```

Not very strong. You can observe the differences between the methods to computer power in action with

```
> tkbinom.power()
```

(Figure 11.8). Observe the small differences among the methods. □

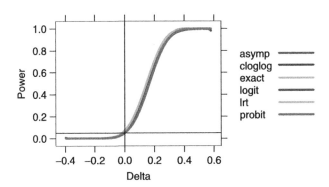

Figure 11.8 Methods to compute power for small sample from the binomial ($n = 25$, $H_0: \pi = \pi_0 = 0.4$ and $H_A: \pi > \pi_0$ with π_A varying between 0.4 and 1.0.

cloglog, logit and so on refer to ways to parameterize π (e.g. $\pi = \exp[\mu]$). Which method should you use? The one that gives you the minimum power for your data.

Sample size

The package binom includes the function cloglog.sample.size(). This function computes sample size for the complementary log parameterization of π:

$$\pi = e^{-\mu}, \quad \mu = e^{\gamma}.$$

Let us obtain sample size with an example.

Example 11.11. Returning to Example 11.10, we ask: How many trials do we need to observe to distinguish between $\pi_0 = 0.4$ and $\pi_A = 0.5$ with $1 - \beta = 0.8$? First we specify the data

```
> pi.0 <- 0.4 ; pi.A <- 0.5 ; alpha <- 0.05 ; beta <- 0.2
```

and then obtain the number of trials

```
> library(binom)
> cloglog.sample.size(pi.A, p = pi.0, power = 1 - beta,
+    alpha = alpha, alternative = 'greater',
+    recompute.power = TRUE)
  p.null p.alt delta alpha      power   n phi
1    0.4   0.5   0.1  0.05 0.8010439 150   1
```

We need to observe 150 trials to distinguish between a hunting success of 0.4 vs. 0.5 at $\alpha = 0.05$ and $1 - \beta = 0.8$. Note the named argument recompute.power = TRUE. Because n is rounded up to the nearest integer, we recompute the power for the integer n. □

11.2.3 Intensities

In (10.9) through (10.11), we established the critical values of l_L and l_H that allowed us to decide about $H_0 : \lambda = \lambda_0$. The sampling density in these cases was χ^2 with the appropriate degrees of freedom. We build on these ideas here.

Power

We consider only the so-called exact method. For lower-tailed hypothesis testing with significance α, we obtain l_L from (10.9). Then the power for the alternative $\lambda = \lambda_A$ is

$$1 - \beta = P(l \leq l_L | 2\lambda_A)/2 \tag{11.27}$$
$$= \text{pchisq(l.L, 2 * lambda.A)}/2$$

where $P(l \leq l_L | 2\lambda_A)$ is the χ^2 distribution with $2\lambda_A$ degrees of freedom. For an upper-tailed test, we first obtain l_H from (10.10) and obtain the power from

$$1 - \beta = 1 - P(l \leq l_H | 2(\lambda_A + 1))/2 \tag{11.28}$$
$$= 1 - \text{pchisq(1 - alpha, 2 * (lambda.A + 1))}/2 \, .$$

For a two-tailed test, we calculate the power for the lower-tailed and upper-tailed hypothesis tests separately (with appropriate substitutions for α and λ_A) and then add them together. Because the distributions involved in lower-tailed power are different from upper-tailed power (they do belong to the same family, though), this is a good time to look at the meaning of power again.

Example 11.12. The arrival rate to an emergency room is 10 persons per hour. We assume that people arrive independent of each other and that the events (arrivals) are uniformly distributed in time. Therefore, the arrivals represent the Poisson density. We wish to test H_0: $\lambda = \lambda_0 = 10$ vs. H_A: $\lambda < \lambda_0$. We are interested in the power to distinguish between λ_0 and $\lambda_A = 9$ at $\alpha = 0.05$. While at it, we are also interested in H_A: $\lambda > \lambda_0$ and in the power of distinguishing between λ_0 and $\lambda_A = 11$.

364 Power and sample size for single samples

Using (11.27) and (11.28) we obtain:

```
> lambda.0 <- 10 ; alpha <- 0.05 ; lambda.A <- c(9, 11)
> l.L <- qchisq(alpha, 2 * lambda.0) / 2
> l.H <- qchisq(1 - alpha, 2 * (lambda.0 + 1)) / 2
> round(c('lower-tailed' =
+   pchisq(l.L, 2 * lambda.A[1]) / 2,
+   'upper-tailed' =
+   1 - pchisq(l.H, 2 * (lambda.A[2] + 1)) / 2), 3)
lower-tailed upper-tailed
       0.001        0.925
```

Figure 11.9 illustrates the results. □

Other approaches to calculate the power use the gamma distribution or the so-called Byar's formula (see documentation for `pois.exact()` in the `epitools` package).

Sample size

Here we may take advantage of the binomial approximation to the Poisson. Let n be the number of unit intervals and n_S the number of events we count during n.

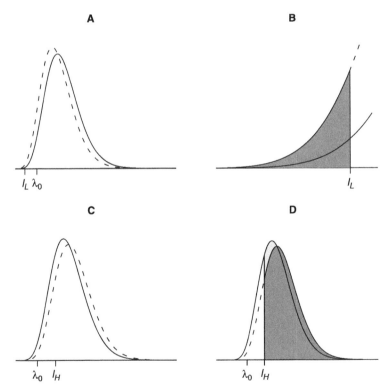

Figure 11.9 Solid and broken curves show the sampling densities of λ_0 and λ_A, respectively. A - The power associated with lower-tailed hypothesis testing. B - The dark polygon shades the power. C - The power associated with upper-tailed hypothesis testing. D - The dark polygon shades the power.

We choose the unit intervals to be small enough so that the probability of more than one event during the unit interval is practically zero. Then

$$\frac{(n_S/n)^{n_S}}{n_S!} \exp[-n_S/n] \approx \binom{n}{n_S} (n_S/n)^{n_S} (1 - n_S/n)^{n-n_S}.$$

As the next example illustrates, we can now use `cloglog.sample.size()` in the package `binom` to compute the needed number of unit intervals.

Example 11.13. The arrival rate of patients to an emergency room is 10 per day. Divide the day into small enough time units such that the probability that more than one patient arrives during the interval is practically zero. Take the interval to be minutes. We have $24 \times 60 = 1440$ time units. How many time units do we need to count events to distinguish between arrival rate of 10 and 11 patients per day with $1 - \beta = 0.8$ and $\alpha = 0.05$? First, the data:

```
> lambda.0 <- 10 ; lambda.A <- 11  ; alpha <- 0.05
> beta <- 0.2 ; n.units <- 24 * 60
> lambda.0 <- lambda.0 / n.units
> lambda.A <- lambda.A / n.units
```

Next, the results:

```
> round((n <- cloglog.sample.size(lambda.A, p = lambda.0,
+   power = 1 - beta,
+   alpha = alpha, alternative = 'greater',
+   recompute.power = TRUE)), 4)
  p.null p.alt delta alpha power     n phi
1 0.0069 0.0076 7e-04  0.05   0.8 93646   1
> n[6] / n.units
         n
1 65.03194
```

So we need to record arrivals (by the minute) for 65 days and 46 minutes. □

11.3 Power and sample size for arbitrary densities

So far, we assumed that the probability density of the population from which we drew a sample was known. Suppose that it is not. Then we need to revert to the bootstrap. Say we wish to test $H_0: \theta = \theta_0$ vs. one of the three possible H_A for some parameter θ (e.g. median). We draw a sample from the population and obtain the sampling density of θ with the bootstrap. To compute power and sample size, we need to provide a sample under θ_A. This brings us to the topics of two-sample tests, power and sample size. So we defer the discussion of bootstrap power and sample size to Chapter 13.

11.4 Assignments

Unless otherwise stated, use $\alpha = 0.05$ and $1 - \beta = 0.8$ where necessary.

Exercise 11.1.

1. We have a model of a normal population with $\mu_0 = 10$ and $\sigma = 2$. Draw a diagram that shows the type II error for $\mu_A = 10.5$, 11 and 11.5. Is the power increasing or decreasing as μ_A increases? Explain.

2. What would be the sample size necessary to distinguish between μ_0 and the three values of μ_A? Is the sample size increasing or decreasing with increasing μ_A? Explain.

Exercise 11.2. Repeat the analysis in Example 11.4 for each species separately. Specifically, for each species, examine the power of distinguishing between the given mean mercury and the nearest lower and upper boundaries on mercury concentration with respect to recommended upper limit on fish meals per month. Write your power results in a table format, displaying the species, its mean concentration, standard deviation, sample size and power "from below" and power "from above."

Exercise 11.3. Determine the sample size necessary to identify the population mean of mercury concentration for all species in Table 11.2 with confidence levels of $\alpha = 0.01$ and $1 - \beta = 0.8$ to within 10 ppm.

Exercise 11.4. Table 28-1 from http://nces.ed.gov/programs/coe/2006/section3/indicator28.asp compares averaged freshman graduation rate for public high school students and number of graduates, by state: 200001, 200102 and 200203. The table is stored in high-school.xls at the book's site. According to the table, 83.9% of high school freshman graduated in 2001–02. In 2002–03, 84.8% graduated. Newspapers around the countries came with headlines such as "High school graduation hits all time high." Suppose that the percent graduating every year is independent of another year. Use the table to answer the following questions for Minnesota:

Suppose that a 1% change in freshmen graduating from one year to the next represents, for all practical purposes, no change. What should be the sample size needed to distinguish a significant difference in percent graduation at $\alpha = 0.05$ and $1 - \beta = 0.8$?

Exercise 11.5. Based on past data, we know that the levels of a certain carcinogen in water range between 50 to 700 ppm. How many water samples should we analyze to estimate the true mean concentration of the carcinogen to within 10 ppm of the true mean with 95% confidence?

Exercise 11.6. Repeat the analysis in Exercise 11.4, except that now treat the graduation as a rate, not as a proportion.

For the next three exercises, *do not* reinvent the wheel! Hack code as much as you can. Search for code you need, either at the book's site or simply look for appropriate functions in R and then type the function name in R's workspace without parentheses. This will usually print the function's code. Modify the code to fit the exercises.

Exercise 11.7.

1. Write a function that computes the power of distinguishing between any pair of null vs. alternative normal models for any level of significance and any sample size for upper-, lower- or two-tailed hypotheses.

2. Write a function that computes the sample size for any level of significance, any desired power and any detectable difference for upper-, lower- or two-tailed hypotheses.

Exercise 11.8. Repeat Exercise 11.7 for the binomial.

Exercise 11.9. Repeat Exercise 11.7 for the Poisson.

12

Two samples

So far, our interest in hypothesis testing was to make inferences about a population parameter from a sample. For example, we examined cases where a sample provided a proportion, p and we wished to infer about the value of the population proportion, π, of some trait.

Often the question is how trait values from two populations compare. For example, can we say that the distribution of the concentration of a pollutant in wells in one region is different from that in another region? Does the distribution of beak length in one species of bird differ from that of another? Can we assert that the proportion of the U.S. adult population that supported the death penalty in 1936 is different from that in 2004? Is the treatment of patients with a particular medicine effective compared to no treatment?

Our approach should be familiar by now: We wish to estimate the value of a parameter in the population. We obtain a sample and compute the sample-based best estimate of the parameter (the statistic). Next, we develop the sampling density of the statistic and from it draw conclusions about plausible values of the population parameter.

As in Chapters 9 and 10, we discuss comparisons of means, proportions and intensities (rates) with small and large samples. We also discuss situations where we do not know the density of the trait in the population. For the most part, we assume that samples are independent. A special situation arises when observations are pairwise dependent, but the observations are independent among themselves. In other words, if x_i and y_i are the ith pair, then we do not assume that they are independent. We do assume that (x_i, y_i) are independent of (x_j, y_j) for $i \neq j$.

Consider two populations, x and y, with the same family of densities, but different parameter values. To simplify the notation, we address a single parameter family of densities, $P(X = x | \theta)$. We obtain a sample from each population

$$\bm{X} := [X_1, \ldots, X_{n_1}], \quad \bm{Y} := [Y_1, \ldots, Y_{n_2}].$$

From the samples, we estimate $\widehat{\theta}_1$ and $\widehat{\theta}_2$. We are interested in the following question: Is the difference between θ_1 and θ_2 large enough so that we can claim that the densities are significantly different? To stay close to the developments in Chapter 10, we construct a test statistic from $\widehat{\theta} := \widehat{\theta}_1 - \widehat{\theta}_2$ and we need the sampling density of $\widehat{\theta}$. Thus, at least in notation, we are back to a single-sample hypothesis test:

$$H_0 : \theta = \theta_0 \text{ vs. one of } H_A : \theta \begin{cases} < \theta_0 \\ \neq \theta_0 \\ > \theta_0 \end{cases} \qquad (12.1)$$

where now θ_0 is a hypothesized difference $\theta_1 - \theta_2$. We will consider situations where either θ_0 is location invariant or devise ways to get around this problem. By location invariance we mean that the statistical properties (in particular the sampling density) of θ_0 remain unchanged regardless of where the difference between θ_1 and θ_2 is located.

12.1 Large samples

In this section we address hypothesis testing for large samples. We discuss means, proportions and rates.

12.1.1 Means

Here we adopt the notation detailed in Table 12.1. We base the inference for means on the difference between them. According to (12.1), we are interested in

$$H_0 : \mu = \mu_0 \text{ vs. one of } H_A : \mu \begin{cases} < \mu_0 \\ \neq \mu_0 \\ > \mu_0 \end{cases} .$$

Because n_1 and n_2 are large (> 30), the sampling density of $\overline{X}_1$ and $\overline{X}_2$ is approximately normal with μ_1, $\sigma_1/\sqrt{n_1}$ and μ_2, $\sigma_2/\sqrt{n_2}$. It turns out that the sampling density of $\overline{X} := \overline{X}_1 - \overline{X}_2$ is normal with

$$\mu_{\overline{X}} = \mu_1 - \mu_2 \, , \quad \sigma_{\overline{X}} = \sqrt{\frac{\sigma_1^2}{n_1} + \frac{\sigma_2^2}{n_2}} \, .$$

With this in mind and from the central limit theorem, we conclude that the properties of the sampling density of $\overline{X} := \overline{X}_2 - \overline{X}_1$ are:

1. The means of the sampling density of the rv $\overline{X}$ are centered around μ; i.e. $\mu_{\overline{X}} = \mu$ (where $\mu := \mu_2 - \mu_1$).

Table 12.1 Population and sample notation for means. Difference refers to the parameter of the first population (sample) subtracted from the second population (sample).

	Population parameter		Sample parameter		
Population	Mean	Variance	Mean	Variance	Sample size
1	μ_1	σ_1^2	$\overline{X}_1$	S_1^2	n_1
2	μ_2	σ_2^2	$\overline{X}_2$	S_2^2	n_2
Difference	μ	σ^2	$\overline{X}$	S^2	

2. The standard deviation of the sampling density is given by

$$\sigma_{\overline{X}} = \sqrt{\frac{\sigma_1^2}{n_1} + \frac{\sigma_2^2}{n_2}}$$

(recall that the standard deviation of the sampling density is called the standard error).

3. When both n_1 and n_2 are large (> 30), the central limit theorem implies that the sampling density of $\overline{X}_1$ and $\overline{X}_2$ is approximately normal. So is their difference.

The standard error of $\overline{X}$ is

$$\text{SE} := \sqrt{\frac{S_1^2}{n_1} + \frac{S_2^2}{n_2}}. \tag{12.2}$$

When both n_1 and n_2 are large (> 30), we use $S_1 \approx \sigma_1$ and $S_2 \approx \sigma_2$ to estimate the standard deviation of the sampling density of $\overline{X}$; i.e.

$$\sigma_{\overline{X}} = \sqrt{\frac{\sigma_1^2}{n_1} + \frac{\sigma_2^2}{n_2}} \approx \text{SE}. \tag{12.3}$$

We can now proceed with hypothesis testing as usual.

Hypothesis testing

We set the null hypothesis to $H_0 : \mu = \mu_0$ where $\mu_0 := \mu_2 - \mu_1$. Our alternative hypothesis, H_A, is one of $\mu \neq \mu_0$, $> \mu_0$ or $< \mu_0$ with significance level α. From the samples we obtain $\overline{X} := \overline{X}_2 - \overline{X}_1$, S_1^2 and S_2^2 and use (12.2) to obtain SE.

For lower-tailed hypothesis testing, we compute x_L with (10.2). If $\overline{X} \leq x_L$ we reject H_0 and conclude that $\mu_2 - \mu_1 < \mu_0$ with the given significance. We can also compute the,

$$p\text{-value} = \Phi\left(\overline{X}\,|\mu_0, \text{SE}\right) \tag{12.4}$$
$$= \text{pnorm}(\text{X.bar}, \text{mu.0}, \text{SE}).$$

If the p-value $\leq \alpha$ we reject H_0.

For an upper-tailed test, we use (10.3) to obtain x_H and

$$p\text{-value} = 1 - \Phi\left(\overline{X}\,|\mu_0, \text{SE}\right) \tag{12.5}$$
$$= 1 - \text{pnorm}(\text{X.bar}, \text{mu.0}, \text{SE})$$

to calculate the p-value. If $\overline{X} > x_H$ (in which case the p-value $< \alpha$) we reject H_0 and conclude that $\mu_2 - \mu_1 > \mu_0$ with the given significance.

For a two-tailed test, the alternative is H_A: $\mu < \mu_0$ or $\mu > \mu_0$ with significance levels α_L and α_H, respectively. To implement the test, we first obtain x_L from (10.2) (using α_L instead of α) and the p-value from 12.4. If $\overline{X} \leq X_L$, or the p-value $\leq \alpha_L$, we reject H_0 and conclude that $\mu_2 - \mu_1$ is smaller than μ_0. The upper-tailed test is similar; e.g. if p-value $< \alpha_H$ we reject H_0 and conclude that $\mu_2 - \mu_1 > \mu_0$. The p-value is obtained with (12.5). Under this arrangement, it is possible that we reject H_0 on one side but not on the other. In such a case, by our rule, we reject H_0.

Example 12.1. We return to the capital punishment data, first introduced in Example 9.13 (United States Department of Justice, 2003). Recall that between 1973 and 2000, there were 7658 cases of capital punishment in the U.S. We wish to test the hypothesis that the mean age at sentencing to death of whites and blacks do not differ at $\alpha_L = 0.025$ and $\alpha_H = 0.025$. Because we have the true population values, we have an opportunity to test how effective a sample is in detecting differences. First, we load the data set and rename it for ease of reference:

```
> load('capital.punishment.rda')
> cp <- capital.punishment
```

Next, we calculate the age at sentencing based on the month and year of birth and the month and year of sentencing (columns 11, 12, 25 and 26, respectively):

```
> age <- cp[, 25] + cp[, 26] * 12 - cp[, 11] - cp[, 12] * 12
> age <- age / 12
```

We must get rid of all the rows in which at least one of the columns 9, 11, 12, 23 or 24 in the data is tagged as NA. These are the columns that give skin color, month of birth, year of birth, month of conviction and year of conviction:

```
> na.data <- is.na(cp[, 9]) | is.na(cp[, 11]) |
+    is.na(cp[, 12]) | is.na(cp[, 23]) | is.na(cp[, 24])
> w <- cp$Race == 'White' & !na.data
> b <- cp$Race == 'Black' & !na.data
```

Now w and b are logical vectors that have TRUE values wherever we need them. So the population data are

```
> x.whites <- age[w] ; x.blacks <- age[b]
```

where x.whites and x.blacks are the age at sentencing of the white and black populations, respectively. We rid the data of NA this way to illustrate combined conditional statements. An easier way to keep only those rows in a data frame where there is no NA in any column is to use complete.cases(). Next, we sample the populations

```
> set.seed(1)
> n.whites <- 50 ; n.blacks <- 50
> X.whites <- sample(x.whites, n.whites)
> X.blacks <- sample(x.blacks, n.blacks)
```

and compute the statistics we need

```
> X.bar.whites <- mean(X.whites)
> X.bar.blacks <- mean(X.blacks)
> var.whites <- var(X.whites) ; var.blacks <- var(X.blacks)
```

Here is what we have thus far:

```
> info <- rbind(mean = c(whites = X.bar.whites,
+    blacks = X.bar.blacks),
+    variance = c(var.whites, var.blacks),
+    'sample size' = c(n.whites, n.blacks))
> round(info, 1)
```

	whites	blacks
mean	30.9	26.5
variance	88.7	40.0
sample size	50.0	50.0

In our notation, $n_1 = 50$, $n_2 = 50$, $\overline{X}_1 = 30.9$ years old (at the time of sentencing), $\overline{X}_2 = 26.5$, $S_1^2 = 88.7$ and $S_2^2 = 40.0$. Because the samples are large enough, we use $\sigma_1 \approx S_1$ and $\sigma_2 \approx S_2$. We have $\mu_2 - \mu_1 = 0$ and our hypotheses are:

$$H_0 : \mu = 0 \quad \text{vs.} \quad H_A : \mu \neq 0 \,.$$

Note that $\mu_2 - \mu_1 < 0$. Therefore, we only need to test the lower tail of the hypotheses. Using (10.2) we obtain

```
> X.bar <- X.bar.blacks - X.bar.whites
> SE <- sqrt(var.whites / n.whites + var.blacks / n.blacks)
> alpha.L <- alpha.H <- 0.025
> round(c(x.L = qnorm(alpha.L, 0, SE),
+   X.bar = X.bar,
+   p.value = pnorm(X.bar, 0, SE)), 3)
    x.L   X.bar p.value
 -3.145  -4.423   0.003
```

and conclude that the age at sentencing of blacks is significantly younger than that of whites. □

Confidence intervals

The construction of confidence intervals for two-sample comparisons for large samples is identical to the single sample (see Chapter 9). To obtain the interval, we estimate $\mu := \mu_2 - \mu_1$ with $\widehat{\mu} = \widehat{\mu}_2 - \widehat{\mu}_1$ and use the SE as defined in (12.3). Accordingly, (9.6) becomes

$$\begin{aligned} I_{1-\alpha}\left(X\,\middle|\,\widehat{\theta}\right) &= \left[\varPhi^{-1}\left(\alpha/2\,\middle|\,\widehat{\mu},\mathrm{SE}\right)\,,\ \varPhi^{-1}\left(1-\alpha/2\,\middle|\,\widehat{\mu},\mathrm{SE}\right)\right] \\ &= \mathtt{c(qnorm(alpha/2, mu.hat, SE),} \\ &\quad \mathtt{qnorm(1 - alpha/2, mu.hat, SE))} \,. \end{aligned} \quad (12.6)$$

Example 12.2. Lead is one of the oldest metals used by humans. It is a cumulative neurotoxin. It impairs brain development in children. In adults, it is associated with elevated blood pressure (hypertension), heart attacks, and premature death. Emissions from vehicles are the largest source of lead exposure in many urban areas in countries that do not enforce the use of unleaded gasoline. Lead toxicity is also a health problem for those involved in its production. Emissions from lead smelters and refineries expose workers to lead.

Data on the maximum concentration of lead in gasoline (grams per liter) from 1992 to 1996 were reported by The World Bank (1996a) and The World Bank (1996b). Let us examine the data:

```
> load('l.rda')
> head(l)
        africa a.lead      europe e.lead
```

374   Two samples

```
1    Algeria      0.63     Austria 0.00
2     Angola      0.77 Belarus Rep 0.82
3      Benin      0.84     Belgium 0.15
4   Botswana      0.44    Bulgaria 0.15
5 Burkina Faso    0.84 Croatia Rep 0.60
6    Burundi      0.84   Czech Rep 0.15
```

The

```
> boxplot(l$a.lead, l$e.lead, names = c('Africa', 'Europe'))
```

reveals one outlier (Figure 12.1). To identify it, we do

```
> identify(rep(2, length(l$e.lead)), l$e.lead,
+   labels=l$europe)
```

We use identify() to label the outlier. Note that the x coordinate of points in a box plot is identified by the group the points belong to (2 in our case). When we call identify(), we want to obtain both the x and y coordinates of the outlier, hence the rep(2, length(e.lead)). The corresponding label of the point is obtained from the vector l$europe. Because Belarus is the only outlier, we trim the mean by a fraction of $1/31$ on both sides. We also trim the data before calculating the standard deviations:

```
> n = c(length(l$a.lead), length(l$e.lead) - 2)
> l.stats <- data.frame(mu.hat, S, n)
> dimnames(l.stats)[[1]] <- c('Africa', 'Europe')
       mu.hat      S  n
Africa 0.6477 0.1892 31
Europe 0.2179 0.1754 29
```

According to (12.6), the confidence interval is

```
> SE <- sqrt(sum(S^2 / n)) ; alpha <- 0.05
> c(low = qnorm(alpha / 2, mu.hat[2] - mu.hat[1], SE),
+   high = qnorm(1 - alpha / 2, mu.hat[2] - mu.hat[1], SE))
    low    high
-0.5221 -0.3375
```

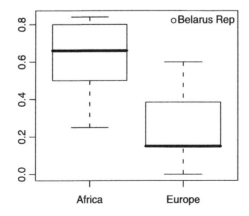

Figure 12.1   Lead in gasoline by continent.

The estimated difference in the mean populations (countries) in Africa and Europe does not cross zero. Therefore, were we to test H_0: $\mu_0 = 0$ vs. H_A: $\mu_0 \neq 0$ with $\alpha = 0.05$, we would have rejected H_0. This does *not* necessarily work the other way: a rejection via hypothesis testing implies that the difference crosses zero only when the rejection region (on the lower tail in this case) of hypothesis testing equals half the value of the confidence coefficient. □

12.1.2 Proportions

In the case of two samples, to implement the normal approximation to the binomial, we require that $n_i p_i$ and $n_i(1 - p_i)$ are both ≥ 5 for $i = 1, 2$.

Hypothesis testing

We have two independent samples and wish to test for $\pi_0 := \pi_2 - \pi_1$. Our hypotheses are H_0: $\pi = \pi_0$ vs. one of the three alternatives. The sampling density of p_i is $\phi\left(p_i \mid \pi_i, \sqrt{\pi_i(1 - \pi_i)/n}\right)$. Therefore, the sampling density of $p_2 - p_1$ is normal. Its mean is π_0 and its variance is

$$\frac{\pi_0 \times (1 - \pi_0)}{n_1} + \frac{\pi_0 \times (1 - \pi_0)}{n_2} = \pi_0 \times (1 - \pi_0)\left(\frac{1}{n_1} + \frac{1}{n_2}\right).$$

To estimate the standard error, we use

$$\bar{p} := \frac{n_{1S} + n_{2S}}{n_1 + n_2}$$

and so

$$\text{SE} = \sqrt{\pi_0 \times (1 - \pi_0)\left(\frac{1}{n_1} + \frac{1}{n_2}\right)} \approx \sqrt{\bar{p}(1 - \bar{p})\left(\frac{1}{n_1} + \frac{1}{n_2}\right)}. \quad (12.7)$$

When $n_1 = n_2$, or when both are large, we do not need to use the continuity correction. If you wish to use the correction, call the appropriate R function (see below). To test hypotheses, replace μ_0 and SE in (10.2) through (10.5) with π_0 and SE from (12.7).

Confidence intervals

We construct the confidence interval for $\hat{\pi} = \hat{\pi}_2 - \hat{\pi}_1$ to estimate the location of $\pi = \pi_2 - \pi_1$. Our rv is $p := p_2 - p_1$. From (9.8),

$$I_{1-\alpha}(p \mid \hat{\pi}) = \left[\Phi^{-1}(\alpha/2 \mid \hat{\pi}, \text{SE}), \Phi^{-1}(1 - \alpha/2 \mid \hat{\pi}, \text{SE})\right]$$
$$= \text{c(qnorm(alpha/2, pi.hat, SE),} \quad (12.8)$$
$$\text{qnorm(1 - alpha/2, pi.hat, SE))}.$$

Example 12.3. In 1936 and in 2004, Gallup Poll published results of a poll concerning the following question, asked of adults in the U.S.: "Are you in favor of the death penalty for a person convicted of murder?" The results are detailed in Table 12.2. It is not clear from the article how many participated in each of the surveys, so we

Table 12.2 Results from Gallup poll.

Survey date	Proportion		
	Favor	Opposed	Undecided
May 2–4, 2004	0.71	0.26	0.03
December 2–7, 1936	0.59	0.38	0.03

assume 500. Divide the answers into those who were in favor as opposed to those who were opposed and undecided and ask: Was the proportion of U.S. adults who supported the death penalty in 2004 larger than in 1936 at $\alpha = 0.1$? Translated to our language, we wish to test

$$H_0 : \pi = 0 \quad \text{vs.} \quad H_A : \pi > 0.$$

To test the hypotheses via the confidence interval, we use 95% confidence coefficient ($\alpha = 0.05$). The data are:

```
> n <- c(500, 500) ; p <- c(0.59, 0.71) ; alpha <- 0.05
> p.bar <- sum(n * p) / sum(n)
> SE <- sqrt(p.bar * (1 - p.bar) * sum(1 / n))
> pi.hat <- p[2] - p[1]
```

(we could be more terse with the data, but we want to relate to our notation). From (12.8),

```
> round(c(low = qnorm(alpha / 2, pi.hat, SE),
+    high = qnorm(1 - alpha / 2, pi.hat, SE)), 2)
 low high
0.06 0.18
```

Because we arranged for the correct α and because the confidence interval does not cross zero, we conclude that we are 95% confident that the proportion of U.S. adults that supported the death penalty in 2004 was larger than in 1936. In R, we use (with and without correction):

```
> correction <- prop.test(n * p, n)
> no <- prop.test(n * p, n, correct = FALSE)
```

Both returned data are lists. To extract the confidence interval from the list, we examine its names:

```
> names(no)
[1] "statistic"   "parameter"   "p.value"     "estimate"
[5] "null.value"  "conf.int"    "alternative" "method"
[9] "data.name"
```

and then produce a report:

```
> CI <- rbind(correction$conf.int, no$conf.int)
> dimnames(CI) <- list(c('correction', 'no'),
+    c('low', 'high'))
> round(CI, 2)
```

```
            low    high
correction  -0.18  -0.06
no          -0.18  -0.06
```

(the switch in sign compared to our results is because the order of subtraction). We conclude that the true difference in proportions between 2004 and 1936 is somewhere between 6 and 18%. □

Contingency tables

Rather than use the normal approximation to the binomial, we can approach the analysis of two binomial samples with a *contingency table*. This approach relies on obtaining a sampling density from the following considerations. Suppose that we have two populations, with π_1 and π_2 reflecting proportions of some trait, e.g. gender, sick vs. healthy, cross-fertilization vs. self-fertilization. We take samples of sizes n_1 and n_2 from the populations. We count n_{1S} and n_{2S} successes in each sample. This leads to the following 2 by 2 table:

success	population 1	population 2	total
yes	n_{1S}	n_{2S}	n_S
no	$n_1 - n_{1S}$	$n_2 - n_{2S}$	$n - n_S$
total	n_1	n_2	n

Here we use the notation $n := n_1 + n_2$ and $n_S := n_{1S} + n_{2S}$. We wish to test

$$H_0 : \pi = 0 \quad \text{vs.} \quad H_A : \pi \neq 0$$

(where $\pi := \pi_2 - \pi_1$) with significance α. We use the data to estimate π_1 and π_2:

$$\widehat{\pi}_1 = \frac{n_{1S}}{n_1}, \quad \widehat{\pi}_2 = \frac{n_{2S}}{n_2}.$$

Under H_0, both samples come from the same population, with proportion estimated by

$$\widehat{\pi} = \frac{n_1}{n_1 + n_2} \widehat{\pi}_1 + \frac{n_2}{n_1 + n_2} \widehat{\pi}_2.$$

We use $\widehat{\pi}$ to obtain the *expected* values in each of the table's cells under H_0:

success	population 1	population 2	total
expected yes	$\widehat{\pi} \times n_1$	$\widehat{\pi} \times n_2$	$\widehat{\pi} \times (n_1 + n_2)$
expected no	$(1 - \widehat{\pi}) \times n_1$	$(1 - \widehat{\pi}) \times n_2$	$(1 - \widehat{\pi}) \times (n_1 + n_2)$
sample size	n_1	n_2	n

Next, we compare the expected values to the observed values. The larger the difference between the expected and observed values, the more we believe that the proportions in the populations differ. It turns out that under H_0, the statistic

$$X^2 := \sum_{i=1}^{4} \frac{(O_i - E_i)^2}{E_i}$$

(where E_i are the expected values and O_i are the observed values) has a χ^2 density with 1 degree of freedom. The latter is determined from the number of column cells -1, i.e. $(2-1)$ times the number of row cells -1, i.e. $(2-1)$. Contingency tables consist of counts. The χ^2 density is continuous. Therefore, we often need to implement the so-called Yates' continuity correction. Our statistic is now

$$X^2 = \sum_{i=1}^{4} \frac{(|O_i - E_i| - 0.5)^2}{E_i}.$$

Example 12.4. We return to the capital punishment data (United States Department of Justice, 2003). We have data for the population of inmates sentenced to death in the US (see Example 9.13 for details). Suppose we have access to inmates' paper files only. We wish to answer the following question: Is the proportion of married black inmates different from the proportion of married white inmates. We have two "populations" of inmates: married and single. In each of these populations, we define black as success and not black as failure. Here is how we prepare the data:

```
> load('capital.punishment.rda')
> attach(capital.punishment)
> color <- ifelse(Race == 'Black', 'Black', 'Other')
> status <- ifelse(MaritalStatus == 'Married',
+    'Married', 'Single')
> x <- data.frame(color, status)
> head(x, 4)
  color status
1 Other Single
2 Black Single
3 Black Single
4 Other Single
```

We draw 400 random cases from x. Because x is a data frame, we draw the sample by first creating an index vector and then using it to select our 400 cases:

```
> set.seed(100)
> idx <- sample(1 : length(color), 400)
> s <- x[idx, ]
```

To tally the results, we do

```
> (s <- table(s))
       status
color   Married Single
  Black      40    136
  Other      45    179
```

Now implementing the contingency-table calculations, we do

```
> n.1S <- 40 ; n.2S <- 45 ; n.1F <- 136 ; n.2F <- 179 ;
> n.1 <- n.1S + n.1F ; n.2 <- n.2S + n.2F ; n <- n.1 + n.2
> pi.1 <- n.1S / n.1 ; pi.2 <- n.2S / n.2
> pi.hat <- n.1 / n * pi.1 + n.2 / n * pi.2
> E <- c(pi.hat * n.1, pi.hat * n.2,
+    (1 - pi.hat)* n.1, (1 - pi.hat) * n.2)
```

```
> O <- c(n.1S, n.2S, n.1F, n.2F)
> (chisq.value <- sum((abs(O - E) - 0.5)^2 / E))
[1] 0.2673797
> (p.value <- 1 - pchisq(chisq.value, 1))
[1] 0.605095
```

The p-value $> \alpha$. Therefore, we conclude that the proportion of blacks in the population of married inmates does not differ from the proportion of blacks in the population of single inmates. Here inmates refer to those who were sentenced to death. □

All of the work we have done on our own can be accomplished in R with chisq.test().

Example 12.5. Continuing with Example 12.4, we obtain

```
> chisq.test(s)

        Pearson's Chi-squared test with Yates'
        continuity correction

data:  table(s)
X-squared = 0.2674, df = 1, p-value = 0.6051
```

Note that chisq.test() wants a table (s in our case). □

The χ^2 test works best for large n. As a rule, none of the cells should include expected counts of less than 5.

12.1.3 Intensities

The development in this section parallels that in Section 12.1.2.

Hypothesis testing

We have two independent samples and wish to test for $\lambda_0 := \lambda_2 - \lambda_1$. Our hypotheses are $H_0: \lambda = \lambda_0$ vs. one of the three alternatives. The sampling densities of l_i are $\phi\left(l_i \,\middle|\, \lambda_i, \sqrt{\lambda_i/n_i}\right)$. Therefore, the sampling density of $l_2 - l_1$ is normal. Its mean is λ_0 and its variance is

$$\frac{\lambda_0}{n_1} + \frac{\lambda_0}{n_2} = \lambda_0 \left(\frac{1}{n_1} + \frac{1}{n_2}\right).$$

To estimate the standard error, we use

$$\bar{l} := \frac{n_{1S} + n_{2S}}{n_1 + n_2}$$

where n_{iS} are the event counts in n_i interval units. So

$$\text{SE} = \sqrt{\lambda_0 \left(\frac{1}{n_1} + \frac{1}{n_2}\right)} \approx \sqrt{\bar{l} \left(\frac{1}{n_1} + \frac{1}{n_2}\right)}. \tag{12.9}$$

The test is acceptable as long as $\lambda_i > 2.5$ (Thode Jr, 1997; Detre and White, 1970). When $n_1 = n_2$, or when both are large, we do not need to use the continuity correction.

To test hypotheses, replace μ_0 and SE in (10.2) through (10.5) with λ_0 and SE from (12.9).

Confidence intervals

We construct the confidence interval for $\hat{\lambda} = \hat{\lambda}_2 - \hat{\lambda}_1$ to estimate the location of $\lambda = \lambda_2 - \lambda_1$. Our rv is $l := l_2 - l_1$. From (9.9) we have

$$I_{1-\alpha}\left(l\,|\,\hat{\lambda}\right) = \left[\Phi^{-1}\left(\alpha/2\,|\,\hat{\lambda}, \text{SE}\right),\ \Phi^{-1}\left(1-\alpha/2\,|\,\hat{\lambda}, \text{SE}\right)\right]$$
$$= \text{c(qnorm(alpha/2, lambda.hat, SE)},\qquad(12.10)$$
$$\text{qnorm(1 - alpha/2, lambda.hat, SE))}\ .$$

Example 12.6. In two different areas we count the number of individuals of a species in $1\,\text{m}^2$ plots. The data are

```
> n <- c(150, 200) ; n.S <- c(40, 20) ; alpha <- 0.05
```

Are the encounter rates significantly different and what is the confidence interval of the difference between the encounter rates in the two areas at $\alpha = 0.05$? We obtain $\hat{\lambda}$, $\bar{l}$ and SE with

```
> lambda.hat <- (n.S/n)[2] - (n.S/n)[1]
> l.bar <- sum(n.S) / sum(n)
> SE <- sqrt(l.bar * sum(1 / n))
```

and from (12.10),

```
> round(c(low = qnorm(alpha / 2, lambda.hat, SE),
+    high = qnorm(1 - alpha / 2, lambda.hat, SE)), 3)
   low   high
-0.254 -0.079
```

The answers to the two questions are yes. □

12.2 Small samples

So far, we examined the cases where the normal density, or its approximation to the binomial and Poisson hold. When sample sizes are small or a particular density's parameters do not conform to the usual requirement (i.e. that np and $n(1-p)$ are both ≥ 5), then we must rely on a different approach. We tackle these issues in this section.

Regarding densities, we have a few generic possibilities (Figure 12.2). We discuss cases (a) and (b) in Section 12.2.3. Cases (c) and (d) are discussed in Section 12.3.1.

12.2.1 Estimating variance and standard error

We wish to estimate $\sigma_1^2 + \sigma_2^2$ from two samples. If the variances are presumed equal, then to obtain an unbiased estimate of the pooled variances we use

$$S^2 = \frac{n_1 - 1}{n_1 + n_2 - 2} S_1^2 + \frac{n_2 - 1}{n_1 + n_2 - 2} S_2^2$$

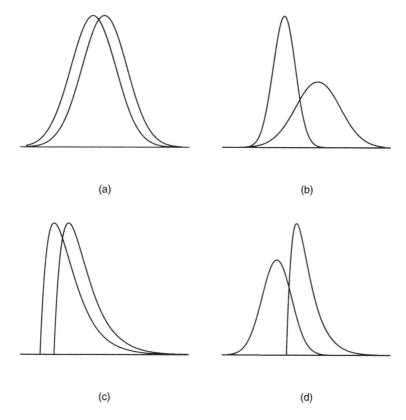

Figure 12.2 Generic possibilities of densities of 2 samples: (a) Means differ, variances equal, both populations are normal. (b) Means differ, variances differ, both populations are normal. (c) Means differ, variances equal, both populations are not normal. (d) Means differ, variances differ, both populations are not normal and are not equal.

where S_1^2 and S_2^2 are the samples variance and S^2 is the pooled variance. The terms

$$\frac{n_1 - 1}{n_1 + n_2 - 2}, \quad \frac{n_2 - 1}{n_1 + n_2 - 1}$$

weigh the contributions of S_1^2 and S_2^2 to the pooled sample variance. The standard error of the difference between $\overline{X}_1$ and $\overline{X}_2$ is then

$$\text{SE} = S \times \sqrt{\frac{1}{n_1} + \frac{1}{n_2}}. \tag{12.11}$$

If the variances of the two samples are not presumed equal, then we estimate the standard errors separately and pool them like this:

$$\text{SE}_1 = \frac{S_1}{\sqrt{n_1}}, \quad \text{SE}_2 = \frac{S_2}{\sqrt{n_2}}, \quad \text{SE} = \sqrt{\text{SE}_1^2 + \text{SE}_2^2}. \tag{12.12}$$

12.2.2 Hypothesis testing and confidence intervals for variance

To implement any of the approaches to obtaining the pooled variance (Section 12.2.1), we need a way to test for the equality of variance. Let S_1^2 and S_2^2 be the rv variances obtained from two samples (of size n_1 and n_2) taken from normal populations with μ_1, σ_1 and μ_2, σ_2. It turns out that the sampling density of $X := S_1^2 / S_2^2$ is $F(X | n_1 - 1, n_2 - 1)$. So we test for

$$H_0 : \rho = \rho_0 \quad \text{vs.} \quad H_A : \rho \neq \rho_0$$

where $\rho := \sigma_2^2 / \sigma_1^2$ and ρ_0 is the variance ratio we wish to test for. The estimated ratio under H_0 is X. For a lower-tailed test

$$\begin{aligned} p_L\text{-value} &= F(X | n_1 - 1, n_2 - 1) \\ &= \mathtt{pf(X, n.1 - 1, n.2 - 1)} \end{aligned}$$

and for an upper-tailed test

$$\begin{aligned} p_H\text{-value} &= 1 - F(X | n_1 - 1, n_2 - 1) \\ &= 1 - \mathtt{pf(X, n.1 - 1, n.2 - 1)} \,. \end{aligned}$$

For a two-tailed test

$$p\text{-value} = 2 \min (p_L\text{-value}, p_H\text{-value}) \,.$$

The corresponding confidence intervals are:

$$\begin{aligned} \text{lower-tailed CI} &= \left[0, X / F^{-1}(\alpha | n_1 - 1, n_2 - 1)\right] \\ &= \mathtt{c(0, X/qf(alpha, n.1 - 1, n.2 - 1))} \,, \end{aligned}$$

$$\begin{aligned} \text{upper-tailed CI} &= \left[X / F^{-1}(1 - \alpha | n_1 - 1, n_2 - 1), \infty\right] \\ &= \mathtt{c(X/qf(1 - alpha, n.1 - 1, n.2 - 1), Inf)} \end{aligned}$$

and

$$\begin{aligned} \text{two-tailed CI} &= \left[X / F^{-1}(1 - \alpha/2 | n_1 - 1, n_2 - 1), X / F^{-1}(\alpha/2 | n_1 - 1, n_2 - 1)\right] \\ &= \mathtt{c(X/qf(1 - alpha/2, n.1 - 1, n.2 - 1),} \\ &\quad\ \mathtt{X/qf(alpha/2, n.1 - 1, n.2 - 1)} \,. \end{aligned}$$

Example 12.7. We generate two small samples from two normal densities and test for the equality of variances ($\rho_0 = 1$) for lower-tailed

```
> set.seed(28) ; n.1 <- 20 ; n.2 <- 25
> X.1 <- rnorm(n.1) ; X.2 = rnorm(n.2, 0, 2)
> alpha <- 0.05 ; X <- var(X.1) / var(X.2)
> p.L <- pf(X, n.1 - 1, n.2 - 1)
```

upper-tailed
```
> p.H <- 1 - pf(X, n.1 - 1, n.2 - 1)
```
and two-tailed tests
```
> p <- 2 * min(p.L, p.H)
```
Note that for convenience we use S_1^2/S_2^2. Were we to test S_2^2/S_1^2, the upper- and lower-tailed tests would have switched significance. To see the results, we do
```
> p.value <- rbind(p.L, p.H, p)
> dimnames(p.value) <- list(c('lower-tailed', 'upper-tailed',
+   'two-tailed'), 'p-value') ; round(p.value, 3)
             p-value
lower-tailed   0.020
upper-tailed   0.980
two-tailed     0.039
```
So for upper- and two-tailed tests we reject H_0 and conclude that the variances are different. Because $S_1^2 / S_2^2 \approx 0.39$, the upper-tailed test is not significant. Indeed, S_1^2 is not $>$ than S_2^2 and we cannot reject H_0.

For confidence intervals we obtain
```
> CI.L <- c(0, X / qf(alpha, n.1 - 1, n.2 - 1))
> CI.H <- c(X / qf(1 - alpha, n.1 - 1, n.2 - 1), Inf)
> CI <- c(X / qf(1 - alpha / 2, n.1 - 1, n.2 - 1),
+   X / qf(alpha / 2, n.1 - 1, n.2 - 1))
> CI <- rbind(CI.L, CI.H, CI)
> dimnames(CI) <- list(c('lower-tailed', 'upper-tailed',
+   'two-tailed'), c('low', 'high')) ; round(CI, 3)
               low   high
lower-tailed 0.000  0.821
upper-tailed 0.190    Inf
two-tailed   0.165  0.952
```
Because neither lower- nor two-tailed confidence intervals cross 1, we conclude that the variances are different. The upper-tailed test indicates that they are not. All of this can be implemented with
```
> var.test(X.1, X.2)

        F test to compare two variances

data:  X.1 and X.2
F = 0.3881, num df = 19, denom df = 24, p-value =
0.03905
alternative hypothesis: true ratio of variances is not equal to 1
95 percent confidence interval:
 0.1654920 0.9517562
sample estimates:
ratio of variances
         0.3881043
```
which confirms our results. □

12.2.3 Means

Here we distinguish between paired and unpaired observations. Recall that we have two samples, X_1 and X_2, of size n_1 and n_2. In the case of unpaired observations, the value for the ith observation from X_1, denoted by X_{1i}, is independent from the value of the $i$the observation from X_2, denoted by X_{2i}. In paired observations, we record two values from a single object. Obviously, in the case of paired observations, $n_1 = n_2$. If you have paired observations, use the paired test for means, otherwise use the unpaired test.

Unpaired observations

To test lower-, upper- and two-tailed hypotheses, use (10.6), (10.7) and (10.8) where $\mu_0 := \mu_2 - \mu_1$ and depending on whether $S_1^2 = S_2^2$ or not, the SE is calculated according to (12.11) or (12.12). The p-values are calculated as usual.

Example 12.8. The data for midterm and final scores in a Statistics course are:

```
> load('scores.rda') ; scores
$midterm
 [1] 66 78 62 99 80 63 82 86 84 70 98 81 66 42 92 74 75 89
[19] 87 84 89 87 76 45 84

$final
 [1] 79 78 65 75 84 94 79 84 79 66 76 76 79 91 88 78 77 87
[19] 86 73 73 84 88 79
```

(one student dropped the class). Do the means on the midterm and final differ at $\alpha = 0.05$? The (edited) results for the test of equality of variances are

```
> alpha <- 0.05
> (v.equal <- var.test(scores$midterm, scores$final))

        F test to compare two variances

F = 3.9807, num df = 24, denom df = 23, p-value = 0.001486
```

Based on the p-value, the variances are different. The (edited) results of the two-sided t-test are

```
> p.v.equal <- v.equal$p.value
> v.equal <- TRUE
> if(p.v.equal <= alpha) v.equal <- FALSE
> t.test(scores$midterm, scores$final, var.equal = v.equal)

        Welch Two Sample t-test

t = -0.7354, df = 35.656, p-value = 0.4669
sample estimates:
mean of x mean of y
 77.56000  79.91667
```

From the *p*-value we conclude that the mean score on the finals did not differ from the mean score on the midterm. □

In addition to the *p*-value of the difference in means, `t.test()` provides confidence intervals on the difference of means.

Paired observations

In a paired design, we record two values for each object in our sample. Examples are before and after, two measurements on an object at different times and so on. A paired design is preferable to a random experiment design because it results in smaller variance. Here is why. Let X_1 and X_2 be paired random variables from a paired normal population of size N. Then there must be some amount of correlation between X_1 and X_2

$$\rho = \frac{\sigma_{X_1 X_2}}{\sigma_{X_1}\sigma_{X_2}} \quad \text{or} \quad \sigma_{X_1 X_2} = \rho \times \sigma_{X_1} \times \sigma_{X_2}$$

where $\sigma_{X_1 X_2}$ is the covariance between X_1 and X_2, defined as

$$\sigma_{X_1 X_2} := \frac{1}{N}\sum_{i=1}^{N}(X_{i1}-\mu_1)(X_{i2}-\mu_2).$$

For each pair of observations we have

$$X_i := X_{i2} - X_{i1}.$$

Therefore,

$$\mu_X = \mu_{X_1 - X_2}$$

and

$$\sigma_X^2 = \sigma_{X_1}^2 + \sigma_{X_2}^2 - 2\sigma_{X_1 X_2}$$
$$= \sigma_{X_1}^2 + \sigma_{X_2}^2 - 2\rho\sigma_{X_1}\sigma_{X_2}.$$

Each pair of measurement is independent of other pairs. However, pairs are dependent. Therefore, if $\rho > 0$, then

$$\sigma_X^2 < \sigma_{X_1}^2 + \sigma_{X_2}^2.$$

This means that we should prefer to test paired comparisons over pooled comparisons because the variance for the difference in paired comparisons is smaller. The difference between paired and unpaired designs is in the way we calculate the mean and the variance. Once these are obtained, the test proceeds as usual.

For a paired sample of size n

$$\overline{X} := \frac{1}{n}\sum_{i=1}^{n} X_i \quad \text{and} \quad S_X^2 = \frac{1}{n-1}\sum_{i=1}^{n}(X_i - \overline{X})^2$$

where again, $X_i = X_{i2} - X_{i1}$ for $i = 1, 2, \ldots, n$.

386 Two samples

Example 12.9. Continuing with Example 12.8, the first score in the midterm is for the student who dropped the class. So the data are:

```
> midterm <- scores$midterm[-1] ; final <- scores$final
```

and the test (let us indulge and be terse here) is

```
> t.test(midterm, final, var.equal =
+        (var.test(midterm, final)$p.value <= alpha))

        Two Sample t-test

t = -0.5725, df = 46, p-value = 0.5698
sample estimates:
mean of x mean of y
 78.04167  79.91667
```

(the output was edited). As in the unpaired test, we do not reject the null hypothesis and conclude that the mean scores were not different on the midterm and final. □

12.2.4 Proportions

Let us go back to go back to contingency tables:

success	population 1	population 2	total
yes	n_{1S}	n_{2S}	n_S
no	$n_1 - n_{1S}$	$n_2 - n_{2S}$	$n - n_S$
total	n_1	n_2	n

For $\pi := \pi_2 - \pi_1$, we wish to test

$$H_0 : \pi = 0 \quad \text{vs.} \quad H_A : \pi \neq 0$$

with significance level of α. The null hypothesis dictates that $\pi := \pi_1 = \pi_2$. Under the null, we estimate π with

$$p := \frac{n_1}{n_1 + n_2} \times \frac{n_{1S}}{n_1} + \frac{n_2}{n_1 + n_2} \times \frac{n_{2S}}{n_2}.$$

Also under the null, the expected values in the four cells are

success	population 1	population 2
yes	$p \times n_1$	$p \times n_2$
no	$(1-p) \times n_1$	$(1-p) \times n_2$

If any of the four expected values are < 5, then we must use Fisher's exact test. Let

$$A := n_S!\,(n - n_S)!\,n_1!\,n_2!\,,$$
$$B := n!\,n_{1S}!\,n_{2S}!\,(n_1 - n_{1S})!\,(n_2 - n_{2S})\,.$$

Under the assumption that the margins in the contingency table are fixed, the probability of obtaining the table is computed according to Fisher's exact test as

$$p\text{-value} = \frac{A}{B}.$$

If p-value $< \alpha$, we reject the null hypothesis.

Example 12.10. After the first examination in a Statistics class, 4 students scored an A and 5 students a C. The examination took place on the 5th class. The A students missed a total of 2 classes among them and the C students missed a total of 14 classes. Did the A and C students differ in the number of classes they missed? We use a significance level of 0.05.

Assume that students miss classes independent of each other and at random class sessions. Then we prepare the table

```
> classes <- rbind(c(3, 8), c(18, 11))
> dimnames(classes) <- list(Classes = c('missed', 'attended'),
+    Grade = c('A', 'C'))
> classes
          Grade
Classes    A  C
  missed   3  8
  attended 18 11
```

and run the test

```
> fisher.test(classes, alternative = 'less')

        Fisher's Exact Test for Count Data

data:  classes
p-value = 0.05275
alternative hypothesis: true odds ratio is less than 1
95 percent confidence interval:
 0.000000 1.018084
sample estimates:
odds ratio
 0.2381634
```

Based on the p-value, we conclude that class attendance did not affect the student grades. □

12.2.5 Intensities

We present only one approach (the so-called conditional test or binomial exact test) to testing hypotheses for two Poisson parameters. Let N_1 and N_2 be counts from two independent populations over n_1 and n_2 interval units with Poisson parameters λ_1 and λ_2. The best estimates of λ_1 and λ_2 are $\widehat{\lambda}_1 := l_1 = N_1 \,/\, n_1$ and $\widehat{\lambda}_2 = l_2 = N_2 \,/\, n_2$. The uppertailed hypotheses are

$$H_0 : \rho = \rho_0 \quad \text{vs.} \quad H_A : \rho > \rho_0$$

where $\rho_0 := \lambda_2 \,/\, \lambda_1$ is given. Here we have $E[N_1] = n_1\lambda_1$ and $E[N_2] = n_2\lambda_2$. The estimates of the expectations are $X_1 = n_1 l_1$ and $X_2 = n_2 l_2$ and their corresponding realizations are x_1 and x_2. Let $X := X_1 + X_2$ with the realization $x = x_1 + x_2$. Then

$$P(X_2 = x_2 \,|\, \rho_0, x) = \binom{x}{x_2} \rho_0^{x_2} (1 - \rho_0)^{x - x_2}$$

where the probability of success is

$$\rho_0 = \frac{n_2 \lambda_2}{n_1 \lambda_1 + n_2 \lambda_2}.$$

So

$$\begin{aligned}p\text{-value} &= P(X_2 \geq x_2 \,|\, \rho_0, x) \\ &= 1 - \texttt{pbinom}(\texttt{x.2} - 1, \texttt{x}, \texttt{rho.0}).\end{aligned}$$

For testing $\lambda_0 := \lambda_1 = \lambda_2$ and

$$H_0 : \rho = \rho_0 \quad \text{vs.} \quad H_A : \rho \neq \rho_0,$$

we have

$$\rho_0 = \frac{n_2}{n_1 + n_2}$$

and the p-value is given by

$$\begin{aligned}p\text{-value} &= 2 \min\left[P(X_2 \geq x_2 \,|\, \rho_0, x), P(X_2 \leq x_2 \,|\, \rho_0, x)\right] \\ &= 2 * \min\left(1 - \texttt{pbinom}(\texttt{x.2} - 1, \texttt{x}, \texttt{rho.0}),\right. \\ &\quad\left. \texttt{pbinom}(\texttt{x.2}, \texttt{x}, \texttt{rho.0})\right).\end{aligned}$$

There are other ways to test for two samples from Poisson densities (Krishnamoorthy and Thomson, 2004, and citations therein).

Example 12.11. We wish to determine if the hunting successes of two prides of lions are different. We follow the first pride for 20 attempts, all of which failed. We follow the second pride for 20 attempts, three of which succeeded. So

```
> x.1 <- 0 ; x.2 <- 3 ; x <- x.1 + x.2 ; rho.0 <- 0.5
> (p.value <- 2 * min(1 - pbinom(x.2 - 1, x, rho.0),
+       pbinom(x.2, x, rho.0)))
[1] 0.25
```

and we conclude that the hunting successes of both prides are equal. □

12.3 Unknown densities

So far, we examined small samples from known densities. As we saw, the addition of uncertainty about the variances (compared to large samples) forced us to use the t-test when the populations were normal. When the populations are binomial or Poisson, we use Fisher's exact test or the binomial exact test (Sections 12.2.4 and 12.2.5). What do we do when the densities are not known and samples sizes are small? Then we use either the rank sum test or the paired signed rank test. Because the tests do not rely on known densities, they are called *nonparametric* tests. We are interested in testing for differences in means. Therefore, we must assume that the densities are symmetric. The rank sum and signed rank tests can be used to test for the differences of medians of two samples instead of means. For medians, we do not need to assume symmetry of the population densities.

12.3.1 Rank sum test

This test is often called the Wilcoxon rank sum test or the Mann-Whitney U test. We shall call it the rank sum test. Here we consider the case where two samples come from symmetric densities with the same spread (equal variance), but different location (Figure 12.2c). The assumption of symmetry allows us to test for the equality of means. Otherwise, the test is applicable for the equality of medians. The assumption of symmetry is not as restrictive as it might seem. Often the differences between the means lead to a symmetric sampling density. Also, the test is robust to slight asymmetries. The rank sum test is the nonparametric counterpart of the t-test.

Consider testing

$$H_0 : \mu = 0 \quad \text{vs.} \quad H_A : \mu \neq 0$$

with samples of size $n_1 = n_2 = n$ and $\mu := \mu_2 - \mu_1$. Under the null hypothesis, both samples come from the same density. Therefore, if we pool the data and rank the $2n$ observations, then we expect the values that come from the first sample to be equally scattered among the values that come from the second sample. If we sum the ranks of two samples from the same populations (under H_0), then the sum of the ranks should be about equal. Here is an example.

Example 12.12. Two samples of young walleye were drawn from two different lakes and the fish were weighed. The data in g are:

```
> X.1 <-c (253, 218, 292, 280, 276, 275)
> X.2 <- c(216, 291, 256, 270, 277, 285)
> sample <- c(rep(1, 6), rep(2, 6))
> w <- data.frame(c(X.1,X.2), sample)
> names(w)[1] <- 'weight (g)'
> cbind(w[1 : 6, ], w[7 : 12, ])
  weight (g) sample weight (g) sample
1        253      1        216      2
2        218      1        291      2
3        292      1        256      2
4        280      1        270      2
5        276      1        277      2
6        275      1        285      2
```

Next, we sort the data keeping track of the group identity

```
> idx <- sort(w[, 1] , index.return = TRUE)
> d <- rbind(weight = w[idx$ix, 1], sample = w[idx$ix, 2],
+    rank = 1:12)
> dimnames(d)[[2]] <- rep('', 12) ; d

weight 216 218 253 256 270 275 276 277 280 285 291 292
sample   2   1   1   2   2   1   1   2   1   2   2   1
rank     1   2   3   4   5   6   7   8   9  10  11  12
```

Finally, we sum the ranks of the observation in the pooled data:

```
> rank.sum <- c(sum(d[3, d[2, ] == 1]),
+   sum(d[3, d[2, ] == 2]))
> rank.sum <- rbind(sample = c(1,2),
+   'rank sum' = rank.sum)
> dimnames(rank.sum)[[2]] <- c('','') ; rank.sum

sample    1  2
rank sum 39 39
```

In this case, the rank sums are equal. □

Suppose that all the observation values in the first sample are smaller than all the observation values in the second sample. Now rank the $2n$ observations as a single sample. Then the first n observations belong to the first sample, the rest to the second. Therefore, in our example, the first sample will have the smallest possible rank sum of

$$1 + 2 + 3 + 4 + 5 + 6 = 21$$

and the second sample will have the largest possible value of rank sum,

$$7 + 8 + 9 + 10 + 11 = 57.$$

The rank sum of each sample can range between 21 and 57. We want the probabilities (density) of all possible values of rank sums. Under H_0, the ranks of the first sample should be equally scattered among the ranks of the two samples pooled. Altogether, the ranks of the first sample can be scattered in

$$\frac{12!}{6!6!} = 924$$

different ways. The rank sum of 21 is achieved in only one possible way: When all of the values of one sample are smaller than those of the other. Similarly, the rank sum of 57 can be achieved in only one possible way. Consequently,

$$P(W_1 = 21 \text{ and } W_2 = 57) = \frac{2}{924} = 0.002$$

where W_1 and W_2 denote the rank sum of sample 1 and sample 2. If H_0 is true and we get a rank sum of 21 or 57, we will reject H_0 for $\alpha = 0.05$ because p-value $= 0.002$. Continuing this way, we determine in how many ways we can produce rank sums ≤ 22, 23 and so on. For example, by enumeration, we can show that

$$P(W_1 \leq 23 \text{ and } W_2 \geq 55) = \frac{8}{924} = 0.009$$

The R function `wilcox.test()` makes the job of computing rank sum easy.

Example 12.13. For the data in Example 12.12, we use the hypotheses that

$$H_0 : \mu = 0 \quad \text{vs.} \quad H_A : \mu \neq 0$$

with $\alpha = 0.05$:
```
> wilcox.test(X.1, X.2)
```
         Wilcoxon rank sum test

data: X.1 and X.2
W = 18, p-value = 1
alternative hypothesis: true location shift is not equal to 0

Here the value of the W statistic is 18 and the p-value is virtually 1. Therefore, we do not reject H_0. □

So far, we tested for $H_0 : \mu = 0$. To test for $H_0 : \mu = \mu_0$ (where μ_0 is some constant), we simply rearrange the null hypothesis to $H_0 : \mu - \mu_0 = 0$.

Example 12.14. Continuing with Example 12.13, we wish to test
$$H_0 : \mu = 50\,g \quad \text{vs.} \quad H_A : \mu < 50\,g$$
at $\alpha = 0.05$:
```
> wilcox.test(X.1, X.2, mu = 50, alternative = 'less')
```
         Wilcoxon rank sum test

data: X.1 and X.2
W = 4, p-value = 0.01299
alternative hypothesis: true location shift is less than 50

Therefore we reject H_0 in favor of H_A and conclude that the difference between the mean weight of the two populations of young walleye is less that 50 g. □

When some of the ranks are equal, we assign each tied rank the average value of the rank. When the proportion of ties in the two samples is inordinately large, say over 25%, then a correction factor needs to be applied (Mosteller, 1973; Wayne, 1990).

Example 12.15. Consider two samples, with 6 observations each:
```
> s1 <- c(3, 4, 5, 7, 3, 8)
> X.1 <- c(3, 4, 5, 7, 3, 8)
> X.2 <- c(10, 6, 9, 1, 2, 7)
> wilcox.test(X.1, X.2)
```
         Wilcoxon rank sum test with continuity correction

data: X.1 and X.2
W = 15.5, p-value = 0.7479
alternative hypothesis: true location shift is not equal to 0

Warning message:
cannot compute exact p-value with ties in:
wilcox.test.default(X.1, X.2)

The warning message refers to the fact that the exact p-value cannot be computed with ties. Exact p-value refers to computation based on the densities of various rank sums computed directly from combinatorial considerations as we did above. □

12.3.2 t vs. rank sum

Which test should we choose: the t-test or the rank sum test? It turns out that when both samples come from a normal density, the t-test performs slightly better—it detects significant differences when they exist—than the rank sum. However, when departures from normal are marked, the t-test is inferior and the resulting p-values cannot be trusted. Also, when the samples are very small (say 5–10 observations), then tests of normality are not reliable and we prefer to use the rank sum test. The upshot? When in doubt, be conservative and use the rank sum.

12.3.3 Signed rank test

When two independent samples come from normal distributions and the samples are small, we use the t-test. When observations are paired and independent (but pairs are dependent), when the samples come from normal populations and when they are small, we use the paired t-test. When the two samples are independent, but they come from similar symmetric distributions with equal variance and different location, we used the rank sum test. We are now ready to discuss the case where the small samples conform to the assumptions we used for the rank sum test with one additional condition: observations are paired. In this case we use the signed rank test. It is the nonparametric counterpart of the paired t-test.

Under the assumption that $\mu = 0$, the test statistic is computed thus:

1. Rank $|X_i|$ and denote the ith rank by R_i.
2. Restore the signs to R_i.
3. Sum the positive ranks and denote it by W_+. Sum the negative ranks and denote it by W_-. Do not use ranks for zero difference and reduce the sample size by the number of zero differences.
4. Denote by W_S the smaller sum. The *signed rank sum* W_S is our test statistic.

From this construction, large W_S indicates large X_i and we consequently reject H_0. The sampling density of W_S is known and thus we can compute p-values and confidence intervals. It should not be confused with the sampling density of the sum rank, W. Here is an example that details the calculation steps of the sign rank statistic, W_S.

Example 12.16. One semester, one of us was particularly interested in comparing the performance of 12 students in a Statistics class. Here are their test scores on the midterm and final and the sequence calculations that we need to obtain the W_S statistic:

```
> load('test.scores.rda')
> (z <- test.scores.rda)
  midterm final diff abs.diff rank signed.rank
1      48    44    4        4  2.5         2.5
2      51    62  -11       11  9.0        -9.0
```

```
3      57   64   -7     7   7.5       -7.5
4      67   62    5     5   4.0        4.0
5      46   64  -18    18  11.5      -11.5
6      67   85  -18    18  11.5      -11.5
7      68   62    6     6   5.5        5.5
8      60   75  -15    15  10.0      -10.0
9      91   95   -4     4   2.5       -2.5
10     86   92   -6     6   5.5       -5.5
11     87   94   -7     7   7.5       -7.5
12     87   84    3     3   1.0        1.0
```

The data do not seem normal

```
> par(mfrow = c(1, 2))
> qqnorm(midterm, main = 'midterm') ; qqline(midterm)
> qqnorm(final, main = 'final') ; qqline(final)
```

(Figure 12.3). Yet, the test scores are paired. So we use the signed rank test with

$$H_0 : \mu = 0 \quad \text{vs.} \quad H_A : \mu \neq 0$$

and with $\alpha = 0.05$. From the last column above,

```
W.plus <- sum(z$signed.rank[z$signed.rank > 0])
W.minus<- - sum(z$signed.rank[z$signed.r < 0])
c(W.plus, W.minus)
[1] 13 65
```

or in our notation,

$$W_+ = 13 \quad \text{and} \quad W_- = 65$$

Now

```
> W <- min(W.plus, W.minus)
> round(c('W' = W, 'p.value' = 2 *
+     psignrank(W, length(z[, 1]), length(z[, 1]))), 3)
      W p.value
 13.000   0.042
```

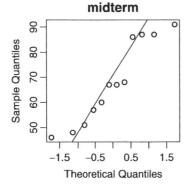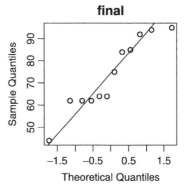

Figure 12.3   Test scores.

394   Two samples

The value of the test statistic $W_S = 13$ and its p-value (obtained with psignrank (13,12)) is 0.042. The mean test scores on the midterm and final for the 12 students were different at $\alpha = 0.05$. So far, we did our own computations. Using R, we obtain

```
> wilcox.test(midterm, final, paired = TRUE)

        Wilcoxon signed rank test with continuity correction

data:  midterm and final
V = 13, p-value = 0.04513
alternative hypothesis: true location shift
is not equal to 0

Warning message:
cannot compute exact p-value with ties in:
wilcox.test.default(midterm, final, paired = TRUE)
```

Our direct computation and those of R are within roundoff errors.   □

When there are too many ties (say above 20%), you should doubt the legitimacy of the test results. For $n > 10$ and for X_i independent and symmetrically distributed around zero, the statistic

$$Z_{W_S} := \frac{W_S - \mu_{W_S}}{\sigma_{W_S}}$$

is approximately normal under H_0 where

$$\overline{W}_S = 0 \quad \text{and} \quad S_{W_S} = \sqrt{\frac{n(n+1)(2n+1)}{6}}.$$

With these approximations, we can test hypotheses and obtain confidence intervals.

12.3.4   Bootstrap

In this section, we discuss the case where nothing is known about the distributions of the samples. In fact the samples may come from populations with any two distributions. We introduce the topic by way of an example.

Example 12.17. We use the data presented in (Efron and Tibshirani, 1993, Table 2.1, p. 11), where two groups of mice were subjected to treatment and control ($n_1 = 7$ and $n_2 = 9$):

```
treatment <- c(94, 197, 16, 38, 99, 141, 23)
control <- c(52, 104, 146, 10, 50, 31, 40, 27, 46)
```

We wish to produce the 95% confidence interval of $\mu := \mu_1 - \mu_2$, the difference between the means of the populations. So we take a sample of size 7 (with replacement) from the treatment group and compute its mean $\overline{X}_1$. Similarly, we take a sample of size 9 with replacement from the control group and obtain $\overline{X}_2$. We now have our first instance of $\overline{X}$. We repeat the process 1 500 times. Thus, we get an approximation of the sampling density of $\overline{X}$. Assuming that this sampling density is approximately normal, we obtain an estimate of $\mu_{\overline{X}}$ and the standard error $\sigma_{\overline{X}}$. Using these, the confidence interval is

$$95\% \text{ CI} = \left[\overline{X} - 1.96 \times \text{SE},\, \overline{X} + 1.96 \times \text{SE}\right].$$

Implementing the bootstrap procedure,

```
> library(simpleboot)
> set.seed(10)
> b <- two.boot(treatment, control, mean,
+    R = 1500, student = TRUE, M = 50)
```

we read the value of $\overline{X}$ from

```
> b$t0[1]
[1] 30.63492
```

Next, we calculate the normally approximated confidence interval

```
> bci <- boot.ci(b)
> bci$normal
     conf
[1,] 0.95 -22.20960 85.36527
```

or in our notation

$$95\% \text{ CI} = [-22.21,\, 85.37].$$

To view the results, we do

```
> hist(b)
> abline(v = b$t0[1], col = 'red', lwd = 2)
> abline(v = bci$normal[2], lty = 2)
> abline(v = bci$normal[3], lty = 2)
```

(see Figure 12.4). Because the confidence interval spans zero, the population means are not judged to be different. □

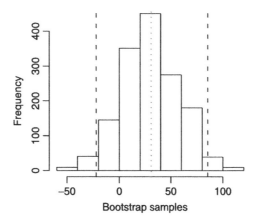

Figure 12.4 Bootstrap frequency and 95% confidence interval for the difference between the means of two populations with unknown distributions.

12.4 Assignments

Unless otherwise specified, use two-tailed tests with $\alpha = 0.05$.

Exercise 12.1. Take two independent samples from populations with the following parameters:
$$\mu_1 = 20, \sigma_1 = 3, n_1 = 45,$$
$$\mu_2 = 30, \sigma_2 = 4, n_2 = 47.$$

1. What is the sampling distribution of $\mu_2 - \mu_1$?
2. What is the mean of this sampling distribution?
3. What is the variance of this sampling distribution?

Exercise 12.2. Walleye are sampled from two different lakes in southern and northern Minnesota. The data summary on fish lengths (mm) are as follows:

population	mean	standard deviation	sample size
1	110	12	54
2	125	25	52

A fisheries biologist claims that with a significance level of 0.05, these results support the hypothesis that fish in warmer waters (southern Minnesota) grow to be longer than fish in colder lakes (northern Minnesota). Do the data support her claim?

Exercise 12.3. Deer feeding in the winter in northern latitudes is a controversial issue. Some claim it reduces mortality. Others claim it does the opposite—strong deer get to eat most of the food and thus deny others of the opportunity to eat and larger winter populations mean larger summer population. Weight serves as an index of mortality. The larger the weight, the smaller the mortality. A single deer from each of 11 isolated populations were weighed by the end of five winters. These populations were not weighed during the winter. A single deer from each of 12 isolated populations were weighed by the end of six winters. These populations were given supplemental food during the winter. Assume that weights between years are independent. The data were as follows:

Supplemental Feeding	mean	standard deviation	sample size
Yes	165	25	72
No	160	22	55

1. What is the 95% confidence interval estimate for the population difference in the mean weight?
2. Use $\alpha = 0.1$ to determine if supplemental feeding results in different deer weight.

Exercise 12.4. For the following exercise, refer to the data in Table 9.3, on page 302.

1. Based on these data, would you conclude that all possible pairs are sufficiently different from each other to justify separation of subspecies? Compare males to males and females to females. Do not compare males to females. Be sure to run the appropriate test.

2. Sexual dimorphism refers to the idea that within a species, males and females may differ in one or more morphological traits. Would you conclude that there is wing chord sexual dimorphism?

Exercise 12.5. Responses to public opinion surveys often depend on subtle differences in the wording of questions. In the paper "Attitude measurement and the gun control paradox" (Public Opinion Quarterly 1977–1978, pp. 427–438), the investigators were interested in how the wording of a question influences the response. They worded the question about gun control in two ways:

A. "Would you favor or oppose a law that would require a person to obtain a police permit before purchasing a gun?"

B. "Would you favor or oppose a law that would require a person to obtain a police permit before purchasing a gun, or do you think that such a law would interfere too much with the right of citizens to own guns?"

The second question should elicit a smaller proportion of "yes" than the first. Here are the data:

```
             n in favor
question A 615     463
question B 585     403
```

Did the wording have an effect on responses?

Exercise 12.6. The sex ratio of reptiles is determined by temperature during incubation. Suppose that 150 eggs of alligators were exposed to temperature t_1. Of these, 90 eggs hatched into females. Another sample, of 200 eggs were exposed to temperature t_2, and 125 of them hatched into females.

1. Does temperature make a difference in the sex ratio (use $\alpha = 0.05$)?
2. Compute the 95% confidence interval for the true difference in the the sex ratio for the "populations" in this experiment.

Exercise 12.7. Kimmo et al. (1998) studied the survival of Willow tits over the winter. They trapped birds in the autumn and then recorded the number of birds resighted in the following season. They interpreted these data as survival rate. Here is a subset of their report (see their Table 1):

```
                     91-92 92-93 95-96
trapped.adults         49    60    25
resighted.adults       37    36    22
trapped.yearlings      45    19    40
resighted.yearlings    17     2    14
```

1. For which of these years (if any) can you compare the survival rate of adults and yearlings with the normal approximation for inference about proportions?
2. Compare the survival of adults to yearlings. Are they significantly different at $\alpha = 0.05$ for a two-tailed test?
3. Repeat (2) for $\alpha = 0.01$.
4. Repeat (1) and (2) for a one-tailed test (set up the test so that it makes biological sense).

5. If you find data that you could use, construct the 95% confidence interval for the survival proportion for adults. Does the survival rate for yearlings fall within this interval?
6. Repeat (5) for a 99% confidence interval.

Be sure to show your calculations. Write your conclusions formally; i.e. given that... we reject (or do not reject) the null hypothesis.

Exercise 12.8. The data for this exercise are from Gholz et al. (1991). The authors' hypothesis was that fertilization with nitrogen increases leaf area. Prior to the experiment, the authors assigned fertilization or control (no fertilization) to randomly selected plots. They had to make sure that the number of trees per plot (of size 1 ha) were about equal. So they determined the following:

```
Trees per ha in fertilized plots:
1024, 1216, 1312, 1280, 1216, 1312, 992, 1120

Trees per ha in unfertilized plots:
1104, 1072, 1088, 1328, 1376, 1280, 1120, 1200
```

Do you believe the authors' statement that the number of trees per plot were approximately equal before the beginning of the experiment at $\alpha = 0.01$?

Exercise 12.9. The data for this exercise are from Frelich and Lorimer (1991). The authors claim that spreading fires do more damage to hardwood than do spot fires. Damage was measured by the percent of trees scarred by fires. The authors used a t-test to justify their claim. This means that they assumed that the data came from a normal distribution. Here are the data:

```
Spreading fires:
21.0, 26.7, 9.2, 6.7, 29.2, 26.7, 6.7, 8.3, 18.4, 4.9
Spot fires:
1.6, 4.6, 1.1, 1.2, 21.1, 11.9, 1.8, 4.7, 7.4
```

1. Based on normal probability plots, do you agree with the authors that the data came from a normal distribution?
2. Do the data conform to the assumptions of the t-test for small samples?
3. Assume that they do. Then set up a null and alternative hypothesis, run the test, and draw explicit conclusions abut the authors' claim. Be sure to use the p-value in drawing conclusions.
4. Compute the confidence interval for $\alpha = 0.05$ and $H_0 : \mu = 0$ What do you conclude about the true difference between μ_1 and μ_2?

Exercise 12.10. Species diversity is known to be related to soil nutrients. Twenty-five plots were divided into two subplots. One subplot was treated with fertilizer, the other was not. By the end of the experiments, the following number of species were determined:

```
> fertilized
 [1] 10 12 10 15 12 10 12 13 13 11 15 12 10  7 14 11 11 13
    13 13 13 13 12  8 13
```

```
> not.fertilized
 [1] 13 13 11 13 14 16 13 14 13 11 13 13 13 16 15 13 13 15
    15 12 12 14 15 13 15
```

Run the appropriate test. Was fertilization associated with change in species number?

Exercise 12.11. Brown-headed cowbirds are known as nest parasites. They leave their eggs in other species nests, where the eggs are "adapted" by the nest owners. To determine if nest parasitism by cowbirds is different in a prairie habitat compared to a forested habitat, 25 nests were selected for observation in a prairie habitat. Of these, 12 were parasitized. In the forest habitat, of 22 nests, 8 were parasitized. Run the appropriate test. What is your conclusion?

Exercise 12.12. A biologist is interested in establishing differences in the fitness of a population of elk in two different habitats, A and B. In one year, in habitat A he counts 12 births for 25 animals (per 100 Ha). In the same year, he counts 8 births for 25 animals (per 100 Ha). Are the birth rates (a measure of fitness) different?

Exercise 12.13. Answer the following briefly:

1. What are the conditions under which the rank sum test for differences between two means from two independent samples is performed?
2. What are the conditions under which the rank sum test for differences between two medians from two independent samples is performed?
3. What are the conditions under which the signed rank test for differences between two means from paired samples is performed?
4. Given the choice, which test would you prefer for testing the difference of means from two samples from symmetric population distributions with equal variance: t-test or rank sum? Why?
5. What are the conditions under which the t-test is performed for the difference between the means from two samples?
6. What are the conditions under which the paired t-test is performed?
7. Given the choice, which one would you prefer, t-test or paired t-test? Why?

Exercise 12.14. You are given the choice of the following tests of hypotheses about the difference of two samples means: Z, t, paired t-test, rank sum, signed rank. Rank the tests from the least to the most specific in terms of the assumptions about the underlying sampling distributions. Explain your choice of ranking.

Exercise 12.15. Refer to Exercise 12.9.

1. Run a formal test of normality on each vector. What are your conclusions with regard to normality of the data?
2. Which test would you use to examine the hypothesis that spreading fires do more damage than spot fires if you doubt the normality of the data? Run it. What are your conclusions?

Exercise 12.16. Write a function that does a two-sample test of significance for the difference between proportions. The arguments should be p_1, p_2, n_1, n_2, α and one or two-sided test.

13

Power and sample size for two samples

In this chapter, we are interested in the question of power and sample size for comparing two samples. The samples may come from populations with normal, binomial or Poisson densities and our estimates of power and sample size refer to differences between means, proportions and rates.

Let us summarize the issues involved with power and sample size. In planning a two-sample study, we must guard against two types of errors. The first is Type I error. It refers to declaring the difference in, for example, proportions significant when in fact it is not. To guard against this error, we set α to be as small as we can tolerate—usually 0.1, 0.05 or 0.01. By increasing sample size, we can also achieve the desired significance, no matter how small the difference is between two proportions. So we need to specify a difference as large as we deem detectable. The second is Type II error. Here we declare the difference between two population parameters (means, proportions or intensities) as significant while in fact it is not. So after we specify the minimum difference that is important to be detected, we need to specify the probability of detecting this difference. This probability, denoted by $1 - \beta$, determines the power of the test. Recall that β is the probability of Type II error. To compute a necessary sample size, we specify the minimum detectable difference between the parameters of interest, the desired significance and the desired power.

13.1 Two means from normal populations

Here we discuss how to obtain the power to distinguish the difference between the means of two populations based on two samples. We shall also see how to obtain sample sizes necessary to distinguish between the means with a given difference, significance and power.

13.1.1 Power

The hypotheses to be tested are $H_0 : \mu = 0$ vs. one of the usual three alternatives for a specified α. Here $\mu := \mu_2 - \mu_1$. To obtain the power to distinguish between two means from normal populations based on two samples from these populations, we must specify a value for the alternative difference between the means, denoted by μ_A. Let

$$\text{SE} := \sqrt{\frac{S_1^2}{n_1} + \frac{S_2^2}{n_2}}$$

where n_i and S_i^2 ($i = 1, 2$) are the respective means and variances of the two samples. Denote by $P(Z < z)$ the probability that the rv Z takes on values less than z where Z is from a standard normal distribution. For the hypotheses $H_0 : \mu = 0$ vs. $H_A : \mu \neq 0$ and for a given alternative μ_A, the two-sided power is given by

$$\begin{aligned} 1 - \beta &= P\left(Z < \frac{\mu_A}{\text{SE}} - z_{1-\alpha/2}\right) + P\left(Z < -\frac{\mu_A}{\text{SE}} - z_{\alpha/2}\right) \\ &= \texttt{pnorm(mu.A/SE - qnorm(1 - alpha/2))} + \\ &\quad \texttt{pnorm(-mu.A/SE - qnorm(alpha/2))}. \end{aligned} \quad (13.1)$$

For $H_A : \mu > \mu_2 - \mu_1$, the power is given by

$$\begin{aligned} 1 - \beta &= P\left(Z < \frac{\mu_A}{\text{SE}} - z_{1-\alpha}\right) \\ &= \texttt{pnorm(mu.A/SE - qnorm(1 - alpha))} \end{aligned} \quad (13.2)$$

and for $H_A : \mu < \mu_2 - \mu_1$, the power is given by

$$\begin{aligned} 1 - \beta &= P\left(Z < -\frac{\mu_A}{\text{SE}} - z_{\alpha}\right) \\ &= \texttt{pnorm(mu.A/SE - qnorm(alpha))}. \end{aligned} \quad (13.3)$$

Example 13.1. Consider the capital punishment data first introduced in Example 2.12. To examine differences of age at sentencing between blacks and whites, we sample the data with $n_1 = 35$, $n_2 = 40$ and find that $\overline{X}_1 = 29.0$, $\overline{X}_2 = 27.6$, $\sigma_1^2 = 64.8$ and $\sigma_2^2 = 61.1$ respectively. The p-value for the difference in the means is $0.112 > 0.025$ and we do not reject the hypothesis that the mean age at sentencing is equal for whites and blacks. How powerful is our ability to distinguish between these two means if in fact the difference between the true (population) means is $|\mu| = |27.6 - 29| = 1.4$? Before answering this question, let us first examine the power profiles according to Equations (13.1), (13.2) and (13.3):

```
1   source('power-normal.R')
2
3   alpha <- 0.05 ; mu.0 <- 0 ; mu.1 <- 29 ; mu.2 <- 27.6
4   mu.A <- seq(-10, 10, length = 201) ; V.1 <- 64.8
5   V.2 <- 61.1 ; n.1 <- 35 ; n.2 <- 40 ; k <- n.2 / n.1
6
7   par(mfrow = c(1, 3))
8   alt <- c('two.sided', 'greater', 'less')
9   for (i in 1 : 3){
10      if(i == 1) ylab = 'power' else ylab = ''
11      p <- power.normal(mu.A = mu.A, mu.0 = mu.0, n.1 = n.1,
```

```
      n.2 = n.2, S.1 = sqrt(V.1), S.2 = sqrt(V.2),
      alt = alt[i])
  plot(p$pwr, xlab = expression(mu[A]),
      ylab = ylab, type = 'l', main = alt[i])
}
```

Let us explain the code. In line 1, we execute a script in which the function power.normal() resides. The code for this function resides in power-normal.R at the book website. In lines 3–5 we specify the data. We set $\mu_0 := \mu_2 - \mu_1 = 0$. In other words, we are looking for the power to distinguish between two populations' means under the null hypothesis that the means are not different at $\alpha = 0.05$. In line 4, we set the alternative difference between the population means as a vector (we wish to examine the power profile). In lines 4 and 5 we specify the variances, sample sizes and the ratio of the sample sizes. Because we wish to plot the profiles for the three H_A, we open a window ready to accept a matrix of plots with one row and three columns (line 7). We store in alt the three power types we wish to examine and plot.

In lines 9–16 we call power.normal() and plot the results. The function can calculate power for a single or two samples from a normal density. For two samples, the function requires the arguments as shown. Here S.1 and S.2 denote the standard deviations of the samples and alt denotes the alternative for which we wish to determine the power. The resulting power profiles are shown in Figure 13.1. To obtain the power for $|\mu_A| = 1.4$ under H_0: $\mu = 0$, we simply call

```
> mu.A <- 1.4
> p <- power.normal(mu.A, mu.0, n.1, n.2, S.1 = sqrt(V.1),
+       S.2 = sqrt(V.2), alt = 'two.sided')
> p$pwr
   mu.A      power
1   1.4  0.1186401
```

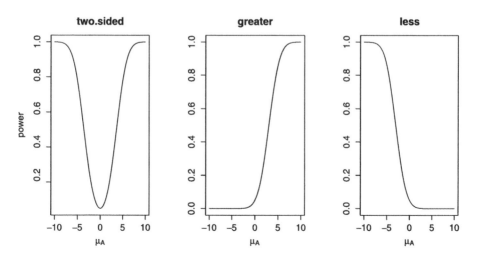

Figure 13.1   Power profiles for distinguishing between the age of sentencing to death of blacks and white inmates in the U.S.

Thus, our ability to distinguish a difference of 1.4 years between the mean ages at sentencing to death of whites and blacks is negligible; i.e. $1 - \beta \approx 0.189$. □

13.1.2 Sample size

To compare two means of two samples from normal populations with

$$H_0 : \mu = \mu_0 \quad \text{vs.} \quad H_A : \mu \neq \mu_0$$

with significance level α and power $1 - \beta$ we need to specify the smallest detectable difference. Recall that $\mu := \mu_2 - \mu_1$. Also recall that because we are dealing with large samples, we use $\sigma_1 \approx S_1$ and $\sigma_2 \approx S_2$ where S_1 and S_2 are the sample-based standard deviations of X_1 and X_2. If we have no idea about the population standard deviations, we may use the range of the data divided by 4 to estimate the variance and then the standard deviations. Let n be the sample size of each of the two samples. Then, for the two-tailed estimate, the smallest sample size we need is

$$n = \frac{\left(\sigma_1^2 + \sigma_2^2\right)\left(z_{1-\alpha/2} + z_{1-\beta}\right)^2}{\mu^2}.$$

Often, because of cost or other concerns, we anticipate that n_2 will be larger than n_1 by a factor k; i.e. $n_2 = k \times n_1$. In such cases, we estimate the needed sample size with

$$n_1 = \frac{\left(\sigma_1^2 + \sigma_2^2/k\right)\left(z_{1-\alpha/2} + z_{1-\beta}\right)^2}{\mu^2},$$

$$n_2 = \frac{\left(k\sigma_1^2 + \sigma_2^2\right)\left(z_{1-\alpha/2} + z_{1-\beta}\right)^2}{\mu^2}.$$

For one-tailed estimates, replace $z_{1-\alpha/2}$ above with $z_{1-\alpha}$.

Example 13.2. We continue with the capital punishment data (Example 12.1). From the samples we had, we specify $\sigma_1^2 \approx 64.8$ for whites' age at sentencing to death and $\sigma_2^2 \approx 61.1$ for blacks. We wish to calculate the sample sizes that we need to obtain a significant difference at $\alpha = 0.05$ with $1 - \beta$ between 0.6 and 0.9 for detectable differences between -5 and 5 years of age. The following script accomplishes the task.

```
alpha <- 0.05 ; mu.0 <- 0 ; V.1 <- 64.8 ; V.2 <- 61.1
mu <- c(seq(-8, 8, length = 161))
pwr <- seq(0.6, 0.9, length = 30)

s <- sample.size.normal(mu, S.1 = sqrt(V.1),
  S.2 = sqrt(V.2), power = pwr, alt = alt[i])

s$size[s$size$mu > -2 & s$size$mu < 2, 3 : 4] <- NA
sm <- matrix(s$size$n.1, ncol = length(mu),
  nrow = length(pwr), byrow = TRUE)

```

```
12  persp(pwr, mu, sm, theta = 30, phi = 30, expand = 0.5,
13    col = "gray90", ticktype = 'detailed', shade = 0.2,
14    xlab = 'power', ylab = 'difference', zlab = 'sample')
```

The function that computes sample size for one or two samples is called sample.size.normal(). It is available from the book website in the link for the file sample-size-normal.R. In lines 1 to 3 we prepare the data. In lines 5 and 6 we call the function with both power and μ vectors (their values are set in lines 2 and 3). sample.size.normal() returns a list with the data and the output. The latter is stored in the list as a data frame. Here are some of its lines:

```
> head(s$size)
    mu power n.1 n.2
1 -8.0   0.6   8   8
2 -7.9   0.6   8   8
3 -7.8   0.6   8   8
4 -7.7   0.6   8   8
5 -7.6   0.6   8   8
6 -7.5   0.6   9   9
```

In line 8 we remove values of n_1 and n_2 from the results for those absolute values of mu that are too close to zero because the sample sizes for these values are either too large, or because a detectable difference between ± 2 years is not important.

To prepare the function output for a 3D plot, we create a matrix from the values of n_1. This matrix has as many columns as the length of the vector mu and as many rows as the length of pwr (the former contains the values of the differences between μ_2 and μ_1 and the latter the values of the power).

Finally, in lines 12 to 14, we call the R function persp() (see Figure 13.2). To obtain the sample size for $\alpha = 0.05$, for a detectable age difference of 2.5 years between the mean ages of blacks and whites at the time of sentencing and for $1 - \beta = 0.8$, we first set the condition for extraction of the results from the s$size data frame:

```
> condition <- round(s$size$mu,2) == 2.50 &
+     round(s$size$power,2) == 0.80
```

and then

```
> s$size[condition, ]
      mu   power    n.1 n.2
3165 2.5 0.7965517 124 124
```

In other words, to detect the desired difference in age with the desired power, we need a sample of 124 whites and blacks. Let us see if this indeed is the case. In sampling the data from the population, we set.seed() to 10 and $n_1 = n_2 = 124$. This gives

```
            whites  blacks
mean          32.7    27.9
variance      98.1    51.5
sample size  124.0   124.0
```

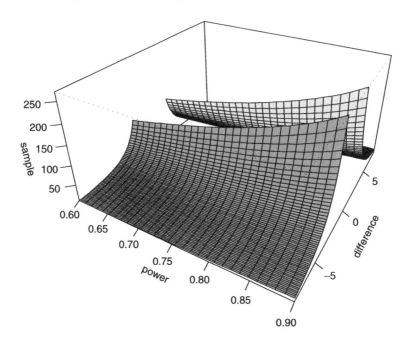

Figure 13.2   Sample size for combinations of power and $\mu := \mu_2 - \mu_1$.

and a p-value of

```
> 1-pnorm(X.bar.1 - X.bar.2, 0, SE)
[1] 7.488974e-06
```

We thus conclude that if 2.5 years of age-difference (of blacks and whites at the time of sentencing to death) indicates, for example, prejudice against blacks (in sentencing young to death), then a sample of 124 will suffice to detect this difference. □

Example 13.2 illustrates an extremely important point. One of the most frequent criticisms of the abuse of statistics is this: You can always establish a significant difference if you use a large enough sample. We know by now that this criticism is valid because the standard deviation of the sampling density (the standard error) decreases as the sample size increases. So if you have a large enough population, you can always establish a significant difference by increasing your sample size (recall our bigot in Example 10.14). *However*, if common sense dictates that the smallest detectable difference $\mu := \mu_2 - \mu_1$ makes sense, then you can calculate the sample size needed to detect this difference and thus avoid abusing statistics. In the case of our capital punishment example, we decide (for whatever reason) that a detectable difference in mean age of sentencing of at least 2.5 years between blacks and white may be practically important. Thus, any sample larger than 124 will amount to "forcing the issue."

13.2 Two proportions

Here, we follow the same sequence as we did in Section 13.1. Unlike the power obtained from comparing two means, we usually do not have repeated experiments. That is, we must distinguish between n_i representing repetitions (as in Section 13.1) and between two experiments, one with n_1 trials and n_{1S} successes and the other with n_2 trials and n_{2S} success.

13.2.1 Power

We are interested in the power to distinguish between proportions from two populations. To obtain it, we must specify the level of significance, the difference between π_1 and π_2 that is important to detect and whether we are testing for one- as opposed to two-tailed hypotheses.

The hypotheses to be tested are $H_0 : \pi = 0$ vs. one of the usual three alternatives for a specified α. Here $\pi := \pi_2 - \pi_1$. As usual, π_1 and π_2 are the probabilities of success in the respective populations. These are estimated with $\pi_i \approx p_i = n_{iS} / n_i$, $i = 1, 2$.

Under the assumption that $\pi_1 = \pi_2$, we have that the mean of π, denoted by $\overline{\pi}$ and its standard error, $\overline{SE}$, are

$$\overline{\pi} = \frac{n_1 \pi_1 + n_2 \pi_2}{n_1 + n_2}, \quad \overline{SE} = \sqrt{\overline{\pi}(1 - \overline{\pi})\left(\frac{1}{n_1} + \frac{1}{n_2}\right)}.$$

We use the samples' proportions of success, p_1 and p_2, to estimate π_1 and π_2. Consequently, the standard error of the sampling distribution of $\pi_2 - \pi_1$ is given by

$$SE \approx \sqrt{\frac{p_1(1 - p_1)}{n_1} + \frac{p_2(1 - p_2)}{n_2}}.$$

For the two-sided power (i.e. $H_A : \pi \neq \pi_2 - \pi_1$), the power is given by

$$1 - \beta = 1 - P\left(Z < \frac{z_{1-\alpha/2} \overline{SE} - |\pi|}{SE}\right) + P\left(Z < \frac{-z_{1-\alpha/2} \overline{SE} - |\pi|}{SE}\right)$$

where $P(Z < z)$ is the probability (area) under the standard normal density that $Z < z$. For $H_A : \pi_2 > \pi_1$, the one-sided "greater than" power is given by

$$1 - P\left(Z < \frac{z_{1-\alpha} \overline{SE} - |\pi|}{SE}\right)$$

and for the "less than" $H_A : \pi_2 < \pi_1$, the power is given by

$$P\left(Z < \frac{-z_{1-\alpha} \overline{SE} - |\pi|}{SE}\right).$$

Example 13.3. Two groups of 40 patients each were selected for a study of the effectiveness of flu shots. Members of the treatment group received a flu shot. Members of the control group received a saline shot. The medical history of both groups was followed for the duration of the flu season. Of the control group, 15 suffered from flu symptoms at least once. Of the treatment group, 10 did. We wish to answer the following:

1. Was the treatment effective?
2. If not, what is the probability that we accept the hypothesis that the treatment was not effective in preventing flu while in fact it was (i.e. type II error, β)?

3. What should have been the number of people in the treatment group that did not suffer from flu symptoms for a power of 0.8; i.e. for a power that will guarantee a small (0.2) type II error?

To answer these questions, we first set the notation:

Treatment: $n_1 = 40$, $n_{1S} = 10$, $p_1 = n_{1S}/n_1 = 0.25$.
Control: $n_2 = 40$, $n_{2S} = 15$, $p_2 = n_{2S}/n_2 = 0.375$.
Hypotheses: $H_0: \pi_1 = \pi_2$, $H_A: \pi_1 < \pi_2$, $k = n_2/n_1 = 1$,
$\pi_1 \approx p_1$, $\pi_2 \approx p_2$, $\bar{p} = \dfrac{n_{1S} + n_{2S}}{n_1 + n_2} = 0.3125$,
$\bar{\pi} \approx \bar{p}$.

Regarding the first question, we have

```
> prop.test(c(10, 15), c(40, 40), alternative = 'g')

        2-sample test for equality of proportions
        with continuity correction

data:  c(10, 15) out of c(40, 40)
X-squared = 0.9309, df = 1, p-value = 0.8327
alternative hypothesis: greater
95 percent confidence interval:
 -0.3189231  1.0000000
sample estimates:
prop 1 prop 2
 0.250  0.375
```

and we do not reject the null hypothesis. Therefore, we conclude that flu shots were not effective.

To answer the second question, we set the data and call bp() (for binomial power), available in bp.R, at the book's site, with a one sided test:

```
> source('bp.R')
> n <- c(40, 40) ;   n.S <- c(10, 15) ; p <- n.S / n
> Power <- bp(p[1], p[2], n1 = n[1], n2 = n[2],
+    alt = 'greater')
> print(c(beta = 1 - as.vector(Power)))
     beta
0.6710638
```

Therefore, the type II error is approximately 0.671. In other words, the probability that we accept the hypothesis that the treatment was not effective in preventing flu while in fact it was is 0.671—not a good state of affairs because we may deny effective treatment.

To answer the third question, we do:

```
> pi.A <- seq(0, p[2], length = 201)
> Power <- bp(pi.A, p[2], n1 = n[1], n2 = n[2],
+    alt = 'greater')
> plot(pi.A, Power, xlab = expression(pi[A]), type = 'l')
```

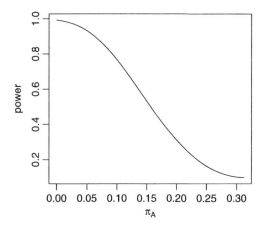

Figure 13.3   One sided power profile for $\pi_2 = 0.375 > \pi_A$ between 0 and π_2.

(see Figure 13.3). Thus we find

```
> c(pi.A = pi.A[72], bp(pi.A[72], p[2], n1 = n[1],
+     n2 = n[2], alt = 'greater'))
     pi.A      Power
0.1331250 0.8090285
> floor(pi.A[72] * n[1])
[1] 5
```

In other words, in the current experiment, we needed no more than five people from the experiment group contracting the flu to obtain a power of approximately 0.8. Such power presents a balance between the probability of denying effective treatment (0.2) and the probability of providing flu shots while they are not effective (0.05). Under such conditions, it might be reasonable to select $\alpha = 0.1$ for then we will decrease the probability of denying effective treatment. □

13.2.2   Sample size

Here we are interested in determining the sample size needed to distinguish between two proportions with a particular power and level of significance.

Let $\rho := n_2 / n_1$. Under the null ($\pi_2 = \pi_1$) and alternative ($\pi_2 \neq \pi_1$) hypotheses, we first obtain the pooled proportion

$$\bar{\pi} := \frac{(\pi_1 + \rho\pi_2)}{1 + \rho}.$$

Next, the standard deviations under the null and under the alternative, where for the alternative we specify $\pi_A := |\pi_2 - \pi_1|$, are

$$\sigma_0 = \sqrt{\bar{\pi}(1 - \bar{\pi})\left(1 + \frac{1}{\rho}\right)},$$

$$\sigma_A = \sqrt{\pi_1(1 - \pi_1) + \frac{\pi_2(1 - \pi_2)}{\rho}}.$$

Then, the two-sided sample size is obtained from

$$n' = \left[\frac{z_{1-\alpha/2} \times \sigma_0 + z_{1-\beta} \times \sigma_A}{\pi_A}\right]^2,$$

$n_2 = $ largest integer closest to $\rho \times n'$,

$n_1 = $ largest integer closest to n'/ρ.

For a one-sided test, use $z_{1-\alpha}$.

Often, cost and other considerations dictate that the sample sizes should be different. This can be achieved by using appropriate values of ρ.

Example 13.4. Continuing with Example 13.3, we wish to determine the sample sizes that are necessary to establish a difference of $0.375 - 0.25$ between the proportion that got sick in the control and treatment groups. We use the standard values of $\alpha = 0.05$ and $1 - \beta = 0.8$ and the same fraction of the total sample allocated to both groups. Then

```
> library(Hmisc)
> ceiling(bsamsize(p.1, p.2))
 n1  n2
435 435
```

In other words, we need 870 people to achieve the desired significance. Suppose that it is twice as expensive to follow members of the treatment group compared to the control group e.g. following a member of the treatment group costs $100 and following a member of the control group costs $50. Then, our desired fraction of allocation to the treatment group is $1/3$ and

```
> ceiling(bsamsize(p.1, p.2, fraction = 1/3))
 n1  n2
312 624
```

The cost for the treatment group is $312 \times \$100 = \$31\,200$. The cost for the control group is $624 \times \$50 = \$31\,200$ for a total cost of $62\,400$. Here we need more people (936) compared to equal sample sizes (870). We may wish to investigate the possibility of allocating the 936 people to both groups in a way that will maximize the power we can achieve. Then

```
> ba <- ballocation(p.1, p.2, 936)
> as.vector(c(936 * ba[4], ba[4]))
[1] 442.2857658   0.4725275
```

Thus, we conclude that instead of allocating 312, we may allocate 443 (of the 936) to the treatment. This will maximize the power we expect to achieve at a cost of $68\,900$ (compared to $62\,450$ when no power-maximizing is considered). □

13.3 Two rates

Let t' denote the time from the occurrence of the last event. Denote by $P(X < 1|t')$ the probability that no event occurred by t'. As t' increases, this probability decreases because the more time passes since the time of last event, the smaller the probability that the event does not occur. It can be shown that if X is Poisson with λ, then

$$P(X < 1|t') = e^{-\lambda t'}.$$

Therefore,
$$P(X \geq 1|t') = 1 - e^{-\lambda t'}.$$

For a sample of size n, the expected number of events is then
$$m := nP(X \geq 1|t') = n\left(1 - e^{-\lambda t'}\right).$$

For two Poisson populations we have $\lambda_1, \lambda_2, t'_1, t'_2, n_1, n_2, m_1$ and m_2. Denote by π the probability of an event from n_1. Suppose we observe n_1 for t_1 time units and n_2 for t_2 time units. Then

$$T_1 = n_1 t_1, \quad T_2 = n_2 t_2, \quad \pi = \frac{\lambda_1 T_1}{\lambda_1 T_1 + \lambda_2 T_2}. \tag{13.4}$$

Events are independent. Therefore, the number of events from n_1 is binomial with parameters π and $m_1 + m_2$. During the time we follow subjects (t'_1 and t'_1), we expect that
$$m = m_1 + m_2 = n_1\left(1 - e^{-\lambda_1 t'_1}\right) + n_2\left(1 - e^{-\lambda_2 t'_2}\right)$$
events will occur.

To proceed, we define $\rho := \lambda_1/\lambda_2$. Then dividing the numerator and the denominator of the expression for π in (13.4) by λ_2, we obtain

$$\pi = \frac{\lambda_1 T_1}{\lambda_1 T_1 + \lambda_2 T_2}$$
$$= \frac{\lambda_1 T_1/\lambda_2}{\lambda_1 T_1/\lambda_2 + \lambda_2 T_2/\lambda_2}$$
$$= \frac{T_1 \rho}{t_1 \rho + T_2}.$$

We wish to test the hypothesis that the rates λ_1 and λ_2 are equal. So we set
$$H_0 : \rho = 1 \quad \text{vs.} \quad \rho > 1$$
which is equivalent to
$$H_0 : \pi = \frac{T_1}{T_1 + T_2} \quad \text{vs.} \quad \pi > \frac{T_1}{T_1 + T_2}.$$

To simplify the notation, we let $\pi_0 := T_1/(T_1 + T_2)$ and $\pi_A > \pi_0$, where π_A is specified. So equivalent to $H_0 : \rho = 1$ we have $H_0 : \pi = \pi_0$, with the alternative specified. Thus, similar to the development in Section 11.1.2, for $\pi_A > \pi_0$, we obtain

$$\text{power} = P\left(Z \leq \frac{(\pi_A - \pi_0)\sqrt{m} - z_{1-\alpha}\sqrt{V_0}}{\sqrt{V_A}}\right) \tag{13.5}$$

where
$$V_0 := \pi_0(1 - \pi_0) \quad \text{and} \quad V_A := \pi_A(1 - \pi_A).$$

For $\pi_A < \pi_0$, we use

$$\text{power} = P\left(Z \le \frac{(\pi_0 - \pi_A)\sqrt{m} - z_{1-\alpha}\sqrt{V_0}}{\sqrt{V_A}}\right). \qquad (13.6)$$

To obtain two-sided power, replace $z_{1-\alpha}$ by $z_{1-\alpha/2}$ and sum the right hand sides of equations (13.5) and (13.6). To verify that our computations are correct, we use Example 14.15 in (Rosner, 2000, page 692).

Example 13.5. The incidence rate of a genetic mutation in population 1 is 375 per 100 000 in one year. In population 2 it is 300 per 100 000 in one year. We take a sample of 5 000 from each population. What is the power of distinguishing between $\lambda_1 = 375 \times 10^{-5}$ and $\lambda_2 = 300 \times 10^{-5}$ at $\alpha = 0.05$?

The expected number of incidences in $t' = 5$ years are

$$m_1 = 5000 \times \left(1 - e^{-375/100\,000 \times 5}\right) = 92.877,$$

$$m_2 = 5000 \times \left(1 - e^{-300/100\,000 \times 5}\right) = 74.44$$

and $m = m_1 + m_2 = 167.32$. Also

$$T_1 = T_2 = 5 \times 5\,000 = 25\,000.$$

Therefore,

$$\pi_0 = \frac{T_1}{T_1 + T_2} = 0.5,$$

$$\pi_A = \frac{25\,000 \times \dfrac{375}{300}}{25\,000 \times \dfrac{375}{300} + 25\,000} = 0.556.$$

We wish to test

$$H_0 : \rho = 1 \quad \text{vs.} \quad H_A : \rho \ne 1.$$

This is equivalent to testing

$$H_0 : \pi = \pi_0 \quad \text{vs.} \quad \pi \ne \pi_0.$$

To determine the power, we use π_A. Here

$$V_0 = 0.5\,(1 - 0.5) = 0.25 \quad \text{and} \quad V_A = 0.556\,(1 - 0.556) = 0.247.$$

Therefore,

$$\begin{aligned} z_1 &:= \frac{(\pi_A - \pi_0)\sqrt{m} - z_{1-\alpha/2}\sqrt{V_0}}{\sqrt{V_A}} \\ &= \frac{(0.556 - 0.5)\sqrt{167.32} - 1.96\sqrt{0.25}}{\sqrt{0.247}} \\ &= -0.526 \end{aligned}$$

and
$$z_2 := \frac{(0.5 - 0.556)\sqrt{167.32} - 1.96\sqrt{0.25}}{\sqrt{0.247}}$$
$$= -3.418.$$

Thus, we obtain
$$P(Z < z_1) + P(Z < z_2) = 0.3.$$

We will not reject a wrong null hypothesis in about 30% of the cases. Not a very good power.

Here is the code for a function that computes two-sample power for the Poisson:

```
1  Poisson.power <- function(t, n, l, alpha = 0.05){
2    q <- qnorm(1 - alpha / 2)
3    T <- t * n
4    rho <- l[1] / l[2]
5    p0 <- T[1] / sum(T) ; v0 <- p0 * (1 - p0)
6    pa <- T[1] * rho / (T[1] * rho + T[2])
7    va <- pa * (1 - pa)
8    m <- sum(n * (1 - exp(-l * t)))
9    A <- ((pa - p0) * sqrt(m) - q * sqrt(v0)) / sqrt(va)
10   B <- ((p0 - pa) * sqrt(m) - q * sqrt(v0)) / sqrt(va)
11   pnorm(A) + pnorm(B)
12 }
```

The function takes the following arguments (except for alpha, all vectors are of size 2):

> t Time period for each sample.
> n Size of each sample.
> l λ_1 and λ_2.
> alpha Significance level α (default value = 0.05).

The function returns the two-sided power $(1 - \beta)$ for a given α. Let us follow the code for the function. In line 2 we obtain the quantile for the appropriate value of α. In line 3, we obtain the values of subject-time for each of the sample. In our example, we have mutation-years. When the "rate" is not with respect to time, the latter represents the number of repetitions of counts for each subject. We then compute the rate ratio in line 4. In lines 5 and 6 we calculate the probability under the null and the alternative hypotheses, respectively. The variances of each sample are calculated in lines 5 and 7. The expected number of incidences (mutations in our example) are calculated in line 8. Lines 9 and 10 calculate the quantiles given in equations (13.5) and (13.6). We need both quantiles because Poisson.power() returns a two-sided power. Line 11 returns the power. In Exercise 13.5 you are asked to generalize the function for one-sided power (greater than and less than). □

Let us discuss the sample size m that will give us a desired power. Rearranging (13.5) and (13.6) for a two-sided test, we obtain

$$m = \left(\frac{z_{1-\alpha/2}\sqrt{V_0} + z_{1-\beta}\sqrt{V_A}}{|\pi_0 - \pi_A|}\right)^2 \quad (13.7)$$

where m is the expected number of events in both populations. Let $k := n_2/n_1$. Then if we specify k, we get the necessary sample sizes for each population from

$$n_1 = \frac{m}{k + 1 - e^{-\lambda_1 t_1'} - k e^{-\lambda_2 t_2'}}, \quad (13.8)$$

$$n_2 = k n_1 \,.$$

Example 13.6. Continuing with Example 13.5, we ask: How many subjects do we need to follow for 5 years to obtain 80% power at significance of 0.05 for a two-tailed test and equal numbers from both populations?

Using (13.7) we write

$$m = \left(\frac{1.96\sqrt{0.5(1-0.5)} + 0.84\sqrt{0.556(1-0.556)}}{|0.5 - 0.556|}\right)^2 = 633.40 \,.$$

So we need to choose n_1 and n_2 such that we anticipate 634 events to occur. From (13.8),

$$n_1 = n_2 = \frac{634}{2 - e^{-375/100\,000 \times 5} - e^{-300/100\,000 \times 5}} = 18\,928.05 \,.$$

We therefore need to follow 18 929 subjects from each population for 5 years.

Here is a function that computes Poisson sample size for two samples:

```
Poisson.sample.size <- function(t, n, e,
  rho = (e[1] / n[1]) / (e[2] / n[2]), alpha = 0.05,
  power = 0.8, k = 1)
{
  q <- qnorm(1 - alpha / 2)
  p <- qnorm(power)
  p0 <- t[1] / sum(t) ; v0 <- p0 * (1 - p0)
  pa <- t[1] * rho / (t[1] * rho + t[2])
  va <- pa * (1 - pa)
  m <- (q * sqrt(v0) + p * sqrt(va)) / (abs(p0 - pa))
  m <- ceiling(m * m)
  d <- k + 1 - exp(-e[1] / n[1] * t[1]) -
    k * exp(-e[2] / n[2] * t[2])
  n1 <- m / d; n2 <- k * n1
  ceiling(c(n1, n2))
}
```

The function computes the sizes of two samples from Poission populations that are necessary to achieve a given power for a given significance level and for a given ratio of the sample sizes. The function takes the following arguments:

> t Time period for each sample.
> n Size of each sample from past data.
> e Incidence count for each sample from past data.
> rho The desired ratio of λ_1 to λ_2. If not provided,
>   the ratio is computed from e and n.
> alpha The desired significance level (default value = 0.05).
> power The desired power (default value = 0.8).
> k The desired ratio n_1 / n_2.

In lines 5 and 6 we compute the quantiles for α and $1 - \beta$. In lines 7 to 9 we compute π_0 and π_A under the null and alternative hypotheses and their variances. The required number of incidences is computed in line 10 (see equation 13.7). To obtain n_1, we first compute the denominator in equation (13.8). In line 14 we compute the necessary n_1 and n_2. □

13.4 Assignments

Exercise 13.1. Download the file walleye.rda from the book's site. It contains the following list of walleye weights from two lakes:

```
$sample.1
 [1] 0.86 1.38 1.43 1.38 1.58 0.62 1.74 2.04 1.37 1.72 2.62
[12] 1.52 2.13 1.27 0.95 1.54 2.30 1.40 1.31 1.85 1.19 1.96
[23] 1.17 1.00 1.25 1.06 2.65 1.28 1.38 0.49 1.54 1.95 1.78
[34] 0.54 1.64 1.85 1.13 1.60 0.40 1.32

$sample.2
 [1] 0.81 2.46 2.05 1.11 1.31 0.97 1.04 1.61 2.09 1.63 1.48
[12] 1.69 1.78 1.89 2.03 1.27 2.34 1.90 2.18 1.59 1.84 1.95
[23] 1.67 1.66 1.78 2.34 1.50 2.02 1.04 1.83 1.14 0.83 1.69
[34] 1.68 2.15 2.40 1.56 1.73 0.65 1.76 2.26 1.23 2.62 1.27
[45] 2.83
```

1. Create power profiles for the difference between the weights under the assumption that $\mu_1 - \mu_2 = 0$ for $\mu_A < \mu_0$, $\mu_A > \mu_0$ and $\mu_A \neq \mu_0$. Set the range of μ_A from -1 to 1.
2. What is the power of distinguishing between weights of the two samples at $\alpha = 0.05$ and a minimum detectable difference of 0.2 kg for $\mu_A < \mu_0$, $\mu_A > \mu_0$ and $\mu_A \neq \mu_0$?

Use Example 13.1 as a guide.

Exercise 13.2. Continuing with the walleye.rda (Exercise 13.1), assume that the samples' variances approximately equal the population variances. Set $\alpha = 0.05$, power between 0.6 and 0.9 and detectable difference between -1 and 1 kg. With these:

1. Draw and interpret a figure for these data similar Figure 13.2.
2. What would be the sample size necessary to detect a difference of 0.2 kg with power = 0.9?

Exercise 13.3. Two separate populations of deer were chosen for a study of the effect of reducing winter mortality due to supplemental feeding. The first population included 38 deer and the second 42. Habitats in the two areas where the populations reside were comparable and so was the weather. The averages of the population weight at the beginning of the winter were not different. The first population received supplemental feeding, the second did not. By the end of the winter, 9 and 12 deer died from starvation in the first and second populations, respectively.

1. Was the feeding effective in reducing winter mortality?
2. What is the probability that we accept the hypothesis that the supplemental feeding was not effective in reducing mortality while in fact it was?
3. What should have been the number of deer in the winter-fed population that survived for a power of 0.8 (with $\alpha = 0.05$?

Exercise 13.4. Continuing with Exercise 13.3, determine the population sizes that are necessary to establish a difference of 0.1 in the winter mortality between the fed and unfed deer populations. Use $\alpha = 0.05$ and $1 - \beta = 0.8$. Assign the same fraction of the total number of deer to the fed and unfed populations.

Exercise 13.5. Write a function that returns the one-sided (less than or greater than) or two-sided power of a test of the difference between λ_1 and λ_2. Use the code for `Poisson.power()` as a guide.

14
Simple linear regression

So far, we were mostly concerned with the following question: What is the density of some trait in a population from which we have samples? The answer to this question boils down to estimating and comparing parameters of some density. For example, in Chapter 9, we learned to estimate the mean of a population from the normal density and in Chapter 12 we learned how to compare samples from two populations.

Here we are concerned with the following question: Given a sample for which we have a pair of (random) values obtained for each object, say X_i and Y_i, is there a relationship between these pairs of values? More specifically, we are interested in the linear relationship

$$y = \beta_0 + \beta_1 x \tag{14.1}$$

where β_0 and β_1 are coefficients. For example, we may ask: Are there relationships between the number of cigarettes people smoke and the incidence of lung cancer? What can we say about the relations between the age of a tree and its height? Does infant mortality increase with lower per capita income? Can we say that more years of education are associated with longer life expectancy?

14.1 Simple linear models

Simple linear models refer to linear functions that describe the relationship between two variables, y and x. Linear functions are generally defined as

Linear function A function f is linear if and only if

$$f(\alpha x) = \alpha f(x), \quad f(x+y) = f(x) + f(y)$$

for α constant.

We will use a special case of linear functions, $f(x) = \beta_0 + \beta_1 x$ where β_0 and β_1 are constants. We call such functions *simple linear*.

14.1.1 The regression line

Equation (14.1) is deterministic—if we know the value of x, then we know exactly the value of y. When additive random effects are involved, we have

$$Y_i = \beta_0 + \beta_1 X_i + \varepsilon_i \tag{14.2}$$

where i denotes a specific pair of values (Y_i, X_i) for the ith object in the population and ε_i is the so-called *error term*. Now if the density of ε is normal with $\mu = 0$ and σ_ε, then

$$\begin{aligned} E[Y_i|X_i] &= E[\beta_0 + \beta_1 X_i + \varepsilon] \\ &= \beta_0 + \beta_1 E[X_i] + E[\varepsilon] \\ &= \beta_0 + \beta_1 X_i \,. \end{aligned} \tag{14.3}$$

The last equality holds because $E[\varepsilon] = 0$ and $E[X_i] = X_i$. Equation (14.3) is referred to as the *regression line*. The coefficient β_0 is its *intercept* and β_1 is its *slope*. Y is said to be the *dependent variable*, (also known as the variate or the response) and X is the *independent variable* (also known as the explanatory variable, the covariate, the treatment or the stimulus). These terms should *not* be interpreted as necessarily cause and effect. To be consistent with traditional mathematical definitions, we shall call Y the dependent variable and X the independent variable.

Example 14.1. The following data were obtained from http://www.cdc.gov/nchs/about/major/nhanes/nhanes2005-2006/exam05_06.htm. It gives measurements of various body parts for 9 950 individuals from a 2005–2006 survey of U.S. adults. The file is in a SAS export format (xpt). To import both it and the variable descriptions, we

```
> bm <- read.xport('body-measurements.xpt')
> bmv <- read.table('body-measurements-variables.txt',
+     sep = '\t', header = TRUE, skip = 2)
```

and

```
save(bm, file = 'bm.rda')
save(bmv, file = 'bmv.rda')
```

The

```
> head(bmv)
     Item.ID                          Label
1       SEQN      Respondent sequence number
2   BMDSTATS  Body Measures Component Status Code
3      BMXWT                     Weight (kg)
4      BMIWT                  Weight Comment
5   BMXRECUM            Recumbent Length (cm)
6   BMIRECUM         Recumbent Length Comment
```

reveals space characters at the end of items ID and variable labels. In the spirit of avoiding for loops like the plague, we get rid of these spaces with:

```
> trimmed <- apply(as.array(bmv[, 2]), 1, function(x)
+     if (substr(x, nchar(x), nchar(x)) == ' ')
+     strtrim(x, nchar(x) - 1) else strtrim(x, nchar(x)))
> bmv[, 2] <- unlist(trimmed)
> head(bmv, 4)
    Item.ID                           Label
1       SEQN         Respondent sequence number
2 BMDSTATS  Body Measures Component Status Code
3      BMXWT                        Weight (kg)
4      BMIWT                    Weight Comment
```

Note that: apply() wants an array and returns a list; nchar() returns the number of characters in a string; strtrim() trims the string to a specified length; unlist() collapses a list to a vector. In Exercise 14.1, you are asked to clean the potentially leading and trailing spaces from Item.ID. Let us look at variables 3, 9 and 16 in bm and label them legibly with bmv:

```
> pairs(bm[, c(3, 9, 16)],
+     labels = bmv[c(3, 9, 16), 2])
> pairs(bm[, c(3, 9, 16)],
+     labels = bmv[c(3, 9, 16), 2])
```

(Figure 14.1). The pair height and upper arm length seem to be linearly related in the sense we use here. The other relationships seem to be polynomial (probably up to power of 2). □

14.1.2 Interpretation of simple linear models

Depending on the sign and value of β_1 in (14.3), we can interpret the model in three ways: As the value of X increases, the values of Y may increase, decrease, or remain unchanged. The next example illustrates these possible relationships.

Example 14.2. The data are about quality of life indicators for various countries for the years 1999 to 2003.[1] We wish to explore the linear relationships (if any) among various economic indicators for the year 2000. We are interested in CO_2 emissions (metric tons per capita), energy use (kg of oil equivalent per capita), gross domestic product (GDP) in US$, fertility rate (births per 1000 woman-year), life expectancy at birth (years), infant mortality rate (mortality per 1000 live births-year) and mortality of children under the age of 5 (mortality per 1000 children under 5). Figure 14.2 summarizes the data (the script is shown in Exercise 14.9). There is positive relationship between CO_2 emission and GDP or energy use. Clearly, life expectancy decreases as fertility rate increases while children's mortality increases with increased infant mortality. In the first case, $\beta_1 < 0$; in second, $\beta_1 > 0$. We also observe that at fertility rate near zero, life expectancy at birth is about 80 years. When infant mortality is nearly zero, so is children's mortality. In the former $\beta_0 > 0$ and in the latter $\beta_0 \approx 0$. □

---
[1]The data were obtained from http://devdata.worldbank.org/data-query/

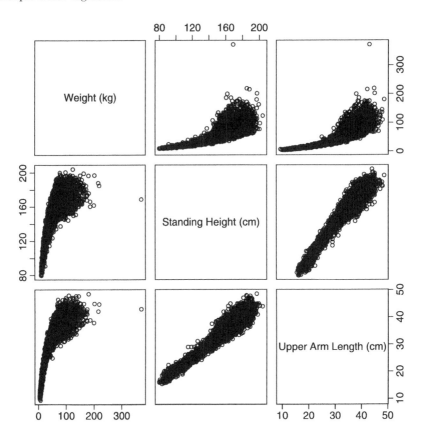

Figure 14.1   Relationships among body measurements.

How do we interpret the regression coefficients? Just like algebraic equalities, units must be identical across equalities. The pair of equalities

$$E[Y|X] = \beta_0 + \beta_1 E[X],$$
units of Y = units of β_0 + units of β_1 × units of X

must always be satisfied.

Example 14.3. We are interested in the linear relationship between height (cm) and weight (kg). Then

$$E[Y|X] = \beta_0 + \beta_1 E[X],$$
cm = units of β_0 + units of β_1 × kg .

To maintain the equality across units, the units of β_0 and β_1 must be cm and cm/kg for then we have

$$\text{cm} = \text{cm} + \frac{\text{cm}}{\text{kg}} \times \text{kg}$$
$$= \text{cm} + \text{cm}$$

and the addition of cm to cm gives cm. □

Simple linear models 421

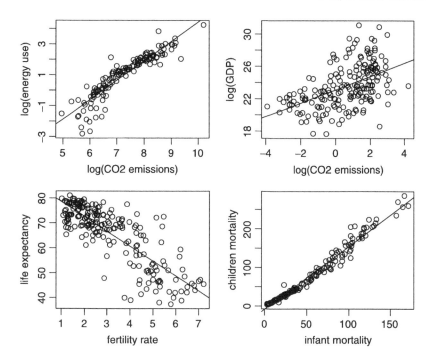

Figure 14.2 Linear relationships between pairs of economic indicators for various countries. Data are for the year 2000.

The idea of units in modeling is extremely important—it allows for correct interpretation of the model.

Example 14.4. Let us interpret the units of β_0 and β_1 in the relationships demonstrated in Figure 14.2. The bottom left shows

$$\text{life expectancy} = \beta_0 + \beta_1 \times \text{fertility rate},$$
$$\text{years} = \text{units of } \beta_0 + \text{units of } \beta_1 \times \frac{\text{births}}{\text{woman-year}}. \quad (14.4)$$

To maintain the equality over units, the units of β_0 must be in years and the units of β_1 must be in

$$\text{units of } \beta_1 = \frac{\text{years} \times \text{woman-year}}{\text{births}} = \frac{\text{years}}{\text{units of } x} = \frac{\text{years}}{\text{births/woman-year}}.$$

We then say that the units of β_1 are years per births per woman-year. □

Example 14.5. Continuing with Example 14.2, we obtain[2] for the line in the bottom left (of Figure 14.2) that

$$E[Y|X] = 84.67 - 6.00 E[X].$$

---
[2]We shall see later how.

Based on the interpretation of the units of β_0 and β_1 in (14.4), we conclude that when fertility rate is zero ($x = 0$), the expected life expectancy is $E[Y|X=0] \approx 85$ years. For each unit of increase in the fertility rate we have

$$E[Y|X+1] - E[Y|X] = 84.67 - 6.00\,(E[X]+1) - [84.67 - 6.00E[X]]$$
$$= -6.00E[X] - 6.00 + 6.00E[X]$$
$$= -6.00$$

or approximately 6 years' decrease in life expectancy for each unit increase in fertility rate. We thus conclude that to increase life expectancy, policy makers should work to decrease fertility rate. Of course, there are other factors that affect life expectancy. □

We talked about interpretation of the model coefficients. But how do we obtain values for these coefficients? We address this issue next.

14.2 Estimating regression coefficients

Based on given sample values of X and Y, we wish to estimate the best values of β_0 and β_1. Best in what sense? After obtaining the coefficient values for the regression lines, we usually use it to obtain the expected value of Y given a value of X, sometimes called the *predicted* value. Therefore, the best values of the line coefficients will be those that minimize the error ε in $Y = \beta_0 + \beta_1 X + \varepsilon$.

Example 14.6. We go back to the data in Example 14.1. To illustrate the ideas, we pick 4 subjects from the data and examine the relationship between $\log(X)$ (standing height) and $\log(Y)$ (upper arm length):

```
> Y <- bm[, 16] ; X <- bm[, 9]
> log.X <- log(X) ; log.Y <- log(Y)
> idx <- c(2174, 3499, 4779, 6309)
> log.X <- log.X[idx] ; log.Y <- log.Y[idx]
```

We will discuss in a moment how to estimate the coefficients of the regression line. For now, we need these coefficients. So we do

```
> model <- lm(log.Y ~ log.X)
```

The call to the linear model (a function named `lm()`) returns an object of class `lm`. The formula `log.Y ~ log.X` tells `lm()` that `log.Y` is the dependent variable and `log.X` the independent. We assign this object to `model`. The call to `coefficients()` with an object of class `lm` retrieves the coefficient values:

```
> round(coefficients(model), 3)
(Intercept)        log.X
     -4.694        1.659
```

We use `round()` with the argument 3 to print up to three decimal digits. Now

$$E[Y|X] = -4.694 + 1.659 E[X] \,. \tag{14.5}$$

Table 14.1   log upper arm length (Y) vs. log height (X) and the expected values of Y according to (14.5).

i	X	Y	$E[Y|X]$	ε	ε^2
1	4.742	3.367	3.174	0.193	0.037
2	4.629	2.912	2.986	−0.074	0.005
3	4.964	3.627	3.542	0.085	0.007
4	4.875	3.190	3.395	−0.204	0.042

If we do not have repetitions for specific values of X, then the best estimate of $E[X_i]$ is X_i. From Table 14.1,

$$\begin{aligned} E[Y_1|X_1] &= -4.694 + 1.659 X[1] \\ &= -4.694 + 1.659 \times 4.742 \\ &= 3.174 \; . \end{aligned}$$

Rather than compute the expected (predicted) values of each point by hand, we use predict(model). Given an object of class lm, predict() returns the expected values of Y based on the coefficients and data that are stored in the object. The errors are given by log.Y - predict(log.Y). You can also access the predicted values with model$fitted.values. To observe the errors, we

```
> plot(log.X, log.Y, xlim = c(4.6, 5.05),
+   xlab='log(height)', ylab='log(upper arm length)')
```

draw the regression line

```
> abline(reg = model)
```

and connect each expected value with a line to its corresponding log.Y value:

```
> for(i in 1:length(log.X))
+   lines(c(log.X[i], log.X[i]),
+     c(log.Y[i], model$fitted.values[i]))
```

In the call to abline(), we specify the named argument reg and assign an object of class lm to it. abline() extracts the coefficients from model and plots the line (Figure 14.3). The points and their values are drawn with

```
> points(log.X, model$fitted.values, pch = 19)
> text(log.X, model$fitted.values,
+   labels = round(model$fitted.values, 3), pos = 4)
> text(log.X, log.Y, labels=round(log.Y, 3),
+   pos = 4)
```

Finally, we add the error values with

```
> for(i in 1 : length(log.X)){
+   if(i == 2 | i == 4) pos = 4 else pos = 2
+   text(log.X[i],
```

```
+            (model$fitted.values[i] + log.Y[i]) / 2,
+          labels=bquote(epsilon[.(i)]==.(round(log.Y[i] -
+          model$fitted.values[i], 3))), pos = pos)
+   }
```

(we discussed `bquote()` following the script on page 138 and in Example 7.27). Figure 14.3 illustrates the relationship among the data, the regression line and the predicted values of Y. □

Because we wish to minimize the error for the whole data and because we are not interested in the sign of the error, but rather its magnitude, we can sum the absolute values of errors ε. However, working with absolute values is mathematically cumbersome. Instead, we work with the sum of the squares of the errors. Thus, we seek those values of β_0 and β_1 that minimize

$$\text{SSE} := \sum_{i=1}^{n} \varepsilon_i^2 \tag{14.6}$$

$$= \sum_{i=1}^{n} [Y_i - (\beta_0 + \beta_1 X_i)]^2$$

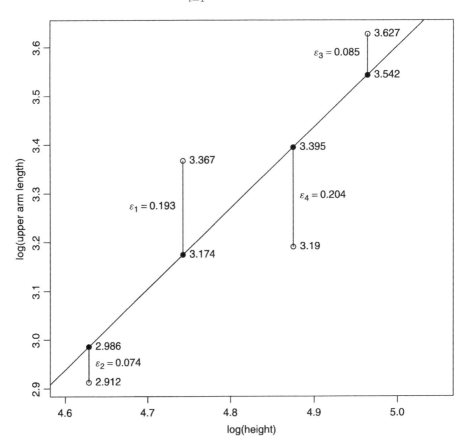

Figure 14.3   The regression line for upper arm length vs. height where $E[Y|X]$ are shown as filled circles. Here $\varepsilon_i = Y_i - E[Y_i|X_i]$. See Table 14.1.

where SSE stands for the sum of squares of the errors and X_i and Y_i are the data. We now define the

Estimated regression line is the line defined by the estimated values β_0 and β_1 such that the SSE is minimized. These estimated values are denoted by $\widehat{\beta}_0$ and $\widehat{\beta}_1$.

Because the minimizing criterion is the SSE, this line is often called the *least-squares regression line*. We interpret β_0 and β_1 as the population coefficients and $\widehat{\beta}_0$ and $\widehat{\beta}_1$ as the sample-derived coefficients. Some authors denote the population coefficients by α and β and their estimates by a and b. Associated with $\widehat{\beta}_0$ and $\widehat{\beta}_1$ are the estimated errors, defined as

The estimated ith residual ($\widehat{\varepsilon}_i$) associated with X_i is

$$\widehat{\varepsilon}_i := Y_i - (\widehat{\beta}_0 + \widehat{\beta}_1 x_i) . \tag{14.7}$$

We are left with the problem of how to obtain $\widehat{\beta}_0$ and $\widehat{\beta}_1$. One possibility is brute force—simply compute many SSE for many pairs of values for the coefficients and then choose the pair that give in the smallest SSE. A much better approach is to use calculus.

Example 14.7. Using the values in Table 14.1 with (14.6), we obtain

$$\text{SSE} = [3.367 - (\beta_0 + \beta_1 \times 4.742)]^2 + [2.912 - (\beta_0 + \beta_1 \times 4.629)]^2$$
$$+ [3.627 - (\beta_0 + \beta_1 \times 4.964)]^2 + [3.190 - (\beta_0 + \beta_1 \times 4.875)]^2 .$$

To minimize SSE, we take the derivative of SSE with respect to β_0 and equate it to zero and with respect to β_1 and equate it to zero. These give two equations with two unknowns whose solutions are the regression coefficients $\widehat{\beta}_0 = -4.742$ and $\widehat{\beta}_1 = 1.659$ (see Exercise 14.2). □

For didactic reasons, we introduced Example 14.7 with more work than necessary. Recall that the variance of X is

$$S_X^2 = \frac{\sum_{i=1}^n (X_i - \overline{X})^2}{n-1}$$

and the covariance of X and Y is

$$S_{XY} = \frac{\sum_{i=1}^n (X_i - \overline{X})(Y_i - \overline{Y})}{n-1} .$$

Then following the procedure in Example 14.7 (this time with symbols instead of numbers), we obtain

$$\widehat{\beta}_1 = \frac{S_{XY}}{S_X^2}, \quad \widehat{\beta}_0 = \overline{Y} - \widehat{\beta}_1 \overline{X} . \tag{14.8}$$

We now define the

Predicted value of Y is denoted by $\widehat{Y}$ and is given by

$$\widehat{Y} = \widehat{\beta}_0 + \widehat{\beta}_1 X .$$

The predicted values are also called the *fitted* values.

From this definition, we conclude that the points $(X_i, \widehat{Y}_i)$ always fall on the regression line. The least squares estimates of the coefficients result in unbiased estimates, i.e.

$$E[\widehat{\beta}_i] = \beta_i, \quad E[S_{\widehat{\varepsilon}}^2] = \sigma_\varepsilon^2.$$

Here, the variance of the estimated errors (residuals), $S_{\widehat{\varepsilon}}^2$, is computed as

$$S_{\widehat{\varepsilon}}^2 := \frac{\sum_{i=1}^n \widehat{\varepsilon}^2}{n-p}$$

where p is the number of estimated parameters (2 for simple linear regression). To simplify our notation, we define

$$\sigma^2 := \sigma_\varepsilon^2, \quad S^2 := S_{\widehat{\varepsilon}}^2.$$

We introduced numerous symbols. For reference, we summarize them in Table 14.2. We finish this section with an example that explains the impact of the log transformation on the statistical model.

Table 14.2 Notation for simple linear regression. The estimated error variance is often called the residual mean square (Residual MS). The value of p in this chapter is always 2.

Parameter or coefficient	Population value	Estimated or predicted value
intercept	β_0	$\widehat{\beta}_0$
slope	β_1	$\widehat{\beta}_1$
error	ε	$\widehat{\varepsilon}$
dependent variable	Y	$\widehat{Y}$
error variance	$\sigma^2 := \sigma_\varepsilon^2$	$S^2 := S_{\widehat{\varepsilon}}^2$
number of estimated parameters	p	

Example 14.8. The data for this example are from the New York City (NYC) Open Accessible Space Information System Cooperative (OASIS); a partnership of more than 30 federal, state and local agencies, private companies, academic institutions and nonprofit organizations. The goal of the project is to enhance the stewardship of open space for the benefit of NYC residents.[3] The data refer to information about 322 trees in NYC. We are interested in tree age, defined as X, vs. diameter at breast height (DBH), defined as Y. We use log transformation on both variables. Figure 14.4 (in Exercise 14.5, you are asked to produce the figure) displays the data (left), the log transformed data and the estimated regression line (right)

$$E[Y|X] = -0.210 + 0.696X.$$

To simplify the notation, let

$$\log(a) := \widehat{\beta}_0, \quad b := \widehat{\beta}_1.$$

---
[3] See http://www.oasisnyc.net

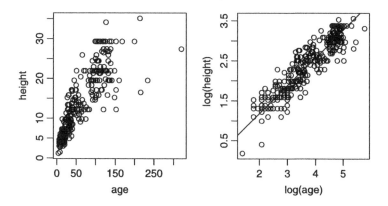

Figure 14.4   Tree height vs. age.

and define $X := \log(\text{age})$ and $Y := \log(\text{height})$. Thus,

$$\log(\text{height}) = \log(a) + b \times \log(\text{age}) \ .$$

Using the log rules, we find

$$a = \exp\left[\widehat{\beta_0}\right]$$
$$= e^{-0.210}$$
$$= 0.810$$

and

$$\text{height} = a \times \text{age}^b \tag{14.9}$$
$$= 0.810 \times \text{age}^{0.696} \ .$$

The units on the left-hand side are m and so must they be on the right-hand side. Therefore,

$$\text{m} = \frac{\text{m}}{\text{years}^{0.696}} \times \text{years}^{0.696} \ .$$

The coefficient a describes the growth rate in m per years$^{0.696}$. The predicted height of a 10-year-old tree is

$$\widehat{\text{height}} = 0.810 \times 10^{0.696}$$
$$= 4.022 \, \text{m} \ .$$

The growth rate (in height) of a tree is not constant. Therefore, it makes no sense to talk about a growth rate in a year. It makes sense to talk about an instantaneous growth rate. That is, the growth rate at a particular age. The instantaneous growth rate is obtained from the derivative of (14.9) with respect to age:

$$\text{growth rate} = a \times b \times \text{age}^{b-1}$$
$$= 0.564 \times \text{age}^{-0.304} \ .$$

A 5-year-old tree grows at a rate of $0.564 \times 5^{-0.304} = 0.346$ m per year while a 10-year-old tree grows at a rate of $0.564 \times 10^{-0.304} = 0.280$ m per year. □

In biology, Equation (14.9), written generally as

$$\widehat{Y} = a \times X^b,$$

is referred to as an *allometric* rule. Such equations apply to numerous other biological measures such as growth of children and metabolic rate as function of weight (see also Example 9.19).

14.3 The model goodness of fit

We fit the regression line and obtain estimates for the coefficients. Our next task is to decide how well the model describes the data. Heuristically, in the context of simple linear regression we say that

Goodness of fit refers to a significance test that compares the linear relationship between the dependent and independent variable to no relationship at all.

One way to view the idea of goodness of fit is to ask: Do we get better estimates of $\widehat{Y}$ based on $\widehat{Y} = \widehat{\beta}_0 + \widehat{\beta}_1 X$ compared to $\widehat{Y} = \widehat{\beta}_0$ (where $\widehat{\beta}_0 = \overline{Y}$)? In other words, does the model improve the predictability of Y based on the values of X compared to no model at all? Here "no model" means that Y is not a function of X.

The overall fit of a model is a separate issue from the significance of each model coefficient. As we shall see, in the case of simple linear regression, the overall model fit and the conclusion that β_1 is different from zero are indistinguishable. This is not the case when there is more than one independent variable: each coefficient is or is not significant and the whole model does or does not improve the predictability of Y compared to no model at all. In the remainder of this section we concentrate on the whole model goodness of fit. In Section 14.4, we discuss the significance of individual coefficients.

14.3.1 The F test

Recall that our least squares estimate of β_0, which we denote by $\widehat{\beta}_0$, is obtained from

$$\widehat{\beta}_0 = \overline{Y} - \widehat{\beta}_1 \overline{X}.$$

Therefore,

$$\widehat{Y} = \widehat{\beta}_0 + \widehat{\beta}_1 X$$
$$= \overline{Y} - \widehat{\beta}_1 \overline{X} + \widehat{\beta}_1 X.$$

For $X = \overline{X}$, we obtain

$$\widehat{Y} = \overline{Y} - \widehat{\beta}_1 \overline{X} + \widehat{\beta}_1 \overline{X} = \overline{Y}.$$

Thus, the predicted value of Y at $\overline{X}$ is $\overline{Y}$. In other words, the regression line always passes through the point $(\overline{X}, \overline{Y})$. Next, consider a typical data point in the context of the regression line (Figure 14.5). The location of each data point (X_i, Y_i), with

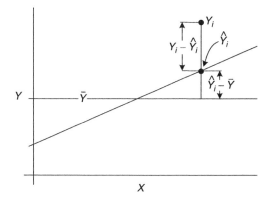

Figure 14.5 The distance of Y_i from the mean $\overline{Y}$ is the sum of the distance of the predicted value, $\widehat{Y}_i$, from the mean and the distance of the point from its predicted value, all along X_i.

respect to its distance from the sample mean of Y, consists of two components: The distance of the predicted value of Y at X_i from $\overline{Y}$ and the distance from the predicted value to the point. In symbols,

$$Y - \overline{Y} = (\widehat{Y} - \overline{Y}) + (Y - \widehat{Y})$$

(see also Figure 14.3). Because we are interested in the magnitude of these distances, not in their sign, we square these distances for all points and sum them. So the sum of squares of distances of all Y_i from $\overline{Y}$ can be partitioned into two sum of squares: one, the distances from the predicted values to the mean and the other, the distances of the points to predicted values. With SS denoting sum of squares, we thus define

$$\text{Regression SS} := \sum_{i=1}^{n} \left(\widehat{Y}_i - \overline{Y}\right)^2,$$

$$\text{Residual SS} := \sum_{i=1}^{n} \widehat{\varepsilon}_i^2$$

(where $\widehat{\varepsilon}_i$ is defined in (14.7)) and

$$\text{Total SS} := \sum_{i=1}^{n} \left(Y_i - \overline{Y}\right)^2$$
$$= (\text{Regression SS}) + (\text{Residual SS}) .$$

These sum of squares are illustrated in the next example.

Example 14.9. The first case in Figure 14.6 reflects the strongest argument we can make that the model fits the data. Here we have large Regression SS and small Residual SS. The slope is distinct from the horizontal line, which represents $\overline{Y}$. The points are bunched close to the regression line. The fourth case is the weakest. Here the Regression SS is small, i.e. the slope is not much different from 0, which is the slope of the mean $\overline{Y}$. Worse yet, the Regression SS is large; the points are scattered

430 Simple linear regression

away from the regression line. For the sake of completeness, we include the script that produces Figure 14.6:

```
1   rm(list = ls())
2   x <- seq(1, 10, length = 21)
3
4   ss <- function(i){
5     r <- c(0.8, 2.5, 0.8, 2.5)
6     b0 <- 1
7     b1 <- c(1, 1, .5, .5)
8     set.seed(i + 1)
9     Y <- b0 + b1[i] * x + rnorm(length(x), 0, r[i])
10    adj <- mean(Y) - 5
11    model <- lm(Y ~ x)
12    m <- c('large reg SS, small res SS',
13       'large reg SS, large res SS',
14       'small reg SS, small res SS',
15       'small reg SS, large res SS')
16    xlab <- '' ; ylab <- ''
17    if (i == 1 | i == 3) ylab = 'Y'
18    if (i == 3 | i == 4) xlab = 'x'
19    plot(x, Y, ylim = c(0, 12), main = m[i],
20       xlab = xlab, ylab = ylab)
21    abline(reg = model)
22    abline(h = mean(Y))
23    RegSS <- sum((predict(model) - mean(Y))^2)
24    ResSS <- sum((Y - predict(model))^2)
25    cbind('regression SS' = RegSS, 'residual SS' = ResSS,
26       'total SS' = RegSS + ResSS, Y = mean(Y))
27  }
28
29  openg(4.5, 4.5)
30  par(mfrow = c(2, 2))
31  b0 <- 1
32  SS <- matrix(ncol = 4, nrow = 4)
33  for (i in 1 : 4) ss(i)
34  saveg('goodness-of-fit', 4.5, 4.5)
```

Because we have four plots to produce, we write the function ss() in lines 4–27. In it, r is the standard deviation of the four residuals (as generated in line 9). The residuals are generated with rnorm() with mean zero and standard deviation one of the four elements of r. Once we generate the data for Y in line 9, we fit a linear model to the data in line 11. Lines 12–26 produce the plots. The plots result from the four calls to ss(). Recall that openg() and saveg() save the plots in a variety of convenient formats as discussed in Section 1.11. □

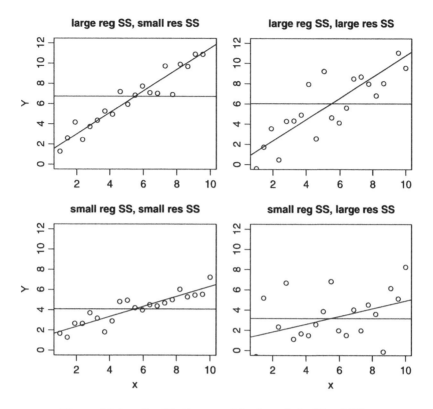

Figure 14.6   Qualitative partitioning of the Total SS.

We wish to develop a statistical test that will quantify the argument in Example 14.9. A good way to do this is to look at the ratio of the Regression SS over the Residual SS. As Figure 14.6 illustrates, we desire large Regression SS and small Residual SS. So the larger the ratio, the stronger claim we have that the slope of the regression line is different from zero. Now that we have a criterion, we look for its sampling density. It turns our that the slightly adjusted ratio (Regression SS) / (Residual SS) has a known sampling density. So let us first modify the ratio and then name the density. With MS denoting mean square, we define

Regression MS is defined as

$$\text{Regression MS} := \frac{\text{Regression SS}}{k}$$

where k is the number of covariates in the regression. We refer to k as the number of degrees of freedom of the Regression SS.

Residual MS Let $p = k + 1$ be the number of estimated model coefficients. Then we define the Residual MS as

$$\text{Residual MS} := \frac{\sum_{i=1}^{n} \widehat{\varepsilon}_i^2}{n - p} \quad (14.10)$$

where n is the sample size. We refer to $n - p$ as the number of degrees of freedom (df) of the residual sum of squares.

To obtain the df, we subtract from the number of observations the number of estimated parameters. In simple linear regression, we have a single covariate with one coefficient ($k = 1$) and the intercept. Therefore

$$\text{Regression MS} = \text{Regression SS}, \quad \text{Residual MS} = \frac{\sum_{i=1}^{n} \hat{\varepsilon}_i^2}{n-2}.$$

We wish to test the hypothesis

$$H_0 : \beta_1 = 0 \quad \text{vs.} \quad H_A : \beta_1 \neq 0.$$

Under this hypothesis, the statistic

$$F = \frac{\text{Regression MS}}{\text{Residual MS}}$$

has an F density with 1 and $n - 2$ df. With confidence level α, if $F_{1,n-2} > F_{1,n-2,\alpha}$ (where $F_{1,n-2,\alpha}$ denotes the critical value), then we reject the null hypothesis in favor of the alternative. We usually report results of the F-test with the so-called analysis of variance (ANOVA) table as detailed in Table 14.3 (ANOVA is discussed in Chapter 15). Instead of comparing F values, we can obtain the p-value directly and if the p-value $< \alpha$, we reject the null hypothesis:

$$p\text{-value} = F_{1,n-2}^{-1}\left(\frac{\text{Regression MS}}{\text{Residual MS}}\right)$$
$$= 1 - \mathtt{pf}(\mathtt{F}, 1, \mathtt{n} - 2)$$

where F is defined in Table 14.3.

Table 14.3 Typical ANOVA table. SS denotes sum of squares and df denotes degrees of freedom.

SS	df	Mean SS	F
Regression SS	1	Regression MS = $\frac{\text{Regression SS}}{1}$	$F = \frac{\text{Regression MS}}{\text{Residual MS}}$
Residual SS	$n - 2$	Residual MS = $\frac{\text{Residual SS}}{n-2}$	

Example 14.10. We return to Example 14.8 (see Figure 14.4). The null hypothesis is that $\beta_1 = 0$ and the alternative is that $\beta_1 \neq 0$. We select $\alpha = 0.05$. From Table 14.4 we read that under the null hypothesis, $F_{1,230} = 1\,967$. The critical value of the statistic is $F_{1,230,.05} = 3.87$. Because $1\,967 > 3.87$, we reject H_0 and surmise that $\beta_1 \neq 0$. From the results, we conclude that $\beta_1 > 0$. Therefore, the model fits the data with positive relationship between log(age) and log(height). Given that the p-value $= 2.2 \times 10^{-16}$ (Table 14.4), we may skip the third step. Because the p-value < 0.05 we reject H_0. □

Table 14.4 ANOVA for the regression of log(height) vs. log(age) (see Example 14.8 and Figure 14.4).

Source	SS	df	Mean SS	F	p-value
Regression	150.5	1	150.50	1967	≈ 0
Residuals	24.5	320	0.08		

14.3.2 The correlation coefficient

Recall that

$$\text{Total SS} := \sum_{i=1}^{n} (Y_i - \overline{Y})^2$$

$$= (\text{Regression SS}) + (\text{Residual SS}) .$$

Now if all the data points fall on the regression line, then Residual SS = 0 and Total SS = Regression SS. In this case, all of the sum of squared deviations of Y_i from $\overline{Y}$ is accounted for by the sum of squared deviations of $\widehat{Y}$ from $\overline{Y}$. If Total SS = Residual SS, then Regression SS = 0. In this case, the Regression SS accounts for none of the sum of the squared deviations of Y_i from $\overline{Y}$. So to measure the goodness of fit, we define the

Squared correlation coefficient (R^2)

$$R^2 := \frac{\text{Regression SS}}{(\text{Regression SS}) + (\text{Residual SS})} . \qquad (14.11)$$

From its definition, we have the following

Properties of R^2
1. $0 \leq R^2 \leq 1$.
2. The larger the value of R^2, the better the goodness of fit.
3. The smaller the value of R^2, the worse the goodness of fit.

The first property follows because all quantities in (14.11) are nonnegative and because Residual SS ≥ 0. We discussed the last two properties prior to the definition of R^2.

Example 14.11. From Table 14.4,

$$R^2 = \frac{150.5}{150.5 + 24.5} = 0.86 .$$

About 86% of the variation in Y is therefore accounted for by the linear regression. □

The variation in Y that is accounted for by R^2 is often referred to as the variation in Y that is *explained* by the regression. "Explain" should *not* be construed as cause and effect.

14.3.3 The correlation coefficient vs. the slope

We defined Pearson's sample correlation coefficient (R) in (8.6) and the population correlation coefficient ρ in (8.7). Writing the expression for Pearson's sample correlation coefficient as in (8.8), and the expression for $\widehat{\beta}_1$ as in (14.8), side by side,

$$\widehat{\rho} = \frac{S_{XY}}{S_X S_Y} \quad \widehat{\beta}_1 = \frac{S_{XY}}{S_X^2},$$

we conclude that

$$\widehat{\beta}_1 = \frac{S_{XY}}{S_x^2} = \frac{S_{XY} S_Y}{S_X^2 S_Y} = \frac{S_Y}{S_X} \frac{S_{XY}}{S_X S_Y} = \frac{S_Y}{S_X} \widehat{\rho}.$$

Therefore, we have the

Interpretation of $\widehat{\rho}$ Because $R^2 \leq 1$, so is $|\widehat{\rho}| \leq 1$. Furthermore, if X and Y are distributed normally, then:
1. If $\widehat{\rho} < 0$, we say that X and Y are negatively correlated—an increase in X is generally associated with a decrease in Y.
2. If $\widehat{\rho} > 0$, we say that X and Y are positively correlated—an increase in X is generally associated with an increase in Y.
3. If $\widehat{\rho} \approx 0$, we say that X and Y are uncorrelated—a change in X is generally not associated with a change in Y.

This interpretation is often useful in exploring linear relationships between X and Y without having to get into detailed regression analysis.

Example 14.12. In Example 14.2, we examined visually the paired relationship between energy use and CO_2 emission, CO_2 emission and GDP, life expectancy and fertility rate and child mortality and infant mortality (Figure 14.2). If we are interested in the qualitative relationships only, then we compute (see Exercise 14.9):

Pair	$\widehat{\rho}$	
log (energy use) vs. log (CO_2 emissions)	0.92	*
log (CO_2 emissions) vs. log (GDP)	0.50	*
life expectancy vs. fertility rate	−0.82	*
child mortality vs. infant mortality	0.99	*

The stars indicate significance (see Section 14.4.5). □

14.4 Hypothesis testing and confidence intervals

Now that we know how to test the overall fit of the model, we need to address the following questions: Are $\widehat{\beta}_0$ and $\widehat{\beta}_1$ different from zero? What are the confidence intervals on the coefficients? The tests we discuss here apply to each coefficient independent of the other. As such, they are *not* appropriate as tests of the whole model. We are also interested in the confidence intervals on $\widehat{Y}$. The larger the intervals, the less confidence we have about the predictions of the model.

14.4.1 t-test for model coefficients

For simple linear regression, the F-test (Section 14.3.1) and the t-test are equivalent. They are not when the regression includes more than one independent variable. In the latter case, the F-test refers to the model goodness of fit and the t-test refers to the significance of each coefficient.

We wish to test
$$H_0 : \beta_1 = 0 \quad \text{vs.} \quad H_A : \beta_1 \neq 0.$$
Under H_0, the sampling density of $\widehat{\beta}_1$ is t_{n-2} with
$$E\left[\widehat{\beta}_1\right] = 0, \quad \text{Var}\left[\widehat{\beta}_1\right] = \frac{\sigma_{\beta_1}^2}{n\sigma_X^2}$$
where $\sigma_{\beta_1}^2$ is the variance of the population coefficient β_1 and σ_X^2 is the population variance of X. Because both these variations are usually not known, we estimate $\text{Var}[\widehat{\beta}_1]$ with
$$S_{\widehat{\beta}_1}^2 = \frac{\text{Residual MS}}{(n-1)\,S_X^2} \quad \text{or} \quad \text{SE}\left[\widehat{\beta}_1\right] = \sqrt{\frac{\text{Residual MS}}{(n-1)\,S_X^2}}. \tag{14.12}$$

The t statistic is then
$$T = \frac{\widehat{\beta}_1}{\text{SE}\left[\widehat{\beta}_1\right]} \tag{14.13}$$
with $n-2$ df. Under the null hypothesis $H_0 : \beta_1 = 0$ vs. $H_A : \beta_1 \neq 0$ and significance level α, we compute the T statistic according to (14.13). If $|T| > t_{n-2, 1-\alpha/2}$, we reject H_0 in favor of H_A. Otherwise, do not reject H_0. As usual, the shortcut to this procedure is to use
$$\begin{aligned} p\text{-value} &= t_{1,n-2}^{-1}(T) \\ &= 1 - \texttt{pt(T, n}-2) \end{aligned}$$
where T is calculated according to (14.13). If the p-value $< \alpha$ then we deem $\widehat{\beta}_1$ significant.

Example 14.13. From Table 14.4, we find that Residual MS $= 0.08$. Therefore
$$\text{SE}\left[\widehat{\beta}_1\right] = \sqrt{\frac{0.0765}{(322-1) \times 0.968}} = 1.569 \times 10^{-2}.$$

The p-value of the test statistic is 2.2×10^{-16}, which is identical to the p-value for the F statistic in Example 14.10. □

14.4.2 Confidence intervals for model coefficients

Confidence intervals serve two purposes. First, they give us an idea about the precision of our estimates of the regression coefficients. Second, we can use them to draw conclusions about the population values of the coefficients.

The SE of $\widehat{\beta}_1$ is given in (14.12). The SE of $\widehat{\beta}_0$ is

$$\mathrm{SE}\left[\widehat{\beta}_0\right] = \sqrt{S^2\left(\frac{1}{n} + \frac{\overline{X}^2}{(n-1)S_X^2}\right)}.$$

Therefore, for significance level α, the confidence intervals are

$$(1-\alpha) \times 100\%\ \mathrm{CI}\left[\widehat{\beta}_0\right] = \widehat{\beta}_0 \pm t_{n-2, 1-\alpha/2} \mathrm{SE}\left[\widehat{\beta}_0\right],$$
$$(1-\alpha) \times 100\%\ \mathrm{CI}\left[\widehat{\beta}_1\right] = \widehat{\beta}_1 \pm t_{n-2, 1-\alpha/2} \mathrm{SE}\left[\widehat{\beta}_1\right]$$

with the appropriate substitutions of qt() for $t_{n-2, 1-\alpha/2}$ in the R vernacular.

Example 14.14. From Table 14.4, $S^2 = 0.08$ and $n = 322$. Also, $\overline{X} = 3.627$ and $S_X^2 = 0.968$. Therefore

$$\mathrm{SE}\left[\widehat{\beta}_0\right] = \sqrt{0.0765 \times \left(\frac{1}{322} + \frac{3.627^2}{321 \times 0.968}\right)}$$
$$= 5.896 \times 10^{-2}.$$

From Example 14.13, $\mathrm{SE}[\widehat{\beta}_1] = 1.569 \times 10^{-2}$. With $\alpha = 0.05$ we obtain for β_0

$$95\%\ \mathrm{CI}[\beta_0] = -0.210 \pm 1.96 \times 5.896 \times 10^{-2}$$
$$= [-0.326, -0.094]$$

and for β_1

$$95\%\mathrm{CI}[\beta_1] = 0.696 \pm 1.96 \times 1.569 \times 10^{-2}$$
$$= [0.665, 0.727].$$

From these results we conclude that both β_i are most likely different from zero. If it turns out that another regression, based on much larger sample size results in, say, $\widehat{\beta}_1$ outside the given confidence interval, then we would conclude that our sample comes from an unusual population of trees. It would be unusual in the sense that the relationship between the log of height and the log of age is different from the population. □

14.4.3 Confidence intervals for model predictions

From interval estimation on model predictions we draw conclusions about the accuracy of the predictions. There are two types of predictions we can make and these relate to two different confidence intervals:

1. For a given value of X not included in the data, we may be interested in the confidence interval of $\widehat{Y}$, the so-called *regression confidence interval*.

2. For a given X, we may be interested in the confidence interval of the average value of $\widehat{Y}$, which we denote by $\overline{\widehat{Y}}$. This is called the *prediction confidence interval*.

In both cases, $\widehat{Y} = \overline{\widehat{Y}} = \widehat{\beta}_0 + \widehat{\beta}_1 X$. However, the respective standard errors (and associated confidence intervals) are different:

$$\text{SE}\left[\widehat{Y}\right] = \sqrt{S^2 \left(1 + \frac{1}{n} + \frac{(X - \overline{X})^2}{(n-1) S_X^2}\right)}, \qquad (14.14)$$

$$\text{SE}\left[\overline{\widehat{Y}}\right] = \sqrt{S^2 \left(\frac{1}{n} + \frac{(X - \overline{X})^2}{(n-1) S_X^2}\right)}. \qquad (14.15)$$

From the equations we conclude that $\text{SE}\left[\widehat{Y}\right]$ will always be larger than $\text{SE}\left[\overline{\widehat{Y}}\right]$. Furthermore, for large n, the standard errors are approximately equal. The sampling density of $\widehat{Y}$ is F with $k+1$ and $n-(k+1)$ df (k is the number of covariates). Therefore, for simple linear regression the df are 2 and $n-2$. With confidence level $1 - \alpha$, the confidence intervals are

$$(1 - \alpha) \times 100\% \text{ CI}\left[\widehat{Y}\right] = \widehat{Y} \pm \sqrt{2 F_{2, 1-\alpha, n-2}} \times \text{SE}\left[\widehat{Y}\right],$$
$$(1 - \alpha) \times 100\% \text{ CI}\left[\overline{\widehat{Y}}\right] = \widehat{Y} \pm \sqrt{2 F_{2, 1-\alpha, n-2}} \times \text{SE}\left[\overline{\widehat{Y}}\right].$$

Example 14.15. Building on Figure 14.4, we compute both confidence intervals: for $\text{SE}\left[\widehat{Y}\right]$ and for $\text{SE}\left[\overline{\widehat{Y}}\right]$ for a sample of five points from the data. This will allow us to see the difference between the regression and prediction confidence intervals. So we load the data, attach them and call the library that contains ci.plot() with

```
> dbh <- read.table('DBH.txt', header = FALSE, sep = '\t')
> names(dbh) <- c('DBH', 'height', 'age')
> attach(dbh)
> library(RcmdrPlugin.HH)
```

Next, we assign the data, take a sample of five points, construct the model and plot the confidence intervals:

```
> X <- log(age) ; Y <- log(height)
> set.seed(1)
> idx <- sample(1 : length(X), 5)
> X <- X[idx] ; Y <- Y[idx]
> model <- lm(Y ~ X)
> ci.plot(model, main = '')
```

(Figure 14.7).   □

438 Simple linear regression

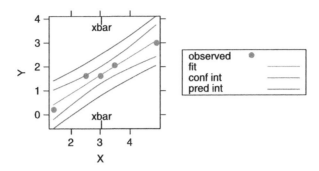

Figure 14.7   95% confidence intervals on $\widehat{Y}$ and $\overline{\widehat{Y}}$ for five random points from the data.

The confidence intervals in Figure 14.7 curve on both sides of the regression line. The further X is from $\overline{X}$, the wider the confidence interval. These trends can also be deduced from the terms $(X - \overline{X})^2$ in (14.14) and (14.15). Most statistical packages compute the prediction confidence interval (for $\widehat{Y}$) by default and this is what we shall use from now on.

14.4.4   t-test for the correlation coefficient

We use this test for small sample sizes (say < 30). We are interested in testing

$$H_0 : \rho = 0 \quad \text{vs.} \quad H_A : \rho \begin{matrix} \neq \\ > \\ < \end{matrix} 0 \; .$$

Under H_0, the test statistic is

$$T = \frac{\widehat{\rho}\sqrt{n-2}}{\sqrt{1-\widehat{\rho}^2}} \tag{14.16}$$

(where T's density is t with $n - 2$ df). To implement the t-test for the correlation coefficient, we compute $\widehat{\rho}$ and the test statistic T according to (14.16). Then for significance level α:

1. For $H_A : \rho \neq 0$, reject H_0 if $|T| > t_{n-2, 1-\alpha/2}$.
2. For $H_A : \rho > 0$, reject H_0 if $T > t_{n-2, 1-\alpha}$.
3. For $H_A : \rho < 0$, reject H_0 if $T < -t_{n-2, 1-\alpha}$.

Example 14.16. The data are for a family of five:

```
         height weight age
father    5.917    200  58
mother    5.250    110  54
son       6.000    190  24
son       6.167    200  20
daughter  5.500    140  19
```

Heights are in fractions of feet and weights in pounds. The regression of weight (Y) vs. height (X) (Figure 14.8) is significant:

```
> X <- height ; Y <- weight
> model <- lm(Y ~ X)
> summary(model)
```

```
              Estimate Std. Error t value Pr(>|t|)
(Intercept)   -435.41      86.22   -5.05  0.01498
X              104.64      14.93    7.01  0.00596

Residual standard error: 11.32 on 3 degrees of freedom
R-Squared: 0.9425
```

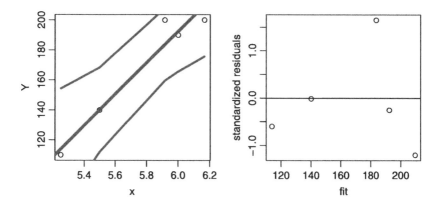

Figure 14.8 Weight (Y) vs. height (X) $\pm 95\%$ confidence lines for a family of five.

Because of the small sample size, it is difficult to judge the model by its residuals. To implement the t-test for the significance of the correlation coefficient, we find $\hat{\rho} = \sqrt{0.9425} = 0.971$ and $T = 7.010$. For $\alpha = 0.05$, $7.010 > t_{3,0.975} = $ qt(0.975, 3) $= 3.182$. Therefore, we reject the null in favor of $\rho \neq 0$. The same conclusions can be reached by observing that

```
> round(1 - pt(7.010, 3), 3)
[1] 0.003
```

Given our expectation that $\rho > 0$, we could choose to test $H_A : \rho > 0$. □

14.4.5 z tests for the correlation coefficient

This test is used for large sample sizes (> 30). Here we discuss significance tests with respect to ρ under the hypotheses $H_0 : \rho = \rho_0$ vs. $H_A : \rho \neq \rho_0$ where $\rho_0 \neq 0$. We present one- and two-sample tests.

One-sample test

We are interested in testing

$$H_0 : \rho = \rho_0 \quad \text{vs.} \quad H_A : \rho \begin{matrix}\neq\\>\\<\end{matrix} \rho_0 \, . \tag{14.17}$$

For $\rho_0 \neq 0$, the density of ρ is skewed. The following transformation, due to Fisher,

$$z_0 = \frac{1}{2} \log\left(\frac{1+\rho_0}{1-\rho_0}\right), \tag{14.18}$$

normalizes the density of ρ_0. Under H_0, the sample-based estimates of the mean and variance of z_0 are

$$\widehat{Z} = \frac{1}{2} \log\left(\frac{1+\widehat{\rho}}{1-\widehat{\rho}}\right), \quad S^2 = \frac{1}{n-3} \tag{14.19}$$

and the test statistic

$$Z := \left(\widehat{Z} - z_0\right) \sqrt{n-3} \tag{14.20}$$

is standard normal. With (14.17) in mind, we compute $\widehat{\rho}$, z_0 according to (14.18) and $\widehat{Z}$ according to (14.19). Next, we compute the test statistic Z according to (14.20). With significance level α:

1. For $H_A : \rho \neq \rho_0$, reject H_0 if $|Z| > z_{1-\alpha/2}$.
2. For $H_A : \rho > \rho_0$, reject H_0 if $Z > z_{1-\alpha}$.
3. For $H_A : \rho < \rho_0$, reject H_0 if $Z < -z_{1-\alpha}$.

Two-sample test

We have two samples of sizes n_1 and n_2 from presumably different populations. For each sample, we calculate $\widehat{\rho}_1$ and $\widehat{\rho}_2$. Are the correlation coefficients for the two populations different? Fisher's transformation provides the necessary machinery. We wish to test

$$H_0 : \rho_1 = \rho_2 \quad \text{vs.} \quad H_A : \rho_1 \begin{cases} \neq \rho_2 \\ > \rho_2 \\ < \rho_2 \end{cases}$$

where ρ_1 and ρ_2 are the respective population correlation coefficients. The test statistic is

$$Z = \frac{\widehat{Z}_1 - \widehat{Z}_2}{\sqrt{1/(n_1-3) + 1/(n_2-3)}} \tag{14.21}$$

where the density of

$$\widehat{Z}_i = \frac{1}{2} \log\left(\frac{1+\widehat{\rho}_i}{1-\widehat{\rho}_i}\right) \quad \text{for} \quad i = 1, 2 \tag{14.22}$$

is standard normal. To implement the test, we compute $\widehat{\rho}_1$ and $\widehat{\rho}_2$ and their respective transformations $\widehat{Z}_1$ and $\widehat{Z}_2$ according to (14.22). Next, we compute the test statistic Z according to (14.21) and with significance level α:

1. For $H_A : \rho_1 \neq \rho_2$, reject H_0 if $|Z| > z_{1-\alpha/2}$.
2. For $H_A : \rho_1 > \rho_2$, reject H_0 if $Z > z_{1-\alpha}$.
3. For $H_A : \rho_1 < \rho_2$, reject H_0 if $Z < -z_{1-\alpha}$.

14.4.6 Confidence intervals for the correlation coefficient

To obtain confidence limits for ρ, we first compute $\widehat{\rho}$ and the transformation $\widehat{Z}$ according to (14.19). Next, let $\widehat{Z}_L$ and $\widehat{Z}_H$ denote the low and high estimated limits of the confidence interval of $\widehat{Z}$ for a given α. Then

$$\widehat{Z}_L := \widehat{Z} - \frac{z_{1-\alpha/2}}{\sqrt{n-3}}, \quad \widehat{Z}_H := \widehat{Z} + \frac{z_{1-\alpha/2}}{\sqrt{n-3}}$$

where z is standard normal and n is the sample size. The confidence interval around $\widehat{Z}$ is then

$$(1-\alpha) \times 100\% \text{ CI}\left[\widehat{Z}\right] = \left[\widehat{Z}_L, \widehat{Z}_H\right].$$

Let $\widehat{\rho}_L$ and $\widehat{\rho}_H$ be the low and high limits of the confidence interval around $\widehat{\rho}$ for a given α. Then

$$\widehat{\rho}_L = \frac{e^{2\widehat{Z}_L} - 1}{e^{2\widehat{Z}_L} + 1}, \quad \widehat{\rho}_H = \frac{e^{2\widehat{Z}_H} - 1}{e^{2\widehat{Z}_H} + 1}$$

and

$$(1-\alpha) \times 100\% \text{ CI}\left[\widehat{\rho}\right] = [\widehat{\rho}_L, \widehat{\rho}_H].$$

Example 14.17. Figure 14.4 details the relationship between the log of age and log of height for 322 trees in New York (see Example 14.8). The regression results are

```
            Estimate Std. Error t value Pr(>|t|)
(Intercept) -0.21019  0.05895   -3.565  0.000419
x            0.69576  0.01569   44.355  < 2e-16
Residual standard error: 0.2766 on 320 degrees of freedom
Multiple R-Squared: 0.8601
```

What is the range of correlations values within which we might find similar regression for all of the trees in New York (assuming that the 322 trees are a random sample of trees in New York)?

We find $\widehat{\rho} = 0.927$ and $\widehat{Z} = 1.640$. For $\alpha = 0.05$,

$$\widehat{Z}_L = 1.640 - \frac{1.96}{\sqrt{319}} = 1.530, \quad \widehat{Z}_H = 1.640 + \frac{1.96}{\sqrt{319}} = 1.749.$$

Therefore,

$$95\% \text{ CI}\,[0.927] = [1.530, 1.749].$$

The 95% confidence interval around $\widehat{\rho}$ is then

$$\widehat{\rho}_L = \frac{\exp[2 \times 1.530] - 1}{\exp[2 \times 1.530] + 1} = 0.910, \quad \widehat{\rho}_H = \frac{\exp[2 \times 1.749] - 1}{\exp[2 \times 1.749] + 1} = 0.941$$

or

$$95\% \text{ CI}\,[0.927] = [0.910, 0.941].$$

This is the range within which we are likely to find the correlation between log age and log height for all trees in New York city. □

14.5 Model assumptions

Our goal at this point is to develop ways to judge the adequacy of the simple linear model in light of its assumptions. Throughout, we wrote the linear equation as $Y = \beta_0 + \beta_1 X + \varepsilon$. This notation implies that Y is a function of X. When we wish to emphasize this fact, we write $Y(X)$, instead of Y. With this notation, we have

Assumptions
1. The mean value of $Y(X)$ is $\overline{Y(X)} = \beta_0 + \beta_1 X$ for any X.
2. $Y(X)$ has a normal density with mean $\overline{Y(X)}$ and standard deviation $\sigma_{Y(X)} = \sigma$, where σ is constant for all X.
3. For the data (X_i, Y_i), $i = 1, \ldots, n$, the values of $\varepsilon_i := Y_i - (\beta_0 + \beta_1 X_i)$ are independent.

From assumption 1 we conclude that $\overline{\varepsilon}_i = 0$ for all i. From assumptions 2 and 3 we conclude that $\text{Var}[\varepsilon_i] = \sigma^2$ for all i. We summarize these conclusions as a

Corollary The density of the *residuals* ε_i is normal with mean 0 and standard deviation σ.

We should emphasize that the assumptions and their corollary all refer to the population (true) regression line, *not* to the estimated regression line. The assumptions and their corollary are illustrated in Figure 14.9 (in Exercise 14.11 you are asked to write a script that produces the figure). The next example illustrates situations where the assumptions are violated.

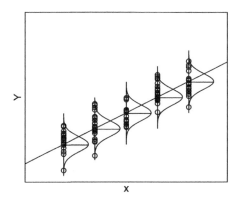

Figure 14.9 For each X, the mean of $\overline{Y(X)}$ is on the line $\beta_0 + \beta_1 X$. The density of $Y(X)$ is normal with mean $\overline{Y(X)}$ and constant standard deviation. These densities are superimposed on the line.

Example 14.18. Figure 14.10 (in Exercise 14.12 you are asked to write a script that produces the figure) illustrates two typical situations in which the assumptions of the model are violated. In the first case (left panel), the variance of $Y(X)$ is not constant across all values of X. In the second (right panel), $\overline{Y(X)} \neq \beta_0 + \beta_1 X$. □

With our understanding of the assumptions about linear regression, we are ready to develop the diagnostics that will allow us to judge if the model is adequate.

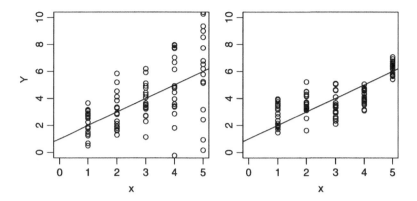

Figure 14.10 Left panel: the variance of the residuals increases with X. Right panel: the mean of the residuals is not zero. This can happen either because the residuals are not independent or because the relation between Y and X is not linear.

14.6 Model diagnostics

Even if the model fits (Section 14.3) and we reject the hypothesis that the slope is not different from zero (Section 14.4), we are not done yet. Why? Because the tests we discussed assume particular sampling densities of the residuals and coefficients under the null hypothesis. If these assumptions are not met, then the tests are moot.

Example 14.19. The data for this example are from the 1994 March–April issue of Academe. They refer to the average salary and overall compensation, broken down by full, associate and assistant professor ranks, for 1161 colleges in the U.S. We wish to examine the relationship between the number of full professors and their average salary. We load the data and the sources for two functions that compute and plot confidence intervals around a lm():

```
> load('aaup.rda')
> source('confidence-interval.R')
> source('see.R')
> X <- aaup[, 5] ; Y <- aaup[, 13]
> test <- (is.na(X) == FALSE & is.na(Y) == FALSE)
> X <- X[test] ; Y <- Y[test]
> see(X, Y,'AAUP-PROFS-VS-SALARY')
```

confidence-interval.R and see.R are at the book's site. A call to see() produces Figure 14.11 and a report of the lm() (edited):

```
            Estimate Std. Error t value Pr(>|t|)
(Intercept) -264.40622   16.40534  -16.12   <2e-16
x              0.69684    0.03053   22.82   <2e-16
```

From the data it seems that salary increases with the number of full professors at a particular university. The results, as reflected by the p-value (Pr(>|t| show that the parameter values are significant for both the intercept and the slope: As expected, ANOVA[4] (Table 14.5), also produced by see(), indicates that the overall model fit is

---
[4] Albeit not necessary, we use the ANOVA table for heuristic reasons.

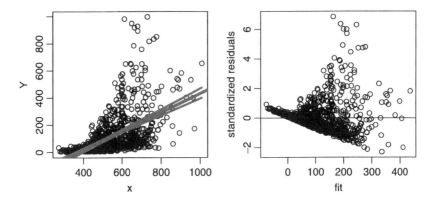

Figure 14.11 Left: Average salary (US$×1 000) for 1 161 campuses in the U.S. The relationship is obviously nonlinear. Also shown are the regression line and its 95% confidence interval. The discrepancy between the number of campuses and the df in Table 14.5 is due to missing data. Right: the standardized residuals (see Section 14.6.2).

Table 14.5 ANOVA table for the regression between number of full professors and their average yearly income on 1 161 U.S. campuses.

Source	SS	df	Mean SS	F	p-value
Regression	7 422 774	1	7 422 774	520.9	≈ 0
Residuals	15 546 487	1 091	14 250		

significant. Furthermore, the value of $R^2 = 0.323$ is significant. Are we then to conclude that the model is adequate—i.e. that we can use the regression line to predict full professors salary based on the number of full professors? The answer is no and the reasons are explained next. □

From the assumptions about the regression line we conclude that residuals are good candidates to develop diagnostics that tell us how well our model and data conform to the assumptions. Recall that the unknown *population* residual for the ith data pair (X_i, Y_i) is

$$\varepsilon_i := Y_i - (\beta_0 + \beta_1 X_i)$$

and the estimated ith residual is

$$\widehat{\varepsilon}_i := Y_i - \left(\widehat{\beta}_0 + \widehat{\beta}_1 X_i\right) .$$

We will use the residuals to examine the conformance of the model to the assumptions. There are many diagnostics to choose from and their nuanced differences are sometimes difficult to discern. In addition, the diagnostics are interrelated and with large samples, some diagnostics give nearly identical results. Diagnostics fall under two broad categories: those that address the effect of observations on specific coefficients and those that address the whole model—all in the context of the current model coefficients and specific observations.

Diagnostics are not discussed in the literature uniformly. Notation and formulas often differ among software implementations, journal articles and textbooks. We follow the conventions used in R. In fact, most of the formulas concerning specific residuals were translated directly from R's code. In this context, the code is the final arbiter.

14.6.1 The hat matrix

As we shall see, the *hat* matrix is essential in presenting various diagnostics. A detailed discussion of the hat matrix is beyond our scope. We will introduce the subject mostly by examples and for simple linear models only. Heuristically, the hat matrix, denoted by H, is a matrix with n rows and n columns. The ith diagonal element of H reflects the multidimensional distance of the ith datum from the multidimensional center of the data. An example of a one-dimensional distance is $(X_i - \overline{X})^2$ where $\overline{X}$ is the center of the data. The diagonal elements of H are denoted by h_i, for $i = 1, \ldots, n$. As we shall see, H has many uses; one of them is in computing the predicted (hat) values. The diagonal elements of the hat matrix reflect the corresponding influence of the ith data point on the model fit.

To explain the hat matrix, we need to develop shorthand notation. For data with n observations, let

$$Y := \begin{bmatrix} Y_1 \\ Y_2 \\ \vdots \\ Y_n \end{bmatrix} \quad X := \begin{bmatrix} 1 & X_1 \\ 1 & X_2 \\ \vdots & \vdots \\ 1 & X_n \end{bmatrix}.$$

Here (X_i, Y_i) are pairs of observations. A column of 1s is added to the vector that represents the independent variable data. This is used to estimate the intercept; hence the name of the matrix X, the so-called *design matrix*. The number of columns of X is denoted by p. The latter is also the number of coefficients in the model. In our case, $p = 2$. We present the data with

$$Y_1 = \beta_0 + \beta_1 X_1 + \varepsilon_1,$$
$$Y_2 = \beta_0 + \beta_1 X_2 + \varepsilon_2,$$
$$\ldots$$
$$Y_n = \beta_0 + \beta_1 X_n + \varepsilon_n,$$

where ε_i, $i = 1, \ldots, n$ are the residuals. Denoting the column of $(\varepsilon_1, \ldots, \varepsilon_n)$ by ε, we write the above in a shorthand notation:

$$Y = X\beta + \varepsilon$$

where $\beta := [\beta_1, \beta_2]$ is the vector of coefficients. Two quantities that appear in many computations of linear models are

$$A := \left(X'X\right), \quad A^{-1} \tag{14.23}$$

where X' denotes the transpose of X and A^{-1} is the inverse of A.

Example 14.20. Consider data with $n = 4$ observations. Then the design matrix and its transpose are

$$X = \begin{bmatrix} 1 & X_1 \\ 1 & X_2 \\ 1 & X_3 \\ 1 & X_4 \end{bmatrix}, \quad X' = \begin{bmatrix} 1 & 1 & 1 & 1 \\ X_1 & X_2 & X_3 & X_4 \end{bmatrix}.$$

Let
$$\mathrm{SX} = \sum_{i=1}^{4} X_i, \quad \mathrm{SSX} := \sum_{i=1}^{4} X_i^2.$$

Then using matrix multiplication rules we get

$$A = \begin{bmatrix} 4 & \mathrm{SX} \\ \mathrm{SX} & \mathrm{SSX} \end{bmatrix}, \quad A^{-1} = \begin{bmatrix} \frac{\mathrm{SSX}}{B} & \frac{-\mathrm{SX}}{B} \\ \frac{-\mathrm{SX}}{B} & \frac{4}{B} \end{bmatrix} \quad (14.24)$$

where
$$B = 4 \times \mathrm{SSX} - (\mathrm{SX})^2$$

For data with n observations, replace 4 everywhere it appears by n. □

With this in mind, we define

The hat matrix (H) is the projection of the data points onto the space spanned by X,
$$H := XA^{-1}X'$$
where X' is the transposed X and A^{-1} is defined in (14.24) for simple linear regression.

Leverage (h) The diagonal elements of H, denoted by h_i ($i = 1, \ldots, n$) reflect the influence of the ith data point on the model fit.

Cutoff criterion for h_i If
$$h_i \geq \min\left[\frac{3p}{n}, 0.99\right]$$
then X_i is judged as having an unusual predictor value.

The predicted mean vector of the n responses is given by
$$\widehat{Y} = X\widehat{\beta} = HY$$

This notation is particularly convenient with multiple linear regression. We use specific numerical results in the next example to clarify some ideas.

Example 14.21. Consider the data with $n = 4$,
$$X = \begin{bmatrix} 1 & 1 \\ 1 & 2 \\ 1 & 3 \\ 1 & 4 \end{bmatrix}, \quad Y = \begin{bmatrix} 1.127 \\ 1.541 \\ 1.846 \\ 2.407 \end{bmatrix}.$$

Then
$$\widehat{\beta} := \begin{bmatrix} \widehat{\beta}_0 \\ \widehat{\beta}_1 \end{bmatrix} = A^{-1}X'Y = \begin{bmatrix} 0.694 \\ 0.415 \end{bmatrix}$$

and
$$H = \begin{bmatrix} 0.7 & 0.4 & 0.1 & -0.2 \\ 0.4 & 0.3 & 0.2 & 0.1 \\ 0.1 & 0.2 & 0.3 & 0.4 \\ -0.2 & 0.1 & 0.4 & 0.7 \end{bmatrix}.$$

Also
$$\widehat{Y} = X\widehat{\beta} = HY = \begin{bmatrix} 1.109 \\ 1.523 \\ 1.938 \\ 2.352 \end{bmatrix}.$$

The diagonal elements are 0.7, 0.3, 0.3 and 0.7. Here $p = 2$ and $n = 4$. Therefore, $3p/n = 1.5$. Consequently, the criterion for judging the effect of each observation on the predicted values is 0.99. None of the 4 values of X have an exceptionally large effect on the fitted values. □

14.6.2 Standardized residuals

Two residual-related measures of interest are:

Standard deviation of residuals Let i be the ith data point $(i = 1, \ldots, n)$, n the number of data points, p the number of fitted parameters and S_X^2 the sample-based variance of X. Then we estimate the standard deviation of the ith residual with

$$\text{SD}\left[\widehat{\varepsilon}_i\right] = \sqrt{\text{Residual MS}} \sqrt{1 - \frac{1}{n} - \frac{(X_i - \overline{X})^2}{(n-1)S_X^2}}$$

where Residual MS is defined in (14.10).

Standardized residuals Denote by $\widehat{\varepsilon}'_i$ the standardized residual of the ith data. We estimate it with

$$\widehat{\varepsilon}'_i = \frac{\widehat{\varepsilon}_i}{\sqrt{\text{Residual MS}}\sqrt{1 - h_i}} \tag{14.25}$$

where h_i is the diagonal element of H.

Overall, the value of the standardized residuals should remain constant over the entire range of X and they should show no trends with increasing values of $\widehat{Y}$. Otherwise, the assumptions of constant variance and independent errors are violated and we may have to reject a model even if its fit and coefficients are significant. Standardized residuals can be used to check major departures from model assumptions. In particular, we can verify the assumption that the residuals are normal (with mean zero) with Q-Q plots. Standardized residuals are not very good at detecting observations that might have large influence on the model fit and estimates of coefficients.

Example 14.22. In Example 14.19, we established that the model fits the data (about professor salaries) and that the slope is significant (Figure 14.11). In Figure 14.11 (right)—also produced by see()—we examine the standardized residuals against $\widehat{Y}$. The plot reflects a major departure from model assumptions: the variance of the studentized residuals is obviously not constant. Can we fix this problem? Yes (almost) with the log transformation. We define $X :=$ log of number of full professors and $Y :=$ log of average salary of professors. The linear model and Figure 14.12 are produced with

```
> log.X <- log(X) ; log.Y <- log(Y) ; model<-lm(log.Y ~ log.X)
> see(log.X, log.Y, 'AAUP-PROFS-VS-SALARY-LOG')
            Estimate Std. Error t value Pr(>|t|)
```

```
(Intercept)  -18.9731     0.7071   -26.83    <2e-16
X              3.6685     0.1133    32.38    <2e-16
```

Residual standard error: 0.8323 on 1091 degrees of freedom
R-Squared: 0.49,   Adjusted R-squared: 0.4895
F-statistic: 1048 on 1 and 1091 DF, p-value: < 2.2e-16

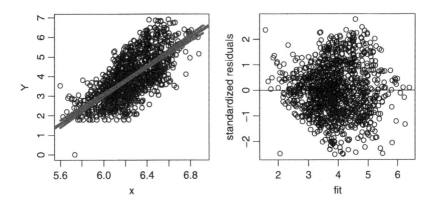

Figure 14.12   Here $X :=$ log of number of full professors and $Y :=$ log of average salary of professors in U.S. \$100. Left: the data and the fitted line with 95% confidence interval of $\widehat{Y}$. Right: A plot of the standardized residuals by predicted values.

Both the F-test and R^2 indicate that the model is an improvement over no model in terms of reducing the variability in Y. From Figure 14.12, we conclude that except for both extreme values of $\widehat{Y}$, the residuals meet the assumptions of linear regression: The trend in the standardized residuals is the result of small variances at both extreme values of $\widehat{Y}$. To verify that the density of the residuals is normal we use

> qqnorm(residuals(model))
> qqline(residuals(model))

Figure 14.13 confirms our previous conclusion: except for the tails, the residuals are normal. □

From the fact that the tails of the Q-Q plot depart from normality we conclude that perhaps we should clip the data at both ends of X. Schools with exceptionally low or high salaries are not typical. Our model is log-linear. This means that extrapolations (predicting Y for X outside its range of data) are risky at best—large values of X may lead us to predict astronomical salaries for full professors (wishful thinking). More generally, one should use caution in extrapolating linear models. They may lead to silly conclusions.

14.6.3 Studentized residuals

If one $\widehat{\varepsilon}$ is very large, then the estimate of the Residual MS will be large. Consequently, all the residuals will be small. To get around this problem, we *studentize* the residuals.

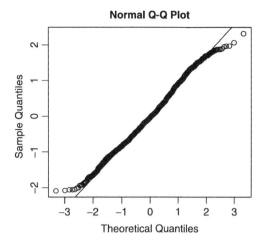

Figure 14.13 Q-Q plot of the standardized residuals of full professor salaries.

Studentized residuals For the ith data point,

$$\widehat{\varepsilon}_i^* := \frac{\widehat{\varepsilon}_i'}{\operatorname{SD}[\widehat{\varepsilon}_i]}$$

where $\widehat{\varepsilon}_i'$ denotes the ith standardized residual (defined in 14.25) and $\widehat{\varepsilon}_i^*$ denotes the ith studentized residual.

The further a data point is from the center of the data (the intersection of the horizontal line $y = \overline{Y}$ and the vertical line $x = \overline{X}$), the larger its influence on the estimate $\widehat{\beta}_1$. The studentized residuals tend to emphasize this attribute of data points. To enhance this property, we use the following residuals.

14.6.4 The RSTUDENT residuals

One way to observe the effect of a data point on its fitted value is this: remove the ith point from the data; compute a new regression line; use the new line to provide a new prediction based on X_i. We denote the prediction based on a model fitted without the ith point by $\widehat{Y}_{(i)}$. Similarly, we denote by Residual MS$_{(i)}$ the residual mean square of the model fitted without the ith observation. Consequently, we have the following definition:

The $\widehat{\varepsilon}_{(i)}^*$ residuals (RSTUDENT) Let $\widehat{\varepsilon}_{(i)}^*$ denote the studentized residual based on a model with the ith point removed. Then

$$\widehat{\varepsilon}_{(i)}^* = \frac{\widehat{\varepsilon}_i}{\sqrt{\operatorname{Residual\ MS}_{(i)}}\sqrt{1 - h_i}}$$

where $\widehat{\varepsilon}_i$ and h_i are the residual and leverage of the ith point and Residual MS$_{(i)}$ is the Residual MS, calculated with the ith point removed, i.e.

$$\operatorname{Residual\ MS}_{(i)} := \frac{\sum_{j=1, j \neq i}^n \widehat{\varepsilon}_j^2}{n - p - 1}$$

Cutoff criterion for $\widehat{\varepsilon}^*_{(i)}$**]** If $|\widehat{\varepsilon}^*_{(i)}| \geq 2$ then X_i is judged as having an unusual predictor value.

The df of $\widehat{\varepsilon}^*_{(i)}$ are $n - p - 1$ because we are using $n - 1$ data points. The residuals $\widehat{\varepsilon}^*_{(i)}$ are often called RSTUDENT or *jackknifed residuals*. The sampling density of $\widehat{\varepsilon}^*_{(i)}$ is t_{n-p}.

Note the difference between $\widehat{\varepsilon}^*_i$ and $\widehat{\varepsilon}^*_{(i)}$. The former is used to observe overall departure of residuals from their expected behavior if the model is correct. The latter is used to identify points with unusually large change in their predicted values (due to their removal from estimates of the coefficients).

Example 14.23. Here we analyze the normal average January minimum temperature in °F for 56 U.S. cities from 1931 to 1960, compared to latitude. The data appeared in Peixoto (1990) and in Hand (1994), pp. 208–210. We load the data and the source for the function that produces confidence intervals on the regression line, fit the model and obtain its summary:

```
> load(file = 'temperature.rda')
> source('confidence-interval.R')
>
> X <- temperature$Lat
> Y <- temperature$JanTemp
> n <- length(X) ; p <- 2
> summary(model <- lm(Y ~ X))

Call:
lm(formula = Y ~ X)

Residuals:
     Min       1Q   Median       3Q      Max
-10.6812  -4.5018  -0.2593   2.2489  25.7434

Coefficients:
            Estimate Std. Error t value Pr(>|t|)
(Intercept)  108.7277     7.0561   15.41   <2e-16
X             -2.1096     0.1794  -11.76   <2e-16

Residual standard error: 7.156 on 54 degrees of freedom
Multiple R-squared: 0.7192,     Adjusted R-squared: 0.714
F-statistic: 138.3 on 1 and 54 DF,  p-value: < 2.2e-16
```

(in the interest of saving space and because they duplicate the results of the *t*-test, we no longer show the results of the ANOVA). The model obviously fits well. In Figure 14.14 (left), we plot the model, the data, the confidence intervals (ci.lm() resides in confidence-interval.R) and the index of data for coastal cities (except Spokane) that apparently belong to a different model (Table 14.6):

```
> coast <- c(5, 6, 41, 52)
> par(mfrow = c(1, 2))
> ci.lm(X, model)
```

```
> points(X[coast], Y[coast], pch = 19)
> identify(X, Y)
[1]  5  6 41 52
```

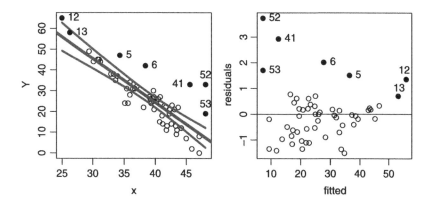

Figure 14.14 Mean January minimum temperature (Y) and latitude (x) of 56 U.S. cities (left). Standardized residuals (right). Corresponding points are identified by numbers. The numbers correspond to the cities in Table 14.6.

Table 14.6 Average January temperature for selected U.S. cities. See Figure 14.14. Influential residuals are identified with stars. The residual columns refer to (in order): RSTUDENT, DFBETAS for intercept, DFBETAS for slope, DFFIT, Cook's distance and diagonals of the hat matrix.

ID	City	Temperature	Latitude	$\widehat{\varepsilon}^*_{(i)}$	$\Delta\widehat{\beta}^*_{0(i)}$	$\Delta\widehat{\beta}^*_{1(i)}$	$\Delta\widehat{y}^*_i$	D_i	h_i
5	Los Angeles, CA	47	34.3						
6	San Francisco, CA	42	38.4	*					
12	Key West, FL	65	25						*
13	Miami, FL	58	26.3						*
41	Portland, OR	33	45.6	*			*		
52	Seattle, WA	33	48.1	*		*	*		
53	Spokane, WA	19	48.1						

On the right side, we plot the standardized residuals and indicate the coastal cities of interest:

```
> residuals.standardized <- rstandard(model)
> plot(model$fitted.values, residuals.standardized,
+    xlab = 'fitted', ylab = 'residuals')
> abline(h=0)
> points(model$fitted.values[coast],
+    residuals.standardized[coast], pch = 19)
> text(model$fitted.values[coast],
+    residuals.standardized[coast], labels = coast,pos = 4)
```

452 Simple linear regression

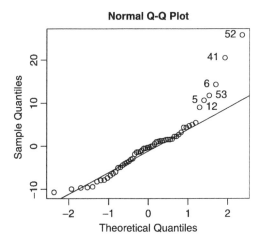

Figure 14.15   Q-Q plot of standardized residuals. Numbered points correspond to the cities in Table 14.6.

Observe that the points that are visually different (black circles) on the scatter plot emerge so in the residuals plot. Other points that are not visually different on the scatter plot emerge as so on the residual plot. For example Spokane (53) does not seem far out on the scatter plot. It does appear far on the residual plot because of its high leverage: it is far from the center of the data and thus influences the regression more than those that are closer to the center of the data. Without the identified points, the residuals seem to be well behaved. To verify that the standardized residuals are normal, we use the Q-Q plot

```
> q <- qqnorm(residuals(model))
> identify(q$x, q$y)
> qqline(residuals(model))
```

(Figure 14.15). The plot confirms our suspicion that the identified points are different from the others. Without them, the Q-Q plot shows that the remaining residuals are normal.

In Figure 14.16 we compare the studentized to the standardized residuals with:

```
> RSTUDENT<-rstudent(model)
> plot(model$fitted.values, RSTUDENT,
+    xlab = 'fitted', ylab = 'standard residuals')
> abline(h = 0) ; abline(h = 2, lty = 2)
> points(model$fitted.values[coast], RSTUDENT[coast],
+    pch = 19)
> points(model$fitted.values, residuals.standardized, cex = 2)
> identify(model$fitted.values, RSTUDENT)
```

The values of small residuals are nearly identical. However, the studentized residuals (with one point removed) pull the small residuals closer to the horizontal zero and push the large residuals further apart (compared to the standardized residuals). The studentized residuals of San Francisco, Portland and Seattle are significantly large. They may belong to a different model. The results here were produced with temperature-RSTANDARD-vs-RSTUDENT.R which resides at the book's site. □

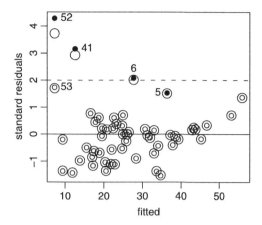

Figure 14.16 Large circles show the standardized residuals and small circles the $\widehat{\varepsilon}^*_{(i)}$. The black circles corresponding to those in Figure 14.14. The broken line indicates the cutoff value of influential $\widehat{\varepsilon}^*_{(i)}$.

14.6.5 The DFFITS residuals

Consider the following. We fit a regression and obtain predictions for each value of X_i, namely $\widehat{Y}_i$. Now remove the ith observation from the data, fit a new model and obtain a new prediction for the ith observation. Denote this new prediction by $\widehat{Y}_{(i)}$. By how much did the prediction of Y for the ith observation change? If it changed substantially, then we know that the ith observation is influential in the model (e.g. with regard to the values of the coefficients). So we can use $\widehat{Y}_i - \widehat{Y}_{(i)}$ to evaluate the amount of change. But we still have a problem. A small change in $\widehat{Y}_i - \widehat{Y}_{(i)}$ for an observation that is near the center of the data is not comparable to a small change in $\widehat{Y}_i - \widehat{Y}_{(i)}$ for an observation that is far from the center of the data. It is "harder" for the former to effect change than the latter. So we must scale the difference $\widehat{Y}_i - \widehat{Y}_{(i)}$ by how far the point is from the center of the data. The scale of choice is $\sqrt{\text{Residual MS}_{(i)}} \times \sqrt{h_{(i)}}$. After some algebraic manipulations, we obtain

Difference in fits (DFFITS) The difference in fits for the ith observation is defined as

$$\Delta \widehat{Y}_i^* = \frac{\widehat{\varepsilon}_i \sqrt{h_i}}{\sqrt{\text{Residual MS}_{(i)} (1 - h_i)}} . \quad (14.26)$$

Cutoff criterion for $\Delta \widehat{Y}_i^*$ If

$$|\Delta \widehat{Y}_i^*| \geq 3 \sqrt{\frac{p}{n-p}} ,$$

then x_i is judged as having large influence on the overall model fit.

The $\Delta \widehat{Y}_i^*$ residuals are often denoted by DFFITS. Large values of $\Delta \widehat{Y}_i^*$ indicate influential observations. A large value must be determined by some criterion. For linear models, the sampling density of $\Delta \widehat{Y}_i^*$ is $F_{p, n-p-1}$ (in our case, the number of coefficients we fit is $p = 2$).

454 Simple linear regression

Example 14.24. Returning to Example 14.21, we calculate

$$\Delta \widehat{\boldsymbol{Y}}^* = \begin{bmatrix} 0.494 \\ 0.131 \\ -7.050 \\ 3.447 \end{bmatrix}$$

Here $p = 2$ and $n = 4$. Therefore, $3\sqrt{2/2} = 3$. The last two observations are influential. □

Example 14.25. Figure 14.17, obtained with

```
> load(file = 'temperature.rda')
> source('confidence-interval.R')
> X <- temperature$Lat ; Y <- temperature$JanTemp
> model <- lm(Y ~ X) ; p <- 2 ; n <- length(X)
> coast <- c(5, 6, 12, 13, 41, 52, 53)
> DFFITS <- dffits(model)
> plot(model$fitted.values, DFFITS,
+    xlab = 'fitted', ylab = 'DFFITS')
> abline(h = 0)
> abline(h = 3 * sqrt(p / (n - p)), lty = 2)
> abline(h = -3 * sqrt(p / (n - p)), lty = 2)
> points(model$fitted.values[coast], DFFITS[coast], pch = 19)
> identify(model$fitted.values, DFFITS)
```

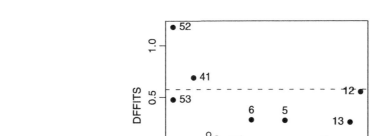

Figure 14.17 The $\Delta \widehat{Y}_i^*$ residuals for the city temperature data.

details the $\Delta \widehat{Y}_i^*$ residuals for the city temperature data. The broken horizontal line is the cutoff value. Comparing $\widehat{\varepsilon}_{(i)}^*$ (Figure 14.16) to $\Delta \widehat{Y}_i^*$, we see that San Francisco (6) is no longer above the cutoff level. □

14.6.6 The DFBETAS residuals

In simple linear regression, we estimate two parameters: $\widehat{\beta}_0$ and $\widehat{\beta}_1$. Remove the ith observation and recompute the regression coefficients. Denote these two coefficients by $\widehat{\beta}_{0(i)}$ and $\widehat{\beta}_{1(i)}$. The influence of removing the ith observation on the estimated coefficients can be estimated with $\widehat{\beta}_{0i} - \widehat{\beta}_{0(i)}$ and $\widehat{\beta}_{1i} - \widehat{\beta}_{1(i)}$. As was the case for $\Delta \widehat{Y}_{(i)}$, not all observations are born equal and we must rescale them. The scale of choice is $\sqrt{\text{Residual MS}_{(i)}} \times \sqrt{(\boldsymbol{A}^{-1})_{ii}}$ where $\boldsymbol{A}$ is given in (14.23) and $(\boldsymbol{A}^{-1})_{ii}$ is the ith element of the diagonal of $\boldsymbol{A}^{-1}$. We now have

Difference in coefficients (DFBETAS) Let $\widehat{\beta}_{0(i)}$ and $\widehat{\beta}_{1(i)}$ be the estimated values of β_0 and β_1 with the ith observation removed. Then

$$\Delta \widehat{\beta}^*_{0(i)} := \frac{\widehat{\beta}_0 - \widehat{\beta}_{0(i)}}{\sqrt{\text{Residual MS}_{(i)}} \sqrt{(\boldsymbol{A}^{-1})_{ii}}},$$

$$\Delta \widehat{\beta}^*_{1(i)} := \frac{\widehat{\beta}_1 - \widehat{\beta}_{1(i)}}{\sqrt{\text{Residual MS}_{(i)}} \sqrt{(\boldsymbol{A}^{-1})_{ii}}}$$

are the standardized differences of the estimates of β_0 and β_1 with the ith observation included and excluded.

Cutoff criterion for $\Delta \widehat{\beta}^*$ If $\left|\Delta \widehat{\beta}^*_{0(i)}\right| > 1$ or $\left|\Delta \widehat{\beta}^*_{1(i)}\right| > 1$ then x_i is influential in estimating the respective regression coefficient.

Example 14.26. Continuing with Example 14.24, we have

$$\Delta \widehat{\boldsymbol{\beta}}^*_0 = \begin{bmatrix} 0.482 \\ 0.097 \\ 0.000 \\ -1.682 \end{bmatrix}, \quad \Delta \widehat{\boldsymbol{\beta}}^*_1 = \begin{bmatrix} -0.396 \\ -0.053 \\ -2.878 \\ 2.764 \end{bmatrix}.$$

The last point has strong influence on the intercept and the two last points have strong influence on the slope. □

Example 14.27. Figure 14.18, produced with

```
> load(file = 'temperature.rda')
> source('confidence-interval.R')
> X <- temperature$Lat ; Y <- temperature$JanTemp
> model <- lm(Y ~ X) ; p <- 2 ; n <- length(X)
> coast <- c(5, 6, 12, 13, 41, 52, 53)
> par(mfrow = c(1, 2))
> DFBETAS <- dfbetas(model)
> DFBETAS.0 <- DFBETAS[,1] ; DFBETAS.1 <- DFBETAS[, 2]
> plot(model$fitted.values, DFBETAS.0,
+   xlab = 'fitted', ylab = 'intercept DFBETA')
> abline(h = 0) ; abline(h = 1, lty = 2)
> abline(h = -1, lty = 2)
> points(model$fitted.values[coast], DFBETAS.0[coast],
+   pch = 19)
> identify(model$fitted.values, DFBETAS.0)
> plot(model$fitted.values, DFBETAS.1,
+   xlab = 'fitted', ylab = 'slope DFBETA')
```

```
> abline(h = 0) ; abline(h = 1, lty = 2)
> abline(h = -1, lty = 2)
> points(model$fitted.values[coast], DFBETAS.1[coast],
+   pch = 19)
> identify(model$fitted.values, DFBETAS.1)
```

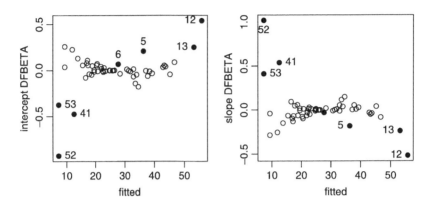

Figure 14.18 The $\Delta\widehat{\beta}^*$ residuals on the intercept and the slop of the regression between latitude and mean minimum January temperature for 53 U.S. cities.

details the $\Delta\widehat{\beta}^*$ for both regression coefficients. None of the temperatures influences the intercept. Only Seattle influences the slope. Why? Because it is far enough to the North to have enough influence on the slope. □

14.6.7 Cooke's distance

This measure indicates how large the influence of the ith point is on the combined values of the model coefficients.

Cook's distance (D) Let $\widehat{\varepsilon}_i$ be the ith residual, h_i the ith diagonal element of $\boldsymbol{H}$, p the number of model coefficients and Residual MS as given in (14.10). Then Cook's distance is defined as

$$D_i = \frac{h_i}{p}\left[\frac{\widehat{\varepsilon}_i}{\sqrt{\text{Residual MS}\,(1-h_i)}}\right]^2.$$

Cutoff criterion for D If $F_{p,n-p}$ for the ith observation is ≥ 0.5, then the distance is considered unusually large.

The sampling density of D_i is $F_{p,n-p}$. D_i is related to $\Delta\widehat{Y}_i^*$ as given in (14.26) thus

$$\Delta\widehat{Y}_i = \frac{\sqrt{\text{Residual MS}}}{\sqrt{\text{Residual MS}_{(i)}}}\sqrt{D_i p}\,.$$

Example 14.28. Figure 14.19, produced with

```
> load(file = 'temperature.rda')
> source('confidence-interval.R')
> X <- temperature$Lat ; Y <- temperature$JanTemp
```

```
> model <- lm(Y ~ X) ; p <- 2 ; n <- length(X)
> coast <- c(5, 6, 12, 13, 41, 52, 53)
> COOK <- cooks.distance(model)
> plot(COOK, xlab = 'observation number',
+   ylab ="Cook's distance", type = 'h', ylim = c(0, 1))
> cutoff <- qf(0.5, p, n - p, lower.tail = FALSE)
> abline(h = cutoff, lty = 2)
> points(coast, COOK[coast], pch = 19) ; identify(COOK)
```

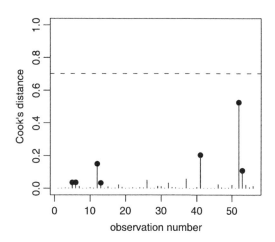

Figure 14.19   Cook's distance for the temperature data. The broken line indicates the cutoff value for influential residuals.

details Cook's distances for the temperature data. We find that $P(X > 0.702) = 0.5$ where X is a random variable from $F_{2,54}$ density. Therefore, the cutoff value for D_i is 0.702. None of the residuals is significantly influential.   □

14.6.8 Conclusions

The data for cities that we suspected to be influential (as identified in Figure 14.13) are detailed in Table 14.6. Spokane did not turn out to be influential and the two Florida cities were influential in the leverages. If we are to adhere to the residual analysis strictly, then Portland and Seattle emerge as the most influential cities. However, statistical procedures in general and residual analysis in particular are not straight jackets. They are used to enhance our understanding of the data.

Both Table 14.6 and Figure 14.12 lend support to the idea that the U.S. coastal cities (e.g. Los Angeles, San Francisco, Portland and Seattle) belong to a different model. Why? Because physical geography tells us that Oceans moderate temperatures along coasts. You might wonder why East Coast cities like New York and Boston do not distinguish themselves, along with the Florida cities, as well as the West Coast cities do. Here again, we must rely on our understanding of the underlying processes in nature. The warm Gulf Stream flows close to Florida and then heads East, leaving northern East Coast cities out in the cold. So they do not align themselves as clearly as the West Coast cities.

458 Simple linear regression

The take-home message is this: Statistical analysis enhances our understanding of the data. It should not replace it.

14.7 Power and sample size for the correlation coefficient

We wish to investigate the power of the test of significance under

$$H_0: \rho = 0 \quad \text{vs.} \quad H_A: \rho = \rho_0 > 0 \tag{14.27}$$

for a given α and given alternative model correlation ρ_0. We use Fisher's transformation (14.19). Under H_0, the mean of $\widehat{Z}$ is 0 and the variance is $1/(n-3)$. Therefore, we reject H_0 if

$$\widehat{Z}\sqrt{n-3} > z_{1-\alpha} .$$

Let z_0 be given as in (14.18). Then we reject H_0 also if

$$\widehat{Z}\sqrt{n-3} - z_0\sqrt{n-3} > z_{1-\alpha} - z_0\sqrt{n-3}$$

or

$$\left(\widehat{Z} - z_0\right)\sqrt{n-3} > z_{1-\alpha} - z_0\sqrt{n-3} .$$

Because $\widehat{Z}$ is a rv, so is $Z = \left(\widehat{Z} - z_0\right)\sqrt{n-3}$. Under H_1, the rv Z is standard normal. Therefore

$$P\left(Z > z_{1-\alpha} - z_0\sqrt{n-3}\right) = 1 - P\left(Z \leq z_{1-\alpha} - z_0\sqrt{n-3}\right)$$
$$= P\left(Z \leq z_0\sqrt{n-3} - z_{1-\alpha}\right) .$$

To obtain power at $1 - \beta$, we set

$$1 - \beta = z_0\sqrt{n-3} - z_{1-\alpha} .$$

Therefore,

Power for the correlation coefficient For the hypotheses (14.27), for the alternative $H_A := \rho_0$ with significance level α, the power $(1 - \beta)$ for sample size n, is given by

$$\text{power} = P\left(Z \leq z_0\sqrt{n-3} - z_{1-\alpha}\right)$$

where Z is a standard normal rv.

To obtain the corresponding sample size, we solve the power for n:

Sample size for the correlation coefficient For the hypotheses (14.27), for the alternative $H_A := \rho_0$ with significance level α and given power $1 - \beta$, the required sample size n, is given by

$$n = \left(\frac{z_{1-\alpha} + z_{1-\beta}}{z_0}\right)^2 + 3 .$$

14.8 Assignments

Exercise 14.1. Refer to Example 14.1.

1. How would you determine if `Item.ID` in `bmv.rda` includes leading or trailing white spaces?
2. Follow the ideas introduced in the example to rid `Item.ID` from potential leading and trailing white spaces.

Exercise 14.2. Use calculus to show that $\hat{\beta}_0 = -4.742$ and $\hat{\beta}_1 = 1.659$ in Example 14.7.

Exercise 14.3. In Example 14.8, we use the square transformation to relate tree diameter at breast height (DBH) to age. The transformation results in an apparent linear relationship. Why?

Exercise 14.4. What are the units of the intercept and slope of the following linear relationships?

1. Y - height (cm), x - weight (kg)
2. Y - basal metabolic rate (Kcal per hour), x - weight (kg)
3. Y - plants per m^2, x - m^2

Exercise 14.5. Write an R script that produces Figure 14.4 and prints the summary of the linear model.

Exercise 14.6. Here are data about the number of classes missed and the corresponding score on the final exam for 120 students in a statistics class (see `exercise-skipping-class.txt`):

```
 1  1  1  1  1  1  1  1  1  1  1  1  1  1  1  1  1  1  1  1
81 86 80 95 87 80 88 89 88 83 94 87 81 72 92 85 85 91 90
 1  1  1  1  1  1  1  1  1  2  2  2  2  2  2  2  2  2  2
89 91 90 85 73 89 85 84 76 74 80 85 76 79 77 69 75 75 77
 2  2  2  2  2  2  2  2  2  2  2  2  2  2  2  2  2  2  2
84 82 76 75 81 80 73 73 79 82 76 82 79 73 79 70 86 89 75
 2  2  2  3  3  3  3  3  3  3  3  3  3  3  3  3  3  3  3
71 80 76 84 70 74 70 66 71 59 79 71 83 73 66 74 64 62 72
 3  3  3  3  3  3  3  3  3  3  4  4  4  4  4  4  4  4  4
67 70 70 66 67 69 77 61 74 72 71 63 67 67 62 72 72 69 75
 4  4  4  4  4  4  4  4  4  4  4  4  4  4  4  4  4  4  4
68 57 62 58 62 61 65 60 66 61 76 69 70 67 75 61 62 74 61
 4  4  4  4  4  4
64 63 63 63 68 64
```

1. Plot the scatter of the data.
2. Add to the plot points that show the mean score for those who missed one class, two classes, three and four.
3. Add to the plot the regression line.
4. On the average, how many points can a student expect to lose from the score on the final exam?

Exercise 14.7. Using the data in Exercise 14.6, plot the expected values of Y and Y − predicted Y.

Exercise 14.8. Import `basal-metabolic-rate.txt'`. The column names identify the order, family, species, mass (M, in g), body temperature (T, in °C), and the basal metabolic rate (BMR, in kcal/hr). Next:

1. Plot the scatter of the numerical data in pairs and identify a pair of columns that indicate potential linear relationship.
2. Plot a scatter of the log of this pair.
3. Add the regression line to this scatter.
4. Identify and label by species name extreme points in the scatter plot.
5. Write the formula that relates your two variables.
6. Does the overall model fit the data?

Exercise 14.9. Explain the following script (see Example 14.2):

```
1   rm(list=ls())
2
3   wb <- read.csv('world-bank.csv', header = TRUE, sep = ',',
4     stringsAsFactors = FALSE)
5   names(wb) <- c('country', 'indicator', '1999', '2000',
6     '2001', '2002', '2003')
7
8   wb.split <- split(wb[,-2],wb$indicator)
9   names(wb.split) <- levels(wb$indicator)
10
11  out <- function(d, x, y, xlab, ylab, trans = FALSE){
12    countries <- intersect(d[[x]][,1], d[[y]][,1])
13    index <- vector()
14    for(i in 1 : length(countries)){
15      index[i] <- which(d[[x]][, 1] == countries[i])
16    }
17    if(trans) xx <- log(d[[x]][index, 3])
18    else xx <- d[[x]][index, 3]
19    index <- vector()
20    for(i in 1 : length(countries)){
21      index[i] <- which(d[[y]][, 1] == countries[i])
22    }
23    if(trans) yy <- log(d[[y]][index, 3])
24    else yy <- d[[y]][index, 3]
25    plot(xx, yy, xlab = xlab, ylab = ylab)
26    model <- lm(yy ~ xx)
27    abline(reg = model)
28    print(summary(model))
29  }
30
31  openg(4,4)
```

```
32  par(mfrow=c(2, 2))
33  # energy use vs CO2 emissions
34  out(wb.split, 4, 2, 'log(CO2 emissions)',
35    'log(energy use)', trans = TRUE)
36  # GDP vs CO2 emissions
37  out(wb.split, 2, 6, 'log(CO2 emissions)', 'log(GDP)',
38    trans = TRUE)
39  # fertility rate vs infant mortality
40  out(wb.split, 5, 8, 'fertility rate', 'life expectancy')
41  # edu vs infant mortality
42  out(wb.split, 11, 12, 'infant mortality',
43    'children mortality')
44  #saveg('world-bank', 4, 4)
45
46  print(c(sqrt(0.8415), sqrt(0.2498), sqrt(0.6801),
47    sqrt(0.9766)))
```

Exercise 14.10. Write a script to obtain $\hat{\rho}$ from Example 14.12.

Exercise 14.11. Write an R script that reproduces Figure 14.9.

Exercise 14.12. Write an R script that reproduces Figure 14.10.

Exercise 14.13. Use Examples 14.23 to 14.28 as guidelines.

1. Run all diagnostics as shown in the examples and identify where British Columbia and Alaska coastal cities fit in the model.
2. Based on the diagnostics, do coastal cities in Alaska and British Columbia stand out? Explain.

15

Analysis of variance

In many ways, analysis of variance (ANOVA) is similar to linear regression. The main difference is in our treatment of the independent variable. Recall that in linear regression, we chose both variables to be numeric (decimal). In ANOVA, it is (they are) factors or occasionally ordered factors. As we shall see in this chapter, in modern regression, the distinction between linear models and ANOVA blurs. Roughly speaking, ANOVA is simply a different way of summarizing results. This and the nature of the ANOVA introduce difficulties in applying the analysis and interpreting the results. We will discuss the major types of ANOVA; however, the topic is large and often requires careful considerations of experimental design, applying the analysis and interpretation.

Like the Australian aborigines (who count one, two, many), ANOVA may be broadly classified as one-way, two-way and many-way. We will not go beyond two-way. One-way ANOVA may be further classified into fixed- and random-effects ANOVA and within these categories, we might have balanced and unbalanced designs. Two-way ANOVA may be classified similarly with the addition of mixed effects (fixed and random). We shall elaborate upon some of these ideas in this chapter.

To run ANOVA often requires a bit of work in preparing data for analysis. Information from different data files need to be coalesced, factors introduced in the right order and so on. We will use mostly large, publicly available data. This will give us a chance to do some heavy data manipulations.

15.1 One-way, fixed-effects ANOVA

One-way ANOVA is a method to analyze data where one variable is a factor (categorical) and the other is numeric. If the levels of the factor variable are fixed, then we have *fixed-effects* ANOVA. For example, we may classify people into various ethnic groups (the factor) and study the relationship between ethnicity and income (the

latter is the dependent variable). Say we have four ethnic groups. For each group we
may have equal number of observations. This is called a *balanced design*. If the number
of observations per group are not equal, we have *unbalanced design*. In fixed-effects
ANOVA we are interested in difference among group means.

15.1.1 The model and assumptions

Let us start with an example which, by the way, will tax R's ability to deal with large
data sets. We also show how to draw maps with R.

Example 15.1. The European Union (EU) maintains large data sets about air
quality. We are interested in comparing the means of the maximum value of atmospheric sulfur dioxide (SO_2) for three cities: Berlin, Madrid and Rome for the year
2005. We download the air quality data from http://dataservice.eea.europa.eu/
dataservice/metadetails.asp?id=949. The files are named Airbase_v1_station.
txt and Airbase_v1_statistic. txt. The former provides information about the
data collection stations and the latter contains the 190 million bytes (Mb) of data.
First, we import the data and save it for later use:

```
> EU.station <- read.table('Airbase_v1_station.txt',
    header = TRUE, sep = ',', stringsAsFactors = FALSE)
> save(EU.station, file = 'EU.station.rda')
> EU <- read.table('Airbase_v1_statistic.txt', header = TRUE,
    sep = ',', stringsAsFactors = FALSE)
> save(EU, file = 'EU.rda')
```

An important note about reading text files Occasionally, fields in text files may
contain a single quote, for example a French name, like *Count d'Money*. This will
confuse read.table() and its allies—they will lose count of the number of fields
in a row!

Importing the statistics data takes a while, but R saves it in a binary file of size 5.9 Mb
only. The station information includes

```
> load('EU.station.rda')
> names(EU.station)
 [1] "station_EUropean_code" "country_iso_code"
 [3] "country_name"          "type_of_station"
 [5] "station_type_of_area"  "station_longitude_deg"
 [7] "station_latitude_deg"  "station_altitude"
 [9] "sabe_country_code"     "station_city"
```

and the columns of interest in the statistics data are:

```
> head(EU[c(3, 13, 15)])
  component_caption statistic_shortname statistic_value
1               S02         Days(c > 125)           0.000
2               S02                  Max4          77.083
3               S02                   Max          91.875
4               S02                  Mean          19.135
5               S02                   P50          14.375
6               S02                   P95          45.833
```

Next, we want to extract the data for the maximum sulfur dioxide (SO_2) for Berlin, Madrid and Rome for 2005. But first, let us see a map where the measurements had been collected. We start by loading the necessary data and packages,

```
> load('EU.station.rda')
> library(maps)
> library(mapdata)
```

identifying the regions we wish to draw and their colors

```
> r <- c('Spain', 'Italy', 'Germany')
> col<-colors()[c(360, 365, 370)]
```

and plotting the map with its coordinates (longitude and latitude):

```
> map('world', regions = r, fill = TRUE, col = col)
> map.axes()
```

(Figure 15.1). To name the countries, we use

```
> text(c(-2, 7, 3), c(36, 43, 51), labels = r)
```

To isolate the coordinates of the stations in the three countries, we do

```
> m <- match(EU.station[, 3], c('ITALY', 'SPAIN', 'GERMANY'),
+     nomatch = 0)
```

match() returns the occurrences of all elements in the right argument (the countries) in the left argument (the third column in EU.station). The returned values are integers. We ask that no matching locations in the third column be set to zero. So now we can extract the desired

```
> long <- EU.station[m > 0, 6]
> lat <- EU.station[m > 0, 7]
```

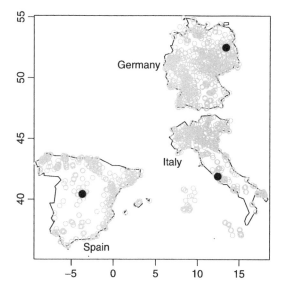

Figure 15.1   Stations (circles) where air quality data had been collected in Spain, Italy and Germany. Black disks locate Berlin, Madrid and Rome.

Finally, we draw the station locations:

```
> points(long, lat, col=c(360))
```

To add the points for our cities, we do a quick search on the Web and then

```
> cities.long <- c(13 + 25 / 60, -(3 + 42/60),
+    12 + 27 / 60)
> cities.lat <- c(52.5, 40 + 26 / 60, 41 + 54 / 60)
> points(cities.long, cities.lat, col = 'black',
+    pch = 19, cex = 1.5)
```

Now to extract the data, we need both EU.rda and EU.station.rda. The stations data associate the station codes with their nearby city. So we need to extract these station codes from the stations data *and* then use these codes to extract the data from the statistics file. First, we extract the stations that correspond to the cities of interest:

```
> m <- match(toupper(EU.station[, 10]),
+    c('MADRID', 'ROMA', 'BERLIN'), nomatch = 0)
> stations <- EU.station[m > 0, 1]
```

Note the use of toupper(). The function changes all of a string's letters to upper case. We do this because in the data, Berlin is written as BERLIN or Berlin. Next, let us extract the data for SO_2 from the stations in the cities of interest:

```
> tmp <- EU[EU$statistic_shortname == 'Max' &
+    EU$component_caption == 'SO2', ]
> m <- match(tmp[, 1], stations, nomatch = 0)
> tmp <- tmp[m > 0, c(1, 8, 15)]
```

(the eighth column includes the year the data were collected). Next, we extract and tighten up the data for 2005

```
> tmp <- tmp[tmp[, 2] == 2005, c(1, 3)] ; head(tmp, 3)
       station_european_code statistic_value
351946                 DE0715A          28.083
351953                 DE0715A          38.000
362262                 DE0742A          30.217
```

The first two letters of the station code include the country; the data are for known cities in each country, for SO_2 for 2005. We add a column for city name like this:

```
> n  <- tapply(substr(tmp[, 1], 1, 2), substr(tmp[, 1], 1, 2),
+    length)
> city <- c(rep('Berlin', n[1]), rep('Madrid', n[2]),
+    rep('Roma', n[3]))
> SO2 <- cbind(tmp, city)
> names(SO2)[1:2] <- c('station', 'Max SO2') ; head(SO2, 3)
        station Max SO2   city
351946 DE0715A  28.083 Berlin
351953 DE0715A  38.000 Berlin
362262 DE0742A  30.217 Berlin
> save(SO2, file = 'SO2.rda')
```

substr() extracts a substring that begins in the second argument (1) and end in the third (2 in our case). In the next example we shall start with the data analysis. □

As you can see, extracting desired data is not a trivial task. In Example 15.1 we have three groups. The number of groups is denoted by k and in R they are represented by a factor variable with factor levels Berlin, Madrid and Roma. So k ($=3$) is the number of levels of the factor variable city. Let us denote the *response* variable (Max.SO2 in our case) by the rv Y. In each group (for each city), we have a value of Y_{ij} where i is the group index ($i = 1, 2, 3$) and j is the measurement index within a group.

Example 15.2. Continuing with Example 15.1, Table 15.1 illustrates the notation we use. Here $k = 3$, $i = 1, \ldots, k$. For $i = 1$, j goes from 1 to 12 and so on. Figure 15.2 illustrates the notation. To produce the figure, we load and plot the data with no axes (axes), x-label with a subscript notation (xlab) and letters larger than usual (cex):

```
> plot(SO2[, 2], SO2[, 3], axes = FALSE,
+      xlab = expression(SO[2]),
+      ylab = 'city', ylim = c(1, 3.2),
+      xlim = c(0, 160), cex = 1.2)
```

We then draw the default x-axis and the y-axis with tick marks at 1, 2, 3 and the appropriate tick labels:

```
> axis(1)
> axis(2, at = c(1, 2, 3),
+      labels = c('Berlin', 'Madrid', 'Roma'))
```

Table 15.1 ANOVA notation for the EU data (Example 15.1)

Factor level	i	Measurement index (j)	Measurement value
Berlin	1	1	28.083
⋮	⋮	⋮	⋮
Berlin	1	12	34.000
Madrid	2	1	48.800
⋮	⋮	⋮	⋮
Madrid	2	50	92.890
Roma	3	1	26.909
⋮	⋮	⋮	⋮
Roma	3	6	27.000

Next, we draw horizontal lines for each city, calculate city means and add the them with appropriate notation (pch):

```
> abline(h = c(1, 2, 3))
> means <- tapply(SO2[, 2], SO2[, 3], mean)
> points(means, 1 : 3, pch = '|', cex = 2)
```

468 Analysis of variance

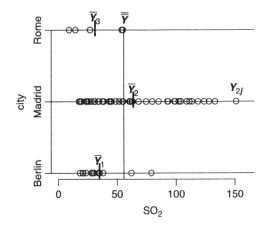

Figure 15.2 Maximum SO_2 air pollution in Berlin, Madrid and Rome.

Here

```
> Y.bar.bar <- mean(SO2[, 2]) ;
> lines(c(Y.bar.bar, Y.bar.bar), c(0, 3.01))
```

we add a vertical line for $\overline{\overline{Y}}$ (we could do this with `abline(v = Y.bar.bar)` but it does not look nice). We use `expression()` to draw the notation. The tricky one is

```
> text(SO2[26, 2], 2,
+      labels = expression(bolditalic(Y[2*j])),
+      pos = 3)
```

You will not get Y_{ij} correctly unless you juxtapose 2 and j with *. □

Now we assume the model:
$$y_{ij} = \mu + \alpha_i + \varepsilon_{ij} \ . \tag{15.1}$$

The corresponding sample variables are detailed in Table 15.2. If the k samples (groups) are taken from a single population, then μ is the mean of the population. In the model, α_i are the differences between the overall population mean (μ) and the mean of each group (sample). Finally, ε_{ij} are the random errors about $\mu + \alpha_i$, with density $\phi(0, \sigma)$. We are therefore assuming that the *variances within each group are all equal* to σ^2. If we want to estimate μ and all α_i, we have to estimate $k + 1$ parameters. This is not possible because we have k observed means and we wish to estimate $k + 1$ parameters. So we shall *assume* that

$$\sum_{i=1}^{k} \alpha_i = 0 \ .$$

We thus have the following definition

One-way ANOVA We say that (15.1) is a one-way ANOVA.

Table 15.2 Population and sample quantities in ANOVA.

Population	sample	
y_{ij}	Y_{ij}	Observation j in group i
μ	$\overline{\overline{Y}}$	total mean
$\mu + \alpha_i$	$\overline{Y}_i$	mean for group i

Note that if we assign α_i in (15.1) to $\beta_1 X_i$ in (14.2), then ANOVA is essentially a linear model. Now if each group's density is normal with variance σ^2, we can obtain a meaningful statistics with known density that allows us to compare the arbitrary (k) number of means. The comparison involves the null hypothesis that *all k group means are equal*.

Example 15.3. Applying these ideas to our SO_2 pollution data (Example 15.2), we say that each value can be predicted by the overall mean plus the group mean ("city effect") plus random-effects within each group (city). □

In summary, model (15.1) assumes:

1. independent samples
2. equal group variances
3. normal error with mean 0 and variance σ^2

The last assumption implies that the data are normal.

Example 15.4. In the case of our SO_2 pollution data, we may be violating the assumption of independent samples. For example, data from nearby monitoring stations taken during the same time may be dependent. The sampling dates were not available and so we assume (perhaps erroneously) that the data are independent. The assumptions of equal variance and normality of errors can be tested (see Sections 12.2.2 and 7.2.4, respectively). □

15.1.2 The F-test

ANOVA involves the F-test. As opposed to the t-test, which applies to pairs of means, the F-test applies to overall comparison of *all* means.

Example 15.5. With the SO_2 pollution data, we test the null hypothesis that the means of maxima of atmospheric SO_2 concentrations (the response variable) were equal in Berlin, Madrid and Rome during 2005. □

Formally, with one-way, fixed-effects ANOVA, we test the null hypothesis (H_0) that the means of all group are equal, or equivalently, that each $\alpha_i = 0$. The alternative hypothesis is that at least one $\alpha_i \neq 0$. Under the null hypothesis, we construct the so-called F-statistic whose sampling density is known to be $F(k-1, n-k)$. Here F is the density (detailed in Section 6.8.5) and $k-1$ and $n-k$ are the degrees of freedom. k is the number of groups (sometimes called treatments or effect) and n

is the total number of observations. To construct the sampling density, consider the deviation (under the null hypothesis) $Y_{ij} - \overline{\overline{Y}}$ which may be written as

$$Y_{ij} - \overline{\overline{Y}} = \underbrace{Y_{ij} - \overline{Y}_i}_{\text{within group deviation}} + \underbrace{\overline{Y}_i - \overline{\overline{Y}}}_{\text{between group deviation}}.$$

Squaring and summing over the appropriate number of observations we obtain

$$\underbrace{\sum_{i=1}^{k}\sum_{j=1}^{n_i}\left(Y_{ij} - \overline{\overline{Y}}\right)^2}_{\text{Total SS}} = \underbrace{\sum_{i=1}^{k}\sum_{j=1}^{n_i}\left(Y_{ij} - \overline{Y}_i\right)^2}_{\text{Within SS}} + \underbrace{\sum_{i=1}^{k}\sum_{j=1}^{n_i}\left(\overline{Y}_i - \overline{\overline{Y}}\right)^2}_{\text{Between SS}}$$

where SS stands for Sum of Squares and n_i denotes the number of observations in the ith group. In our notation, the last equation is written as

$$\text{Total SS} = \text{Within SS} + \text{Between SS}.$$

Now the means of Within SS and Between SS, denoted by Within MS and Between MS are

$$\text{Within MS} = \frac{1}{n-k}\sum_{i=1}^{k}\sum_{j=1}^{n_k}\left(Y_{ij} - \overline{Y}_i\right)^2,$$
$$\text{Between MS} = \frac{1}{k-1}\sum_{i=1}^{k}\sum_{j=1}^{n_k}\left(\overline{Y}_i - \overline{\overline{Y}}\right)^2 \quad (15.2)$$

(for computations, more efficient formulas are used).

Example 15.6. Let us implement these computations for the SO_2 data (Example 15.1) in R. First, we load the data

```
> load('SO2.rda')
```

Next, we determine n_i for $i = 1, 2, 3$ and the total mean $(\overline{\overline{Y}})$ with

```
> (n <- tapply(SO2[, 2], SO2[, 3], length))
Berlin Madrid   Roma
    12     50      6
> (Total.mean <- mean(SO2[, 2]))
[1] 55.68776
```

The group means are

```
> (Group.means <- tapply(SO2[, 2], SO2[, 3], mean))
  Berlin   Madrid     Roma
34.86800 63.63320 31.11533
```

To simplify scripts for later computations, we add the group means to SO2

```
> (SO2 <- data.frame(SO2, Group.means = c(
+    rep(Group.means[1], n[1]), rep(Group.means[2], n[2]),
+    rep(Group.means[3], n[3]))))
       station Max.SO2    city Group.means
351946 DE0715A  28.083  Berlin     34.86800
351953 DE0715A  38.000  Berlin     34.86800
362262 DE0742A  30.217  Berlin     34.86800
...
```

The Total SS is

```
> (Total.SS <- sum((SO2[, 2] - Total.mean)^2))
[1] 79462.8
```

The Within SS is

```
> (Within.SS <- sum(tapply((SO2[, 2] - SO2[, 4])^2,
+     SO2[, 3], sum)))
[1] 67481.92
```

and the Between SS is

```
> (Between.SS <- sum(tapply((SO2[, 4] - Total.mean)^2,
+     SO2[, 3], sum)))
[1] 11980.87
```

The mean squares are

```
> (Within.MS <- Within.SS / (sum(n) - length(n)))
[1] 1038.183
> (Between.MS <- Between.SS / (length(n) - 1))
[1] 5990.437
```

The computations can be done a bit more efficiently but this way we can see explicitly what is going on. □

Now the ratio Between MS / Within MS has a known sampling density; it is F with $k - 1$ and $n - k$ degrees of freedom (see Section 6.8.5). Obviously, as Between MS grows for fixed Within MS, the contribution of the variance between the groups to the total variance compared to the variance within the groups grows. So when the F-statistic (= Between MS / Within MS) grows, it will reach a value large enough for us to reject the hypothesis that all group means are equal (our H_0).

The null and alternative hypotheses are

$H_0 :=$ all group means are equal to the total mean vs.

$H_A :=$ at least one of the means is different from the total mean .

Note that H_0 implies that $\sum \alpha_i = 0$. Next, we calculate the statistic

$$F = \frac{\text{Between MS}}{\text{Within MS}}$$

where Between MS and Within MS are calculated according to (15.2). Then, if for significance level α,

$$p\text{-value} = 1 - \text{pf}(F, k - 1, n - k) < \alpha$$

(where pf() is the F *distribution* in R), reject the null hypothesis in favor of the alternative.

Let us pause for a moment to discuss the difference between balanced and unbalanced design by example.

Example 15.7. Let A, B and C be the treatments. The response values for each treatment were generated from a normal density with means 10, 20 and 30 and

SD = 15 (see `balanced-vs-unbalanced-design.R` in the book's site). Here are the results for balanced and unbalanced design

```
> source('balanced-vs-unbalanced-design.R')
    Within.MS Between.MS      F p.value
SS     199.97      687.8   3.44    0.05
n       10.00       10.0  10.00      NA
    Within.MS Between.MS      F p.value
SS     200.33     430.73   2.15    0.14
n        5.00      15.00  10.00      NA
    Within.MS Between.MS      F p.value
SS     195.65     613.65   3.14    0.06
n       15.00      10.00   5.00      NA
```

In short, for each of the three analyses, 30 samples were drawn from the same population but group allocation differed. The balanced design indicates significant F value and is easiest to interpret. In the other two cases, a firm conclusion is more difficult because the unclear effect of sample sizes. □

Let us run ANOVA on the SO_2 air pollution.

Example 15.8. Back to Example 15.6. First, we load the data

```
> load('SO2.rda')
```

The sample sizes are

```
> (n <- tapply(SO2$'Max SO2', SO2$city, length))
Berlin Madrid   Rome
    12     50      6
```

So we have unbalanced number of observations for each group. We will deal with such analysis later. For now, let us use a balanced design:

```
set.seed(101) ; n.c <- cumsum(n)
i.1 <- sample(1 : n.c[1], n[3])
i.2 <- sample((n.c[1] + 1) : n.c[2], n[3])
i.3 <- (n.c[2] + 1) : n.c[3]
d <- SO2[c(i.1, i.2, i.3), ]
```

To test if the variances are equal, we follow Section 12.2.2:

```
> (vars <- tapply(d[, 2], d[, 3], var))
   Berlin   Madrid     Roma
 221.8475 669.1904 376.5391
```

Now that we have the variances and sample sizes, we lower-tail test all pairs of ratios:

```
> (vars.ratio <- c(vars[1]/vars[2],
+      vars[1] / vars[3], vars[3] / vars[2]))
   Berlin    Berlin      Roma
0.3315163 0.5891752 0.5626786
> (p.L <- c(pf(vars.ratio[1], 5, 5),
+      pf(vars.ratio[2], 5, 5),
+      pf(vars.ratio[3], 5, 5)))
   Berlin    Berlin      Roma
0.1254588 0.2878244 0.2716446
```

None of the ratios is significant and we conclude that all variances are equal. Thus we meet one of the ANOVA assumptions. To test for normality of the data we use the Q-Q plot

```
> par(mfrow = c(1, 3))
> qqnorm(d[d[, 3] == 'Berlin', 2], main = 'Berlin')
> qqline(d[d[, 3] == 'Berlin', 2])
```

(similarly for Madrid and Rome) to obtain Figure 15.3 and conclude that the data for each group are marginally normal. So we proceed with the ANOVA:

```
> a <- aov(d$'Max SO2' ~ d$city, data = d)
> summary(a)
           Df Sum Sq Mean Sq F value Pr(>F)
d$city      2 2035.8  1017.9  2.4091 0.1238
Residuals  15 6337.9   422.5
```

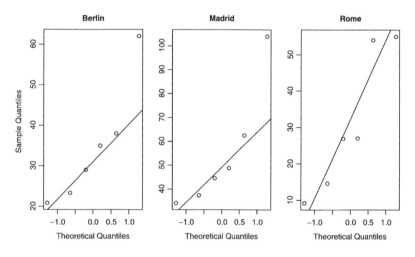

Figure 15.3   Q-Q plots for SO_2 atmospheric pollution.

R uses the column name of the "treatment" to name the row for the Between MS, hence the row name city. The Within MS are named Residuals. From the p-value we conclude that the means of SO_2 pollution in these three cities were not different for 2005. If you wish, you can produce a nice graphical output of your ANOVA like this:

```
> library(granova)
> granova.1w(SO2$Max.SO2, SO2$city)
```

(Figure 15.4). Except for contrasts, the graph is self explanatory (see granova()'s help). We shall meet contrasts soon. □

R's output of aov() is standard. Such output is called ANOVA table and should always be reported with the analysis. We know how to test for means equality. If we reject H_0, we follow up by more detailed analysis to discover which of the means are responsible for rejecting H_0 (we shall pursue this in the next section).

Next, let us see how to run ANOVA with unbalanced design.

474 Analysis of variance

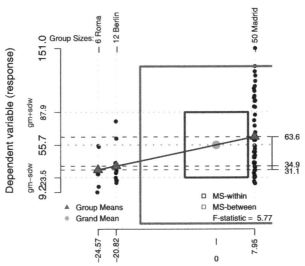

Figure 15.4   Plot of ANOVA output.

Example 15.9. We repeat the analysis in Example 15.8, but this time, for unbalanced design:

```
> load('SO2.rda')
> (n <- tapply(SO2$'Max SO2', SO2$city, length))
Berlin Madrid   Roma
    12     50      6
> d <- SO2
```

Repeat of the analysis on the differences of variances among the groups reveals that they are different. So we log-transform the response variable:

```
> d[, 2] <- log(d[, 2])
> (vars <- tapply(d[, 2], d[, 3], var))
    Berlin    Madrid      Roma
0.1919883 0.3424642 0.5029848
> (vars.ratio <- c(vars[1]/vars[2],
+      vars[1] / vars[3], vars[3] / vars[2]))
    Berlin    Berlin      Roma
0.5606085 0.3816980 1.4687225
> (p.L <- c(pf(vars.ratio[1], 5, 5),
+      pf(vars.ratio[2], 5, 5), pf(vars.ratio[3], 5, 5)))
    Berlin    Berlin      Roma
0.2703705 0.1570159 0.6582729
```

Now paired comparisons reveal that the variances are equal. Because the model is unbalanced, interpreting the sums of squares is difficult. Instead of working with aov() to obtain the ANOVA table (as in Example 15.8), we go through linear regression explicitly and then get the table through ANOVA on the regression results:

```
> model <- lm(d$'Max SO2' ~ d$city)
> anova(model)
Analysis of Variance Table

Response: d$"Max SO2"
          Df  Sum Sq Mean Sq F value   Pr(>F)
d$city     2  5.0599  2.5300  7.6818 0.001012
Residuals 65 21.4075  0.3293
```

The differences are significant with group means

```
> tapply(SO2[, 2], SO2[, 3], mean)
  Berlin   Madrid     Roma
34.86800 63.63320 31.11533
```

The means indicate that as far as atmospheric SO_2 is concerned, you probably would not have wanted to live in Madrid in 2005 (Rome is nice). To verify that we conform to the assumptions of linear models (and therefore ANOVA), we can

```
> par(mfrow = c(2, 2))
> plot(model)
```

Figure 15.5 indicates that we meet the usual assumptions of linear models (and therefore of the ANOVA). You can obtain the same diagnostics with

```
> plot(aov(d$'Max SO2' ~ d$city))
```

Now compare Figure 15.6 obtained with

```
> library(granova)
> granova.1w(d$'Max SO2', d$city)
```

to Figure 15.4. □

The results for anova() in Example 15.9 are identical to those obtained from aov(). This is so because aov() is just a wrapper for lm() and provides for automatic printing of the ANOVA table. However, as we shall soon see, when dealing with two-way ANOVA, it is often nice to explicitly run the linear model first and then produce the ANOVA table with anova().

If we find that the F-test is not significant, we are done. However, if the test is significant, we may wish to pursue a more detailed analysis.

15.1.3 Paired group comparisons

Suppose we find that the F-test is significant (if it is not, do *not* pursue the paired comparisons). We reject H_0 in favor of H_A. But as H_A states, at least one group is different. We wish to investigate which one or ones. But there is a potential problem. If, once we see the ANOVA results we decide to search for significant group mean differences (this is called *post-hoc* analysis), then for many comparisons, some might be significant by chance alone. To guard against this possibility, we choose to use the

476 Analysis of variance

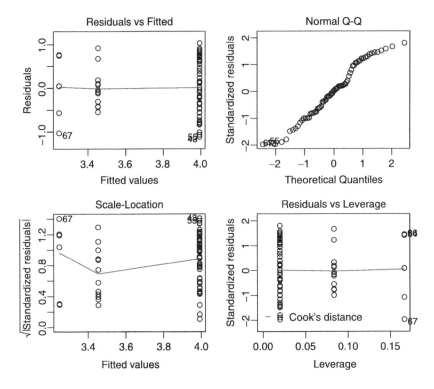

Figure 15.5 ANOVA diagnostics.

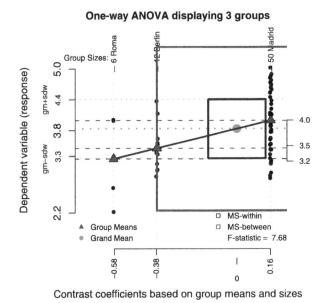

Figure 15.6 Plot of ANOVA output.

Bonferroni test (there are others), which amounts to adjusting α downward as the number of comparisons increases. Paired comparisons and their adjustments are the topics of this section.

Comparing pairs of groups–the least significant (LSD) method

We wish to compare group i to group j for $i \neq j$ and $i, j = 1, \ldots, k$. Based on the ANOVA assumptions (of groups normality) we are after the sampling density of $\overline{Y}_i - \overline{Y}_j$. Under the null hypothesis that two group means are equal, the appropriate sampling density is

$$\phi\left(0, \sqrt{\sigma^2 \left(\frac{1}{n_i} + \frac{1}{n_j}\right)}\right)$$

where n_i and n_j are the respective group sample sizes and σ^2 is the assumed equal variance of the groups. Because σ^2 is not known, in the usual t-test we estimate it with the *pooled* variance

$$S_P^2 = \frac{(n_i - 1) S_i^2 + (n_j - 1) S_j^2}{n_i + n_j - 2}$$

which, under the null hypothesis of equal variance is our Within MS in (15.2). When we have only two samples, the df are $n_i + n_j - 2$. However, we estimate the pooled variance from k samples and therefore must revise the df to $n_1 - 1 + \cdots + n_k - 1 = n - k$. Because we are estimating S^2, our test statistic is

$$T_{ij} = \frac{Y_i - Y_j}{\sqrt{S^2 \left(\frac{1}{n_i} + \frac{1}{n_j}\right)}}$$

where S^2 is the *total* variance (*not* the pooled variance S_P^2) and the sampling density of T_{ij} is t_{n-k}. In summary, for each pairs of groups among the k group we have

$$H_0 : \alpha_i = \alpha_j \quad vs. \quad H_A : \alpha_i \neq \alpha_j$$

with α level of significance. As usual, we reject H_0 if the p-value $= 1 - $ pt(abs(T$_{ij}$), $n - k) < \alpha/2$. This test is often referred to as the least significant difference (LSD). The next example demonstrates how to apply the LSD test, how to deal with infinity (Inf) in R and how to interpret bar plots.

Example 15.10. The data are from a 2004–2005 survey by the US Center for Disease Control (CDC). It was obtained from http://www.cdc.gov/nchs/about/major/nhanes/nhanes2005-2006 We are interested in the demographics survey; a file named demo_d. It is available for download in SAS format and can be imported with read.xport() in the package foreign (we shall skip this step). The file demo_d.short.rda, available from the book's site, includes a subset of the variables in demo_d.rda. The latter contains the full data set, as imported from the CDC site. So, we load:

```
> load('demo_d.short.rda')
```

We wish to run ANOVA on mean yearly household income by ethnicity[1] groups:

```
> names(demo_d.short[, c(4, 10, 11)])
[1] "Race/Ethnicity"        "Household Income from"
[3] "Household Income to"
```

The second and third variables include yearly household income categories from zero to $24 999 in increments of $5 000, then from $25 000 to $74 999 in increments of $10 000. All incomes are pooled for $75 000 and above. Thus, the corresponding last column includes `Inf`—an infinitely large number that R knows how to deal with arithmetically. We want to remove cases with `NA` with `complete.cases()`, to save on typing rename the data frame to i (for income) and rescale:

```
> i <- demo_d.short[, c(4, 10, 11)]
> i <- i[complete.cases(i), ]
> i[, 2] <- i[, 2] / 1000
> i[, 3] <- i[, 3] / 1000 + 0.001
```

Next, we extract ethnicity and the mean of the income categories with units of $1 000 and reassign the data to i:

```
> i <- data.frame(i[, 1], i[, 2] + (i[, 3] - i[, 2])/2)
```

Before going on, we want to examine what we have:

```
> par(mfrow = c(1, 2))
> barplot(table(i$income), las = 2,
+     xlab = 'mean yearly household income category (in $1,000)',
+     ylab = 'count')
```

(Figure 15.7, left). Note that R deals with `Inf` correctly with no extra effort. To remove the "infinite" income (effectively so in some cases), we do

```
> i <- i[i$income <= 75, ]
```

and repeat the `barplot()` (Figure 15.7, right). Now the bar-plot is a count in categories and the data represent a random sample from the U.S. population. So we see that as far as the lower household income "brackets" are concerned, the infinite income category is the largest. For the ANOVA and LSD test, we use the data without the `Inf` income. This will give us an idea about the distribution of income between 0 and $75 000 and help us analyze the data without having to resort to medians.

Let us see what we have:

```
> attach(i)
> par(mar = c(15, 4, 1, 2))
> plot(i, las = 2, ylim = c(0, 80), xlab = '',
+     ylab = 'household yearly income, from (in $1,000)')
```

(Figure 15.8). The `mar` parameter sets the distance of the plotting region (in lines) from the bottom, left, top and right margins of the drawing region. We set it so that we can see the full ethnicity names.

---

[1] Adhering to the biological definition of race, we do not subscribe to the CDC's classification of people of different skin color as being of different races, but we use their terminology to avoid confusion.

One-way, fixed-effects ANOVA 479

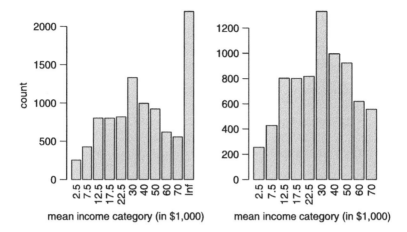

Figure 15.7 Mean yearly household income category with $75 000 and above category assigned infinite income (left) and with it removed (right).

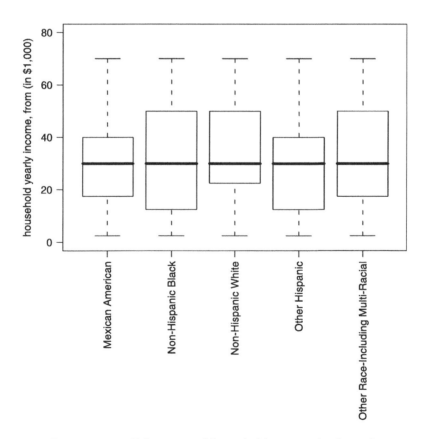

Figure 15.8 Ethnicity and household income (in $1 000).

It is worthwhile to keep in mind the figure when observing the results of the LSD test. Before diving into the ANOVA, we usually need to verify that the data conform to the ANOVA assumptions. The assumption of normality is no problem, for we have large samples. We shall skip the test for equality of variances. The ANOVA

```
> model <- aov(income ~ ethnicity, data = i)
> summary(model)
             Df  Sum Sq Mean Sq F value Pr(>F)
ethnicity     4   56192   14048    40.3 <2e-16
Residuals  7524 2620386     348
```

tells us that the mean incomes by ethnicity are significantly different. But from Figure 15.8, perhaps not all of them. At this point, we run diagnostics on the ANOVA with

```
> par(mfrow = c(2,2))
> plot(model)
```

(output not shown) which leads us to conclude that the data meet the ANOVA assumptions.

To run the LSD test, we load

```
> library(agricolae)
```

obtain the degrees of freedom and the Within MS

```
> df<-df.residual(model)
> MS.error<-deviance(model)/df
```

and run the test for $\alpha = 0.05$ (output edited):

```
> MS.error<-deviance(model)/df
> LSD.test(income, ethnicity, df, MS.error, group = FALSE,
+     main = 'household income\nvs. skin color/ethnicity')

Study: household income
vs. skin color/ethnicity

LSD t Test for income
                           ......
Alpha                      0.050
Error Degrees of Freedom   7524.000
Error Mean Square          348.270
Critical Value of t        1.960

Treatment Means
                              ethnicity income std.err
1                      Mexican American  30.85  0.3648
2                    Non-Hispanic Black  30.93  0.4168
3                    Non-Hispanic White  36.56  0.3806
4                        Other Hispanic  30.01  1.1182
5 Other Race - Including Multi-Racial   35.62  1.0881
```

Comparison between treatments means

```
   tr.i tr.j   diff pvalue
1    1    2 0.08106 0.8855
2    1    3 5.70735 0.0000
3    1    4 0.84168 0.4909
4    1    5 4.77370 0.0000
5    2    3 5.62629 0.0000
6    2    4 0.92275 0.4517
7    2    5 4.69264 0.0000
8    3    4 6.54903 0.0000
9    3    5 0.93365 0.3794
10   4    5 5.61538 0.0002
```

From the results, we observe that the difference between 1 and 2 (Mexican American and Non-Hispanic Black) is *not* significant. The difference (of yearly mean income) between 1 and 4 (Mexican American and Other Hispanic) is not significant either as is the difference between 2 and 4 and 3 and 5. In short two sets of income groups emerge: Mexican American, Non-Hispanic Black and Other Hispanic seem to have significantly lower income (according to our definition) than Non-Hispanic White and Other Race –Including Multi-Racial. In retrospect, we can see these results by observing Figure 15.8. □

The Bonferroni test

If, after we analyze the data, we repeat the tests enough times, some of them may be significant by chance alone. For example, with 5 groups and the LSD tests, there are

$$\binom{5}{2} = 10$$

possible t-tests. With a significance level $\alpha = 0.1$, if we repeat the tests many times, one of them will be falsely significant and we must guard against this possibility. The Bonferroni test, one of many *multiple comparisons tests*, addresses this issue by keeping α at a fixed level.

Our null hypothesis is

$$H_0 : \alpha_i = \alpha_j \quad \text{vs.} \quad H_A : \alpha_i \neq \alpha_j$$

where $i \neq j$ refer to two among k groups. To apply the test, we obtain the Within MS (or residual MS as it is called in R) from the ANOVA results and compute the test statistics

$$T_{ij} = \frac{\overline{Y}_i - \overline{Y}_j}{SE}$$

where

$$SE = \sqrt{\text{Within MS} \times \left(\frac{1}{n_i} + \frac{1}{n_j}\right)}.$$

Next, we adjust α thus:
$$\alpha' := \frac{\alpha}{\binom{k}{2}}.$$

Now for a two-tailed test, if p-value $= 1 - \mathrm{pt}(\mathrm{abs}(\mathrm{T}_{ij}), n-k) < \alpha'/2$ we reject H_0. The application of one-tailed test is done similarly, but with α', instead of $\alpha'/2$. The Bonferroni test assumes that the $\binom{k}{2}$ comparisons are independent. Often they are dependent and the test is therefore conservative.

Example 15.11. Let us use the data frame i as obtained in Example 15.10. Here are a few random records:

```
> set.seed(56) ; idx <- sample(1 : length(i[, 1]), 5)
> i[idx, ]
                                   ethnicity income
3915                      Non-Hispanic Black   50.0
7039 Other Race - Including Multi-Racial     22.5
2997                        Mexican American   70.0
7128                      Non-Hispanic Black    2.5
3370                      Non-Hispanic White   17.5
```

We wish to compare all possible pairs and thus need to adjust the value of α accordingly. Along the way, let us familiarize ourselves a little more with the output of aov(). After this:

```
> n <- tapply(income, ethnicity, length)
> k <- length(n)
> df <- sum(n) - k
> means <- tapply(income, ethnicity, mean)
```

we have a vector of n_i (corresponding to the number of records for each of the five levels of the factor ethnicity), the number of degrees of freedom and the group means. From the ANOVA

```
> (a <- aov(income ~ ethnicity))
Call:
   aov(formula = income ~ ethnicity, contrasts = con)

Terms:
                  ethnicity Residuals
Sum of Squares     56191.7 2620385.8
Deg. of Freedom          4       7524

Residual standard error: 18.662
Estimated effects may be unbalanced
```

we obtain that Within MS $= 18.662^2$. To verify this, we can use the output a:

```
> Within.MS <- sum(a$residuals^2) / a$df.residual
> sqrt(Within.MS)
[1] 18.662
```

One-way, fixed-effects ANOVA

We want to test the significance of all paired comparisons. Specifically, for Mexican American vs. Non-Hispanic white (first and third levels of ethnicity) we get:

```
> SE <- sqrt(Within.MS * (1 / n[1] + 1 / n[3]))
> T.1.3 <- as.numeric((means[1] - means[3]) / SE)
> alpha <- 0.05
> adjust <- choose(5, 2)
> alpha <- alpha / adjust
> c(T = T.1.3, df = df, p.value = 1 - pt(abs(T.1.3), df),
+   alpha = alpha)
         T          df     p.value       alpha
 -10.56739  7524.00000     0.00000     0.00500
```

Even after adjusting α to 0.005, the mean yearly household income for these ethnic groups are significantly different (recall that the data do exclude those with "infinite" income). □

The Bonferroni adjustment to α applies when we resort to multiple pairs of tests, but only (if at all) if we use a "shot gun" approach; that is, we decide to apply the tests in search for significance. If you establish mull hypotheses before running the (multiple) tests, then LSD suffices. Otherwise and if you do wish to be conservative, then you should apply the Bonferroni test. The function Bonferroni() (available from the book's site) implements the paired tests.

Example 15.12. Continuing with i in Example 15.11, we obtain

```
> source('Bonferroni.R')
> (b <- Bonferroni(i))
   i j       T   df p-value alpha
1  1 2  -0.1440 7524  0.4427 0.005
2  1 3 -10.5674 7524  0.0000 0.005
3  1 4   0.6889 7524  0.2455 0.005
4  1 5  -4.4658 7524  0.0000 0.005
5  2 3 -10.2370 7524  0.0000 0.005
6  2 4   0.7526 7524  0.2258 0.005
7  2 5  -4.3702 7524  0.0000 0.005
8  3 4   5.3870 7524  0.0000 0.005
9  3 5   0.8791 7524  0.1897 0.005
10 4 5  -3.6796 7524  0.0001 0.005
```

where i, j refer to the levels of ethnicity

```
> levels(i[, 1])
[1] "Mexican American"
[2] "Non-Hispanic Black"
[3] "Non-Hispanic White"
[4] "Other Hispanic"
[5] "Other Race - Including Multi-Racial"
```
□

Another popular multiple comparisons test is the so-called Tukey honest significant differences (HSD).

484 Analysis of variance

Example 15.13. With the same data as in Example 15.12 and the same linear model

```
> a <- aov(income ~ ethnicity, data = i)
> hsd <- TukeyHSD(a)
> par(mar = c(5, 20, 3, 2), cex.main = 1)
> plot(hsd, las = 2)
```

(Figure 15.9). With the figure, you can quickly determine paired significances. Those intervals that cross zero indicate that the differences between the means of i and j are *not* significant. □

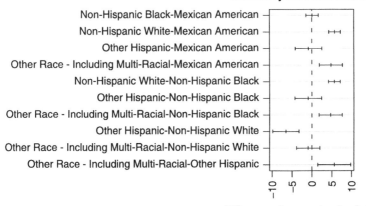

Figure 15.9   Tukey's HSD test. Paired confidence intervals that do not cross the zero line indicate significant differences of the relevant means.

As we have just seen, after comparing groups, the next logical step is to pool groups into sets and examine the significance of differences (in means) between pairs of sets.

15.1.4 Comparing sets of groups

As in section 15.1.3, we first discuss how to construct a statistic for comparing the means of sets of groups and then discuss the adjustments that need to be made for multiple comparisons. The adjustments need to be performed because some comparisons may be significant by chance alone and we must guard against that.

Linear contrasts

To facilitate the construction of sets of groups (e.g. as suggested in Example 15.10), we use the following definition:

Linear contrast (L) We say that any linear combination of group means where the coefficients add to zero is a linear contrast.

In notation, this

$$L = \sum_{i=1}^{k} c_i \overline{Y}_i \text{ such that } \sum_{i=1}^{k} c_i = 0 \qquad (15.3)$$

is a linear contrast.

Example 15.14. We continue where we left off in Example 15.10 (i is the data frame that is ready for us to use). Here is one way to construct a linear contrast. We obtain the number of observations for various ethnicities with

```
> (n <- tapply(income, ethnicity, length))
              Mexican American
                          2273
            Non-Hispanic Black
                          2129
            Non-Hispanic White
                          2515
                Other Hispanic
                           260
Other Race - Including Multi-Racial
                           352
```

Next, we find the proportions in the two sets (let us call them White and Nonwhite) as detailed by the indices of n; first for White

```
> w <- n[c(3, 5)]
> (L.1 <- w / sum(w))
            Non-Hispanic White
                     0.8772236
Other Race - Including Multi-Racial
                     0.1227764
```

and then for Nonwhite

```
> nw <- n[c(1, 2, 4)]
> (L.2 <- nw / sum(nw))
  Mexican American Non-Hispanic Black      Other Hispanic
        0.48755899         0.45667096          0.05577006
```

So if we take our linear contrasts as L.1 and −L.2, we obtain

```
> round(sum(c(L.1, -L.2)), 3)
[1] 0
```

which is, by definition, a linear contrast with c_i ($i = 1, \ldots, 5$). Let us arrange the contrasts according to the factor levels as they appear in i:

```
> contr <- c(L.1, -L.2)
> (contrasts <- contr[c(3, 4, 1, 5, 2)])
              Mexican American
                    -0.48755899
            Non-Hispanic Black
                    -0.45667096
```

```
                  Non-Hispanic White
                         0.87722358
                     Other Hispanic
                        -0.05577006
Other Race - Including Multi-Racial
                         0.12277642
```
□

We denote by μ_L and σ_L^2 the *population* mean and variance of the linear contrast. We assume that $\mu_L = 0$ and wish to estimate it from the sample. We use the total variance, S^2, c_i and n_i to estimate σ_L^2 and thus obtain the contrasted

$$\text{SE} = \sqrt{S^2 \sum_{i=1}^{k} \frac{c_i^2}{n_i}} \qquad (15.4)$$

(recall that for a constant, $\text{Var}(aX) = a^2 \text{Var}(X)$). Intuitively speaking, the linear contrast we just constructed is a way to rearrange the ANOVA into sets of groups where in each set, we weigh the within group variance by the relative sample size of each group within its set of groups. The arrangement ensures that we conform to the ANOVA assumption that the sum of these "within set" variances is zero. You are free to construct any linear contrast you please. However, at some point, you will need to interpret the result after applying the contrast. In our case, the contrast allows us to interpret the ANOVA results as if the design is balanced.

The sampling density of the linear contrast is t and the test goes like this: For significance level α, we wish to test

$$H_0 : \mu_L = 0 \quad \text{vs.} \quad \mu_L \neq 0 \qquad (15.5)$$

(the application to one sided tests should be clear by now). Then the statistic is

$$T_L = \frac{L}{\sqrt{S^2 \sum_{i=1}^{k} c_i^2/n_i}} \,. \qquad (15.6)$$

The density of the statistic is t_{n-k}. Therefore, if the

$$p\text{-value} = 1 - \text{pt}(\text{abs}(\text{T.L}), \text{n} - \text{k}) < \alpha/2\,,$$

we reject H_0 in favor of H_A.

Example 15.15. In Example 15.14, we obtained the contrast vector. To implement T_L, we need the SE according to (15.4):

```
> (v <- var(income))
[1] 355.5496
> (SE <- sqrt(v * sum(contrasts^2 / n)))
[1] 0.4475265
```

and L according to (15.3):

```
> cbind(group.means = means, contrast = contrasts)
                              group.means    contrast
Mexican American                 30.85130 -0.48755899
Non-Hispanic Black               30.93236 -0.45667096
Non-Hispanic White               36.55865  0.87722358
Other Hispanic                   30.00962 -0.05577006
Other Race - Including Multi-Racial  35.62500  0.12277642
> (L <- sum(contrasts * means))
[1] 5.602641
```

The statistic and its corresponding p-value are:

```
> T.L <- L / SE ; df <- sum(n) - length(n)
> p.value <- 1 - pt(abs(T.L), df)
> c(T.L = T.L, df = df, p.value = p.value)
      T.L         df    p.value
 12.51913 7524.00000    0.00000
```

and we conclude that the mean income (as we define in Example 15.10) between the sets White and Nonwhite is different. □

Another way to construct contrasts is to weigh the deviation of the group means from the total mean by the groups' sample size (see Figures 15.4 and 15.6).

There is a subtle point in constructing linear contrasts: do we do it before analyzing the data or after. If we do it after analyzing the data, we can construct (depending on the value of k) a very large number of contrasts and by chance alone, some of them may turn out to be significant. This problem is addressed next.

Multiple comparisons for linear contrasts

After running ANOVA, the results may suggest several paired comparisons of sets of groups (we do not need to run the test here if we are testing for hypothesis that were set up during the design of the study). In this case, we need to penalize ourselves with regard to the significance test because some comparisons may be significant by chance alone.

The notation and hypotheses are as in (15.3), (15.4) and (15.5). Because of the multiple comparisons, the sampling density of the test statistic (15.6) changes to $\sqrt{(k-1)F_{k-1,n-k}}$. Consequently, if

$$|T_L| > \sqrt{(k-1)F_{k-1,n-k,1-\alpha}}$$

we reject the null hypothesis. This is the so-called Scheffé's test.

Example 15.16. In Example 15.15 we compared only two sets of ethnic groups (White vs. Nonwhite). Suppose that these two sets are part of all possible sets of ethnicities. We found that

```
      T.L           df
 12.51913         7524
```

with df = n - k and k = 5. For α = 0.05 we have

```
> (critical.values <- sqrt((k - 1) *
+     qf(1 - alpha, k - 1, sum(n) - k)))
[1] 3.1
```

and we reject the null hypothesis. □

There are many other multiple comparison tests. The differences among them boil down to how conservative and how general they are. The Scheffé's test is general and applies to unbalanced ANOVA. A quick Web search for *"multiple comparison tests"* will introduce you to the sometimes confusing plethora of multiple comparison tests.

15.2 Non-parametric one-way ANOVA

If the assumptions of the data for the continuous variable in one-way ANOVA fail, or if the response is ordinal (as opposed to decimal), then we must rely on non-parametric ANOVA. Non-parametric in the sense that we do not need to make assumptions about the underlying density of the observations and consequently, we do not need to rely on some density parameters to obtain the sampling density of the statistic of interest (differences in means in the case of fixed-effects ANOVA).

15.2.1 The Kruskal-Wallis test

Just as one-way ANOVA generalizes the *t*-test to multiple comparison, the Kruskal-Wallis test generalizes the paired Wilcoxon rank-sum test (Section 12.3.1).

Example 15.17. In Example 15.10, we treated the yearly household income (vs. ethnicity) as a continuous variable. We want to use the original income categories, as reported by the CDC as a rank, so that the Kruskal-Wallis test applies. As discussed in Example 15.10, the CDC data categorize income thus:

```
> (income <- read.table(
+     'CDC-demographics-income-categories.txt',
+     header = TRUE, sep = '\t'))
   code             income
1     1       $ 0 to $ 4,999
2     2   $ 5,000 to $ 9,999
3     3   $10,000 to $14,999
4     4   $15,000 to $19,999
5     5   $20,000 to $24,999
6     6   $25,000 to $34,999
7     7   $35,000 to $44,999
8     8   $45,000 to $54,999
9     9   $55,000 to $64,999
10   10   $65,000 to $74,999
11   11      $75,000 and Over
12   12          Over $20,000
13   13         Under $20,000
```

```
14    77           Refused
15    99           Don not know
16     .           Missing
```

We load the data, extract the ethnicity and income categories into d, assign NA to the records with income category larger than ''11'' and remove all NA from the data:

```
> load('demo_d.short.rda')
> d <- demo_d.short[, c(4, 9)]
> idx <- which(as.numeric(d[, 2]) > 11) ; d[idx, 2] <- NA
> d <- d[complete.cases(d), ]
```

The data

```
> head(d)
       Race/Ethnicity Annual Household Income
1 Non-Hispanic White                        4
2 Non-Hispanic Black                        8
3 Non-Hispanic Black                       10
4 Non-Hispanic White                        4
5 Non-Hispanic Black                       11
6 Non-Hispanic White                       11
```

are now ready for the Kruskal-Wallis test. □

The Kruskal-Wallis test works like this: We compute the average rank to each treatment (e.g. average rank of each ethnic group in Example 15.17) and compare them with the null hypothesis that all average ranks are equal. To obtain the test statistic, let R_i denote the sum of the ranks for each of the treatments ($i = 1, \ldots, k$) and $n := \sum n_i$ be the total number of observations (n_i is the number of observations for each group). If *none* of the R_i are equal (*no* ties), then the test statistic is

$$\chi' = \frac{12}{n(n+1)} \sum_{i=1}^{k} \frac{R_i^2}{n_i} - 3(n+1) . \tag{15.7}$$

If there are $j = 1 \ldots g$ tied groups, denote by m_j the number of observations in the jth set of tied treatments and adjust χ' thus:

$$\chi = \frac{\chi'}{1 - \frac{\sum_{j=1}^{g} (m_j^3 - m_j)}{n^3 - n}} .$$

The density of the statistics χ is $\chi^2(k-1)$ where $k-1$ are the degrees of freedom. Now with ties, if p-value $= 1 - $ pchisq(chi, k $- 1) < \alpha$, reject the null hypothesis—at least one of the mean ranks is significantly different from the rest. If there were noties, use χ' instead of χ. The procedure applies to groups with more than five observations.

Example 15.18. Using d in Example 15.17, we first examine the counts with

```
> par(mfrow = c(1, 2), mar = c(15, 4, 1, 2))
> barplot(table(d$'Race/Ethnicity'), las = 2, ylim = c(0, 4000))
> barplot(table(d$'Annual Household Income'), las = 2,
+     names.arg = income$income[as.numeric(income$code) <= 11],
+     ylim = c(0, 2500))
```

(Figure 15.10). Note how we pluck the category names for the income based on
 d$'Annual Household Income'
from income with
income$income[as.numeric(income$code) <= 11]

To run the analysis, we simply say

```
> kruskal.test(as.integer(d[, 2]) ~ as.factor(d[, 1]))

        Kruskal-Wallis rank sum test

data:  as.integer(d[, 2]) by as.factor(d[, 1])
Kruskal-Wallis chi-squared = 551.4508, df = 4,
p-value < 2.2e-16
```

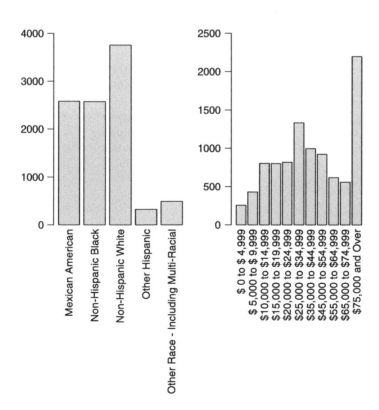

Figure 15.10   U.S. income and ethnicity in the 2004–2005 CDC survey.

and conclude that there is significant overall difference in the mean of the incomes rank among ethnic groups. □

The advantage of the Kruskal-Wallis test compared to the F-test in one-way ANOVA is that we make *no* normality assumption.

As we said, the Kruskal-Wallis test applies when the minimum number of repetitions is five. Otherwise, use the exact test; i.e. calculate quantiles (from the density) or probabilities (from the distribution). Here is an example of how to use the exact test.

Example 15.19. Suppose that we have a sample with 3 groups with 4, 4 and 5 repetitions for each group, respectively. From the data, we calculate $\chi' = 4.668$ with no ties, using (15.7). Then

```
> library(SuppDists)
> (p.value <- 1 - round(pKruskalWallis(4.668, 3, 13,
+     sum(c(1/4, 1/4, 1/5))),2))
[1] 0.09
```

(see help for pKruskalWallis()). For $\alpha = 0.05$ we do not reject the null hypothesis and conclude that there is no significant difference among the mean ranks of the three groups. □

15.2.2 Multiple comparisons

As was the case for one-way ANOVA (Section 15.1.3), we may be interested in paired comparisons. To implement the so-called *Kruskal-Wallis multiple comparisons*, for a significance level α, we adjust α to

$$\alpha' = \frac{\alpha}{k(k-1)} \tag{15.8}$$

where k is the number of groups. To compare the mean rank of group i to group j, we compute the statistic

$$Z = \frac{\overline{R}_i - \overline{R}_j}{\sqrt{\frac{n(n+1)}{12}\left(\frac{1}{n_i} + \frac{1}{n_j}\right)}}$$

where n is the total number of observations and n_i and n_j are the number of observations for groups i and j, respectively. The sampling density of this statistic is standard normal. So for two sided test, if p-value $= 1 -$ pnorm(abs(z)) $< \alpha'$, where α' is obtained from (15.8), then we reject the null hypothesis and conclude that the mean rank for group i and group j are significantly different.

Example 15.20. Continuing with Example 15.18,

```
> library(pgirmess)
> kruskalmc(as.integer(d[, 2]), as.factor(d[, 1]))
Multiple comparison test after Kruskal-Wallis
p.value: 0.05
Comparisons
```

	difference
Mexican American-Non-Hispanic Black	TRUE
Mexican American-Non-Hispanic White	TRUE
Mexican American-Other Hispanic	FALSE
Mexican American-Other Race - Including Multi-Racial	TRUE
Non-Hispanic Black-Non-Hispanic White	TRUE
Non-Hispanic Black-Other Hispanic	FALSE
Non-Hispanic Black-Other Race - Including Multi-Racial	TRUE
Non-Hispanic White-Other Hispanic	TRUE
Non-Hispanic White-Other Race - Including Multi-Racial	FALSE
Other Hispanic-Other Race - Including Multi-Racial	TRUE

(output edited). From the output, we identify for which pairs the mean ranks of the income categories are different. For those incomes that are significantly different, the full output of kruskalmc() allows you to determine which groups' mean rank in the i, j pair was lower. □

15.3 One-way, random-effects ANOVA

In the fixed-effects ANOVA, we were interested in differences in means among groups. Occasionally, we are interested in the proportion that the variation in each group contributes to the total variation in the sample (and by inference, in the population).

Example 15.21. To study changes in ecosystems due to, say, the effect of herbivores on plant communities, ecologists often set up exclosures. These are fenced areas that herbivores cannot enter. Then over a period of years, the plant communities within and outside the exclosures are studied. Say we set up five exclosures in a mixed hardwood forest[2] and after ten years, sample the biomass of birch in and outside the exclosures. Then there will be a variation of biomass within the exclosures, outside the exclosures and between the exclosures and outside of them.

In medical studies, researches often repeat measurements on a single subject (e.g. blood pressure). If the study classifies subjects, then it becomes important to compare the variation among the classification to the variation within subjects. □

The one-way random-effects ANOVA model is

$$y_{ij} = \mu + \alpha_i + \varepsilon_{ij} \tag{15.9}$$

where: y_{ij} is the population value of the ith replicate for the jth individual; α_i is a rv that accounts for the between-subject variability (assumed to be normal with mean 0 and variance σ_α^2); ε_{ij} is a rv which accounts for within-subject variability (assumed to be normal with mean 0 and variance σ^2). ε_{ij} are often referred to as *noise*; they are independent and identically distributed rv. In other words, repeated measure on the jth individual are $\phi(0, \sigma)$. What makes the model random-effects is the fact that α_i is a rv, so that the mean for the jth individual will differ from other individuals.

---

[2]These are northern latitude forests that include a mix of deciduous and pine trees.

In the ANOVA, we are interested to test the hypothesis that there is no between individual variation. In other words, we wish to test

$$H_0 : \sigma_\alpha^2 = 0 \quad \text{vs.} \quad H_A : \sigma_\alpha^2 > 0 .$$

To test the hypothesis, we need a sampling density of some statistic of S_α^2 (the sample approximation of σ_α^2). It turns out that

$$E[\text{Within MS}] = \sigma_\alpha^2, \quad E[\text{Between MS}] = \sigma^2 + n\sigma_\alpha^2$$

where

for unbalanced design: $n = \left(\sum_{i=1}^k n_i - \frac{\sum_{i=1}^k n_i^2}{\sum_{i=1}^k n_i} \right) \Big/ (k-1)$, (15.10)

for balanced design: $n = n_i$.

Here n_i is the number of replications for individual i. In the balanced design all n_i are equal. As was the case for fixed-effects ANOVA, the sampling density of the statistic

$$F = \frac{\text{Between MS}}{\text{Within MS}} \qquad (15.11)$$

is F with $k-1$ and $n-k$ degrees of freedom, where n depends on the design as detailed in (15.10). Here Within MS and Between MS are unbiased estimators of $\sigma^2 + n\sigma_\alpha^2$, computed according to:

$$\text{Between MS} = \frac{\sum_{i=1}^k \left(\overline{Y}_i - \overline{\overline{Y}} \right)^2}{k-1},$$

$$\text{Within MS} = \frac{\sum_{i=1}^k \sum_{j=1}^{n_i} \left(Y_{ij} - \overline{Y}_i \right)^2}{n-k}, \qquad (15.12)$$

where

$$\overline{Y}_i = \sum_{i=1}^k \sum_{j=1}^{n_i} \frac{Y_{ij}}{n_i}, \quad \overline{\overline{Y}} = \frac{\sum_{i=1}^k \sum_{j=1}^{n_i} Y_{ij}}{n'}, \quad n' = \sum_{i=1}^k n_i .$$

The estimator, $\hat{\sigma}^2$, of σ^2 (the within group variance) is given by (15.12). The estimator, $\hat{\sigma}_\alpha^2$, of σ_α^2 (the between group variance) is given by

$$\hat{\sigma}_\alpha^2 = \max \left(\frac{\text{Between MS - Within MS}}{n}, 0 \right)$$

where, depending on the design (balanced or unbalanced), n is calculated according to (15.10).

Once we compute F according to (15.11) and subsequent formulas, we obtain the p-value with $1 - \text{pf}(F, k-1, n-k)$. As usual, if p-value $< \alpha$ (where α is the significance level), then we reject H_0 and conclude that $\sigma_\alpha^2 > 0$. In other words, we have sufficient evidence to claim that between variance is larger than zero. Putting the conclusion differently, we say that in spite of within individual (measurement, noise) variance, we still detect a significant difference among individuals.

The fundamental difference between random and fixed effects ANOVA is this. In the case of fixed-effects, each measurement is independent of the other. For example, for each group, we might have a number of repetitions, but the repetitions are independent of each other. In random-effects, the repeated measures are *not* independent. If a person has high blood pressure and we measure the person's blood pressure twice, the first and second measurements are going to be related by the fact that the person suffers from high blood pressure. The question we are asking in random-effects ANOVA is: "Is the measurement error of the same object (the individual) is so large that we cannot distinguish among individual differences?" If the answer is yes, the within variance may be due to, for example, measurement error. However, it simply may be the case that the blood pressure of a person with high blood pressure fluctuates so much, that we cannot distinguish it from say a person with normal blood pressure, which raises an important question: In repeated measurements, should high blood pressure be defined by mean, variance or both?

Here is another example that illustrates the difference between fixed and random-effects ANOVA.

Example 15.22. In Example 15.21, we discussed a typical exclosure study in ecology. Suppose we have n exclosures, equally divided among wetlands, uplands and riverine habitats. We take a single sample from inside the exclosures, where the response is biomass density of some species. The samples *are* independent and a test on the difference in mean biomass in this case will be a fixed-effects ANOVA. If we take multiple samples within each exclosure, then within an exclosure the samples *are* dependent and we need to implement random-effects ANOVA. If we are interested in differences in both habitat type and within exclosure repetitions, then we have what is called *mixed-effects* ANOVA. □

Here is an example that implements the computations for random-effects ANOVA.

Example 15.23. To verify that all works well, we use the blood pressure data from Rosner (2000), p. 556. Five subjects make the groups and there are two repetitions of blood pressure measures for each subject. We use the log of the response. The balanced random-effects formulas are implemented in `re.1w()` where the data must be presented exactly as shown:

```
> source('random-effect-ANOVA.R')
> group <- rep(1 : 5, 2)
> repl <- c(rep(1, 5), rep(2, 5))
> response <- log(c(25.5, 11.1, 8, 20.7, 5.8, 30.4, 15, 8.1,
+      16.9, 8.4))
> cbind(group, repl, response)
     group repl response
[1,]     1    1 3.238678
[2,]     2    1 2.406945
[3,]     3    1 2.079442
[4,]     4    1 3.030134
[5,]     5    1 1.757858
[6,]     1    2 3.414443
[7,]     2    2 2.708050
```

```
[8,]     3    2 2.091864
[9,]     4    2 2.827314
[10,]    5    2 2.128232
```

To visually examine the data, we do

```
> library(lattice)
> trellis.device(color = FALSE, width = 4, height = 4)
> xyplot(response ~ replication | group, type = 'b', cex = 0.8)
```

(Figure 15.11). Thus we conclude that there is no definitive trend in the first vs. the second measure of individual blood pressure. The random-effect ANOVA is implemented with

```
> re.1w(group, repl, response)
          source df Mean.SS      F p.value
1 Between.MS (model)  4   0.664 22.146   0.002
2  Within.MS (error)  5   0.030
```

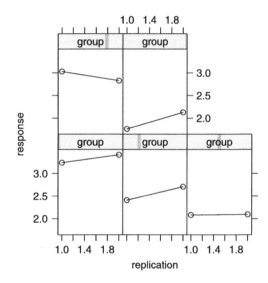

Figure 15.11   Repeated measures by group.

The script for re.1w() resides in random-effect-ANOA.r in the book's site. The Between MS is large enough (compared to the Within MS) and the subject effect overwhelms the fluctuations in blood pressure for two separate measurements within subjects.   □

15.4 Two-way ANOVA

So far, we discussed a single factor variable. Often, we may have two or more. In the former, we have the so-called two-way ANOVA; in the latter multi variable ANOVA, the so-called MANOVA. We shall discuss two-way, fixed-effects mixed-effects and nested ANOVA.

15.4.1 Two-way, fixed-effects ANOVA

We start with an example.

Example 15.24. In Example 15.1, the fixed-effect was a city and the numeric variable was atmospheric concentration of SO_2 in those cities. We may add another factor, say country. So for each country there may be a number of cities and for each city, there would be a number of measurements. If we view the data as two-effect, city and country, then it is two-way fixed-effects ANOVA. However, we may view the data as countries and cities within countries. This leads to *nested* ANOVA.

In Example 15.10, the fixed-effect was ethnicity and the numeric variable was the average of a yearly household income category. With another factor in the data is, for example, marital status allows us to investigate the effect of the gender of the head of the household and ethnicity on income. For each ethnic group, the head of the household may be married or not and for each ethnic-marital status combination we have income. This will constitute nested ANOVA. □

One the central goals in two-way ANOVA is to examine if there exist significant differences in means for one factor, controlling for the effect of the other.

Example 15.25. For the atmospheric concentration of SO_2, we may ask: Are there significant differences in the means among cities after accounting for potential differences (in means) due to country? For example, different countries may have different regulations and after accounting for these, do we still have different means among the cities? □

If the effect of one factor on the numeric value depends on the level of the other factor, then we say that there is an *interaction effect*. The effect of each factor, separate from the other is called the *main effect* of the factor. One of the major tasks in two-way ANOVA is to explore the main and interaction effects.

15.4.2 The model and assumptions

To introduce the notation, let us begin with an example.

Example 15.26. We go back to the EU air pollution data introduced in Example 15.1. We want to compare the means of yearly mean concentration of atmospheric CO (in mg/m^3) by six EU countries, each by three area types where the collecting station is located (rural, suburban and urban). Isolating the data with the desired variables is tricky and we must strive to make the script run fast because the data file is large with

```
> load('EU.rda')
> length(EU[, 1]) * length(EU[1, ])
[1] 21293352
```

data items. The problem is that EU contains station codes and pollutant related data while

```
> load('EU.station.rda')
```

contains the station codes, its country and the area in which the stations are located. Extracting the data is further complicated by the fact that some station codes appear in EU but not in EU.station (apparently an error in the data) and some codes are in EU.station but not in EU (not necessarily an error). First, we want to remove all the NA from EU.station:

```
> stations <- EU.station[complete.cases(EU.station), ]
```

Next, we need to turn factor variables into character variables. This is necessary because if we do comparisons of factor values from two different data frames (by say the index of the values), we are not sure we get what we want—recall that factor levels are really numeric (more specifically integers), but they are represented by legible levels—it will be worth your while to remember this point. So we do

```
> stations$country_name <- as.character(stations$country_name)
> stations$station_type_of_area <-
+    as.character(stations$station_type_of_area)
```

From stations, we want data about:

```
> a <- c('BELGIUM', 'FRANCE', 'GERMANY', 'ITALY', 'SPAIN',
+    'UNITED KINGDOM')
> b <- c('rural', 'suburban', 'urban')
```

We want to extract from stations all the record that correspond to a and b. We do this with:

```
> length(stations[, 1])
[1] 5988
> stations <- stations[is.element(stations[, 3], a), ]
> length(stations[, 1])
[1] 4456
> stations <- stations[is.element(stations[, 5], b), ]
> length(stations[, 1])
[1] 4364
```

Of the original number of records, we end up with 4 364 stations that conform to our country and area criteria. To verify that we got what we want, we do:

```
> unique(stations[, 3])
[1] "BELGIUM"        "GERMANY"         "SPAIN"
[4] "FRANCE"         "UNITED KINGDOM"  "ITALY"
> unique(stations[, 5])
[1] "suburban" "rural"    "urban"
```

So now we have in stations only the data that we need, e.g.

```
> set.seed(2)
> stations[sample(1 : length(stations[, 1]), 5), c(1, 3, 5)]
     station_european_code   country_name station_type_of_area
1327                DE0999A        GERMANY                rural
3783                GB0219A UNITED KINGDOM             suburban
3170                FR0945A         FRANCE                urban
1253                DE0905A        GERMANY                urban
5015                IT1204A          ITALY                urban
```

Our next task is to subset EU, the data frame that holds the pollutants data and add to it columns that include the country and area in which the sampling station resides. This is done through a series of steps (see Exercise 15.3) like this: First,

```
> # isolate CO data
> EU.CO <- EU[EU[, 3] == 'CO' & EU[, 14] == 'annual mean',
+     c(1, 15)]
```

Next,

```
> country <- area <- vector(length = length(EU.CO[, 1]))
> for(i in 1 : length(a)){
+     s <- stations[stations[, 3] == a[i], 1]
+     idx <- is.element(EU.CO[, 1], s)
+     country[idx] <- a[i]
+ }
```

We create the area, long and lat vectors similarly to obtain:

```
> set.seed(4)
> EU.CO[sample(1 : length(EU.CO[, 1]), 5), ]
       station    CO country  area  long   lat
720373 FR0586A 0.098  FRANCE rural  6.05 49.25
143151 BE0235A 0.409 BELGIUM urban  4.45 50.41
476317 DE1142A 0.640 GERMANY urban 11.32 50.98
459921 DE1087A 0.568 GERMANY urban 11.97 51.20
977127 IT0187A 1.395   ITALY urban 11.61 44.84
```

To see a map of the distribution of the stations (Figure 15.12), we load

```
> library(maps)
> library(mapdata)
```

set the regions and colors to be plotted:

```
> r <- c('Belgium', 'France', 'Germany', 'Italy', 'Spain', 'UK')
> col<-c('grey95', 'grey90')
```

draw the map, its axes and the station locations

```
> m<-map('world', regions = r, fill = TRUE, col = col,
+     xlim = c(-15, 20), ylim = c(35,62))
> map.axes()
> points(EU.CO$long,EU.CO$lat, col = 'grey60')
```

and add the countries capitals

```
> for(i in 1 : 6) map.cities(country = r[i], capitals = 1,
+     cex = 1.25)
```

(for unclear reasons, London has to be added with text()). □

In general, in two-way ANOVA we have two sets of groups, one with r groups and the other with c groups. Then there are observation values, indexed by k for the ith and jth groups. In terms of population vs. sample quantities, we shall stick to the lower case and Greek notation vs. upper case notation.

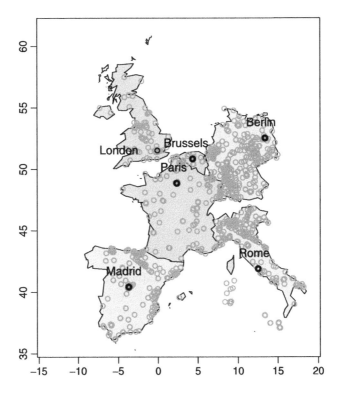

Figure 15.12 Rural, suburban and urban stations by country for which CO measurements were obtained.

Example 15.27. Let us identify the general notation for the CO data frame we obtained in Example 15.26.

```
> (tb <- table(EU.CO[, c(3, 4)]))
               area
country         rural suburban urban
  BELGIUM           0      141   216
  FRANCE          174      747  1239
  GERMANY         351     1668  4086
  ITALY            86      631  2629
  SPAIN           230     1041  2320
  UNITED KINGDOM   21      126  2172
```

So for country we have $r = 6$ groups, for area we have $c = 3$ groups. The value of $Y_{3,2,1}$ can be extracted from

```
> length(Y.3.2 <- EU.CO[EU.CO[, 3] == 'GERMANY' &
+     EU.CO[, 4] == 'suburban', 2])
[1] 1668
```

with

```
> Y.3.2[1]
[1] 0.555
```

where $i = 3$, $j = 2$ and $k = 1$. Note that one of the entry cells (Belgium, rural) has no data. This needs special handling in ANOVA. To stay on course, we will drop Belgium from further considerations. □

Given the general notation, we can now write the two-way, fixed-effects ANOVA model as
$$y_{ijk} = \mu + \alpha_i + \beta_j + \gamma_{ij} + \varepsilon_{ijk} \tag{15.13}$$

Example 15.28. Continuing with Example 15.27, for the "population" of CO measurements, we have

y_{ijk} the kth CO value for the ith country and jth area
μ the overall population mean
α_i the effect of country
β_j the effect of area
γ_{ij} interaction effect between country and area
ε_{ijk} error term from a normal density with mean zero and variance σ^2 □

In addition to the assumption that ε is normal with mean zero and variance σ^2, we assume that
$$\sum_{i=1}^{r} \alpha_i = \sum_{j=1}^{c} \beta_j = 0, \quad \sum_{j=1}^{c} \gamma_{ij} = 0 \text{ for all } i, \quad \sum_{i=1}^{r} \gamma_{ij} = 0 \text{ for all } j.$$

The assumptions imply that the density of y_{ijk} is normal with mean $\mu + \alpha_i + \beta_j + \gamma_{ij}$ and variance σ^2. Consequently, we can derive the sampling densities of the various statistics of Y_{ijk}.

15.4.3 Hypothesis testing and the F-test

A two-way ANOVA with two effects may be represented in a table with i representing row entries and j column entries.

Example 15.29. In the case of the CO data (Example 15.27), the ith entry refers to one of the countries and the jth to one of the area types in which the CO had been measured. If so desired, the table may be rotated and the role of i and j switched. However, as we shall sea, this will lead to different ANOVA results. We can easily find the number of replication for each cell in the table, including the marginals with

```
> replications(CO ~ country + area + country : area,
+    data = EU.CO)
$country
country
       BELGIUM          FRANCE         GERMANY           ITALY
           357            2160            6105            3346
         SPAIN  UNITED KINGDOM
          3591            2319

$area
area
   rural suburban    urban
     862     4354    12662
```

```
$'country:area'
                area
country         rural suburban urban
  BELGIUM           0      141   216
  FRANCE          174      747  1239
  GERMANY         351     1668  4086
  ITALY            86      631  2629
  SPAIN           230     1041  2320
  UNITED KINGDOM   21      126  2172
```

Note that in replications(), the formula CO ~ country + area + country : area produces the marginal counts for country and area, *and* the counts for each country and for each area type. Thus, country : area represents the effect of a specific combination of country : area on the mean CO. This is why we call the term country : area the *interaction effect*. □

It is customary to denote the mean of the interaction by $\bar{y}_{ij}$ and the mean for the ith row by $\bar{y}_{i\cdot}$. Here, $\cdot$ indicates "over all j". Similarly, we denote the mean for the jth column by $\bar{y}_{\cdot j}$ where here, $\cdot$ indicates "over all i". To indicate the overall mean, we write $\bar{y}_{\cdot\cdot}$. We can now write the deviation of each individual observation from the total mean thus:

$$y_{ijk} - \bar{y}_{\cdot\cdot} = y_{ijk} - \bar{y}_{\cdot\cdot}$$
$$- \bar{y}_{ij} + \bar{y}_{ij} - \bar{y}_{i\cdot} + \bar{y}_{i\cdot} - \bar{y}_{\cdot j} + \bar{y}_{\cdot j} - \bar{y}_{\cdot\cdot} + \bar{y}_{\cdot\cdot}.$$

Rearranging, we obtain

$$y_{ijk} - \bar{y}_{\cdot\cdot} = \underbrace{\left(y_{ijk} - \bar{y}_{ij}\right)}_{\text{within group}} + \underbrace{\left(\bar{y}_{i\cdot} - \bar{y}_{\cdot\cdot}\right)}_{\text{row effect}} + \underbrace{\left(\bar{y}_{\cdot j} - \bar{y}_{\cdot\cdot}\right)}_{\text{column effect}} +$$

$$\underbrace{\left(\underbrace{\left(\bar{y}_{ij} - \bar{y}_{i\cdot}\right)}_{\text{column effect in } i\text{th row}} - \underbrace{\left(\bar{y}_{\cdot j} + \bar{y}_{\cdot\cdot}\right)}_{\text{overall column effect}}\right)}_{\text{interaction effect}}.$$

Each term represents a difference that is interpretable. The within group difference is often called the *error term*. Now we can sum the square of the differences over appropriate indices, divide by the degrees of freedom and thus obtain mean sum of squares. The ratios of the mean sum of squares for a sample are known to have the F sampling density with the appropriate degrees of freedom and we are ready to test the hypotheses detailed in Table 15.3. The hypotheses in the table (see also (15.13) and Example 15.28) are hierarchical in the following sense: Row effects are tested first. The test gives the same result as one-way ANOVA with a single (row variable). Next, the column effect is tested after the row effect is removed. Third, the interaction effect is tested after accounting for both row and column effects. Removing, or accounting for an effect means that the variability due to the effect (row, column) is removed before moving on to compute the next effect. In the next example, we examine how R implements these ideas.

502 Analysis of variance

Table 15.3   Hypothesis testing in two-way ANOVA according to (15.13).

Test	H_0	H_A
row effect	all $\alpha_i = 0$	at least one $\alpha_i \neq 0$
column effect	all $\beta_j = 0$	at least one $\beta_j \neq 0$
interaction effect	all $\gamma_{ij} = 0$	at least one $\gamma_{ij} \neq 0$

Example 15.30. Returning to the EU CO data, recall that one cell in the factor table (county : area) for Belgium is empty and that the data are unbalanced. (Example 15.26). So, we

```
> load('EU.CO.rda')
> EU.CO <- EU.CO[EU.CO$country != 'BELGIUM', ]
> attach(EU.CO)
```

country still include an unwanted level (BELGIUM). To remove it we

```
> country <- as.character(country[country != 'BELGIUM'])
> country <- as.factor(country) ; levels(country)
[1] "FRANCE"        "GERMANY"       "ITALY"
[4] "SPAIN"         "UNITED KINGDOM"
```

We wish to explore the effect of interaction between country and area:

```
> interaction.plot(area, country, CO, type = 'b')
```

(Figure 15.13). Of all the countries, Italy stands out: CO increases as one moves from rural to urban areas. Because of the scale of the y-axis, there seem to be no interactions for the remaining countries. A preliminary run revealed, through diagnostics (similar to Figure 15.14), that the ANOVA assumptions are grossly violated

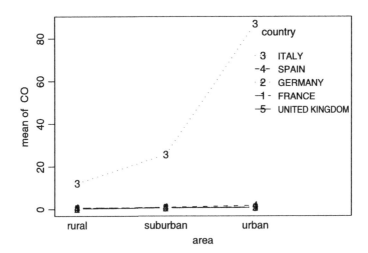

Figure 15.13   Interaction (in atmospheric CO concentration) between the station area and some EU countries.

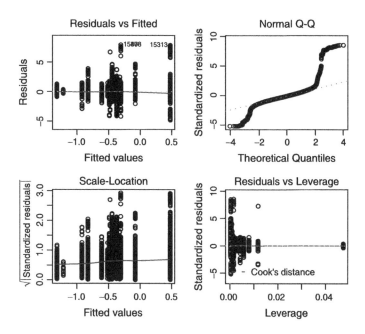

Figure 15.14   Diagnostics of the log transformed atmospheric CO data.

(see Exercise 15.6). So we log transformed CO (we need to add a trace amount to get rid of zeros):

```
> lCO <- log(CO + 0.01)
> full.model <- lm(lCO ~ country + area + country : area)
> anova(full.model)
Analysis of Variance Table

Response: lCO
                Df  Sum Sq Mean Sq F value    Pr(>F)
country          4  1904.2   476.1 552.375 < 2.2e-16
area             2   750.0   375.0 435.141 < 2.2e-16
country:area     8   178.7    22.3  25.914 < 2.2e-16
Residuals    17506 15087.1     0.9
>
> par(mfrow = c(2, 2))
> plot(full.model)
```

(Figure 15.14). The residuals fit the ANOVA assumptions, but the Q-Q plot is suspect. Data are plenty, so we shall continue anyway (you can run nonparametric ANOVA if you wish; see Exercise 15.5).

To explore how R proceeds, let us compare one to two-way ANOVA, without interaction. First, one-way:

```
> summary(aov(CO ~ country))
            Df    Sum Sq Mean Sq F value    Pr(>F)
country      4  14195554 3548889  118.94 < 2.2e-16
```

```
Residuals     17516 522624090          29837
> summary(aov(CO ~ area))
              Df      Sum Sq   Mean Sq  F value    Pr(>F)
area           2      841728    420864   13.756 1.073e-06
Residuals  17518 535977917     30596
```

and then two-way:

```
> summary(aov(CO ~ country + area))
              Df      Sum Sq   Mean Sq  F value    Pr(>F)
country        4    14195554   3548889 119.0198 < 2.2e-16
area           2      397983    198991   6.6736  0.001267
Residuals  17514 522226108     29818
```

Note that the Mean SS for country is the same in the two-way ANOVA as it is for the one-way ANOVA on country. For area, the one-way SS is larger than for two-way. The corresponding p-values behave accordingly (in opposite directions). This is so because in the two-way ANOVA, when the row effect is country, the remaining variability in the model due to area is computed after removing the row effect. In Exercise 15.5, you are asked to verify these results with the area acting as row effect.

Once we obtain the ANOVA model (e.g. full.model), we can further investigate the model with (output edited):

```
> full.model <- aov(CO ~ country + area + country : area)
> model.tables(full.model, type = 'means')
Tables of means
Grand mean

14.81267

 country
     FRANCE  GERMANY    ITALY    SPAIN UNITED KINGDOM
     0.8619    0.851     73.4    1.431         0.7554
rep 2160.0000 6105.000 3346.0 3591.000     2319.0000

 area
     rural suburban   urban
     8.164    7.442   17.77
rep 862.000 4213.000 12446.00

 country:area
                area
 country        rural suburban urban
   FRANCE           1        1     1
   rep            174      747  1239
   GERMANY          0        1     1
   rep            351     1668  4086
   ITALY           12       26    87
   rep             86      631  2629
```

```
SPAIN                     0     1      2
rep                     230  1041   2320
UNITED KINGDOM            0     1      1
rep                      21   126   2172
```

(we run `aov()` before `model.tables()` because the latter wants the former's output). Observe:

- Of the six countries, Italy is by far the most polluted (by two orders of magnitudes) while the U.K. is the least.
- After accounting for the country effect, suburban areas are the cleanest, followed by rural and then by urban. The latter is polluted more than twice over the rural and suburban areas.

Even without much further analysis it becomes quite obvious who is "responsible" for the significant results. □

Keep in mind that interpreting the sum of squares in unbalanced ANOVA is difficult at best. If you wish to fully interpret ANOVA output—including significance and magnitudes of Mean SS—use contrasts to account for the differences in mean values and in the number of cases per factor level. All this, with the assumption that the error variances (residual error) *are* equal. A quick way to check the validity of this assumption is to obtain a linear model with `lm()`, run the diagnostics (`plot()`) on the model and verify that the residuals behave themselves (equal spread). You can run `plot()` on a model obtained from `aov()` directly because the latter is just a wrapper to `lm()` that produces the ANOVA table.

15.5 Two-way linear mixed effects models

It turns out that it is easier to discuss and apply two-way mixed-effects ANOVA through mixed linear models than directly through ANOVA. Recall that in the two way ANOVA model (15.13), μ is the overall mean, α_i and β_j are the two effects means, γ_{ij} is the interaction effect and ε_{ijk} is the error term. In the random-effects one way ANOVA model (15.9), we drop β_j and α_i is a rv. In two-way mixed-effects ANOVA the model is

$$y_{ij} = \mu + \alpha_i + \beta_{ij} + \varepsilon_{ij}$$

where y_{ij} is the value of the response for the jth of n_i observations in the ith group with $i = 1, \ldots, M$; α_i represent the fixed-effect means which, under the null hypothesis, are equal for all groups; β_{ij}, $j = 1, \ldots, R$ are the random effect means for group i (β are rv); and ε_{ij} are the error for observation j in group i. β_{ik} are assumed normal with mean zero and variance σ_k^2 and covariance $\sigma_{kk'}$. Hence, the random effects are *not* assumed independent; in fact, in most cases they will be dependent. Although not necessary, we will use the restrictive assumption that ε_{ij} are independent with variance σ^2.

Example 15.31. We

```
> load('EU.CO.rda')
```

remove Belgium and attach the data frame

```
> EU.CO <- EU.CO[EU.CO$country != 'BELGIUM', ]
> attach(EU.CO)
```

506 Analysis of variance

To get rid of the BELGIUM factor level, we do

```
> country <- as.character(country[country != 'BELGIUM'])
> country <- as.factor(country)
```

and log-transform the response

```
> logCO <- log(CO + 0.001)
```

To examine interactions, we do

```
> par(mar = c(4, 4, 1, 2))
> interaction.plot(area, country, logCO, type = 'b')
```

to obtain Figure 15.15 (compare to Figure 15.13). From the figure we see clear interaction effects. Mixed-effects linear models are implemented in the package nlme. So we

```
> library(nlme)
```

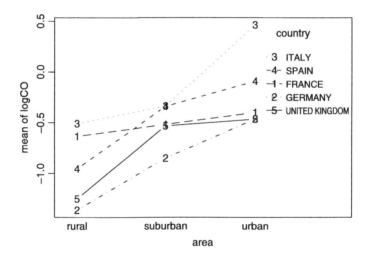

Figure 15.15   Interactions between countries and areas for log(CO).

To use the library effectively, we create a grouped data frame. This is a usual data frame that also specifies how the data are grouped:

```
> grouped <- groupedData(formula = logCO ~ area | country)
> grouped$country <- factor(grouped$country)
> grouped$area <- factor(grouped$area)
> head(grouped)
Grouped Data: logCO ~ area | country
      logCO  area country
1 -1.0328245 rural GERMANY
2 -1.0188773 rural GERMANY
3 -1.0328245 rural GERMANY
4 -1.0133524 rural GERMANY
5 -0.9808293 rural GERMANY
6 -1.0133524 rural GERMANY
```

The grouping formula says "logCo is the response variable, area is one effect and it is conditioned on (or grouped by) country." After grouping, we have to explicitly make country and area factors again. The data frame displays information about the grouping model and it therefore can be used directly in calls to the linear model by simply specifying the grouped data frame. A nicer way (than with interaction.plot()) to examine interactions is with

```
> library(lattice)
```

which allows us to open a graphics window with

```
> trellis.device(color = FALSE, width = 4.5, height = 4)
```

and use

```
> xyplot(logCO ~ area | country,
+     panel = function(x, y){
+        panel.xyplot(x, y)
+        panel.lmline(x, y, lty = 2)
+     }
+ )
```

Within the xyplot() function, we can draw into each panel by making use of the named argument panel. Here, we insert into each panel a linear fit to the data with a call to panel.lmline(). As it is, the plot is unsatisfactory. We need to adjust the size of the strip titles and the x-axis ticks' text. This we do with

```
> update(trellis.last.object(),
+     par.strip.text = list(cex = 0.75),
+     scales = list(cex = 0.6))
```

Thus, we obtain Figure 15.16.

Notice that except for France, all countries exhibit a positive trend as one moves from rural to suburban to urban areas. This trend is most pronounced in Italy. Next, we want to compare the confidence intervals of mean log(CO) by country, for each area. The function lmList() fits a linear mixed-effects model for each group (levels in country). It wants a data frame, *not* a grouped data frame. So we do

```
> country.list <- lmList(logCO ~ -1 + area | country, data =
+     as.data.frame(grouped))
```

(-1 removes the intercept, which we do not want here) and then plot the intervals

```
> i <- intervals(country.list)
```

To modify the axes and strip labels, we

```
> dimnames(i)[[3]] <- c('rural', 'suburban', 'urban')
> p <- plot(i) ; p$ylab <- 'log(CO)' ; p
```

(Figure 15.17). We now see that the means in urban areas in Germany, France and the UK are perhaps not that different whereas Spain and Italy stand on their own, each. The distinctions are not as clear in rural areas, but again, Italy and France are the "leaders". Finally, if you want to live in the suburbia, go to Germany. □

508 Analysis of variance

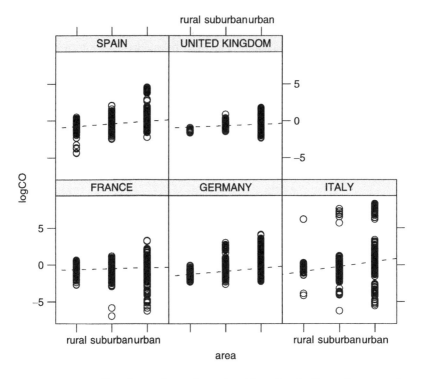

Figure 15.16   Random-effects (area) within fixed-effects (country).

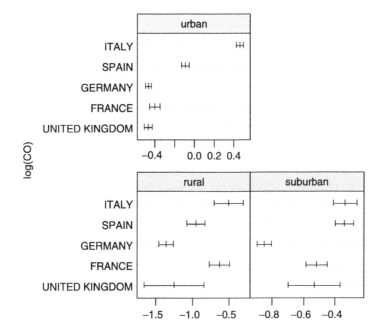

Figure 15.17   Mean log(CO) by area for each country.

15.6 Assignments

Exercise 15.1. Verify with direct calculations that the Within MS in example 15.11 is 18.662.

Exercise 15.2. Interpret the results of the Bonferroni test in Example 15.12 with respect to ethnicity.

Exercise 15.3. Use some of the steps shown in Example 15.26 to produce the data frame EU.CO as shown in the example.

Exercise 15.4. Use EU.CO.rda to show with R that when running two-way ANOVA, column effect is smaller compared to a one-way ANOVA on the column variable only (see Example 15.30).

Exercise 15.5. Run nonparametric ANOVA on the data in Example 15.30. Compare the conclusions of the ANOVA in the example to those you obtain from the nonparametric analysis.

Exercise 15.6. In Example 15.30 we claim that the diagnostics of the original data (before the log transformation) violate the assumptions of two-way ANOVA. Verify this claim by running the diagnostics.

Exercise 15.7. Use Example 15.30.

1. Produce an interaction effect plot.
2. Run ANOVA with interaction effects.
3. Which country is most "responsible" for the interaction effect?
4. Remove Italy from the data and run ANOVA with interaction effect only and a full model. Is the interaction significant without Italy for both models?

16
Simple logistic regression

Here we introduce logistic regression models. We discuss the reasons for using such models. We shall then see how to fit such models to data. Finally, we discuss model diagnostics. In a nutshell, logistic models are used when the response variable is binary (presence absence, yes no and so on). The independent variables (covariates) maybe decimal, integers, factored or ordered factors. In multiple logistic regression, any mix of covariate type is acceptable. We discuss only two-variable logistic regression.

Data that are appropriate for analysis with logistic regression are quite common. For example, you may measure an important habitat variable, say the extent of canopy shading on a forest floor and a response variable which records the absence or presence of a certain plant species. The logistic regression then provides answer to the following: Given a particular amount of solar radiation that reaches the soil surface, what is the probability that we may detect the presence of a particular plant species? In exposure studies, one may be interested in the probability of getting sick given different levels of exposure to a toxic (or sickening) agent.

16.1 Simple binomial logistic regression

Often, response (dependent) variables are categorical while the covariates (independent) variables may be categorical, discrete or continuous.

Example 16.1. Consider a sample of 10 random rows from a data set about fish in two Minnesota rivers:

```
> load('fish.rda')
> idx <- sample(dimnames(fish$adults)[[1]], 10)
> fish$adults[idx, c(1, 4 : 6, 8 : 10)]
```

```
         river water.temp air.temp   habitat depth velocity BCS
2027        YM       21.5     23.5 shoreline    21        3   0
451         OT       13.0     22.0   raceway    49       95   0
1735        YM       10.0      7.0    riffle    69       51   0
1046        OT       21.5     13.5 backwater    74       30   0
596         OT        2.0     -3.0      pool    99       30   0
1765        YM       27.0     31.5 shoreline    37        8   0
57          OT       21.0     20.0   raceway    65       24   0
1829        YM       26.0     25.5    riffle    58       74   0
912         OT       20.0     22.0 deep pool   146        8   0
1875        YM       22.0     23.0      pool    53        4   0
```

Here we take a random sample of 10 rows from the data frame fish$adults. We assign the row numbers (in dimension 1 of the data frame) to an index idx. We then pick those sampled lines and a subset of the columns. The last column indicates the presence (1) or absence (0) of a fish species, coded as BCS (Blackchin shiner, *Notropis heterodon*). The presence of individuals of this species may depend on a host of covariates—air-temperature, water temperature, water depth and velocity. These are variables of continuous type. But there are other variables that might determine the presence of individuals—the river (OT for Otter Tail or YM for Yellow Medicine) and habitat type. These are categorical variables. We are interested in establishing relationship between the probability of detecting individuals of the species and the covariates. □

The data in Example 16.1 are not all continuous and we cannot use classical linear regression to establish relationship between where we find individuals of the species (a binomial random variable) and habitat and environmental variables. We can, however, recast the dependent variable in a probability framework. Probability is a real number, in a closed interval between zero and one. We can transform the probability such that the transformed values may take any real number. Let us pursue this idea.

Consider the binomial rv $Y = 0$ or 1. The latter denotes (by definition) success and the former failure. Denote the probability of success by π and the number of trials by n. Suppose that n_S of these were success. Then the best estimate π is

$$\widehat{\pi} = p = \frac{n_S}{n} \ .$$

Example 16.2. In Example 16.1, Y takes on the value of 0 when no individuals of the species were recorded and 1 when they were. There are 2 152 records in the data and in two of them individuals of Blackchin shiner were found. Therefore, $p = 2/2\,152$. □

We now define

Odds ratio The ratio of successes to failure, i.e.

$$\text{Odds ratio} = \frac{\pi}{1-\pi} \ .$$

To estimate the population odds ratio, we use the ratio $p/(1-p)$. The following transformation maps the range of the odds ratio from $[0, \infty]$ to $[-\infty, \infty]$:

Logit transformation of π is defined as

$$\lambda(\pi) = \log\left(\frac{\pi}{1-\pi}\right).$$

Here log is for the basis of e.[1] The logit transformation is 1 to 1, i.e. for each unique value of π there is a unique value of λ. The opposite is also true. We can therefore talk about the inverse logit, defined next.

Logistic transformation is inverse of the logit transformation, i.e.

$$\lambda^{-1}(\pi) := \pi(\lambda) := \frac{e^\lambda}{1+e^\lambda} = \frac{1}{1+e^{-\lambda}}.$$

The logit transformation is useful because it maps π, which takes on values between 0 and 1, to $\lambda(\pi)$ which takes on any real value. This fact is illustrated in Figure 16.1 which may be reproduced with

```
> logit <- function(p){log(p / (1 - p))}
> p <- seq(.01, .99, length = 101)
> plot(logit(p), p, type = 'l', xlim = c(-4, 4))
> x <- seq(-6, 6, length = 101)
> lines(x, pnorm(x), lwd = 3)
```

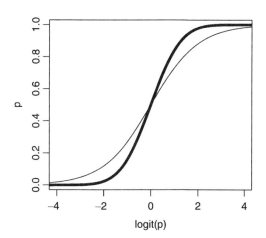

Figure 16.1   The logit transformation (thin curve) compared to the normal distribution (thick curve).

The logit transformation is particularly useful when used in likelihood functions (see Section 5.8).

Consider the rv Y to have a binary outcome with 1 denoting success and 0 failure. In Example 16.1, success is recording individuals of a species in a location in a river

---

[1] Some use ln to distinguish between logarithm with respect to base e and log—the logarithm with respect to base 10. We shall use log to denote logarithm with respect to the basis $e = 2.718282\ldots$.

and failure is not recording any individuals of the species. Suppose that the outcome depends on the value of some explanatory (covariate) rv X. Now because the odds ratio depends on the value of X, we define

Log odds ratio ($\lambda(x)$) The function
$$\lambda(X) = \log\left(\frac{P(Y=1|X)}{P(Y=0|X)}\right) = \log\left(\frac{P(Y=1|X)}{1-P(Y=1|X)}\right).$$

Here, $P(Y=1|X)$ expresses the fact that the probability that $Y=1$ depends on the value of X. λ is a function of X because Y is a constant whereas X on the right hand side is a variable. Now assume that X and Y are related via the linear model
$$\lambda(X) = \beta_0 + \beta_1 X$$
where β_0 is the intercept and β_1 is the slope of the regression. When $X=0$, we obtain
$$\log\left(\frac{P(Y=1|X=0)}{1-P(Y=1|X)}\right) = \beta_0.$$
Therefore,
$$\frac{P(Y=1|X=0)}{1-P(Y=1|X=0)} = e^{\beta_0}.$$
In other words, e^{β_0} is the odds ratio of obtaining $Y=1$ for $X=0$. Often, we scale the data such that $X=0$ is a reference point. Then we say that β_0 is the *baseline log odds* or the *reference log odds*. From this we get that the probability of $Y=1$ for the reference group is
$$P(Y=1|X=0) = \frac{e^{\beta_0}}{1+e^{\beta_0}} = \frac{1}{1+e^{-\beta_0}}.$$
When X increases by one unit, we have
$$\lambda(X+1) - \lambda(X) = \beta_0 + \beta_1(X+1) - (\beta_0 + \beta_1 X)$$
$$= \beta_1.$$
Therefore, we interpret β_1 as the log odds ratio per unit increase in X.

Example 16.3. The data relate to death sentences in the U.S. from 1973 on (United States Department of Justice, 2003); see Example 9.13. It include 7 568 cases of convicts sentenced to death. Two variables of interest are years of education and skin color. We wish to answer the following question: What is the proportion of blacks in the population of convicts sentenced to death as a function of the years of education? Is it true that blacks constitute the same proportion among those with 7 or less years as they are among those with say 12 years?

We load the data and remove the cases for which either Race or Education are missing:

```
> load('capital.punishment.rda')
> length(capital.punishment[, 1])
[1] 7658
```

Simple binomial logistic regression

```
> idx <- complete.cases(capital.punishment[,
+     c('Race', 'Education')])
> cp <- capital.punishment[idx, ]
> length(cp[, 1])
[1] 6495
```

Over 1000 records are missing, so we treat the clean data as a sample. Next, we observe the levels of `skin`:

```
> skin <- cp$Race
> education <- cp$Education
> levels(skin)
[1] "Asian"          "Black"            "Native"
[4] "Other"          "Pacific Islander" "White"
```

and change the level `Black` to `TRUE` and all other levels to `FALSE`:

```
> levels(skin)[-2] <- FALSE
> levels(skin)[2] <- TRUE
> levels(skin)
[1] "FALSE" "TRUE"
```

To run the logistic regression on the data, we use

```
> library(Design)
```

and prepare the data for the logistic regression model `lrm()`:

```
> ddist <- datadist(skin, education)
> options(datadist = 'ddist')
```

`datadist()` lets `lrm()` know about the data through setting `options()`. Thus

```
> (blacks <- lrm(skin ~ education, x = TRUE, y = TRUE))

Frequencies of Responses
FALSE  TRUE
 3740  2755

        Obs  Max Deriv  Model L.R.    d.f.       P
       6495     8e-11       24.48        1       0

             Coef     S.E.    Wald Z   P
Intercept  0.34261  0.13351    2.57   0.0103
education -0.06123  0.01241   -4.94   0.0000
```

(output edited). You can achieve the same results by running the generalized linear model `glm()` directly. `Design`, however, includes several utility functions that assist in obtaining results specific to logistic regression. For example,

```
> plot(blacks, xlab = 'education',
+      ylab = 'log odds of skin color')
```

516 Simple logistic regression

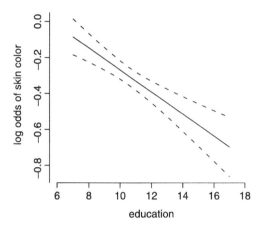

Figure 16.2 Log odds ratio (of being back) vs. years of education ±95% confidence intervals.

produces Figure 16.2. We will discuss the details of the model output soon. For now, we observe that there were 3 740 observations labeled as FALSE (other) and 2 755 labeled as TRUE (blacks). The generalized likelihood ratio (denoted as Model L.R.) tests for the overall significance of the model. We will discuss this statistic soon. For now, we note that it has a χ^2 density with 1 degrees of freedom and that its value is 24.48. The p-value of Model L.R. is virtually zero and we conclude that the model fit is significant. The logistic model is

$$\widehat{\lambda}(X) = 0.343 - 0.061 X$$

with both coefficients significant (p-values of 0.01 and 0.00). When years of education are seven or less (recorded as 7 in the data),

$$\widehat{P}(Y=1|X=7) = \frac{1}{1+e^{0.343-0.061\times 7}} = 0.479 \ .$$

Blacks with seven or less years of education constitute almost half of the population of inmates sentenced to death.[2] The log odds of this probability (i.e. of the probability of being black in the population of inmates who were convicted to death) decreases by

$$\widehat{\lambda}(X) = 0.343 - 0.061 \times 8 - (0.343 - 0.061 \times 7) = 0.061$$

per additional year of education. More explicitly, for each additional year of education,

$$\frac{\widehat{P}\,(Y=1|X)}{1-\widehat{P}\,(Y=1|X)} = e^{-0.061} = 0.94 \ ;$$

the odds of blacks in the population "increases" by a factor of 0.94 (or decreases by 6%) per additional year of education. □

---

[2] Strictly speaking, one needs to keep in mind that the population is pooled over all years of data.

The next example illustrates the idea of using a reference value for a variable in the regression. It also illustrates how to implement logistic regression for data that are presented in a summary table only—a common practice in publications. We also discuss how to interpret the results.

Example 16.4. The data were reported in Kline et al. (1995). A subset of it was analyzed in Fleiss et al. (2003), p. 293 and Table 11.1. The data are about risk factors that affect miscarriage of dead fetuses. One of the reasons for miscarriage is a condition called trisomy, where one of the 23 pairs of chromosomes gains an extra chromosome. For example, trisomy of the 21st chromosome leads to Down's syndrome. We are interested in the relationship between trisomy and maternal age. First, the data as presented in Fleiss et al. (2003):

```
> trisomy.table <- read.table('trisomy-and-maternal-age.txt',
+     header = TRUE, sep = '\t')
> dimnames(trisomy.table)[[1]] <- trisomy.table$age
> trisomy.table <-
+     trisomy.table[, 2 : length(trisomy.table[1, ])]
> save(trisomy.table, file = 'trisomy.table.rda')
> trisomy.table
      coded trisomic normal total proportion fitted
15-19  -2.5        9     70    79      0.114  0.107
20-24  -1.5       26    157   183      0.142  0.145
25-29  -0.5       42    163   205      0.205  0.194
30-34   0.5       37    130   167      0.222  0.254
35-39   1.5       33     59    92      0.359  0.325
40-44   2.5       12     18    30      0.400  0.405
```

The columns are: first–age class (in years), `coded`–coded age class, `trisomic`–incidence of trisomic fetuses among the miscarriages that were studied, `normal`–incidence of non-trisomic fetuses among the premature miscarriages. We discuss the last column in a moment. The coded age gives the midpoint of age intervals, divided by 5 where 30 is the reference age. As Fleiss et al. (2003) pointed out, this is a good way to summarize the data because the intercept of the regression refers to age 30.

Next, we create the data as it might look in a case by case.

```
> trisome <- vector(); age <- vector()
> for(i in 1 : length(trisomy.table[, 1])){
+     trisome <- c(trisome, rep(TRUE, trisomy.table[i, 2]),
+         rep(FALSE, trisomy.table[i, 3]))
+     age <- c(age, rep(trisomy.table[i, 1],
+         trisomy.table[i, 4]))
+ }
> trisomy <- data.frame(age, trisome)
> save(trisomy, file = 'trisomy.rda')
> head(trisomy, 4)
  age trisome
1 -2.5    TRUE
2 -2.5    TRUE
3 -2.5    TRUE
4 -2.5    TRUE
```

518 Simple logistic regression

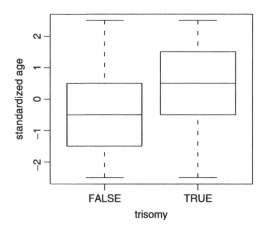

Figure 16.3   FALSE refers to non-trisomic fetuses, TRUE to trisomic fetuses.

These are the data we use for fitting a logistic regression. As mothers' age increases, so does the incidence of trisomy (Figure 16.3), which was produced with

```
> load('trisomy.rda')
> x <- trisomy$age
> y <- as.factor(trisomy$trisome)
> plot(y, x, xlab = 'trisomy', ylab = 'standardized age')
```

Logistic regression provides much more information than an analysis that might rely on the results in Figure 16.3. Fitting the model we get (some output was deleted):

```
> library(Design)
> ddist <- datadist(x, y)
> options(datadist = 'ddist')
> (model <- lrm(y ~ x, x = TRUE, y = TRUE, se.fit = TRUE))
Frequencies of Responses
FALSE  TRUE
  597   159
          Coef    S.E.     Wald Z  P
Intercept -1.254  0.09086  -13.80  0
x          0.347  0.06993    4.96  0
```

The model is then
$$\widehat{\lambda}(X) = -1.254 + 0.347X \ . \tag{16.1}$$

Both coefficients are significant (P=0). At the reference age of $X = 30$, the probability of trisomy among miscarriaged fetuses is

$$\widehat{P}(Y = 1 | X = 30) = \frac{1}{1 + e^{-(-1.254)}} = 0.222 \ .$$

Recall that
$$\lambda(X) = \log\left(\frac{P(Y = 1|X)}{1 - P(Y = 1|X)}\right) \ .$$

Therefore

$$\frac{\widehat{P}(Y=1|X)}{1-\widehat{P}(Y=1|X)} = e^{\widehat{\lambda}(X)}$$
$$= e^{-1.254+0.347X}.$$

So for increase in 1 unit of coded age, we have

$$\frac{\widehat{P}(Y=1|X+1)}{1-\widehat{P}(Y=1|X+1)} - \frac{\widehat{P}(Y=1|X)}{1-\widehat{P}(Y=1|X)} = e^{-1.254+0.347(X+1)-(-1.254+0.347X)}$$
$$= e^{0.347} = 1.41.$$

In other words, the odds of experiencing a miscarriage of a trisomic fetus increase by a factor of 1.41 for every 5 years of the mother's age. For one year the change in this odds is $e^{0.347/5} = 1.072$. Figure 16.4, produced with

```
> plot(model, xlab = 'standardized age',
+     ylab = 'log odds of trisomy')
```

illustrates the results. □

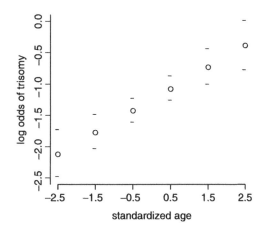

Figure 16.4   log odds ratio of having a miscarriage of a trisomic fetus vs. mother's standardized age (each unit repents 5 years and 0 is at age 30) ±95% confidence interval.

16.2 Fitting and selecting models

In this section we discuss how we compute the regression coefficients and the criteria by which we decide that the model fits the data. We will examine the significance of the model's coefficients. Finally, we will answer the question: Does a model with intercept only suffice or do we need to add a slope?

16.2.1 The log likelihood function

We discussed likelihood, log likelihood and maximum likelihood estimators (MLE) in Sections 5.8 and 9.1.1. Here we apply the ideas to estimating the parameters of the logistic regression. To simplify the discussion, suppose we wish to fit a model to data and consider only two observations: (X_1, Y_1) and (X_2, Y_2) where Y is binomial and X is continuous. Recall that our model is

$$P(Y = 1|X) = \frac{1}{1 + e^{-(\beta_0 + \beta_1 X)}},$$
$$P(Y = 0|X) = 1 - P(Y = 1|X).$$

Take the first pair of data values (X_1, Y_1) and substitute them in the pair of equations above to get

$$P(Y_1|X_1) = \frac{1}{1 + e^{-(\beta_0 + \beta_1 X_1)}},$$
$$P(1 - Y_1|X_1) = 1 - P(Y_1|X_1).$$

Here X_1 and Y_1 are known; β_0 and β_1 are not. Substitutions for the second observation lead to similar expressions. Our task is to come up with a *likelihood function* that reflects the contribution of (X_1, Y_1) and (X_2, Y_2) to the likelihood that both observations occur. We define the contribution of each observation to the likelihood function as[3]

$$\left[\frac{1}{1 + e^{-(\beta_0 + \beta_1 X_i)}}\right]^{Y_i} \left[1 - \frac{1}{1 + e^{-(\beta_0 + \beta_1 X_i)}}\right]^{1 - Y_i}$$

for $i = 1, 2$. Because the observations are presumed independent, their combined contribution to the likelihood function is their product, which we write as

$$\mathcal{L}(\beta_0, \beta_1 | X) = \prod_{i=1}^{2} \left[\frac{1}{1 + e^{-(\beta_0 + \beta_1 X_i)}}\right]^{Y_i} \left[1 - \frac{1}{1 + e^{-(\beta_0 + \beta_1 X_i)}}\right]^{1 - Y_i}.$$

Note that $\mathcal{L}$ is a function of the coefficients, not the data because the data are known. In the general case

$$\mathcal{L}(\beta_0, \beta_1 | X) = \prod_{i=1}^{n} [P(Y_i|X_i)]^{Y_i} [1 - P(Y_i|X_i)]^{1 - Y_i}.$$

We are free to choose any values for β_0 and β_1. Because $\mathcal{L}(\beta_0, \beta_1 | X)$ expresses the likelihood that the observation values occur, we choose the values of β_0 and β_1 such that this likelihood is maximized. The values of coefficients that maximize $\mathcal{L}(\beta_0, \beta_1 | X)$ also maximize $\log[\mathcal{L}(\beta_0, \beta_1 | X)]$. Therefore, we will work with the log likelihood function— this turns products into sums. To simplify the notation, we write

$$L(\beta_0, \beta_1 | X) := \log[\mathcal{L}(\beta_0, \beta_1)].$$

Using log rules, we obtain

$$L(\beta_0, \beta_1 | X) = \sum_{i=1}^{n} \left[Y_i \log\left(\frac{1}{1 + e^{-(\beta_0 + beta_1 X_i)}}\right)\right]$$
$$+ \sum_{i=1}^{n} \left[(1 - Y_i) \log\left(1 - \frac{1}{1 + e^{-(\beta_0 + beta_1 X_i)}}\right)\right]. \quad (16.2)$$

---

[3] The definition of a likelihood function is not unique.

We denote the values of β_0 and β_1 that maximize $L(\beta_0, \beta_1|X)$ (the MLE) by $\widehat{\beta}_0$ and $\widehat{\beta}_1$. In terms of probabilities, we write equation (16.2) (after some algebraic manipulations) as

$$L(\beta_0, \beta_1|X) = \sum_{i=1}^{n} \left[Y_i \log\left(\frac{P(Y_i|X_i)}{1 - P(Y_i|X_i)}\right) + \log(1 - P(Y_i|X_i)) \right]. \quad (16.3)$$

To find the maximum of $L(\beta_0, \beta_1)$ with respect to the coefficients, we must use numerical techniques (see Section 9.1.5). To maintain meaningful notation, we denote any value that is derived from the MLE with $\widehat{}$. For example, when $P(Y = 1|X)$ is computed from the MLE $\widehat{\beta}_0$ and $\widehat{\beta}_1$, we acknowledge this fact with $\widehat{P}(Y = 1|X)$. Once we determine the MLE of the coefficients, we are left with the tasks of implementing statistical inference—we need to test the significance of the coefficients and the significance of the overall model—and model diagnostics. For example, it is conceivable that we do find MLE for $\widehat{\beta}_0$ and $\widehat{\beta}_1$ and yet, the model does not improve significantly our prediction of $P(Y|X)$ compared to no model at all.

16.2.2 Standard errors of coefficients and predictions

To fix ideas, let us recall and introduce the following notation and definitions:

(Y_i, X_i) $i = 1, \ldots, n$ denote the data.
$\widehat{\beta}_0$ and $\widehat{\beta}_1$ The maximum likelihood estimates (MLE) of β_0 and β_1.
$\widehat{P}(X)$ The MLE estimate of model probability for a given X,

$$\widehat{P}(X) = \frac{1}{1 + e^{-(\widehat{\beta}_0 + \widehat{\beta}_1 X)}}.$$

$\widehat{P}_i$ The estimated model probability for a particular X_i, defined as $\widehat{P}_i := \widehat{P}(X_i)$.
w_i Weights, defined as $w_i := \widehat{P}_i \left(1 - \widehat{P}_i\right)$.
$\overline{X}_w$ The weighted average of X_i,

$$\overline{X}_w = \frac{\sum_{i=1}^{n} w_i X_i}{\sum_{i=1}^{n} w_i}.$$

SS_w The weighted sum of squares,

$$SS_w := \sum_{i=1}^{n} w_i \left(X_i - \overline{X}_w\right)^2.$$

We compute the standard errors for $\widehat{\beta}_0$ and $\widehat{\beta}_1$ (see Fleiss et al. 2003) with

$$SE\left(\widehat{\beta}_0\right) = \sqrt{\frac{1}{\sum_{i=1}^{n} w_i} + \frac{\overline{X}_w^2}{SS_w}}, \quad SE\left(\widehat{\beta}_1\right) = \frac{1}{\sqrt{SS_w}}.$$

The covariance between the parameter estimates is given by

$$Cov\left(\widehat{\beta}_0, \widehat{\beta}_1\right) = \frac{\overline{X}_w}{SS_w}.$$

With these expressions, we obtain estimates of the log odds for particular values of X with
$$\lambda\left(\widehat{P}(X)\right) = \widehat{\beta}_0 + \widehat{\beta}_1 X$$
with standard errors
$$\text{SE}\left(\lambda\left(\widehat{P}(X)\right)\right) = \sqrt{\text{SE}\left(\widehat{\beta}_0\right)^2 + 2X\text{Cov}\left(\widehat{\beta}_0, \widehat{\beta}_1\right) + X^2\,\text{SE}\left(\widehat{\beta}_1\right)^2}.$$
The standard error of $\widehat{P}(X)$ is then
$$\text{SE}\left(\widehat{P}(X)\right) = \widehat{P}(X)\left[1 - \widehat{P}(X)\right]\text{SE}\left(\lambda\left(\widehat{P}(X)\right)\right). \tag{16.4}$$
These equations provide confidence intervals for $P(X)$.

Example 16.5. We continue with Example 16.4, where we named the logistic regression output `model`. There, we examined the log odds ratio vs. age. Here, we use `model` to produce the probability of miscarriage vs. age. We also wish to plot the probabilities with their 95% confidence interval. This allows us to compare predictions to data and verify (if at all) that the data are limited by the 95% confidence interval.

In Figure 16.4, we see the predicted values with 95% confidence intervals. These are plotted for the log odds ratios on the y-axis. To produce the probabilities and their 95% confidence interval, we apply the logistic transformation to the model, i.e.
$$\widehat{P}(Y=1|X) = \frac{1}{1+e^{-(-1.254+0.347X)}}$$
for X between -2.5 and 2.5. First, a sample of the data:

```
> load('trisomy.rda')
> X <- trisomy$age
> Y <- as.factor(trisomy$trisome)
> set.seed(22) ; idx <- sample(1 : length(X), 5)
> cbind(X = X[idx], Y = Y[idx])
       X Y
[1,] -1.5 1
[2,] -0.5 1
[3,]  2.5 1
[4,] -0.5 1
[5,]  1.5 2
```

Here is the regression curve on a probability scale:

```
> plot(model$x, 1 / (1 + exp(-model$linear.predictors)),
+      type = 'l', ylim = c(0, 1),
+      xlab = 'maternal standardized age',
+      ylab = 'probability of trisomy')
```

(Figure 16.5) and the 1.96 SE from (16.4) on both sides:

```
> se <- 1.96 * model$se.fit
> lines(model$x, 1 / (1 + exp(-(model$linear.predictors +
+      se ))), lty = 2)
> lines(model$x, 1 / (1 + exp(-(model$linear.predictors -
+      se))), lty = 2)
```

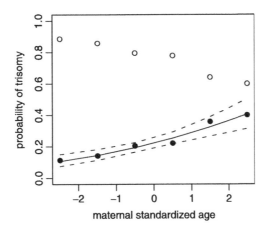

Figure 16.5 Regression curve (predicted values) and the 95% confidence interval on the predictions (broken curves). The y-axis corresponds to probability of trisomy in miscarriaged fetuses (black disks) and $1 -$ this probability (circles).

To obtain empirical probabilities, we need to count the number of TRUE, number of FALSE and divide each by the number of observations. So we split the data into a list (and observe some of the TRUE records):

```
> tri <- split(X, Y)
> head(tri$'TRUE')
[1] -2.5 -2.5 -2.5 -2.5 -2.5 -2.5
```

(recall that the TRUE records indicated the cases of trisomy in miscarriaged fetuses). Next, we count the number of TRUE and FALSE for each center of standardized mother's age,

```
> n.true <- tapply(tri$'TRUE', as.factor(tri$'TRUE'),
+       length)
> n.false <- tapply(tri$'FALSE', as.factor(tri$'FALSE'),
+       length)
```

Thus we obtain the empirical probabilities:

```
> n <- n.true + n.false
> true <- n.true / n
> false <- n.false / n
```

and add them to the plot:

```
> points(unique(model$x), true, pch = 19)
> points(unique(model$x), false)
```
□

To test the significance of the model coefficients, we use the Wald-Z statistic. The statistic is computed for each model coefficient ($\widehat{\beta}_i$, $i = 0, 1$ in our case):

$$\text{Wald-Z} = \frac{\widehat{\beta}_i}{\text{SE}(\widehat{\beta}_i)}.$$

It has a normal density.

Example 16.6. From Example 16.4 we obtain

$$\widehat{\beta}_0 = -1.254 \quad SE(\widehat{\beta}_0) = 0.091 \ .$$

Therefore, Wald-Z $= -13.80$, with a p-value $= 0$. Similarly, for $\widehat{\beta}_1$ we obtain Wald-Z $= 4.96$ with p-value $= 0$. □

16.2.3 Nested models

So far, we assessed the univariate (individual) significance of the model coefficients β_0 and β_1. We now address the issue of the general model adequacy. The central issue here has to do with the choice of a model. What criteria should we use in choosing one model as opposed to another. Because we are dealing with simple logistic regression, we need to distinguish between two models: One with the intercept only and one with intercept and slope. The ideas here extend directly to multivariate models. To proceed, we cast the model adequacy assessment in terms of hypothesis testing. The null hypothesis is that fitting the model with β_0 only suffices—we do not need β_1. Formally, we test

$$H_0 : \lambda(P(X)) = \beta_0 \quad \text{vs.} \quad H_A : \lambda(P(X)) = \beta_0 + \beta_1 X \ .$$

The model under H_0 is obtained by setting $\beta_1 = 0$. Therefore, we say that the model under H_0 is *nested* in the model under H_A. For this reason, H_0 is said to be a *nested hypothesis* of H_A. Implementing the log likelihood (16.3), we obtain $L(\beta_0)$ under H_0 and $L(\beta_0, \beta_1)$ under H_A.[4] Maximizing $L(\beta_0)$, we get the MLE for β_0 and $P(Y)$, which we denote by $\widehat{\beta}_0^0$ and $\widehat{P}^0(Y)$. Similarly, maximizing $L(\beta_0, \beta_1)$, we get MLE for β_0, β_1 and $P(Y)$. We denote these MLE by $\widehat{\beta}_0^A$, $\widehat{\beta}_1^A$ and $\widehat{P}^A(Y)$.

To compare the models, it makes sense to look at the ratio of the maximum likelihood values, $\mathcal{L}(\widehat{\beta}_0^A, \widehat{\beta}_1^A)/\mathcal{L}(\widehat{\beta}_0^0)$. The larger the ratio, the more likely the model under H_A is relative to H_0. To obtain inference about this ratio, we need the sampling density of this ratio. It turns out that for a large sample size, the sampling density of twice the log of the ratio, which we denote by

$$G(H_A : H_0) := 2 \times \log \left[\frac{\mathcal{L}(\widehat{\beta}_0^A, \widehat{\beta}_1^A)}{\mathcal{L}(\widehat{\beta}_0^0)} \right] \tag{16.5}$$

is χ^2 with degrees of freedom equal the number of degrees of freedom of the nesting model less the number of degrees of freedom of the nested model. For H_A we fit two coefficients and therefore we have two degrees of freedom. The model under H_0 has one degree of freedom. Therefore, $G(H_A : H_0)$ has one degree of freedom. G in (16.5) is one example of the *generalized likelihood ratio*. It is conveniently written as

$$G(H_A : H_0) = 2 \left[L\left(\widehat{\beta}_0, \widehat{\beta}_1\right) - L\left(\widehat{\beta}_0\right) \right] \ .$$

Example 16.7. For the trisomy model introduced in Example 16.4, we find

$$\mathcal{L}\left(\widehat{\beta}_0\right) = 1.307\,197 \times 10^{-169} \ , \quad \mathcal{L}\left(\widehat{\beta}_0, \widehat{\beta}_1\right) = 4.425\,755 \times 10^{-164} \ .$$

---
[4] From here on we drop from the notation the dependency on X.

Therefore,
$$\frac{\mathcal{L}\left(\widehat{\beta}_0, \widehat{\beta}_1\right)}{\mathcal{L}\left(\widehat{\beta}_0\right)} = 338\,568.3\,, \quad 2 \times \log\,(338\,568.3) = 25.46\,.$$

The *p*-value of $\chi^2_{1,.05}$ is $4.516\,505 \times 10^{-07}$. Therefore, the probability that we get such a large generalized likelihood ratio by chance alone is so small (smaller than for $\alpha = 0.05$), that we are compelled to reject H_0 in favor of H_A. We thus conclude that the model (16.1) is a significant improvement over the model with intercept only. In other words, age is associated with increased incidence of trisomy among miscarriages. □

16.3 Assessing goodness of fit

In the previous section, we learned how to assess the overall adequacy of the model. Here we are interested in

Goodness of fit Comparing observed outcomes to predicted outcomes based on the logistic model.

We say that a model *fits* if:

- the distance between observed outcomes and predicted outcomes is small and
- the contribution of each pair of observed outcome and fitted outcome to the summary measure is asymmetric and small relative to the error structure of the model.

The fitted (also called predicted) values are calculated from the model. To proceed, we introduce the concept of subsets of equal X_i values. Such subsets are called *covariate patterns*. We need such subsets because they allow us to compare empirical probabilities to predicted probabilities. Empirical probabilities are derived from the proportions of 1 (or whatever signifies a response) to the total number of observations in the subset.

Example 16.8. In a study of habitat relationship between Nashville warbler and canopy cover in a forest, we have 200 observations. Each observation consists of the pair (X_i, Y_i) where Y_i is 0 (absent) or 1 (present) and X_i is the percentage cover. For 50 observations, $X_i = 65\%$; for 60, $X_i = 72\%$; the remaining 90 values are unique. Therefore, we have a total of 92 covariate patterns □

To fix ideas, we denote the number of patterns by J. Each pattern includes m_j observations, for $j = 1, \ldots, J$. We let $\boldsymbol{X}_j$ be a vector of covariate observations that belong to the jth pattern. $\boldsymbol{X}_j$ has m_j elements, all have a single value, X_j. In Example 16.8, $m_1 = 50$ observations with $\boldsymbol{X}_1 = 65\%$ cover; $m_2 = 60$ observations with $\boldsymbol{X}_2 = 72\%$; and $m_3 = \cdots = m_{92} = 1$. In all, we have $J = 92$ covariate patterns. To each covariate pattern there correspond m_j values of Y, some of which are 1 others are 0. From the definition of covariate patterns, we conclude that J can have a minimum value of 1 and a maximum value of n. In the former, all values of $\boldsymbol{X}$ are equal, in the latter they are all different. Based on how the covariate patterns distribute themselves between these two extremes, there are different statistics that are appropriate for evaluation of the goodness of fit.

Numerous approaches to assessing the fit have been proposed (see Hosmer et al., 1997; Hosmer and Lemeshow, 2000). Here, we consider only the Pearson χ^2 statistic, the deviance statistic and the area under the so-called Receiver Operator Characteristic (ROC) curve. For the jth covariate pattern, we have m_j observations with $\widehat{P}(Y=1|\boldsymbol{X}_j)$. Let n_j be the number of observations in the jth covariate pattern for which $Y=1$ and $\widehat{n}_j$ be its estimated value. Similarly, let $P_j := P(Y=1|\boldsymbol{X}_j)$. Then the best estimate of n_j is

$$\widehat{n}_j = m_j \widehat{P}_j = m_j \frac{1}{1+e^{-(\widehat{\beta}_0+\widehat{\beta}_1 X_j)}}.$$

16.3.1 The Pearson χ^2 statistic

We begin with a couple of definitions.

The Pearson residual The jth Pearson residual is defined as

$$r_j = \frac{n_j - m_j \widehat{P}_j}{\sqrt{m_j \widehat{P}_j \left(1 - \widehat{P}_j\right)}} \tag{16.6}$$

$$= \frac{n_j - \widehat{n}_j}{\sqrt{\widehat{n}_j \left(1 - \frac{\widehat{n}_j}{m_j}\right)}}.$$

The Pearson χ^2 statistic is defined as

$$C = \sum_{j=1}^{J} r_j^2.$$

C is χ^2 distributed with $J-2$ degrees of freedom. Here $J-2$ corresponds to J patterns fit over 2 coefficients (β_0 and β_1).

Example 16.9. In Example 16.4, the data, as reported by Fleiss et al. (2003) are already broken into patterns by pooling mothers' ages into 5-year intervals. In the example we obtained $\widehat{\beta}_0 = -1.254$ and $\widehat{\beta}_1 = 0.347$. The results in our notation are detailed in Table 16.1, which was produced from the following script:

```
load('trisomy.table.rda')
m.j <- trisomy.table$total
x.j <- trisomy.table$coded
n.j <- trisomy.table$trisomic
beta.0 <-  -1.254 ; beta.1 <- 0.347
coef <- c(beta.0, beta.1)
n.hat.j <- m.j * (1 / (1 + exp(-(beta.0 + beta.1 * x.j))))
r.j <- (n.j - n.hat.j) /
    (sqrt(n.hat.j * (m.j - n.hat.j) / m.j))
pearson.chi.sq <- sum(r.j^2)
d.j <- sqrt(2 * (n.j * log(n.j / n.hat.j) +
    (m.j - n.j) * log((m.j - n.j) / (m.j - n.hat.j))))
```

```
13  deviance.chi.sq <- sum(d.j^2)
14  df <- length(m.j) - length(coef)
15  pearson.p.value <- 1 - pchisq(pearson.chi.sq, df)
16  deviance.p.value <- 1 - pchisq(deviance.chi.sq, df)
```

The script is a straightforward application of the relevant equations and therefore does not need elaboration.
We find that $C = 1.613$. With 6 patterns and 2 coefficients to fit, we have 4 degrees of freedom and we obtain a large p-value. Consequently, the Pearson residuals, taken as a whole, do not violate the assumption of the model. □

Table 16.1 Pearson and deviance χ^2 residuals.

x_j	m_j	n_j	$\widehat{n}_j$	Pearson	Deviance
−2.5	79	9	8.455	0.198	0.197
−1.5	183	26	26.532	−0.112	0.112
−0.5	205	42	39.665	0.413	0.410
0.5	167	37	42.320	−0.946	0.960
1.5	92	33	29.847	0.702	0.696
2.5	30	12	12.137	−0.051	0.051
χ_4^2				1.613	1.629
p-value				0.806	0.804

16.3.2 The deviance χ^2 statistic

Another residual is defined thus:

Deviance residual For the jth covariate pattern and for $n_j - \widehat{n}_j \geq 0$, the deviance residual is defined as

$$d_j = \sqrt{2\left[n_j \log \frac{n_j}{m_j \widehat{P}_j} + (m_j - n_j) \log \frac{m_j - n_j}{m_j\left(1 - \widehat{P}_j\right)}\right]} \qquad (16.7)$$

$$= \sqrt{2\left[n_j \log \frac{n_j}{\widehat{n}_j} + (m_j - n_j) \log \frac{m_j - n_j}{m_j - \widehat{n}_j}\right]}.$$

For $n_j - \widehat{n}_j < 0$, the deviance residual is $-d_j$. For $n_j = 0$,

$$d_j = -\sqrt{2m_j \left|\log \frac{m_j}{m_j - \widehat{n}_j}\right|}$$

and for $n_j = m_j$,

$$d_j = \sqrt{2m_j \left|\log \frac{m_j}{\widehat{n}_j}\right|}.$$

The deviance χ^2 statistic is defined as

$$D = \sum_{j=1}^{J} d_j^2 .$$

D has χ^2 density with $J - 2$ degrees of freedom. Here $J - 2$ corresponds to J patterns fit over 2 coefficients (β_0 and β_1).

Example 16.10. Returning to Table 16.1, we find that $D = 1.629$ with 4 degrees of freedom. The corresponding p-value is large. Consequently, the deviance residuals, taken as a whole, do not violate the assumption of the model. □

Both the Pearson and the deviance χ^2 statistics cannot be applied when m_j is small. The denominator of the Pearson residual is the standard deviation of the residual in the numerator. It can be shown that the deviance residual is also the result of division by the approximate standard error of the residual. Thus, we expect the mean of these residuals to be 0 and their standard deviation to be 1. This allows us to compare and interpret their magnitudes.

16.3.3 The group adjusted χ^2 statistic

The results in this section fall under the name the Hosmer-Lemeshow tests. When $J \approx n$, we have $m_j \approx 1$. In some cases, there may be too few data in a particular pattern to obtain a reasonable estimate of $\hat{n}_j$. In such cases, neither the C nor the D statistics provide a correct p-value. A way around this problem is to break the data into fewer groups. This may result in more than one pattern in a group. What criteria should we choose to group the covariate patterns? Hosmer and Lemeshow (2000) suggested a probability criterion. The idea is to compute all of

$$\widehat{P}(Y = 1|X_i) := \frac{1}{1 + e^{-(\hat{\beta}_0 + \hat{\beta}_1 X_i)}}$$

and then sort them. We can then group the observations based on the probability quantiles. More than one covariate pattern or a fraction of a covariate pattern may be included in a group. This needs to be taken into account in computing the statistic C. To distinguish between this refinement and the statistic C, we denote the residuals based on grouped patterns by C'. Let n'_k be the number of observations in the kth group. Also, denote by J_k the number of patterns in the kth group. The average value of $\widehat{P}(Y = 1|X)$ for X in the kth group is

$$\overline{P}(Y = 1|X_k) := \sum_{j=1}^{J_k} \frac{m_j}{n'_k} \widehat{P}(Y = 1|X_j).$$

Then the observed and estimated number of $Y = 1$ in the kth group are

$$O_k := \sum_{j=1}^{J_k} n_j, \quad \widehat{O}_k = n'_k \overline{P}(Y = 1|X_k).$$

Let g be the number of groups. Then the statistic C' is computed as follows:

$$C' = \sum_{k=1}^{g} \frac{\left(O_k - \widehat{O}_k\right)^2}{n'_k \overline{P}(Y = 1|X_k)\left(1 - \overline{P}(Y = 1|X_k)\right)}.$$

As long as $J \approx n$ and the number of observations in each group is > 5, C' has a χ^2 density with $g - 2$ degrees of freedom.

16.3.4 The ROC curve

The ROC curve refers to Receiver Operator Characteristic. The idea is borrowed from signal processing, where one is interested in identifying a signal with background noise. To introduce the concept we define:

Sensitivity The proportion of $Y = 1$ correctly identified by a test.
Specificity The proportion of $Y = 0$ correctly identified by a test.

In the context of a logistic regression model, we compute $\widehat{P}_i := \widehat{P}(Y = 1|X_i)$ for all of our observations. We then use $\widehat{P}_i$ to predict Y_i based on a cut-off probability. We wish to choose a cut-off probability such that both sensitivity and specificity are maximized. However, if we choose the cut-off probability to increase sensitivity, we sacrifice in specificity. The point is then to choose an optimal cut-off probability. ROC curves facilitate finding this probability. We plot changes in sensitivity and specificity for a sequence of cut-off probabilities and then choose the probability where both are at their joint possible maximum. Note that $1-$ specificity is the proportion of $Y = 0$ that are identified as 1. The ability of the test to classify $Y = 0$ or $Y = 1$ correctly is measured by the area under the ROC curve. Area of 1 reflects a perfect test; area of 0.5 reflects a worthless test. We have

Rule of thumb regarding ROC If
  ROC $\approx$ 0.5 - no discrimination is possible;
  $0.7 \leq$ ROC < 0.8 - discrimination is acceptable;
  ROC ≥ 0.8 - discrimination is excellent.

Note that even a model that fits the data poorly might have good ROC-based discrimination.

Example 16.11. We use the data introduced in Example 16.1, but for a different fish species: the Spotfin shiner (*Notropis spilopterus*), abbreviated to SFS. Here is a random sample of 10 records from the desired columns of the data frame:

```
> load('fish.rda')
> idx<-sample(dimnames(fish$adults)[[1]], 10)
> col<-c(1, 4, 6, 8, 9, 66)
> fish$adults[idx, col]
      river water.temp        habitat depth velocity SFS
1323     YM         25      shoreline  13.2   18.930   6
1711     YM         10      shoreline  11.0   36.000   0
1694     YM         13         riffle  40.0   93.000   0
793      OT         22           pool  78.0   22.000   1
678      OT          1   side channel  22.0   16.000   0
1559     YM         17        raceway  46.0   47.000   0
226      OT         21      shoreline  60.0   28.000   0
2002     YM         21         riffle  28.0   15.000   4
15       OT         26           pool 101.0   46.000   0
1187     YM         10         riffle  25.4   71.299   0
```

We are interested in the relationship between SFS and water depth. In 2 152 samples from both the Yellow Medicine and Otter Tail rivers (YM and OT) in northwestern

530 Simple logistic regression

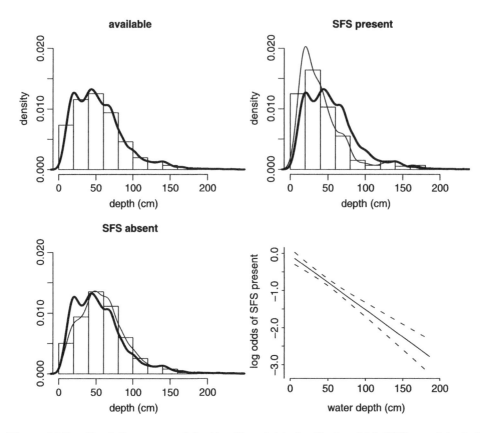

Figure 16.6 Top left: measured depths. Top right: depths in which SFS was detected and measured depths (thick curve). Bottom left: depths in which SFS was not detected and measured depths (thick curve). Bottom right: logistic regression between SFS's presence/absence and water depth.

Minnesota, SFS was present in 665. Let us start with exploring some interesting features in the data. We produce the first three panels in Figure 16.6 thus. First, assignments:

```
> X <- fish$adults$depth
> Y <- ifelse(fish$adults$SFS > 0, 1, 0)
```

Next, the top left histogram:

```
> par(mfrow = c(2, 2))
> hist(X, main = 'available', xlab = 'depth (cm)',
+      ylab = 'density', freq = FALSE,
+      xlim = xlim, ylim = ylim)
```

(we set the limits on the x- and y-axes so that all histograms will be on the same scale). The function density() fits an empirical density to the histogram. It takes a smoothing argument, bw (for band-width). With little experimenting, we set BW to 5 and plot the thick line that shows the density of the measured depths:

```
> BW <- 5 ; ylim <- c(0, 0.02) ; xlim <- c(0, 200)
> lines(density(X, bw = BW), lwd = 3)
```

In the top-right panel we plot the histogram of the depths in which SFS was present and superimpose on it the smoothed density of all sampled depths:

```
> hist(X[Y == TRUE], main = 'SFS present',
+     xlab = 'depth (cm)', ylab = 'density', freq = FALSE,
+     xlim = xlim, ylim = ylim)
> lines(density(X, bw = BW), lwd = 3)
> lines(density(X[Y == TRUE]))
```

Thus we can compare the density of the sampled depths to that of the depths in which we found SFS. The bottom left panel was produced similarly with

```
> hist(X[Y == FALSE], main = 'SFS absent',
+     xlab = 'depth (cm)',   ylab = '', freq = FALSE,
+     xlim = xlim, ylim = ylim)
> lines(density(X, bw = BW), lwd = 3)
> lines(density(X[Y == FALSE]))
```

It compares the density of depths where SFS was not found to that of the sampled depths. From this exploratory analysis we conclude that individuals of SFS tend to be found in shallow waters.

Next, using the R package Design, we fit a logistic regression model and plot it into the bottom right panel of Figure 16.6.

```
> library(Design)
> ddist <- datadist(X, Y)
> options(datadist = 'ddist')
> model <- lrm(Y ~ X, x = TRUE, y = TRUE, se.fit = TRUE)
> plot(model, xlab = 'water depth (cm)',
+     ylab = 'log odds of SFS present')
```

Here is an edited summary of the model:

```
> model

Model L.R.        d.f.          P
    91.83           1           0

            Coef     S.E.    Wald Z  P
Intercept -0.04712 0.092949  -0.51   0.6122
X         -0.01470 0.001655  -8.88   0.0000
```

The generalized likelihood ratio (16.5) is $G(H_A : H_0) = 91.83$. It is distributed according to χ^2 with one degree of freedom and therefore is significant (p-value is practically zero). So we reject the null hypothesis that the model with the intercept only suffices in favor of the alternative hypothesis that the model includes both the intercept and the slope. From the coefficients, their standard error and Wald-Z statistics we conclude that the intercept, $\widehat{\beta}_0 = -0.047$ is not different from zero. The slope, $\widehat{\beta}_0 = -0.015$ is.

Next, we examine the plot of sensitivity and specificity (left panel, Figure 16.7) which is produced as follows. We start with $\widehat{P}$:

```
> p.hat <- 1 / (1 + exp(-model$linear.predictors))
```

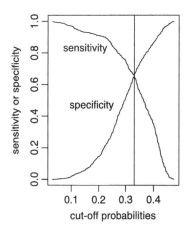 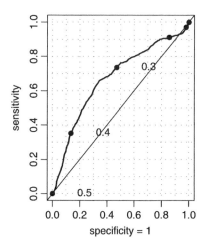

Figure 16.7 Left panel: the vertical line is at the optimal cut-off probability; i.e. the best compromise between sensitivity and specificity. Right panel: the ROC curve.

Next, we compute the sensitivity and specificity vectors. First, we create 200 probability cut values:

```
> BY <- (max(p.hat) - min(p.hat)) / 200
> p.cuts <- seq(min(p.hat), max(p.hat), by = BY)[-1]
```

For each of these values, we create a table and use the table counts to compute the sensitivity and specificity verctors:

```
> sen <- spe <- vector(length = length(p.cuts))
> for(i in 1 : length(p.cuts)){
+     tb <- table(Y, p.hat >= p.cuts[i])
+     sen[i] <- tb[2,2] / (tb[2, 1] + tb[2, 2])
+     spe[i] <- tb[1, 1] / (tb[1, 1] + tb[1, 2])
+ }
```

The calculation requires that we count the zeros and ones above each cut and we accomplish this with `table()`. Here is one example of the table:

```
> tb
y    FALSE TRUE
  0   1486    1
  1    664    1
```

We use the `for` loop above for heuristic purposes. We are now ready to plot the sensitivity and specificity vectors, along with the vertical line that indicates where they are approximately equal. We also label the vectors:

```
> par(mfrow=c(1, 2))
> plot(p.cuts, sen, type = 'l', xlab='cut-off probabilities',
+      ylab = 'sensitivity or specificity')
> lines(p.cuts, spe)
> abline(v = p.cuts[(spe > sen) & (sen > (spe - 0.01))])
> text(locator(), label = c('sensitivity', 'specificity'),
+      pos=c(2, 2))
```

The ROC curve is shown in the right panel of Figure 16.7, which was produced with

```
> library(verification)
> roc.plot(Y, p.hat, main = '', xlab = 'specificity = 1',
+      ylab = 'sensitivity', cex = 2)
> (area<- roc.area(Y, p.hat))
```

The area under the curve is `area$A` = 0.671. This indicates marginal discrimination. □

There are numerous other goodness of fit measures (see Hosmer and Lemeshow, 2000).

16.4 Diagnostics

Once we fit the model, we need to examine whether the fit conforms to the model assumptions. Good fit does not mean that the model is "correct". Residuals may reveal observations that are too far from the model's predictions. Validation—the process by which we test the model on data that were not used in fitting the model—might reveal further shortcomings of the model.

16.4.1 Analysis of residuals

Our outlook so far had been the whole model. Here we are interested in examining particular observations. Are some of them unique with respect to the value of their residual? If so, how much do they influence the MLE of the regression coefficients? Which ones are they? After looking for exceptional residuals, can we still assert that the model fits our assumptions? Albeit not comprehensive, our treatment of analysis of residuals is enough to get you started. For details, consult the documentation for the function `residuals.lrm()` in the R package `Design` and Hosmer et al. (1997). Recall our discussion of covariate pattern in Section 16.3. There, we introduced the Pearson residual, r_j and the deviance residual, d_j (see equations (16.6) and (16.7) and the latter's details). In analyzing residuals, we are interested in two important indicators of a residual: its influence on the values of the regression coefficients and the change in the coefficients when the model is refit without the observation that belongs the residual.

In linear regression, the influence of an observation on the MLE of the regression coefficients is expressed through *leverage* values. These values are proportional to the distance of a point, x_j, from the mean of the data. In logistic regression, there is a comparable approximation to the idea of leverage in linear regression. The approximation is given by

$$n_j - \widehat{n}_j \approx (1 - h_j)\, n_j \quad \text{or} \quad \widehat{n}_j \approx h_j n_j \,. \tag{16.8}$$

Here h_j is the leverage of the residual of the jth pattern. It is a function of the model's coefficients. Because we are using $\widehat{n}_j$ to estimate the unknown population n_j, a large value of h_j indicates that the particular covariate pattern has a large influence on the MLE of the coefficients. Ideally, we wish to have a model where all residuals contribute equally to the coefficient values of the model. To incorporate the potential effect of each residual on the coefficient values, we standardize the residuals. Thus we have the following definition.

Standardized Pearson residual The standardized Pearson residual for the jth covariate pattern is

$$r_{Sj} := \frac{r_j}{\sqrt{1-h_j}} \ . \tag{16.9}$$

A large value of r_{Sj} indicates that observations in the particular pattern have high leverage.

To examine the effect of observations in a particular covariate pattern, we remove them and refit the model. An approximation of the standardized change in the coefficients due to removal of the jth pattern is given by

$$\widehat{\Delta\beta}_j := \frac{r_{Sj}^2 h_j}{1-h_j} \ . \tag{16.10}$$

Skipping the proof, it turns out that the corresponding change in the Pearson χ^2 statistic and in the deviance statistic due to the jth covariate pattern are

$$\Delta C_j = r_{Sj}^2 \ , \quad \Delta D_j = d_j^2 + \frac{r_j^2 h_j}{1-h_j} \ . \tag{16.11}$$

Large values of one or both of these statistics (ΔC_j and ΔD_j) with respect to a covariate pattern j indicate that the jth pattern fits poorly and has large influence on the MLE of the coefficients. Table 16.2 summarizes the diagnostic measures that we discussed thus far. Because not much is known about the sampling densities of the statistics in Table 16.2, we rely on graphical techniques for diagnostics. In running residual analysis, you should, as a rule, plot each of the diagnostics in the second group in Table 16.2 against $\widehat{P}_j$. If possible, you may be able to identify leverage and influence by plotting each of these against h_j.

Example 16.12. We continue with Example 16.11, where we examined the relationship between a fish species and its affinity to habitats with a specific water depth. Figure 16.8 illustrates some of the diagnostics we discussed thus far. For ΔC and ΔD, the residuals that correspond to $Y = 1$ decrease and for $Y = 0$ increase with $\widehat{P}$. Poorly fit points for $Y = 1$ appear at the top left of the figure, with distinct distances from the remaining points. Poorly fit points for $Y = 0$ appear at the top right with distinct distances from the remaining points.

From the standardized Pearson residuals (top left panel in Figure 16.8), we see that for $Y = 1$, there seem to be about 5 points that fit the data poorly. All observations that correspond to $Y = 0$ fit the data. Note that here $m_j = 1$, so $J = n$, where n is the number of observations. The ΔD residuals show results similar to the ΔC residuals. As stated, the density of both statistics is χ^2. Since the number of covariate patterns is the number of points, we have $m_j = 1$ and therefore 1 degree of freedom. At 95% confidence, observations that poorly fit would have values > than 4. There are 2 152 observations, 59 of the standardized Pearson residuals are above 4.

For the standardized deviance residuals (top right panel of Figure 16.8), 41 points are above 4. We do not expect such a small number of misbehaving residuals to invalidate the model. A plot of $\Delta\beta$ illustrates the mirror image of the influence of the points that correspond to $Y = 1$ vis a vis $Y = 0$ on the MLE of the model coefficients. Because of the large number of observations, we do not expect any one point (covariate pattern) to have large enough leverage and large enough Pearson residual to noticeably

Table 16.2 Summary of diagnostics. The first group represents basic diagnostics. The second group represents derived diagnostics (from the first group).

Diagnostic	Notation	Equation	Interpretation
Pearson residual	r_j	16.6	Used in computing the Spearman's χ^2 statistic for overall model fit.
Deviance residual	d_j	16.7	Used in computing the deviance statistic for overall model fit.
Leverage value	h_j	16.8	Large value indicates jth pattern strong influence on the MLE of the regression coefficients.
Standardized Pearson residual	r_{Sj}	16.9	Large value indicates that the jth pattern fits poorly.
Change in Pearson χ^2 statistic	ΔC_j	16.11	Indicates the decrease in C due to removal of the jth pattern from the fit. Large value indicates that the jth pattern fits poorly.
Change in deviance statistic	ΔD_j	16.11	Indicates the decrease in D due to removal of the jth pattern from the fit. Large value indicates that the jth pattern fits poorly.
Influence on MLE of coefficient values	$\widehat{\Delta \beta}_j$	16.10	Indicates the influence of removing the jth pattern from the fit on the MLE model coefficients. Large value indicates strong influence.

influence the MLE of the coefficients. Indeed, the range of values in both bottom panels in Figure 16.8 are small. Yet, good practice requires examination of these diagnostics.

Figure 16.8 was produced as follows. First, we load the data and assign the variables:

```
> rm(list = ls())
> load('fish.rda')
> x <- fish$adults$depth
> y <- ifelse(fish$adults$SFS > 0, 1, 0)
```

Next, we fit the model and calculate $\widehat{P}$:

```
> library(Design)
> ddist <- datadist(x, y)
> options(datadist = 'ddist')
> model <- lrm(y ~ x, x = TRUE, y = TRUE, se.fit = TRUE)
> p <- 1 / (1 + exp(-model$linear.predictors))
```

We continue with the residuals analysis thus:

```
> hat <- residuals.lrm(model, type = 'hat')
> pearson <- residuals.lrm(model, type = 'pearson')
> deviance <- residuals.lrm(model, type = 'deviance')
> standard.pearson <- pearson / (sqrt(1 - hat))
> delta.C <- standard.pearson^2
> delta.D <- deviance^2 / (1 - hat)
> delta.beta <- residuals.lrm(model, type = 'dfbeta')
> delta.beta.0 <- delta.beta[, 1]
> delta.beta.1 <- delta.beta[, 2]
> delta.C.lim <- c(min(delta.C), max(delta.C))
> delta.D.lim <- c(min(delta.D), max(delta.D))
> delta.beta.0.lim <- c(min(delta.beta.0), max(delta.beta.0))
> delta.beta.1.lim <- c(min(delta.beta.1), max(delta.beta.1))
```

Finally, we plot the top left panel of Figure 16.8:

```
> par(mfrow=c(2, 2))
> plot(p[y == 1], delta.C[y == 1],
+    xlab = expression(italic(hat('P'))),
+    ylab = expression(paste(Delta,italic('C'))),
+    ylim = delta.C.lim, cex = 2)
> points(p[y == 0], delta.C[y == 0])
```

The remaining plots were produced similarly. Note the use of `expression()` to annotate the axes. □

16.4.2 Validation

Validation refers to the process of applying the model to data for which the model had not been fitted. Ideally, one would like to fit the regression to data from a population and then validate it for data from another population. Short of this, we can simply exclude part of the data, fit the model and then validate it on the excluded data. There are numerous variations on this theme. When data are scant, we could exclude small part of the data and use bootstrap methods to build the model and test it.

The main method for validating is simply following the residual analysis outlined in the previous section. If the model is valid, then residuals should reflect a fit that are no worse than the data on which the model fit was based. For furhter details, see `validate.lrm(Design)`, Miller et al. (1991) and Hosmer and Lemeshow (2000).

16.4.3 Applications of simple logistic regression to 2 × 2 tables

Analysis of 2 × 2 tables is quite common.

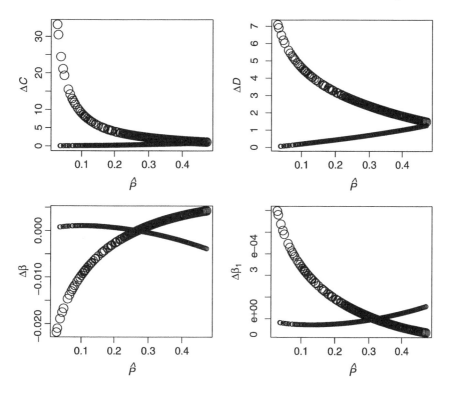

Figure 16.8 The axes notations refer to the diagnostics in (16.11) and (16.10). The $\Delta\beta$ are shown for β_0 and β_1. Large circles identify points for which $Y = 1$ and small circles identify points for which $Y = 0$.

Example 16.13. We are interested in the cross classification between low birth weight ($Y = 0$ or 1) and the mother's smoking status ($X = 0$ or 1). The data, introduced in Hosmer and Lemeshow (2000), are summarized in the following 2×2 table:[5]

```
> load('bwt.rda') ; attach(bwt)
> table(low, smoke)
   smoke
low FALSE TRUE
  0    86   44
  1    29   30
```

or

	low birth weight	smoke		
		1	0	total
	1	30	29	59
	0	44	86	130
	total	74	115	189

---

[5]`bwt.rda` was imported from the original data file—see `bwt.R` at the book's site.

Smoke represents the number of women who smoke (1 yes, 0 no) and low birth weight represents the number of low birth weight babies. □

The odds ratio is used as a measure of association in contingency tables.

Odds ratio (ρ) Let $X = 0$ or 1 be an independent dichotomous variable and $Y = 0$ or 1 a dependent dichotomous variable. The ratio of the odds for $X = 1$ to $X = 0$, called the *odds ratio*, is

$$\rho := \frac{\pi(1)/(1-\pi(1))}{\pi(0)/(1-\pi(0))} .$$

Example 16.14. Continuing with Example 16.13, we have

$$\widehat{\rho} = \frac{(30/74)/(44/74)}{(29/115)/(86/115)} = 2.02 .$$

In this case, the odds ratio represents the risk of low birth weight for a baby born to a smoking mother compared to a mother who does not. In other words, smoking mothers are over twice as likely to give birth to low weight babies compared to non smoking mothers. □

Next, we wish to cast the odds ratio in a logistic regression framework. For $X = 1$ and $Y = 1$, we have

$$\lambda(X) = \beta_0 + \beta_1 X \quad \text{or} \quad \pi(1) = \frac{1}{1 + e^{-(\beta_0 + \beta_1)}} .$$

For $X = 0$ and $Y = 1$ we have

$$\lambda(X) = \beta_0 + \beta_1 \times 0 \quad \text{or} \quad \pi(1) = \frac{1}{1 + e^{-\beta_0}} .$$

Thus we obtain Table 16.3. Implementing the definition of ρ to the results in Table 16.3 and with a little bit of algebra, we have

$$\widehat{\rho} = e^{\beta_1} .$$

Table 16.3 2 × 2 contingency table in terms of logistic regression.

Dependent variable (Y)	Independent variable (X)	
	$X = 1$	$X = 0$
1	$\pi(1) = \dfrac{1}{1+e^{-(\beta_0+\beta_1)}}$	$\pi(0) = \dfrac{1}{1+e^{-\beta_0}}$
0	$1 - \pi(1) = \dfrac{1}{1+e^{\beta_0+\beta_1}}$	$1 - \pi(0) = \dfrac{1}{1+e^{\beta_0}}$
Total	1.0	1.0

Example 16.15. We implement the logistic regression to the data summarized in Example 16.13:

```
> X <- smoke
> Y <- low
> library(Design)
> ddist <- datadist(X, Y)
```

```
> options(datadist = 'ddist')
> (model <- lrm(Y ~ X, x = TRUE, y = TRUE))
Logistic Regression Model

lrm(formula = Y ~ X, x = TRUE, y = TRUE)


Frequencies of Responses
  0   1
130  59

       Obs  Max Deriv  Model L.R.   d.f.        P
       189     4e-08        4.87      1     0.0274
        C       Dxy        Gamma    Tau-a       R2
    0.585      0.17        0.338    0.073    0.036
    Brier
    0.209

             Coef   S.E.   Wald Z  P
Intercept  -1.087  0.2147  -5.06   0.0000
X           0.704  0.3196   2.20   0.0276
```

The p-value for the likelihood ratio is 0.03, so we deem the model significant. Recall that the test statistic here $G(H_A : H_0) = 4.87$ where H_0 is the model with the intercept only and H_A is the model with the intercept and slope. This test statistic has a χ^2 density with 1 degree of freedom. The value of $\widehat{\beta}_1 = 0.704$ is significant ($p = 0.03$). Therefore, the odds ratio is

$$\widehat{\rho} = e^{0.704} = 2.02\,,$$

as required. □

16.5 Assignments

Exercise 16.1. Confirm the statement in Example 16.2: "There are 2 152 records in the data and in two of them individuals of Blackchin shiner were found. Therefore, $p = 2/2\,152$."

Exercise 16.2.

1. What is the domain and range of the odds ratio?
2. What is the domain and range of the logit transformation?

Exercise 16.3. Suppose that the probability of finding a plant species in a particular plot is π.

1. What are the values of the logit transformation for $\pi = 0, 0.25, 0.5, 0.75, 1$?
2. What are the values of the logistic transformation for $\pi = 0, 0.25, 0.5, 0.75, 1$?

Exercise 16.4. In an exposure study, we find that when people are exposed to Radon levels of 0, 10, 20 and 30 Bq (e.g. Example 9.5), the probabilities of developing lung cancer are 0.0001, 0.0002, 0.0003 and 0.0004 (numbers represent fictitious data). Compute and interpret the log odds ratio for these data.

540 Simple logistic regression

Exercise 16.5. Use Example 16.3 as a guideline.

1. What is the proportion of blacks in the population of inmates sentenced to death in the U.S. that have ten years of education?
2. Twelve years of education?

Exercise 16.6. We discussed the CDC demographics data in Example 15.10. The data are in demo_d.short.rda, at the book's site.

1. Load the data and clean it from all NA. Keep the columns Gender, Household Income from and Household Income to only.
2. Let $Y = 0$ for males and 1 for females.
3. Assign to X the mean of Household Income from and Household Income to and \$75 000 to Inf and rescale the data to \$1 000 (divide income by \$1 000).
4. Fit a logistic model to gender vs. mean income and print the results.
5. Plot the model with confidence intervals.
6. Interpret the results.

17

Application: the shape of wars to come

In this chapter, we present a complete analysis of two recent wars, the Iraq war, between the U.S. and Iraqi (government at first and militant organizations later) and the so-called Second Intifada between Israel and militant Palestinian organizations. Our purpose is to illustrate how various ideas presented in the book may be applied to real and current problems to produce publishable manuscripts. The focus here is the data and its interpretation; not as much R and statistics. So we shall not discuss scripts and neither shall we explain the code that produced the analysis and figures. However, Examples 2.16, 8.1, 7.24 and 8.22 refer to the war in Iraq. Examples 5.2, 5.17, 6.9 and 9.8 refer to the Second Intifada. All of these, including the data are available from the book's site.

17.1 A statistical profile of the war in Iraq

We define the War in Iraq (WI) as the period between 03/21/2003 and 10/10/2007—a total of 1 665 days. We obtained data about the date, location and cause of death of every single soldier belonging to the Coalition forces. We also obtained a summary, by month, of the number injured Coalition soldiers. Both the empirical Probability Density (PD) of injuries per month and deaths per week followed the negative binomial PD, indicating that injury and death rates varied over time. Their occurrence remained random (following the negative binomial PD) in spite of a variety of military strategies and social and political policies. Further analysis confirmed this. Our results refute the often made claim that increased activity by the Coalition forces in one place had been compensated by increased activity by Iraqi militant organizations in other places. We found no temporal dependence among the number of deaths in various location across Iraq.

With the significant fit of the negative binomial PD to the data about injuries and deaths we conclude that: (1) One could expect that 95% of the deaths per week would be ≤ 37. (2) One could expect that 95% of the injuries per month would be ≤ 974. Both values may serve as guidelines for organizations responsible for planning of trauma-treatment. We conclude that unless one is willing to use extremely excessive force—as the Russians did during the Second Chechen War (1999–2000)—no realistically large military force can win a war against committed small militant organizations.

17.1.1 Introduction

The War in Iraq (WI), between the U.S. and some militant organizations (MO) in Iraq have exhibited common characteristics with other recent armed conflicts—a small number of individuals clash with large invading armies. Such wars may indicate a change in future warfare from large wars between armies to what some call "the war on terrorism". Similar recent conflicts had been: (i) the First Intifada (1987–1993); (ii) the Second Intifada (2000–2003) –two periods of heightened belligerence between Israel and some Palestinian MO; (iii) the war in Afghanistan between NATO forces and Afghani MO (2001–present); (iv) the First Chechen War (1994–1996); and (v) the Second Chechen War (1999–2000) – both between the Russian military and Chechen MO. Wars between invading or occupying armies and local populations are not new. However, technological advances and instant communication make, on the one hand, such conflicts ever more deadly and on the other, open to global public opinion and scrutiny. See also Geller and Singer (1998); Gelpi et al. (2005/6); Scotbennett and Stam (2006); Alvarez-Ramirez (2006).

These wars exhibit statistical properties that are worthwhile investigating. For example, if we can characterize some of their aspects with well-known probability densities (PD), then emergency service providers may plan for expected magnitudes of disasters (such as the number of deaths and injuries). Also, some PD arise from well known underlying mechanisms. We can then draw conclusions about the random (statistical) processes that underlie such wars and thereby judge the efficacy of diplomatic and military efforts to change the outcome of such conflicts (in particular with respect to the cost in human lives, injuries and suffering). Of the wars listed above, the WI is one of the few where detailed data are available. Thus, we pursue its statistical profile. Detailed data about the Second Intifada are also readily available (Section 17.2). We shall analyze and compare its statistical profile to that of the WI.

17.1.2 The data

The data about deaths begin on 03/21/2003. We stopped updating it on 10/10/2007, 1 665 days since the beginning of the WI. Data were obtained from http://icasualties.org/oif, last visited on 10/10/2007. A list of the countries participating in the Coalition forces may be found at http://www.globalsecurity.org/military/ops, last visited on 10/28/2007. Both sources are often cited in the press. See for example *The Economist*, October 27th–November 2nd, 2007, p. 34; G. Kutler, *Orbis*, 2005, 49: 529–544; 2006, 50: 559–572 and 2007, 51: 511–527. To facilitate date-arithmetic, we added to the raw data a column that lists the Julian day that corresponds to the date of death. The Julian count starts on 1/1/1960. We also classified the reported cause of death to major (Hostile and Non-hostile) and minor. Here are the first three records of the data about deaths:

```
   ID       Date             Rank Age  Srv.Branch
1   1 2003-03-21 2nd Lieutenant  30  U.S. Marine
2  14 2003-03-21 Lance Corporal  22  U.S. Marine
3  13 2003-03-21          Major  34   Royal Navy
  Major.Cause.of.Death Minor.Cause.of.Death             Where
1              Hostile    hostile fire Southern  Iraq
2              Hostile    friendly fire            Um Qasr
3              Hostile helicopter crash            Um Qasr
        Hometown      State Country   Julian
1   Harrison Co. Mississippi   U.S.~  15785
2 Guatemala City   Guatemala   U.S.~  15785
3       Plymouth     England      UK  15785
```

and here are the first three records of the data about injuries:

```
        Date Injured Julian
1 2003-03-03     202  15767
2 2003-04-03     340  15798
3 2003-05-03      55  15828
```

Regrading deaths, we include only those where the cause, as reported by the U.S. Department of Defense, was due to hostile activities. We shall not report summary statistics such as deaths and injuries by nations, causes and so on. These are readily available elsewhere. From here on and unless otherwise specified, *deaths* refer to those (of soldiers that belonged to the Coalition forces) caused by hostile activities, categorized as Hostile in the data. Injuries refer to those soldiers belonging to the Coalition forces.

17.1.3 Results

The run sequences of deaths and injuries (Figure 17.1) may reflect periodicities. Yet, the autocorrelation function did not reveal any significant lags. We shall address this point in a moment. In all, 3 353 soldiers were reported dead because of hostile activities. The deadliest places are identified in Figure 17.2 and their geographic locations in Figure 17.3. We shall isolate the deadliest locations for further analysis soon.

Death, injury rates and their PD

The cumulative sum of injuries and deaths (Figure 17.4) are quite instructive. During the period discussed and based on the slopes of the regression lines in Figure 17.4, the average death rate was 2.02 per day and the average injury rate was 17.80 per day. In both cases, the linear model fit produced $R^2 \approx 0.99$. Of course both R^2 for deaths and injuries are meaningless unless the residuals are sequentially independent. The autocorrelation functions of the residuals (Brockwell and Davis, 1991) of both rates revealed significant autocorrelations among the residuals. In fact, the residuals exhibit four distinct periods of consistent alternating decline and increase in both injury and death rates when compared to the overall average rates (Figure 17.5). From the beginning of the WI (3/21/2003) and for about a year (until 3/29/2004), both death and injury rates steadily declined. Next, for about eight months (until

544 Application: the shape of wars to come

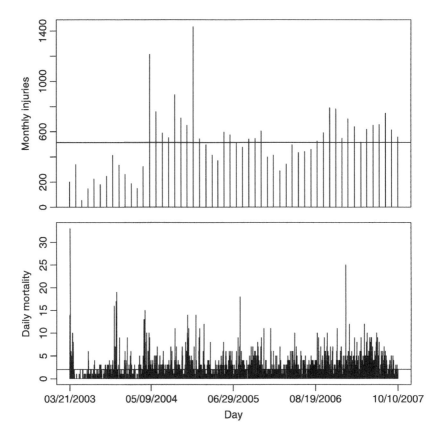

Figure 17.1 Monthly injuries and daily deaths among the Coalition forces. Horizontal lines indicate means.

12/09/2004), both rates increased. For the next 20 months (until 8/26/2006), there was steady decline in both rates. Finally, we see a period of a year of increase in the rates. The last month of the period indicates that perhaps the war was entering its next phase of decline in these rates. Both the monthly injury rate and the weekly death rate seem to be in perfect synchrony—more deaths per week had been associated with more injuries per month and fewer deaths had been associated with fewer injuries.

The analysis of the residuals leads to two conclusions. First, any claim of success or failure in the WI that is based on short-term observations of increase or decrease in the death or injury rates is likely to be premature. Second, the fluctuations in the rates may have been produced by the warring factions adapting their strategies to each other's with time-delays, which are not necessarily constant.

Do deaths and injuries follow some well known PD? To pursue the answer, we first constructed the empirical PD from the data (monthly injuries and weekly deaths). Next, using the total injuries (27 753) and deaths (3 355), we used the empirical density to obtain the expected monthly injuries and expected weekly deaths. These are the points in Figure 17.6. These points were obtained by constructing a density histogram of the injury (death) data and then multiplying the total injuries (deaths) by each density value. The sticks were obtained by calculating the mean, $\overline{X}$ and size, S, of

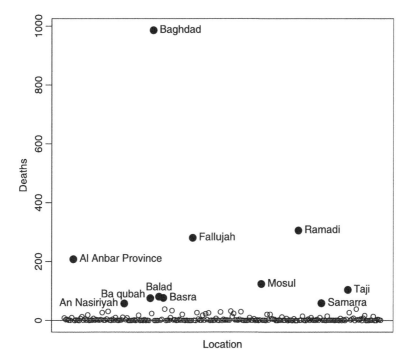

Figure 17.2   Deaths by location (as of 10/10/2007).

the injury (death) data. These are the parameters of the negative binomial PD. The S parameter is given by

$$S := \frac{\overline{X}^2}{V - \overline{X}}$$

where V is the variance of the data. See for example Evans et al. 2000. Statistical Distributions. Wiley, 3rd edition. Albeit counting processes, both empirical densities were over dispersed compared to the Poisson; the mean of injury per month was smaller than its variance (514 vs. 60 577). Such was the case for deaths per week (16 vs. 114). Over dispersion in counting processes arise from two main sources: the Poisson intensity parameter varies over time or there are several Poisson PD with different intensity values underlying the counts.

The Poisson PD process arises when events are scattered uniformly over time— their timing is unpredictable—with some frequency. The negative binomial arises under conditions similar to the Poisson, except that the variance of the former is larger than its mean. For both injuries and deaths, the fit of the theoretical PD (sticks in Figure 17.6) to the empirical PD (points in Figure 17.6) was such that we could not reject the hypotheses that the counts were drawn from count rates that obey the negative binomial PD.[1] As the residuals disclose (Figure 17.5), part of the over dispersion resulted from the time-varying intensity parameter of the Poisson.

---

[1] For the injuries, the results from the Pearsons's Chi-squared test with simulated p-values (based on 2 000 replicates) were $\chi^2 = 19.41$, p-value $= 0.10$. For deaths they were $\chi^2 = 3.87$, p-value $= 1.000$ (degrees of freedom are not needed).

546 Application: the shape of wars to come

Figure 17.3 The 11 deadliest locations (as of 10/10/2007) and their provinces; compare with Figure 17.2.

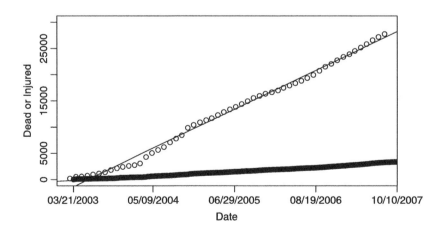

Figure 17.4 Cumulative sum of injuries (upper sequence of points) and deaths (lower sequence of points) with best fit linear models.

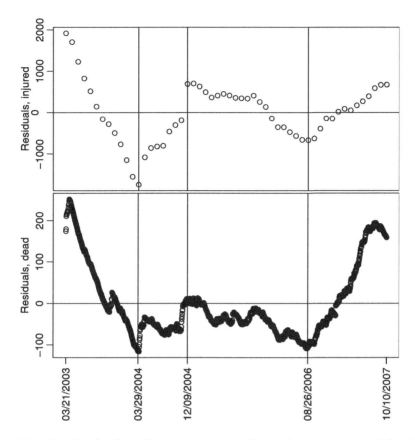

Figure 17.5   Residuals of the linear regressions of cumulative injury and death rates (see also Figure 17.4).

There is also a spatial contribution to the over dispersion—different locations may have had different rates (see next section).

In short, the count of deaths per week and injuries per month remained random (with the negative binomial PD) in the face of various efforts by the U.S. and its allies to try a variety of military strategies and political and social policies. With the given theoretical negative binomial PD, we can answer useful questions. For example, in planning emergency services, one may ask: Let X be the weekly death (or injury) rate in all of Iraq. What is the value of X such that 95% of the weekly deaths or injuries will be expected to be $\leq X$? With the negative binomial, the answer is 37 deaths per week. It was 974 injuries per month.[2] Breaking down this analysis by locations (such as provinces), one can achieve a refined planning of trauma treatment policies that meet future needs. Similar results can be achieved with bootstrap methods. (Efron and Tibshirani, 1993). However, it is nice to obtain significant fits to theoretical PD for then comparisons with other wars are simplified.

---

[2]In R, the result is achieved with the statement qnbinom(.95, size, , mu) where size = 2.63 and mu = 14.12.

548 Application: the shape of wars to come

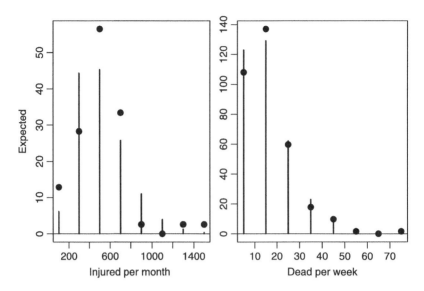

Figure 17.6 Empirical expected values (dots) and theoretical (sticks) for the negative binomial PD.

Chronology by location

News media suggested that Iraqi MO adapt to Coalition military strategy. Wherever Coalition forces concentrate, the MO shift their activities to other places. Let us see if there is credence to this claim. We start by observing the cumulative deaths by the 11 deadliest locations (Figure 17.7). The bottom panel shows the full scale on the Dead-axis. From it, three distinct groups of locations emerge (the top panel identifies them by name). Baghdad stands on its own—a high death rate with some short periods of lull here and there. Then come Fallujah, Ramadi and Al Anbar province. There, death rates were constant for the first few months and then increased sharply, particularly in Fallujah. In between these three and the lower bunch of six locations (Ba qubah, Samarra, An Nasiriyah, Basra, Balad and Taji) stands Mosul. In the latter, the death rate increased sharply, but then as the Kurds established their semi-autonomy, the death rate slowed. The zoomed in view in the top panel allows us to identify in detail periods of sharp increases in death rates by location. The fact that no clear overlaps of these sharp increases emerge, leads us to conclude that various locations ebbed and swelled in death and injury rates at different times. The question of how to quantify potential synchronizations in the increase and decrease of death rates among locations is addressed next.

Let S be the set of all daily dates, between 03/21/2003 and 10/10/2007. We denote by $\#$ the cardinality (number of elements) in a set. So $\#S = 1\,665$. Now denote by A_i, $i = 1, \ldots, 11$, the set of dates in which at least one death was reported from location i (the locations are identified by name in Figures 17.2, 17.3 and 17.7 and for the time being, their corresponding index is not important). These sets represent the dates on which at least one death occurred. Data from dates in which Iraqi MO attempted attacks and no deaths (but possibly injuries) occurred were not available to us. The proportion of death-dates over the whole period for location i is given by $P(A_i) := \#A_i/\#S$ (here $:=$ denotes equality by definition). Similarly, $P(A_i \cup A_j) :=$

A statistical profile of the war in Iraq 549

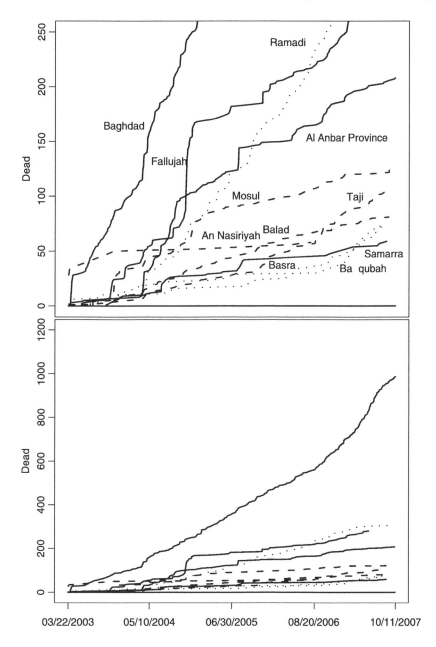

Figure 17.7 Cumulative deaths in the 11 deadliest locations. Bottom - full scale; top - zoomed on the Dead scale.

$\#(A_i \cup A_j)/\#S$ and $P(A_i \cap A_j) := \#(A_i \cap A_j)/\#S$. The proportion of deaths at i, conditioned on deaths at j is given by:

$$P(A_i|A_j) = \frac{P(A_i \cap A_j)}{P(A_j)} \ .$$

Large P indicate large proportions of events co-occurrence at locations i and j.

We are not done yet for we need to integrate results from all locations and derive some statistics that indicate whether the chronology of deaths in locations had been different from random. We pursue this issue next. All of the $P(A_i|A_j)$ can be presented in an 11×11 matrix where rows are indexed by i and columns by j. The matrix need *not* be symmetric. The relative magnitude of the sum of row i (denoted by $P(A_i|A.)$), compared to all other rows, is interpreted as the amount of co-occurrence of deaths at i given deaths in all other locations. Note that $P(A_i|A.)$ is *not* a proper marginal sum because we have chosen a subset of locations (as shown in Figures 17.2 and 17.3). The relative magnitude of the sum of say, column j (compared to all other columns), reflects the "dependence" (co-occurrence) of deaths in all other locations on deaths at location j. So constructing such a matrix might shed light on synchronizations with respect to dates of deaths at locations. However, we must have some yardstick of "randomness" to compare the results to—a null model so to speak.

To achieve such a yardstick, consider the set A_j fixed with respect to both its cardinality and dates, for each $\#A_i$ fixed. If the events at location i are unrelated to those at location j, then we can compute $P(A_i|A_j)$ for fixed A_j, fixed $\#A_i$ (for $i = 1, \ldots, 11$), but random dates for location i within the range of S. Repeating such a simulation, say 1 000 times, for each pair A_i, A_j, we obtain the probability density of a process that allows us to answer the following question:

> A fixed number of events occur at location j on fixed dates. If a fixed number of events occurred at i, but on random dates, what would be the probability density of $P(A_i|A_j)$?

All dates are within the range of S. We can now compare our empirical $P(A_i|A_j)$ to the randomly generated PD of $P(A_i|A_j)$ and determine if the former had been random. A value significantly lower than expected indicates that events at i had been negatively associated with events at j. A value significantly higher than expected indicates positive association. Insignificant values indicate dissociation. Figure 17.8 illustrates the fact that *none* of the conditional proportions were significant. In other words, deaths in any location had been independent of deaths paired with any other location. Thus, we find no evidence for the claim that increased activity by the Coalition forces in one location resulted in compensated increased activity somewhere else. Because all paired conditional proportions were insignificant, it makes no sense to pursue analysis of the marginal proportions.

17.1.4 Conclusions

In this section, we allow ourselves a few speculations. We are *not* military experts and our conclusions should be taken with a grain of salt. Our results may be useful to planners of medical (in particular trauma) treatment facilities. It seems that in spite of various policies, both civilian and military, the underlying random processes of death and injuries in time (for deaths and injuries) and place (for deaths) remained unchanged. The belligerents adapted to each other's military strategies with time-delays. This may have produced alternating periods of increase and decrease (above overall averages) in death and injury rates among the Coalition forces. Such periods may result in false beliefs in military successes (or failures).

Given the ineffectiveness of various efforts by the Coalition forces to control the magnitude of death and injury rates, it seems that no amount of realistic force can win

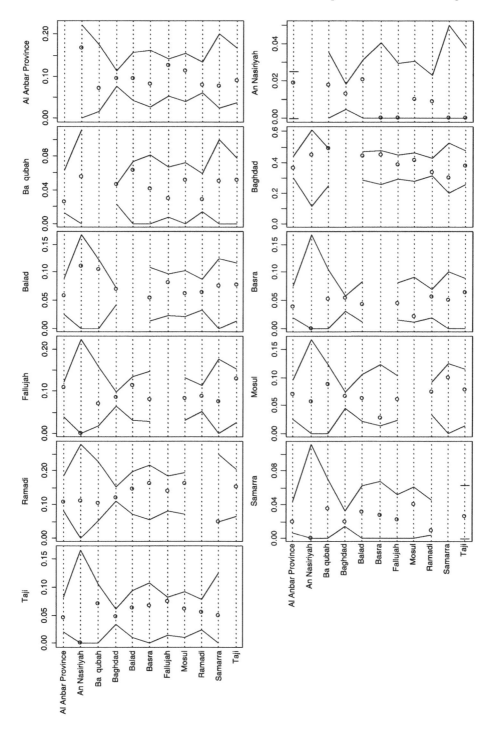

Figure 17.8 Probability density of 1000 repetitions of $P(A_i|A_j)$ (points) with 95% confidence intervals.

552   Application: the shape of wars to come

a war such as the WI unless... One is willing to brutally obliterate the infrastructure upon which MO rely and in the process cause tremendous strife to the population at large. Such was the case in the Second Chechen War.

This, along with our analysis of the Second Intifada, lead us to the following conclusions. To maintain activities, MO need resources. Explosives are expensive, people need to be trained, transported, paid and so on. Perhaps an economic confrontation might be more effective in resolving such wars (where by effectiveness we mean fewer deaths and injuries); more so than military force. Regardless of military efforts, the fact that small groups of MO can garner such power against large military forces means that the grievances of MO (right or otherwise) should be addressed in ways other than brute force.

17.2 A statistical profile of the second Intifada

We analyze the statistical properties of the so-called Second Intifada (SI). We call an explosion triggered by a member (or members) of a Palestinian Militant Organization (PMO) an event. Data about the number of deaths and injuries due to events between 9/27/2000 and 10/4/2003 (1 102 days), the period of the SI, are analyzed. During this period, 278 events had occurred, 763 people died and 3 647 were injured. Of the PMO that claimed responsibility for events, Hammas was the deadliest and Al Aqsa Martyrs Brigades executed more events than any other PMO.

The fortnight death and injury rates fit the negative binomial probability density (PD). Residual analysis revealed cycles in the swell and ebb of the rates of death, injury and event. Because of adjustments by Israelis, the ratio of injuries to death per event increased over time. The barrier that Israel constructed between the Palestinian population in the West Bank and the Israeli population in Israel proper was associated with a decrease in injury and death per event. Using the negative binomial, we find that 95% of the events were expected to result in ≤ 28 deaths per fortnight and ≤ 157 injuries per fortnight. Knowledge of such values should be used in planning medical facilities and treatment.

The statistical properties of the SI resembled those of the war in Iraq (WI), where the negative binomial PD fit the death and injury rates and cycles of increase and decrease in the death and injury rates were identified. In both the SI and the WI, the cycles indicate adjustments of each side to the other's strategy. The analysis cast doubt on the ability of large armies to win such wars unless they are ready to implement extreme violence, as the Russian did in the Second Chechen War (1999–2000).

17.2.1 Introduction

Two recent wars, the so-called Second Intifada (SI), 2000–2003, between Israel and some Palestinian Militant Organizations (PMO) in the West Bank and the Gaza Strip and the War in Iraq (WI), 2003-present, between the U.S. and its allies (called the Coalition) and Iraqi Militant Organizations (IMO) have common characteristics: A small number of individuals clash with large armies. Such wars may portend future warfare and their statistical properties are worthwhile investigating. For example, death and injury rates of members of the Coalition forces in the WI—due to hostile activities—followed the negative binomial PD (see Section 17.1.3). Using this fact, we established that in the WI, one may expect that 95% of the deaths per week would be ≤32 and 95% of the injuries per month would be ≤974. Such information may be useful in planning emergency services. Also, some PD arise from well known mechanisms. We can then draw conclusions about the random (statistical) processes that underly such wars. Here we pursue a statistical profile of the SI. In Section 17.1, we investigated the statistical profile of the WI. To many, the topic is emotional. To avoid semantics from interfering with the analysis, we shall stick to neutral definitions. At the core of our analysis are random events over time. An event is defined as an explosion which in some cases kills its initiator or initiators. The explosion occurs at a particular time and may cause any number of injuries or deaths (including none). Thus, with these random events we associate a count (of deaths or injuries). We use the word death to indicate fatalities, casualties and other such synonyms. Injuries are defined as those people who were injured during an event, regardless of whether they did or did not die later. Deaths and injuries are event related and they include both Israelis and Palestinians. We define the *barrier* as the fence (wall) that was built by Israel and physically separate the Palestinians in the West Bank from Israelis. We are referring to *neither* its legality *nor* its location.

17.2.2 The data

Beginning on 9/27/2000, the Israeli Foreign Ministry has posted event related data: dates, the number injured, the number dead and a short description of the event.[3] The description included the PMO that claimed responsibility for the event. Occasionally, the description detailed the number of dead or injured children and women. Using these postings as a starting point, we cross checked the data with archives of the *New York Times*, *Washington Post*, *Lost Angeles Times* and the *London Times*. By 10/4/2003 (1 102 days later), a total of 278 events had been reported. Here are the first three and last three records of the data:

```
      Date      Dead  Injured  Organization.1
1     9/27/2000   1     0        None
2     9/29/2000   1     0        None
3     10/1/2000   1     0        None
276   9/25/2003   1     6        None
277   9/26/2003   2     0        IJ
278   10/4/2003  19    60        IJ
```

Organization.1 refers to one (of potentially up to three) organizations that claimed responsibility for a single event.

17.2.3 Results

The Second Intifada marked a period of high frequency of events. For our purpose, it lasted between 9/27/2000 and 10/4/2003 (a total of 1102 days).

Overview

Of the 278 events, two organizations claimed responsibility for 19 identical events and three organizations claimed responsibility for one identical event. Therefore, it seems

---
[3] http://www.israel-mfa.gov.il/mfa, last visited on 10/10/2003.

554 Application: the shape of wars to come

Table 17.1 Acronyms and frequency of events by organization. The frequency is based on claimed responsibility.

Acronym	Events	Organization
None	145	None
AAMB	57	Al Aqsa Martyr Brigades
Hammas	43	Hammas
IJ	30	Islamic Jihad
PFLP	13	Popular Front for the Liberation of Palestine
Tanzim	8	Tanzim
AQ	1	Al Quida
Hezbollah	1	Hezbollah
Total	298	

that there was little confusion about "who did what". Table 17.1 lists the organizations (and their acronyms to be used). Claims were counted each time an organization announced responsibility for an event. Therefore, the number of claims exceeds the number of events by 20 (9% of all events). Except for the event by Al Quida (in Kenya) and Hezbollah (from Lebanon), all of the events occurred either within Israel proper or in the occupied territories (West Bank and the Gaza Strip). Also, all of the events in Israel and the occupied territories ended with the initiator or initiators dead.

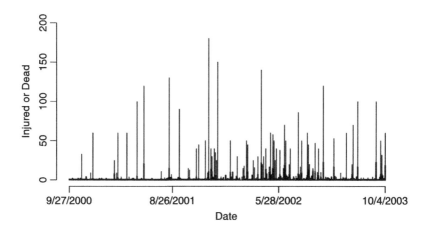

Figure 17.9 Injured (thin) and dead (thick) run sequence during the Second Intifada.

The chronology of the events reveal potential cycles (Figure 17.9). However, neither death nor injury rates showed significant autocorrelations for various time-lags. We shall return to this point soon. During the period reported, 763 people died and 3 647 were injured. Because published reports rarely include those who later died

Table 17.2 Summary statistics for the number of deaths per event by organization that claimed responsibility. SE denotes standard error (standard deviation/$\sqrt{n}$) and N denotes the number of events.

Claimed by	Total dead	Mean	SE	N
None	220	1.52	0.20	145
AAMB	169	2.96	0.47	57
Hammas	288	6.70	0.95	43
IJ	128	4.27	1.03	30
PFLP	25	1.92	0.40	13
Tanzim	10	1.25	0.16	8

from injuries, these numbers represent the minimum death toll. From Table 17.2 we learn that Hammas topped the list in the total number deaths for the events it claimed responsibility for and the mean number of deaths per event, followed by IJ and AAMB. The total deaths in Table 17.2 exceeds 763 because of the 19 events where two organizations claimed responsibility and a single event where three did. From Table 17.3 we learn that Hammas topped the list in the number of injuries per event it claimed responsibility for, followed by IJ and AAMB. For the same reason as above, the sum of the total number of injuries exceeded 3 647. For the reasons detailed, the sum of the totals in Tables 17.2 and 17.3 should not be used to report the total number of deaths and injuries. The question whether the differences in deaths per event—among the organizations that claimed responsibility—are significant will be addressed after we examine the PD of deaths and injuries per event.

Table 17.3 Summary statistics for the number injuries per event, by organization that claimed responsibility. SE denotes standard error and N denotes the number of events.

Claimed by	Total injured	Mean	SE	N
None	724	4.99	1.57	145
AAMB	856	15.02	3.13	57
Hammas	1 688	39.26	6.67	43
IJ	707	23.57	5.13	30
PFLP	100	7.69	3.73	13
Tanzim	11	1.370	0.63	8

Event, death and injury rates

The cumulative sums of the number injured and dead diverged over the entire period (Figure 17.10, left panel): the ratio of injured over dead increased. This may be attributed to a more alert population and extra security measures. For example, security guards at entrances to public places may prevent an event from occurring in

556 Application: the shape of wars to come

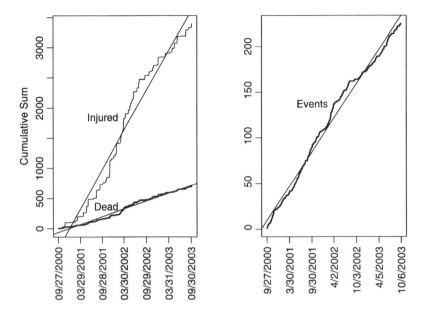

Figure 17.10 Cumulative sums of dead and injured and events. A linear fitted line is drawn to emphasize the changing trends in the rates.

a densely occupied area, where people close to the detonation point are likely to die, but others, behind walls, counters and other barriers are not likely to be injured. For the same reason, detonation outdoors is likely to kill fewer people because they are scattered, but injure more because there are no barriers.

As the best fit linear models illustrate (the straight lines in Figure 17.10), the averages of the death, injury and event rates were 0.748, 3.590 and 0.203 per day, respectively (all with $R^2 > 0.97$). Of course the high R^2 are meaningless unless the residuals are random, which is *not* the case (Figure 17.11). For both the death and injury rates we can clearly identify three periods: Initial decline (until the beginning of 2002), a surge (until June 2002) and a final decline. Associated with the final decline (but *not* necessarily its cause) is the beginning of the construction (in August 2002) of the *barrier*. As was the case of the WI, the analysis of the residuals leads to two conclusions. First, short-term observations of increase or decrease in death, injury or event rates should not lead to long-term predictions. Second, the fluctuations in the rates might have been produced by the warring factions adapting their strategies to each other's with time-delays. Unlike the WI, the alternating periods of increase and decrease in death and injury rates can be associated with changes in strategies, such as increasing the number of guards in public places and physical separation between the populations. Regarding events (Figure 17.10, right panel), there had been short periods of lulls in the rate of events. These are identified in Figure 17.12.

Do deaths and injuries follow some well known PD? To pursue the answer, we first constructed the empirical PD from the data (all rates are on fortnight basis). Next, using the total injuries (3 397), deaths (713) and events (225), we used the empirical densities to obtain the expected fortnight rates. These are the points in Figure 17.13. In R, the points in Figure 17.13 were obtained by constructing a density histogram

A statistical profile of the second Intifada 557

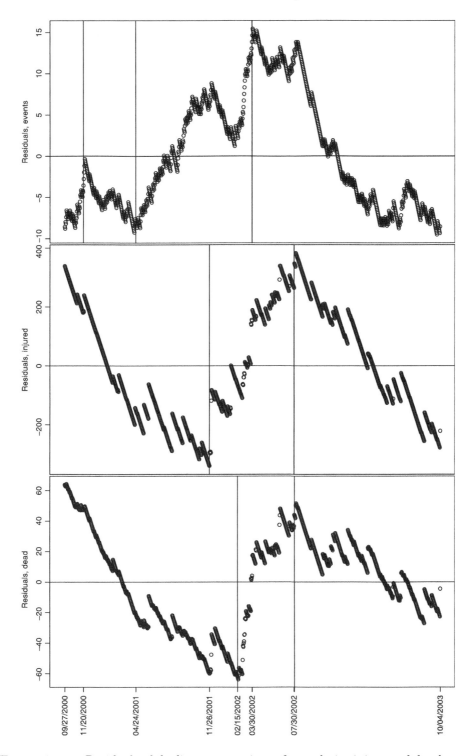

Figure 17.11 Residuals of the linear regressions of cumulative injury and death rates (see Figure 17.10).

558 Application: the shape of wars to come

Figure 17.12 Frequency of events.

of the injuries (death) data and then multiplying the total injuries (death) by each density value. The sticks were obtained by calculating the mean, $\overline{X}$ and size, S, of the injuries and death bi-weekly rates. These are the parameters of the negative binomial PD. The S parameter is given by

$$S := \frac{\overline{X}^2}{V - \overline{X}}$$

where V is the variance of the data. For events, we use the Poisson PD with mean events per two-weeks was the intensity parameter (Evans et al., 2000). Albeit counting processes, the empirical densities of injuries and deaths were over dispersed compared to the Poisson. Results for the fortnight rates were:

	Mean	Variance
Injuries	45.712	3 075.013
Dead	9.507	86.559
Events	3.068	3.398

Over dispersion in counting processes arise from two main sources: the Poisson intensity parameter varies over time or there are several Poisson PD with different intensities underlying the counts. The Poisson PD process arises when events are scattered uniformly over time—their timing is unpredictable—with some constant rate.

Although the mean and variance of events were roughly equal, as we have seen (Figures 17.10 and 17.12), one cannot expect the events to behave according to the Poisson process. The negative binomial arises under conditions similar to the Poisson, except that the variance of the former is larger than its mean. For both injuries and deaths, the fit of the theoretical PD (sticks in Figure 17.13) to the empirical PD (points in Figure 17.13) was such that we could *not* reject that hypotheses that the counts were drawn from count rates that obey the negative binomial PD.[4] As the

---

[4] $\chi^2 = 1.854$ and p-value $= 1$ for injuries; $\chi^2 = 7.505$ and p-value $= 0.943$ for deaths; and $\chi^2 = 46.593$ and p-value $= 0.000$ for events. The p-values for the results from the Pearsons's Chi-squared were simulated (degrees of freedom are not needed).

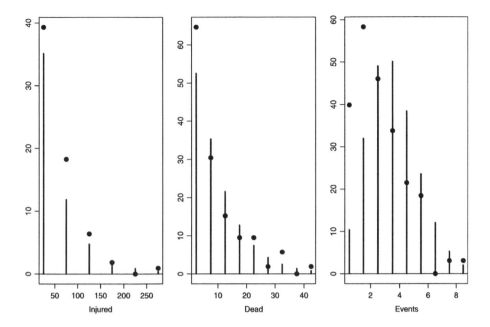

Figure 17.13 Empirical expected values (dots) and theoretical (sticks) for the negative binomial PD for injuries and deaths and Poisson for events.

residuals disclose (Figure 17.11), part of the overdispersion resulted from time-varying rates.

In short, injury, death and event rates seemed to have surged and subside as the warring sides were adjusting to each other's strategies. The SI effectively ended when Hammas declared a cease fire. Israeli daily newspapers[5] consistently attributed the decline in these rates in the last phase of the SI to the *barrier*. This is in sharp contrast to the WI, where no single factor (except perhaps for the delays in adjustment to each other's strategies) could be associated with temporal changes in death and injury rates.

Chronology by organization

During the period summarized, different organizations became active at different times. Figures 17.14 and 17.15 lead to the following conclusions regarding unclaimed events:

- Between 9/27/2000 and 3/1/2001, except for IJ on two occasions, organizations did not claim responsibility for events. During this period, 43 events occurred.
- During this period, all of the events ended with mostly one death.
- In April 2002, the frequency of unclaimed events went down dramatically (but as the right panel of Figure 17.10 illustrates, not the frequency of events).

---

[5] e.g. Ma'ariv, Ha'aretz, Yediot Aharonot and the *Jerusalem Post* kept reporting about sharp decreases in the frequency of attacks in locations in Israel proper next to where the barrier had been erected.

560 Application: the shape of wars to come

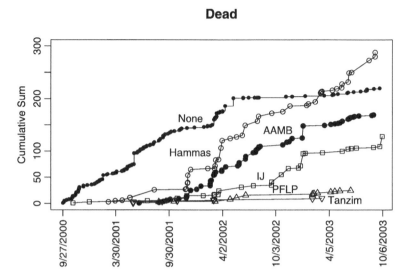

Figure 17.14 Cumulative sum (by date) of the number of dead per event by organizations that claimed responsibility for the events.

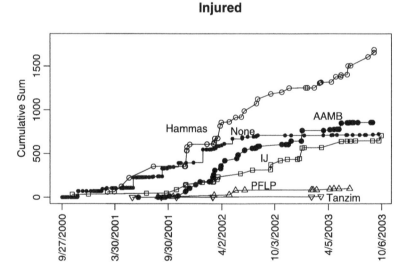

Figure 17.15 Cumulative sum (by date) of the number of injuries per event by organizations that claimed responsibility for the events.

In short, the organizations tended to claim responsibility for events that ended with more than one or two dead. This indicates that events with multiple deaths were regarded as a success by the claiming organization.

Regarding the claimed events, we conclude from Figures 17.14 and 17.15:

- Of the organizations that claimed responsibility, AAMB claimed more events than any other organization (see also Table 17.2).
- The events claimed by Hammas caused more deaths and injuries than any other organization.
- Other than the events claimed by Hammas, the events claimed by AAMB and IJ resulted in most deaths.
- The number of events claimed by PFLP and Tanzim were marginal.

Most events were claimed and except for three, a single organization claimed an event. We can therefore assume that the claims were true. Obviously, executing an event requires both people and resources. Therefore, we conclude that AAMB invested the most during the Second Intifada, followed by Hammas. This conclusion does not allow us to reject the hypothesis that AAMB, an organization associated with the Palestinian Liberation Organization and therefore with the Palestinian Authority, was the most funded compared to other organizations that regularly claimed responsibility for events.

Also, IJ was the first organization to claim responsibility for an event (on 11/2/2000). The first event claimed by Hammas occurred 122 days later, on 3/4/2001. The first event claimed by AAMB occurred 313 days after the first claimed event, on 9/11/2001. This emphasizes the fact that in addition to executing more events than any other organizations, AAMB also executed them at the highest frequency compared to all other organizations.

17.2.4 Conclusions

The similarities between the WI and the SI are striking and the differences are instructive. In both wars we identify alternating periods of ebb and swell in the death and injury rates (and events in the case of SI). These periods may have been related to delays in the warring sides adjusting to each other's change of strategy. However, unlike the WI, we can identify particular strategies (e.g. the *barrier*) that the other side needed to adapt to. In both cases, the fluctuations indicate long-term changes (lasting on the order of months). Reliance on short terms changes to predict the tide of these wars is therefore *not* warranted.

An interesting difference in the fluctuations of death and injury rates between the SI and the WI is the synchrony in fluctuations in the latter. In the case of the SI, the synchrony is not complete (see Figure 17.10). This reflects the fact that changing strategies by Israel and the PMO resulted in ever increasing ratio of injury to death rates. In the case of the SI, the *barrier* is clearly associated with a decline in death, injury and event rates. However, without the Hammas declaring cease fire that effectively ended the SI, it is possible that the MPO would have adapted and the cycle would have started again. This emphasizes the clichés that "wars do not solve anything."

Death and injury rates in both wars fit the negative binomial PD. In the case of the WI, this leads one to expect 95% of deaths to be ≤ 37 per week and 974 per month respectively. In the case of the SI, the corresponding values are $\leq 28/2 = 14$ deaths per week and $157 \times 2 = 314$ injuries per month. These results are important for organizations that are in charge of planning emergency services, for they allow response to anticipated rates of deaths and injuries.

We close with a personal opinion: Both wars indicate that even mighty military forces cannot overcome small groups of local MO that are ready to use any means to cause deaths and injuries. The exception is the Second Chechen War (1999–2000) in which force used without restrain to achieve a goal (it remains to be seen for how long) and which had been conducted without public scrutiny.

References

Adams, D. H. and R. H. McMichael. 1999. Mercury levels in four species of sharks from the Atlantic coast of Florida. *Fisheries Bulletin* **97**:372–9.

Adams, G. D. and Fastnow, 2000. A Note on the Voting Irregularities in Palm Beach. URL http://madison.hss.cmu.edu;http://www.statsci.org/index.html. Florida.

Agresti, A. and B. A. Coull. 1998. Approximate is better than "exact" for interval estimation of binomial proportions. *American Statistician* **52**:119–126.

Alvarez-Ramirez, J. et al. 2006. Fractality and time correlation in contemporary war. Chaos, *Solitons & Fractals* **34**: 1039–49.

Anscombe, F. J. 1948. The transformation of Poisson, binomial and negative binomial data. *Biometrika* **35**:246–54.

Becker, R. A., J. M. Chambers and A. R. Wilks. 1988. *The New S Language*. Chapman & Hall, London.

Berger, J. O. 1985. *Statistical Decision Theory and Bayesian Analysis*. Springer-Verlag.

Bliss, C. I. and R. A. Fisher. 1953. Fitting the negative binomial distribution to biological data. *Biometrics* **9**:176–200.

Box, G. E. P. and G. M. Jenkins. 1976. *Time Series Analysis: Forecasting and Control*. Holden-Day.

Brockwell, P. J. and Davis, R. A. (1991) *Time Series: Theory and Methods*. 2nd edn. Springer.

Buckland, S. T., D. R. Anderson, K. P. Burnham, J. L. Laake, D. L. Borchers and L. Thomas. 2001. *Introduction to Distance Sampling: Estimating Abundance of Biological Populations*. Oxford University Press.

Bui, T. D., C. K. Pham and H. T. Pham et al. 2001. Cross-sectional study of sexual behaviour and knowledge about HIV among urban, rural and minority residents in Viet Nam. *Bulletin of the World Health Organization* **79**:15–21. URL http://www.scielo.br/pdf/bwho/v79n1/v79n1a05.pdf.

Chakravarti, Laha and Roy. 1967. *Handbook of Methods of Applied Statistics*. Wiley, New York.

Chambers, J., W. Cleveland, B. Kleiner and P. Tukey. 1983. *Graphical Methods for Data Analysis.* Wadsworth.

Chambers, J. M. 1998. *Programming with Data.* Springer, New York. URL http://cm.bell-labs.com/cm/ms/departments/sia/Sbook/.

Chambers, J. M. and T. J. Hastie. 1992. *Statistical Models in S.* Chapman & Hall.

Cleveland, W. 1985. *The Elements of Graphing Data.* Wadsworth.

Dalgaard, P. 2002. *Introductory Statistics with R.* Springer.

Davison, A. C., and D. V. Hinkley. 1997. *Bootstrap Methods and Their Application.* Cambridge University Press.

DeNavas-Walt, C., R. W. Cleveland and B. H. Webster, Jr., 2003. *Income in the United States: 2002.* Technical report, US Department of Commerce. Economics and Statistics Administration. US Census Bureau, Washington, D. C.

Dennett, D. C. 1995. *Darwin's Dangerous Idea.* Simon & Schuster, New York.

Detre, K. and C. White. 1970. The comparison of two Poisson-distributed observations. *Biometrics* **26**:851–4.

DiCiccio, T. J., and B. Efron. 1996. Bootstrap confidence intervals (with Discussion). *Statistical Science* **11**:189–228.

Dobson, A. J. 1983. *An Introduction to Statistical Modelling.* Chapman and Hall.

Dobson, A. J., K. Kuulasmaa, E. Eberle and J. Scherer. 1991. Confidence intervals for weighted sums of Poisson parameters. *Statistics in Medicine* **10**:457–62.

e-Digest of Environmental Statistics, 2003*a*. Table 13 Atmospheric inputs from UK sources to the North Sea: 1987–2000. Technical report, Department for Environment, Food and Rural Affairs, UK. URL http://www.defra.gov.uk/environment/statistics/index.htm.

e-Digest of Environmental Statistics, 2003*b*. Table 16 Reported source of pollution by enumeration area: 2001. Technical report, Advisory Committee on Protection of the Sea (ACOPS), Department for Environment, Food and Rural Affairs, UK. URL http://www.defra.gov.uk/environment/statistics/index.htm.

Efron, B. 1987. Better bootstrap confidence intervals (with Discussion). *Journal of the American Statistical Association* **82**:171–200.

Efron, B. and R. J. Tibshirani. 1993. *An Introduction to the Bootstrap.* Chapman & Hall.

Evans, M., N. Hastings and B. Peacock. 2000. *Statistical Distributions.* Third edition. Wiley, New York.

Fleiss, J. L., B. Levin, and M. C. Paik. 2003. *Statistical Methods for Rates and Proportions.* Third edition. Wiley, New York.

Focazio, M. J., Z. Szabo, T. F. Kraemer, A. H. Mullin, T. H. Barringer and V. T. dePaul, 2001. *Occurrence of Selected Radionuclides in Ground Water Used for Drinking Water in the United States: A Reconnaissance Survey, 1998.* Technical report, US Geological Survey, US Geological Survey, Office of Ground Water, Mail Stop 411, 12201 Sunrise Valley Drive, Reston, Virginia 20192. URL http://water.usgs.gov/pubs/wri/wri004273/pdf/wri004273.pdf.

Fox, J. 2002. *An R and S-Plus Companion to Applied Regression.* Sage Publications, Thousand Oaks, CA, USA. URL http://www.socsci.mcmaster.ca/jfox/Books/Companion/.

Frelich, L. E. and C. G. Lorimer. 1991. Natural disturbance regimens in hemlock-hardwood forests of the upper great lakes region. *Ecological Monographs* **61**:145–64.

Gallup Europe, 2003. *Iraq and Peace in the World, Flash Eurobarometer Number* 151. Technical report, European Commission, European Commission, Directorate General, Press and Communication, Public Opinion Analysis sector B-1049 Brussels. URL http://europa.eu.int/comm/public{_}opinion/flash.

Geller, D. S. and Singer, J. D. 1998. *Nations at War*. Cambridge Studies in International Relations, No. 58.

Gelman, A., J. B. Carlin and H. S. Stern. 1995. *Bayesian Data Analysis*. Chapman & Hall.

Gelpi, C. et al. 2005/6. Success matters: Casualty sensitivity and the war in Iraq: *International Security* **30**: 7–46.

Gholz, H. L., W. P. Cropper, Jr, S. A. Vogel, K. McKelvey, K. C. Ewel, et al. 1991. xx. Ecological Monographs **61**:33–51.

Hand, D. J. 1994. *A Handbook of Small Data Sets*. Chapman & Hall, London.

Hosmer, D. W., T. Hosmer, S. Lemeshow, S. le Cessie and S. Lemeshow. 1997. A comparison of goodness-of-fit tests for the logistic regression model. *Statistics in Medicine* **16**:965–980.

Hosmer, D. W. and S. Lemeshow. 2000. *Applied Logistic Regression*. Second edition. Wiley, New York.

International Program Center, 2003. International Data Base. URL http://www.census.gov/ipc/www/. Population Division of the US Bureau of the Census.

Jaynes, E. T. 2003. *Probability Theory: The Logic of Science*. Cambridge University Press, Cambridge, UK.

Johnson, N. L., S. Kotz and N. Balakrishnan. 1994. *Continuous Univariate Distributions*. Second edition. John Wiley & Sons, New York.

Kaye, D. H. 1982. Statistical evidence of discrimination. *American Statistical Association* **77**:772–83.

Keeling, C. D., T. P. Whorf and the Carbon Dioxide Research Group, 2003. *Atmospheric CO_2 concentrations (ppmv) derived from in situ air samples collected at Mauna Loa Observatory, Hawaii*. Technical report, Scripps Institute, La Jolla, California.

Kimmo, L., M. Orell, S. Rytkonen and K. Koivula. 1998. Time and food dependence in Willow tit winter survival. *Ecology* **79**:2904–16.

Kline, J., B. Levin, A. Kinney, Z. Stein, M. Susser and D. Warburton. 1995. Cigarette smoking and spontaneous abortion of known karyotype: Precise data but uncertain inferences. *American Journal of Epidemiology* **141**:417–27.

Kolmogoroff, A. 1956. *Foundations of the Theory of Probability*. Chelsea.

Kotz, S., N. Balakrishnan and N. L. Johnson. 2000. *Continuous Multivariate Distributions*. Second edition. Wiley, New York.

Krebs, C. J. 1989. *Ecological Methodology*. Harper & Row, New York.

Krishnamoorthy, K. and J. Thomson. 2004. A more powerful test for comparing two Poisson means. *Journal of Statistical Planning and Inference* **119**:23–35.

Krivokapich, J., J. S. Child, D. O. Walter and A. Garfinkel. 1999. Prognostic value of dobutamine stress echocardiography in predicting cardiac events in patients with known or suspected coronary artery disease. *Journal of the American College of Cardiology* **33**:708–16.

Lader, D., and H. Meltzer, 2002. *Smoking Related Behaviour and Attitudes, 2001*. Technical report, Office for National Statistics, London. URL http://www.statistics.gov.uk.

Lapsley, M. and B. Ripley, 2004. RODBC. URL http://r-project.org. Package.
Limpert, E., W. A. Stahel and M. Abbt. 2001. Log-normal distributions across the sciences: keys and clues. *BioScience* **51**:341–52.
McLaughlin, M., 1999. Common Probability Distributions. URL http://www.causascientia.org/math_stat/Dists/. Regress + Appendix A.
McNeil, D. R. 1977. Interactive Data Analysis. Wiley.
MFA, 2004. Victims of Palestinian Violence and Terrorism since September 2000. URL http://www.mfa.gov.il/mfa/go.asp?MFAH0ia50. Israel Ministry of Foreign Affairs.
Miller, M. E., S. L. Hsu and W. M. Tierney. 1991. Validation techniques for logistic regression models. *Statistics in Medicine* **10**:1213–26.
Mosteller, F. 1973. *Study of Statistics*. Addison Wesley, Redding, Massachusetts.
National Cancer Institute, D., 2004. Surveillance, Epidemiology and End Results (SEER) Program, SEER*Stat Database: Incidence - SEER 9 Regs Public-Use, Nov 2003 Sub (1973–2001). URL http://www.seer.cancer.gov. Surveillance Research Program, Cancer Statistics Branch, released April 2004, based on the November 2003 submission.
NPS, 2004. All Pakrs Summary Report. URL http://www2.nature.nps.gov/mpur/Reports/reportlist.cfm. US National Park Service.
Orians, G. H. 1980. *Marsh-nesting Blackbirds*. Princeton University Press, Princeton, New Jersey.
Papoulis, S. 1965. *Probability, Random Variables and Stochastic Processes*. McGraw-Hill.
Patten, M. A. and P. Unitt. 2002. Diagnostability versus mean differences of sage sparrow subspecies. *The Auk* **119**:26–35.
Peixoto, J. L. 1990. A property of well-formulated polynomial regression models. *American Statistician* **44**:26–30.
Pielou, E. C. 1977. *Mathematical Ecology*. John Wiley & Sons.
Press, W. H., S. A. Teukolsky, W. T. Vetterling and B. P. Flannery. 1992. *Numerical Recipes in C*. Second edition. Cambridge University Press.
Raina, R., M. M. Lakin, A. Agarwal, R. Sharma, K. K. Goyal, D. K. Montague. 2003. Long-term effect of Sildenafil citrate on erectile dysfunction after radical prostatectomy: 3-year follow-up. *Urology* **62**:110–15.
Ripley, B. D. 1987. *Stochastic Simulations*. John Wiley & Sons.
Rosner, B. 2000. *Fundamentals of Biostatistics*. Fifth edition. Duxbury, Pacific Grove, California.
Ross, S. M. 1993. *Probability Models*. Fifth edition. Academic Press.
Sample, B. E., M. S. Alpin, R. A. Efroymson, G. W. Suter, II and C. J. E. Welsh, 1997. *Methods and Tools for Estimation of the Exposure of Terrestrial Wildlife to Contaminants*. Technical report, Environmental Sciences Division, U.S. Department of Energy. URL http://www.esd.ornl.gov/programs/ecorisk/documents/tm13391.pdf.
Scotbennett, D. and Stam, A. C. (2006) Predicting the length of the 2003 U.S.–Iraq War, *Foreign Policy Analysis* **2**: 101–16.

Shapiro, S. S. and M. B. Wilk. 1965. An analysis of variance test for normality (complete samples). *Biometrika* **52**:591–611.

Stephens, M. A. 1974. EDF Statistics for goodness of fit and some comparisons. *Journal of the American Statistical Association* **69**:730–7.

The R Development Core Team, 2006a. R Data Import/Export.

The R Development Core Team, 2006b. The R Environment for Statistical Computing and Graphics.

The World Bank, 1996a. *Elimination of Lead in Gasoline in Latin America and the Caribbean*. Technical report, Report No. 194/97EN, Energy Sector Management Assistance Programme, Washington, D.C.

The World Bank, 1996b. *Phasing-out Lead From Gasoline: World-Wide Experience and Policy Implications (Annex A update, July 1997)*. Technical report, The World Bank, Washington, D.C.

Thode Jr, H. C. 1997. Power and sample size requirements for tests of differences between two Poisson rates. *The Statistician* **46**:227–30.

Tukey, J. W. 1977. *Exploratory Data Analysis*. Addison-Wesley, Reading, Massachusetts.

Ulm, K. 1990. A simple method to calculate the confidence interval of a standardized mortality ratio. *American Journal of Epidemiology* **131**:373–5.

United Nations. 2003. *World Population Prospects: The 2002 Revision (ST/ESA/ SER.A/224)*. Population Division of the Department of Economic and Social Affairs of the United Nations Secretariat, United Nations, New York.

United States Department of Justice, B. o. J. S., 1995. Survey of Campus Law Enforcement Agencies, 1995: [United States] [Computer file]. Technical report, Conducted by U.S. Dept. of Commerce, Bureau of the Census. ICPSR ed., Ann Arbor, MI: Inter-university Consortium for Political and Social Research [producer and distributor]. URL http://www.ojp.usdoj.gov/bjs.

United States Department of Justice, B. o. J. S., 2003. Capital Punishment in the United States, 1973–2000 [Computer file]. Compiled by the U.S. Dept. of Commerce, Bureau of the Census. Technical report, Conducted by U.S. Dept. of Commerce, Bureau of the Census. ICPSR ed., Ann Arbor, MI: Inter-university Consortium for Political and Social Research [producer and distributor]. URL http://www.ojp.usdoj.gov/bjs.

Velleman, P. and D. Hoaglin. 1981. *The ABC's of EDA: Applications, Basics, and Computing of Exploratory Data Analysis*. Duxbury, Belmont, CA.

Venables, W. N. and B. D. Ripley. 1994. *Modern Applied Statistics with S-Plus*. Springer-Verlag.

Venables, W. N. and B. D. Ripley. 2000. *S Programming*. Springer. URL http://www.stats.ox.ac.uk/pub/MASS3/Sprog/.

Venables, W. N., and B. D. Ripley. 2002. *Modern Applied Statistics with S*. Fourth Edition. Springer. URL http://www.stats.ox.ac.uk/pub/MASS4/.

Venables, W. N., D. M. Smith and the R Development Core Team, 2003. *An Introduction to R*. www.r-project.org, 1.8.0 edition.

Vollset, S. E. 1993. Confidence intervals for a binomial proportion. *Statistical Medicine* **12**:809–24.

Wayne, D. 1990. *Nonparametric Statistics*. PWS-Kent, Boston, Massachusetts.

White, C. R. and R. S. Seymour. 2003. Mammalian basal metabolic rate is proportional to body mass$^2/3$. *Proceedings of the National Academy of Sciences* **100**:4046–4049.

WHO, 2004. *World Population Prospects: The 2002 Revision*. URL `http://www3.who.int/whosis/`. Source: Population Division of the Department of Economic and Social Affairs of the United Nations Secretariat (2003). New York: United Nations. WHO Statistical Information System (WHOSIS).

R index

%*%, 145
\, 17
\t, 59
χ^2
    qchisq(), 303
~, 89, 139
*, 15
+, 11, 15
-, 15
.(), 139
..., 40, 141
.First(), 37, 39
.Last(), 37
.RData, 37, 38
/, 15
:, 139
;, 11
<, 21
<<-, 32
==, 21
>, 11, 21
<=, 21
>=, 21
[], 6
#, 11
$, 54
&, 21, 372

|, 21
{}, 39

abline()
    col, 306
    h, 109, 140, 186, 206, 270, 297, 356
    lty, 254, 306, 395
    lwd, 254, 306
    reg, 220
    v, 254, 270, 297, 306, 356, 395
abline(), 109, 140, 186, 206, 215, 220, 254, 270, 306, 352, 356, 395
acf()
    lag.max, 277
acf(), 277
ad.test(), 225
agricolae, 480
airquality, 56
all, 300
alpha, 354, 363, 414
alt, 403, 408
alternative, 354, 363, 387, 391
angle, 189, 297
ANOVA
    anova(), 475
    aov(), 473, 475, 505
    interaction.plot(), 502

ANOVA (*continued*)
    null, 482
    summary(), 480
anova(), 475
aov()
    data, 473
aov(), 473, 482, 505
apply(), 68, 76
array, 86
array()
    dim(), 51
array(), 52
arrows()
    angle, 189, 297
    code, 189, 206, 297
    length, 189, 206, 297
arrows(), 189, 206, 297
as.character(), 27, 28, 50, 264, 497
as.data.frame(), 54
as.Date(), 66
as.integer(), 28, 49
as.matrix(), 125
as.numeric(), 27, 160
as.vector(), 410
assign(), 32
asymp, 354
Asymptotic, 300
asypow, 361
at, 186, 287, 297
attach(), 55, 84, 378
attr(), 26, 84
attribute, 84
    length, 28
    mode, 27
attributes(), 26, 51
axes, 186, 206, 238, 287, 297, 467
axis()
    at, 186, 287, 297
    font, 186
    labels, 186
    vfont, 186
axis(), 186, 238, 287, 297, 467

ballocation(), 410
barplot()
    col, 47, 80
    las, 47, 80
    main, 47, 80
    names.arg, 80
    ylab, 47
barplot(), 47, 78, 80, 478
base, 9, 10

bca, 243
bill.length, 220
binconf()
    all, 300
    Asymptotic, 300
    Exact, 300
    method, 300
    Wilson, 300
binconf(), 300, 303, 327
binom
    binom.power(), 354, 361
    cloglog.sample.size(), 362
binom, 354, 361
binom.power()
    alpha, 354
    alternative, 354
    asymp, 354
    method, 354
    n, 354
    p, 354
    two.sided, 354
binom.power(), 354, 361
blue, 191, 254
bmr.rda, 305
Bonferroni(), 483
boot, 242
boot()
    bca, 243
    boot(), 242
    summary(), 242
boot(), 242
boot.ci(), 395
boxplot()
    las, 275
    main, 275
    names, 374
boxplot(), 275, 374
bp()
    alt, 408
    greater, 408
bp(), 408
bquote(), 139, 240
breaks, 29, 48, 82, 84
bsamsize(), 410
bwt.rda, 537
byrow, 123

c(), 14, 53, 109, 139, 194, 269, 295, 320, 342,
    373, 389
capital.punishment, 63, 515
capital.punishment.rda, 63, 298, 378
cardiac, 60

cardiac.rda, 266
casualties.rda, 233
cbind(), 20, 52, 68, 123, 146, 216, 290, 356, 389
CDC, 477
CDC-demographics-income-categories, 489
cdfpoe3(), 242
ceiling(), 67, 159, 410
cex, 90, 160, 165, 467
cex.main, 297
character, 27
character(), 27
chisq.test(), 379
choose(), 171
class, 84
cloglog.sample.size()
    alpha, 363
    alternative, 363
    p, 363
    power, 363
    recompute.power, 363
cloglog.sample.size(), 362
code, 189, 206, 297
col, 40, 47, 80, 82, 141, 168, 186, 191, 209, 254, 297, 306, 354, 405, 465
college.crime.rda, 289
color
    blue, 191, 254
    gray90, 82
    grey80, 215
    grey90, 141, 186, 209, 297
    red, 168, 191, 254, 306
color, 495
colors(), 80
combinat, 123, 124
combinat()
    nCm(), 124
    permn(), 123
combn(), 107
complete.cases(), 372, 478, 497, 515
confidence interval
    binconf(), 303
    t.test(), 385
continue, 37
coplot(), 35
cor()
    pairwise.complete.obs, 273
    use, 273
cor(), 273
cor.test()
    method, 275
    spearman, 275

cor.test(), 274, 275
correct, 376
counts, 84
csv, 56, 85
cumsum(), 109
cut()
    breaks, 48
    include.lowest, 159
cut(), 48, 66, 67, 159

data
    airquality, 56
    bill.length, 220
    bmr.rda, 305
    capital.punishment, 63, 515
    capital.punishment.rda, 63, 298, 378
    cardiac, 60
    cardiac.rda, 266
    casualties.rda, 233
    CDC-demographics-income-categories, 489
    college.crime.rda, 289
    complete.cases(), 478
    csv, 56, 85
    data.frame(), 256
    demo_d, 477
    demo_d.short.rda, 477, 489
    discoveries, 90
    distance, 82
    edu, 242
    elections-2000, 56
    EU.rda, 464, 496
    EU.station.rda, 464, 497
    faculty.rda, 256, 260
    file, 256
    fish.rda, 512
    graduation, 76
    head(), 256
    Iraq-casualties, 66
    Iraq.cnts.rda, 277
    l.rda, 374
    load(), 256, 259, 272, 298, 372
    match(), 465
    midterm, 224
    null, 518, 529, 537
    PlantGrowth, 84
    pop.var.names, 78
    read.table(), 464
    rtest, 61
    save(), 256
    score, 47

data (*continued*)
  south, 239
  state.region, 76
  terror, 159
  terror.by.Hamas, 141, 184
  test.scores.rda, 393
  us.income.rda, 259
  wells.info.rda, 272
  wells.nucleotides.rda, 272
  who, 59
  who.by.continents.and.regions, 86
  who.ccodes, 57, 78
  who.pop.2000, 57, 78
  who.pop.var.names, 57
  who.fertility.mortality.rda, 264
data, 473
data export
  write.ftable(), 63
data import
  DSN, 61
  foreign(), 212
  read.csv(), 47, 85
  read.dta(), 60, 212
  read.ftable(), 63
  read.table(), 47, 54, 59, 76
  read.xport(), 477
  RODBC, 60, 61
  scan(), 59, 87
data(), 56, 84, 239
data.frame()
  stringsAsFactors, 50
data.frame(), 50, 54, 125, 165, 256, 264, 268, 272, 378, 389
datadist(), 515, 531
Date, 141
dbeta(), 203
dbinom(), 148, 150, 152, 215, 216, 238
dchisq(), 194
ddouble.exp(), 190
decreasing, 202, 256, 280
demo_d, 477
demo_d.short.rda, 477, 489
densfun, 195
density, 40, 84, 141
density(), 231, 530
density, continuous
  dbeta(), 203
  dchisq(), 194
  ddouble.exp(), 190
  dexp(), 141, 184
  df(), 198
  dgamma(), 201

  dlnorm(), 199
  dnorm(), 186, 206, 209, 213, 222, 234, 240, 254, 287, 289
  dt(), 196
  dunif(), 179
  plnorm(), 199
  pnorm(), 331
  qnorm(), 297
  qt(), 327
density, discrete
  dbinom(), 148, 150, 152, 215, 216, 238
  dgeom(), 139, 143
  dhyper(), 171
  dnbinom(), 168
  dpois, 160
  dpois(), 219, 239
  pois.approx(), 301
Design
  datadist(), 515, 531
  lrm(), 515
  residuals.lrm(), 533
Design, 515, 531
detach(), 34, 55
detailed, 405
dev.copy()
  device, 40
  file, 40
  pdf, 40
  postscript, 40
  win.metafile, 40
dev.copy(), 40
dev.off(), 40
deviance(), 481
device, 40
dexp(), 141, 184
df(), 198
df.residual(), 480
dgamma()
  scale, 201
dgamma(), 201
dgeom(), 139, 143
dhyper(), 171
diff(), 233
digits, 16, 37
dim(), 26, 51
dimnames(), 26, 160, 264, 374, 377, 390
discoveries, 90
distance, 82
distribution, continuous
  pt(), 197
  pbeta(), 203
  pchisq(), 194, 333, 363, 379

pdouble.exp(), 190
pexp(), 187, 189
pf(), 198, 382
pgamma(), 201
pnorm(), 215, 218, 235, 266, 342, 345, 371
psignrank(), 394
punif(), 179, 187
qnorm(), 295
distribution, discrete
    ecdf(), 168
    pbinom(), 150, 152, 218, 388
    pgeom(), 143
    pnbinom(), 168
distribution, empirical
    ecdf(), 203
dlnorm(), 199
dnbinom(), 168
dnorm(), 12, 186, 206, 209, 213, 222, 234, 240, 254, 269, 287, 289
dotchart(), 86
double(), 14
dpois(), 160, 219, 239
DSN, 61
dt(), 196
dunif(), 179

ecdf(), 168, 203, 266
edu, 242
elections-2000, 56
else, 29
epitools
    expand.table(), 63
    pois.byar(), 301
    pois.daly(), 301
    pois.exact(), 301, 304, 364
epitools, 301, 304, 364
equidistant, 84
etc, 37
EU.rda, 464, 496
EU.station.rda, 464, 497
eval(), 88
exact, 300
example(), 10
expand, 405
expand.grid(), 107
expand.table(), 63
expression()
    italic(), 215
expression(), 88, 109, 139, 168, 179, 206, 207, 215, 238, 297

F, 22

factor
    is.ordered(), 49
    ordered, 49
factor(), 47, 49
faculty.rda, 256, 260
FALSE, 13, 22, 78
file, 40, 256, 264
fill, 465
fish.rda, 512, 529
fisher.test()
    alternative, 387
fisher.test(), 387
fitdistr()
    densfun, 195
    start, 195
fitdistr(), 162, 194
floor(), 409
font
    italic, 186
    serif, 186
font, 186
for(), 29, 109, 189, 225, 230
foreign
    read.dta(), 60, 212
foreign, 60, 212
foreign(), 212
freq, 40, 141
function, 27
function(), 209, 231, 293, 513

geoR, 103
gl(), 230
glm(), 515
gpclib, 103
granova
    granova.1w(), 473
granova, 473
granova.1w(), 473
gray90, 80, 82, 354, 405
greater, 408
grey80, 215
grey90, 40, 141, 186, 209, 297
group, 481
groupedData(), 507
gstat, 103
gtools
    permutations(), 125
gtools, 125

h, 109, 140, 186, 206, 216, 270, 297, 356
h()
    ..., 40
    col, 40

h() (*continued*)
    density, 40
    grey90, 40
    l, 191
    main, 40
    type, 191
    xlab, 40, 306
    xlim, 306
    ylab, 40
    ylim, 306
h(), 40, 141, 168, 184, 191, 194, 201, 213, 233, 269, 289, 306
head(), 47, 56, 159, 256, 374
header, 48, 76
height, 39, 40, 495
help(), 6, 8
help.search(), 10, 225
hist()
    ..., 141
    attr(), 84
    breaks, 82, 84
    class, 84
    col, 82, 141
    counts, 84
    density, 84, 141
    equidistant, 84
    freq, 40, 141
    intensities, 84
    main, 84, 141
    mids, 84
    xlab, 84, 141
    xlim, 84
    xname, 84
    ylab, 82, 84, 141
    ylim, 84
hist(), 35, 40, 82, 84, 395
history(), 13
Hmisc
    ballocation(), 410
    binconf(), 300, 303
    bsamsize(), 410
    latex(), 264
Hmisc, 264, 300, 303, 327, 410

identify()
    labels, 86, 256, 275, 374
identify(), 36, 86, 256, 258, 275, 290, 374
if, 29
if(), 29, 209
ifelse(), 30, 109, 264
include.lowest, 159
index.return, 256, 280, 389

Inf, 23, 477
Injured, 141
integer(), 14, 27
intensities, 84
interaction.plot(), 502
intersect(), 69, 108
intervals(), 507
Iraq-casualties, 66
Iraq.cnts.rda, 277
is.array(), 51
is.character(), 27, 50
is.double(), 49
is.element(), 69, 497, 498
is.integer(), 27, 49
is.list(), 66
is.logical(), 27, 28
is.matrix(), 51
is.na(), 18, 22, 372
is.nan(), 23
is.numeric(), 27, 49
is.ordered(), 49
is.vector(), 51
italic, 109, 186
italic(), 139, 179, 215, 238, 297

Julian, 141, 159
julian(), 67

k, 414
Killed, 141
kruskal.test(), 491
kruskalmc(), 492
ks.test(), 224

l, 109, 179, 186, 191
l.rda, 374
labels, 86, 186, 194, 256, 258, 275, 297, 374
lag.max, 277
lapply(), 76
las, 47, 80, 275
latex(), 264
lattice
    trellis.device(), 495
    xyplot(), 89, 495
lattice, 89, 495
legend(), 36
length, 13, 28, 179, 186, 189, 206, 297, 345, 405
length(), 15, 18, 160
sum(), 191
LETTERS, 48, 68, 124, 165
letters, 68, 107
levels(), 71, 515

library(), 89, 125, 212, 220, 239, 242, 264, 294, 300, 354, 395, 410, 491, 495
lines()
    col, 168, 191
    lty, 168
    lwd(), 168
    lwd, 191
    s, 292
    type, 292
lines(), 36, 141, 160, 168, 184, 191, 194, 206, 222, 234, 240, 269, 289, 352
list
    hist(), 84
    length(), 53
    names(), 66
list(), 52, 90, 160, 195, 216, 225, 240, 261, 264, 377
lm(), 220, 505
lmline(), 507
lmline(x, y, lty = 2), 507
lmList, 507
lmomco
    cdfpoe3(), 242
    parpe3(), 242
    quape3(), 242
lmomco, 242, 284, 293
load(), 34, 82, 159, 184, 233, 256, 259, 272, 289, 298, 372, 374, 378
locator(), 36, 194, 533
log(), 225, 289
logical, 27
logical(), 27, 29
lrm()
    x, 515
    y, 515
lrm(), 515
ls(), 38
LSD.test()
    group, 481
LSD.test(), 481
lty, 168, 194, 254, 306, 395
lwd, 140, 179, 191, 254, 260, 306
lwd(), 168

M, 395
main, 40, 47, 80, 84, 141, 160, 186, 275, 393
map(), 465, 498
map()
    col, 465
    fill, 465
    regions, 465
    world, 465

map.axes(), 465, 498
mapdata, 465, 498
mapply(), 76, 226, 230, 261
maps
    map(), 498
maps, 465, 498
mar, 275, 287, 478
match(), 465
matrix()
    byrow, 52, 123
    dim(), 51
    dimnames, 52
    ncol, 52, 123, 305, 329
    nrow, 52, 305, 329
matrix(), 51, 123, 305, 329, 343
max, 7
max(), 159, 269
mean()
    na.rm, 19, 86, 264, 268, 319
    trim, 260, 290
mean(), 8, 19, 86, 168, 201, 212, 213, 224, 260, 264, 268, 290, 319, 329, 372
median(), 259
merge(), 68
method, 275, 300, 354
mfrow, 79, 82, 139, 168, 179, 194, 215, 393, 403
mids, 84
midterm, 224
min, 7
min(), 394
MLE
    fitdistr(), 162
    optim(), 162
mle()
    mle-class, 294
mle(), 293
mle-class, 294
mode
    character, 27
    function, 27
    logical, 27
    numeric, 27
mode, 13, 27
mode()
    numeric, 13
mode(), 26, 27
Model L.R., 516
model.tables()
    type, 505
model.tables(), 505

n, 354

NA, 18, 22, 86, 264
na.rm, 19, 86, 264, 268, 319
names, 374
names(), 20, 66, 76, 86, 268, 272, 376, 389
names.arg, 80
NaN, 23
nCm(), 124
ncol, 123, 305, 329
nlme, 506
no.dimnames(), 33, 107, 123
noquote(), 54, 107, 123
normOrder(), 220
nortest, 225
nqd(), 40, 123
nrow, 305, 329
numeric, 27
numeric(), 13, 27

odbcClose(), 61, 62, 264
odbcConnect(), 61, 264
openg()
    height, 39
    pointsize, 39
    width, 39
openg(), 39
optim(), 162
options
    stringsAsFactors, 268
options()
    continue, 37
    datadist, 515
    digits, 16, 37
    prompt, 37
    show.signif.stars, 37
options(), 16, 37, 39, 268, 515

p, 354, 363
package
    agricolae, 480
    base, 9, 10
    binom, 354, 361
    boot, 242
    combinat, 123, 124
    Design, 515, 531
    detach(), 34
    epitools, 63, 301, 304, 364
    foreign, 60, 212
    foreign(), 212
    geoR, 103
    gpclib, 103
    granova, 473
    gstat, 103
    gtools, 125
    Hmisc, 264, 300, 303, 327, 410
    lattice, 89, 495
    lmomco, 242, 284, 293
    load(), 34
    mapdata, 465, 498
    maps, 465, 498
    nlme, 506
    nortest, 225
    pgirmess, 492
    RODBC, 60, 61, 264
    simpleboot, 395
    splancs, 33, 103
    stats4, 293
    SuppDists, 220, 491
    survival, 33
    UsingR, 239
    verification, 533
    xgobi, 36
    xtable, 33
package, 9
packages, 33
paired, 394
pairs(), 35, 88, 272, 281
pairwise.complete.obs, 273
panel, 507
par()
    mar, 275
    mfrow, 79, 82, 139, 168, 179, 194, 215, 393, 403
par(), 79, 82, 139, 168, 179, 194, 215, 233, 269, 275, 393, 403
par.strip.text, 90
parpe3(), 242
paste()
    sep, 18
paste(), 18, 40, 72, 80, 207, 238
pbeta(), 203
pbinom(), 150, 152, 218, 388
pch, 86, 160, 165, 191, 467
pchisq(), 194, 333, 363, 379
pdf, 40
pdouble.exp(), 190
permn(), 123
permutations()
    repeats.allowed, 125
    v, 125
permutations(), 125
persp()
    col, 405
    detailed, 405
    expand, 405
    gray90, 405

phi, 405
shade, 405
theta, 405
ticktype, 405
xlab, 405
ylab, 405
zlab, 405
persp(), 405
perspective(), 35
pexp(), 187, 189
pf(), 382, 471
pgamma(), 201
pgeom(), 143
pgirmess
    kruskalmc(), 492
pgirmess, 492
phi, 405
pKruskalWallis(), 491
PlantGrowth, 84
plnorm(), 199
plot functions
    abline(), 109, 140, 186, 215, 254, 270, 306, 352, 356, 395
    acf(), 277
    arrows(), 189, 206, 297
    axes, 467
    axis(), 186, 238, 287, 297, 467
    barplot(), 78, 478
    boxplot(), 275, 374
    coplot(), 35
    density(), 530
    dotchart(), 86
    expression(), 139, 168, 179, 215, 238, 297, 467
    h(), 168, 184, 191, 194, 201, 213, 233, 289, 306
    hist(), 35, 82, 395
    identify(), 36, 86, 256, 275, 290, 374
    interaction.plot(), 502
    italic(), 215, 238, 297
    legend(), 36
    lines(), 36, 141, 160, 168, 184, 191, 194, 234, 240, 289, 352
    locator(), 36, 194, 533
    map(), 465
    map.axes(), 465
    pairs(), 35, 88, 272
    persp(), 405
    perspective(), 35
    plot(), 35, 88, 109, 139, 160, 179, 186, 191, 194, 206, 259, 356
    plotmath(), 88, 139

points(), 36, 86, 165
polygon(), 36, 186, 297, 354
qqline(), 289, 393
qqnorm(), 289, 393
roc.area(), 533
roc.plot(), 533
text(), 194, 297
update(), 507
xyplot(), 89, 507
plot()
    aov(), 475
    axes, 186, 206, 238, 287, 297
    box, 275
    cex, 160, 467
    cex.main, 297
    col, 191
    expression(), 88, 109
    h, 140, 216
    italic, 109
    l, 109, 179, 186
    lty, 194
    lwd, 140, 179, 260
    main, 160, 186
    mar, 287, 478
    pch, 160, 191, 467
    run-sequence, 289
    s, 150, 168, 292
    stick, 140
    type, 109, 140, 150, 168, 179, 186, 216, 259, 292
    xlab, 109, 160, 179, 467
    xlim, 179
    ylab, 109, 160, 179
    ylim, 109, 160, 179, 259
    ylog, 356
plot(), 7, 35, 88, 109, 139, 160, 179, 186, 191, 194, 206, 259, 356
plotmath(), 88, 139
pnbinom(), 168
pnorm, 224
pnorm(), 209, 215, 218, 235, 266, 269, 331, 342, 345, 371
points()
    cex, 165
    pch, 86, 165
points(), 36, 86, 165, 269
pointsize, 39, 40
pois.approx
    pt, 301
pois.approx(), 301
pois.byar(), 301
pois.daly(), 301

pois.exact(), 301, 304, 364
Poisson.power(), 413
Poisson.sample.size()
    alpha, 414
    k, 414
    power, 414
    rho, 414
Poisson.sample.size(), 414
polygon()
    col, 186, 209, 297, 354
    gray90, 354
    grey90, 186
polygon(), 36, 186, 209, 297, 354
pop.var.names, 78
pos, 55, 194
postscript, 40
power
    bp(), 408
    large sample, 341
    Poisson.power(), 413
    power.normal(), 403
power, 363, 414
power.normal()
    alt, 403
power.normal(), 403
power.t.test(), 361
print(), 6
probs, 266, 306, 330
prompt, 37, 39
prop.test()
    correct, 376
prop.test(), 376, 408
psignrank(), 394
pt, 301
pt(), 197
punif(), 179, 187

q(), 4
qbinom(), 163
qchisq(), 303, 328, 333, 364
qexp(), 188, 189
qf(), 382
qnorm(), 221, 295, 297, 300, 320, 321, 329,
        342, 345, 373
qqline(), 222, 225, 289, 393
qqnorm()
    main, 393
qqnorm(), 222, 225, 289, 393
qt(), 197, 327
quantile
    probs, 266, 306, 330
    qbinom(), 163

qchisq(), 328, 333, 364
qexp(), 188, 189
qf(), 382
qnorm, 221
qnorm(), 300, 320, 321, 329, 373
qt(), 197, 327
qunif(), 188
quantile(), 266, 269, 276, 306, 330
quape3(), 242
qunif(), 188

R, 395
random
    choose(), 171
    rbeta(), 223
    rbinom(), 163, 238, 303
    rchisq(), 194
    rdouble.exp(), 190
    rexp(), 188, 225, 229
    rgamma(), 201
    rmultinom(), 165
    rnbinom(), 168
    rnorm(), 213, 214, 254, 270, 382
    rpois(), 240, 300
    rt(), 197
    runif(), 7, 19, 202, 329
    sample(), 228, 234, 255, 257, 298, 306,
        321, 329, 378, 472, 511, 512
    set.seed(), 165, 168, 189, 201, 213,
        225, 240, 257, 274, 293, 298,
        321, 329, 372, 378, 382, 395, 405,
        472
Random.user(), 10
range(), 261
rbeta(), 223
rbind(), 52, 160, 215, 219, 226, 269, 295,
        373, 377
rbinom(), 163, 238, 303
rchisq(), 194
rdouble.exp(), 190
re.1w(), 495
read.csv()
    sep, 86
read.csv(), 47, 85
read.dta(), 60, 212
read.ftable(), 63
read.table(), 464
read.table()
    header, 48, 59, 76
    sep, 47, 59, 76
read.table(), 47, 54, 59, 76, 268
read.xport(), 477

recompute.power, 363
red, 168, 191, 254, 306
reg, 220, 221
regions, 465
rep(), 160, 389
repeat, 29
repeats.allowed, 125
replace, 304, 306, 329
replications()
    data, 501
replications(), 501
reshape(), 65
residuals.lrm(), 533
rexp(), 188, 225, 229
rgamma()
    scale, 201
rgamma(), 201
rho, 414
rm(), 38
rmultinom(), 165
rnbinom(), 168
rnorm(), 213, 214, 222, 254, 270, 382
roc.area(), 533
roc.plot(), 533
RODBC
    odbcClose(), 61, 62, 264
    odbcConnect(), 61, 264
    sqlFetch(), 61, 264
    sqlQuery(), 62
    sqlTables(), 61, 62, 264
RODBC, 60, 61, 264
round(), 19, 145, 152, 160, 224, 266, 268,
    269, 295, 321, 342, 373, 394
rpois(), 240, 300
Rprofile, 37
rt(), 197
rtest, 61
runif()
    max, 7
    min, 7
runif(), 7, 19, 109, 202, 329

s, 150, 168, 292
sample
    Poisson.sample.size(), 414
    sample.size.normal(), 405
sample()
    replace, 304, 306, 329
    size, 304, 329
sample(), 228, 234, 255, 257, 298, 304, 306,
    321, 329, 378, 472, 511, 512
sample.size.normal(), 405

sapply(), 47, 76
save()
    file, 256, 264
save(), 86, 256, 264, 268
saveg()
    height, 40
    pointsize, 40
    width, 40
saveg(), 40
scale, 201
scan(), 59, 87
score, 47
sd()
    na.rm, 264, 268
sd(), 213, 224, 234, 264, 268, 319, 329
sep, 18, 47, 76, 86
seq()
    length, 179, 186, 189, 206, 345, 405
seq(), 179, 186, 189, 194, 201, 206, 216, 222,
    238, 240, 254, 266, 269, 287, 345,
    405
serif, 186
set.seed(), 109, 165, 168, 189, 191, 194,
    201, 213, 214, 222, 225, 240, 257,
    274, 293, 298, 321, 329, 372, 378,
    382, 395, 405, 472, 498
shade, 405
shapiro.test(), 225, 226
show.signif.stars, 37
signed rank test, 393
simpleboot
    boot.ci(), 395
    two.boot(), 395
simpleboot, 395
sink(), 12, 13
size, 304, 329
sort()
    decreasing, 202, 256
    index.return, 256, 389
sort(), 67, 202, 256, 280, 389
source(), 11, 12, 472
south, 239
spearman, 275
special values
    F, 22

special values (*continued*)
    FALSE, 22
    Inf, 23
    NA, 22
    NaN, 23
    T, 22
    TRUE, 22
splancs, 33, 103
split(), 66, 77, 230, 272, 523
sqlFetch(), 61, 264
sqlQuery(), 62
sqlTables(), 61, 62, 264
sqrt(), 17, 212, 219, 225, 234, 319, 342, 373
stack(), 64, 71
start, 195
state.region, 76
stats4
    mle(), 293
stats4, 293
stick plot, 140
stringsAsFactors, 50, 268
strsplit(), 72
student, 395
substr(), 72, 467
sum(), 15, 18, 47, 145, 146, 152, 160, 191, 300, 390
summary(), 242, 266
SuppDists
    normOrder(), 220
    pKruskalWallis(), 491
SuppDists, 220, 491
survival, 33
Sys.Date(), 50
Sys.time(), 50
system.time(), 231

T, 22
t(), 80, 107
t.test()
    var.equal, 384
t.test(), 302, 384
table(), 48, 67, 159, 378, 499
tablename, 63
tail(), 47
tapply(), 76, 86, 467, 482, 523
    levels(), 71
terror, 159
terror.by.Hamas, 141, 184
test
    ad.test(), 225
    Bonferroni(), 483
    chisq.test(), 379

    cor.test(), 274, 275
    exact, 491
    fisher.test(), 387
    ks.test(), 224
    LSD, 481
    null, 484
    pnorm, 224
    prop.test(), 408
    shapiro.test(), 225, 226
    signed rank, 393
    t.test(), 302, 384
    var.test(), 383, 384
    wilcox.test(), 390, 394
test.scores.rda, 393
text()
    expression(), 206
    labels, 194
    pos, 194
text(), 194, 206, 297
theta, 405
ticktype, 405
tkbinom.power(), 362
tolower(), 72
toupper(), 72, 466
trellis.device()
    color, 495
    height, 495
    width, 495
trellis.device(), 495, 507
trim, 260, 290
trisomy.rda, 518
TRUE, 13, 22, 78
ts(), 33, 88
TukeyHSD(), 484
two.boot()
    M, 395
    R, 395
    student, 395
two.boot(), 395
two.sided, 354
type, 109, 140, 150, 168, 179, 186, 191, 216, 259, 292, 505
typeof(), 15, 26

union(), 69
unique(), 64, 497, 523
unlist(), 66, 123, 226
unsplit(), 66
unstack(), 64
update()
    par.strip.text, 507
    scales, 507

update(), 507
us.income.rda, 259
use, 273
UsingR, 239
utf8ToInt, 68

v, 125, 254, 270, 297, 306, 356, 395
var
    na.rm, 264
var(), 8, 201, 229, 230, 261, 264, 372
var.equal, 384
var.test(), 383, 384
vector()
    is.na(), 18
    length, 13
    mode, 13
vector(), 13, 109, 202, 229, 233, 240
verification, 533
    roc.area(), 533
    roc.plot(), 533
vfont, 186
vfont(), 186

Warning message, 150
wells.info.rda, 272
wells.nucleotides.rda, 272
while, 29
who.by.continents.and.regions, 86
who.ccodes, 57, 78
who.pop.2000, 57, 78
who.pop.var.names, 57

who.fertility.mortality.rda, 264
width, 39, 40, 495
wilcox.test()
    alternative, 391
    paired, 394
wilcox.test(), 390, 394
Wilson, 300
win.metafile, 40
windows(), 34
world, 465
write.ftable(), 63

x11(), 34
XGobi, 36
xlab, 40, 84, 109, 139, 141, 160, 179, 306, 405, 467
xlim, 84, 179, 306
xname, 84
xtable, 33
xyplot()
    cex, 90
    panel, 507
    par.strip.text, 90
xyplot(), 89, 495, 507

ylab, 40, 47, 82, 84, 109, 141, 160, 179, 405
ylim, 84, 109, 160, 179, 259, 306
ylog, 356

zlab, 405

General index

2×2 tables, 536
$C_{n,k}$, 124
F, 197, 302
    definition, 197
F distribution, 197
F-statistic, 469
F-test, 469
N, 270, 299
N_S, 299
$P(Z < z)$, 402
$P^{-1}(p, bm\theta)$ 188
$P_{n,k}$, 122
R, 271
R_S, 275
S, 262, 318
$S[X]$, 212
S^2, 262, 285
$V(\boldsymbol{X})$, 228, 262
W, 392
W statistic, 391
W_S, 392
Z, 192, 195, 207
Δ, 347
Γ, 197, 200
$\Phi(x)$, 207
α, 200, 296, 317, 328
β, 317, 343, 401
$\boldsymbol{\theta}$, 284

$\mathcal{L}$, 161
χ^2, 164, 303, 328, 363, 377
    applications, 195
    expectation, 194
    variance, 194
$\delta(x)$, 143
$\in$, 137, 161, 314
λ, 141, 154, 184, 188, 189, 205, 225, 239, 291, 300
$\log$, 161
$\mathbb{R}$, 137, 284
$\mathbb{R}^2$, 271
$\mathbb{Z}_+$, 137, 163, 227
$\mathbb{Z}_{0+}$, 137, 143, 161
μ, 12, 192, 199, 205, 206, 468
μ_l, 241
μ_p, 238
$\notin$, 143, 314, 322
ν, 193
$\overline{A}$, 314
$\overline{X}$, 205, 228
$\phi(x)$, 207
π, 139, 205, 236, 290, 323
ρ, 272, 385
σ, 12, 200, 206
σ^2, 192, 199, 262, 468
$\sim$, 221
θ, 161

ε_{ij}, 468
$\varnothing$, 98
$\widehat{P}(Y=1,|X)$ 521
$\widehat{\alpha}$, 200
$\widehat{\gamma}_1$, 192
$\widehat{\gamma}_2$, 192
$\widehat{\lambda}$, 252, 328
$\widehat{\mu}$, 192, 199, 212
$\widehat{\pi}$, 303, 377
$\widehat{\sigma}$, 200, 212
$\widehat{\sigma}^2$, 192, 199, 262
$\widehat{}$, 161, 521
^{222}Rn, 290
e^{β_0}, 514
l, 205, 239
n_S, 299
p, 205
p-value, 224, 330, 371, 379, 384, 516
p-values, 226, 331
t, 195, 301, 326, 332
    applications, 197
    expectation, 197
    variance, 197
t-test, 469
x_H, 320, 322, 329, 330
x_L, 322, 329, 330
*, 20
*, 468
+, 20
-, 20
., 56
/, 20
H_0, 314
LaTeX table, 264
^, 20
3D, 405

A, 125
aborigines, 463
accused, 316
action, 314
action level, 290
adenine, 125
adjust, 323
Afghanistan, 542
Africa, 276, 374
age, 81, 298
age at sentencing, 319, 402
age class, 517
age-adjusted rate, 291
AIDS, 316
air quality, 464

Al Anbar province, 548
Al Aqsa Martyrs Brigades, 552
alleles, 164
alternative hypothesis, 314, 320
amino acids, 125
An Nasiriyah, 548
Analysis of residuals, 533
Anderson-Darling, test, 223
anesthetics, 315
animals, 214
ANOVA, 463
    F-statistic, 469
    F-test, 469
    α_i, 469
    μ, 468
    assumption, 464, 468, 469
    balanced design, 464, 471
    between-subject, 492
    Bonferroni, 477
    contrasts, 487
    df, 477
    diagnostics, 480
    fixed-effect, 463
    fixed-effects, 488
    group, 469
    group mean, 469
    LSD, 477
    main effect, 496
    nested, 496
    non-parametric, 488
    notation, 467
    null hypothesis, 469
    one-way, 463
    one-way random-effects, 492
    Paired comparisons, 477
    post-hoc, 475
    random-effects, 469
    response variable, 469
    Sum of Squares, 470
    table, 473
    two-way, 463, 495
    ubalanced design, 471
    unbalanced design, 464
    Within MS, 470
    Within SS, 470
    within-subject, 492
Anscombe, 1948, 167
approximation, 239
arbitrary densities, 304
arbitrary density, 313, 329
arbitrary parameter, 329
arbitrary parameters, 304

area, density, 185
argument, 4, 7, 31
arithmetic operators, 20
armed conflicts, 542
Armenia, 78
armies, 542
array()
    attribute, 51
    construction, 52
    dimension, 51
arrays, 51
arrival rate, 239
assumption, 313, 464
astronomy, 153
asymptotic, 286, 301
asymptotic properties, 286
Atlantic sharpnose, 350
Australian, 463
Austria, 78
autocorrelation, 543, 554
autocorrelation function, 277

Ba qubah, 548
Baghdad, 177, 548
Balad, 548
balanced design, 464, 471
bar plot, 77
bar plots, 477
barrier, 553
basal metabolic rate, 304
Basra, 548
Bayes estimators, 284
Bayesian, 111
beak length, 369
Becker et al, 1988, 3
Becquerel, 290
Belarus, 374
Belgium, 500
bell curve, 12
Bell Laboratories, 3
belligerents, 550
Berger, 85, 202
Berlin, 464
Bernoulli, 138
    distribution, 142
    experiment, 147
    trial, 106, 142
Bernoulli experiment, 108, 299
best, 286
best estimate, 369
best fit, 556
best-fit line, 220

beta, 201
    applications, 202
    function, 196, 198, 202
    function, incomplete, 198
    function, regularized, 202
    gamma, 200
    mean, 202
    ratios, 202
    regularized function, 198
    variance, 202
between-subject, 492
bias, 284
bias-correct, 243
bigot, 325
bill lengths, 220
binary outcome, 513
binomial, 205, 236, 288, 290, 299–302, 313, 323, 327, 380
    coefficient, 148, 171
    density, 149
    distribution, 149
    expected value, 151
    mean, 163
    MLE, 163
    Poisson approximation, 155
    variance, 151, 163
biology, 153
birch, 492
bird tagging, 317
birds, 111
birth, 167, 372
birth rate, 86, 239
birth weight, 537
bivariate, 55
Blackchin shiner, 512
blacks, 321, 372, 402
Bliss and Fisher, 1953, 167
blond, 299
blood pressure, 373, 495
BMR, 304
body mass, 304
Bonferroni, 477, 482
bootstrap, 235, 242, 304, 313, 326, 329, 365, 394
    confidence interval, 304
    confidence intervals, 329
    implementation, 304
    repetitions, 305
Box and Jenkins, 76, 276
box plot, 275
box plots, 251, 275
braces, 39

brute force, 552
Buckland et al. (2001), 81
burden of proof, 314
Bush, 46
Byar's formula, 364

C, 125
Calliope, 120
calories, 265
cancer, 356
cancer rate, 239
cancer, crude rate, 292
canopy shading, 511
capital punishment, 63, 298, 321, 322, 372, 378, 402
cardinality, 548
casualties, 233, 252, 277
catch, 328
catch rate, 332
categorical, 77, 195
categorical data, 45
cautious action, 316
CDC, 477, 488
central limit theorem, 232, 241, 283, 290, 295, 299, 326, 342, 370
central moment, 241
central processing unit, 231
Chakravarti, 67, 223
Chambers and Hastie, 1992, 3
Chambers, 1983, 251
Chambers, 1992, 56
Chambers, 1998, 3
chance experiment, 106
character, 15
characters, 17
Chebyshev's rule, 251, 267
Chechen, 542
chi-square, 193
children, 553
children mortality, 262
chromosomes, 517
chronology of deaths, 550
chronology of the events, 554
Cleveland 1985, 85
climate change, 87
clumps, 167
clutch size, 327
CO, 502
CO_2, 87
Coalition forces, 541, 543
Cod, 278
codons, 125

coefficient of variation, 265
coercion, 27
color, 80
column bind, 52
column effect, 501
combination, 124
combinations, 123
comment, 11
complement, 314
compound event, 177
conditional probability, 115, 116
conditional proportions, 550
confidence coefficient, 294, 296, 323, 328, 376
confidence interval, 243, 294, 296, 300, 395
    t, 301
    arbitrary densities, 304
    arbitrary parameters, 304
    asymptotic, 301
    binconf(), 327
    bootstrap, 304
    exact, 302
    large sample, 295
    normal, 296
    Poisson, 300
    proportions, 299
    ratio, 298
    small sample, 301
    two samples, 373
    two samples, intensities, 380
    two samples, proportions, 375
    variance, 382
    Wilson method, 302
confidence intervals, 242, 326, 329, 507, 522
consistency, 286
construct, 229
contingency tables, 195, 386, 538
    two samples, intensities, 379
    two samples, proportions, 377
continuation line, 11
continuity correction, 323, 352, 375
continuous, 49
contradictory hypotheses, 314
control, 394, 410
convicts, 319
Cornwall, 291
correlation, 251
correlation coefficient, 269
correlation coefficient, properties, 273
count, 161
countable, 137
counting, 153
counting process, 545

counts, 239, 300, 378
court, 314
court of law, 316
covariance, 521
covariate patterns, 525
covariates, 511
coverage probability, 294
CPU, 231
CPUE, 357
crime, 288
critical value, 320, 322
cross classification, 537
crude rate, 291
cumsum(), 472
customize, 36
cut-off probability, 529
CV, 265
cycles, 552, 554
cytosine, 125

daily maximum temperature, 277
daily visits, 314
Dalgaard, 2002, 3
darts, 315
data
    bivariate, 55
    categorical, 45, 77
    CDC, 477
    factor, 47
    multivariate, 55
    numerical, 49, 77
    ordinal, 48
    SO_2, 469
    tables, 55
    univariate, 55
    WHO, 57
data frame, 54, 264
data frames, 26
data import, 85
data subset, 18
database, 55, 264
date of birth, 298
date-arithmetic, 542
dates, 50
DBF, 60
DBMS, 58
death, 155, 167, 372
death penalty, 319, 369, 375
death rate, 86, 541
death sentence, 514
deaths, 304
deaths per week, 547

decay, 290
decision making, 151, 352
deduction, 251
default values, 10
definition
    F, 197
    χ^2, 193
    p-values, 331
    H_0, 314
    age-adjusted rate, 291
    beta density, 201
    beta distribution, 202
    binomial density, 149
    binomial distribution, 149
    continuous density, 180
    continuous distribution, 177
    crude rate, 291
    discrete density, 138
    discrete distribution, 141
    double exponential density, 189
    double exponential distribution, 190
    Euclidean product, 271
    exponential density, 181
    exponential distribution, 180
    gamma, 200
    geometric density, 140
    hypergeometric, 169
    independent sample, 227
    kurtosis, 192
    likelihood function, 161
    logistic transformation, 513
    logit transformation, 513
    lognormal, 198
    mean squared error, 285
    MLE, 162, 284
    mode, 260
    multinomial, 164
    negative binomial, 166
    normal density, 191, 205
    normal distribution, 206
    null hypothesis, 314
    odds ratio, 512
    One-way ANOVA, 468
    Pearson's population ρ, 272
    Pearson's sample R, 271
    Poisson, 154
    power, 341, 344
    probability, 111
    quantiles, 221
    random variable, 128
    range, 261
    sample, 227

588 General index

definition (*continued*)
    sample variance, 261
    sampling density, 228
    skewness, 192
    standard error, 233
    standard normal, 192
    Standardized Pearson residual, 534
    statistic, 228
    Type I error, 316
    Type II error, 316
    uniform density, 180
    uniform distribution, 179
    variance, 146
degrees of freedom, 164, 193, 301, 302, 326, 328
delays, 561
demographics, 477
Dennett, 95, 97
densities, 283
density, 138, 283
    area, 185
    continuous, 180
    discrete, 138
    empirical, 160
    family, 284
    sampling, 283
density, continuous
    F, 197, 302
    χ^2, 164, 303, 328, 363, 377
    t, 195, 301, 326
    beta, 201
    chi-square, 193
    double exponential, 189
    expected value, 183
    exponential, 141, 181, 225, 229, 260, 292
    gamma, 200
    Laplace, 189
    lognormal, 198
    normal, 12, 186, 191, 205
    Pearson type III, 242
    properties, 182
    standard deviation, 185
    uniform, 180
    variance, 185
density, discrete
    binomial, 149, 205, 288, 290, 300, 327
    empirical, 140
    expected value, 145
    geometric, 138, 140
    histogram, 140
    hypergeometric, 169
    multinomial, 163
    negative binomial, 166, 252, 541
    Poisson, 154, 205, 218, 239, 240, 252, 288, 291, 328
    properties, 144
    standard deviation, 147
    variance, 146
density, sampling
    t, 332
    construct, 229
    intensities, 239
    intensity, 239
    mean, 232, 234, 301
    proportion, 235
    statistic, 229
    variance, 229, 241, 242, 292
Department of Defense, 233, 252, 543
Department of Justice, 63
depth, 531
detectable difference, 341, 342, 349, 404
detectable distance, 356
deviance χ^2 statistic, 527
Deviance residual, 527
deviance residual, 533
df, 194, 477
diagnostics, 480, 511, 534
dimension, 51
dimension names, 160
dimension vector, 51
diploid, 164
Dirac delta, 143
disasters, 542
discrete, 49
    density, 138
discrimination, 324
disjoint, 112, 113
disjoint events, 105
distance, 81
distribution, 188
    continuous, 177
    discrete, 141
    empirical, 191
    estimated, 191
    inverse, 187, 188
    properties, 144
distribution, continuous
    F, 197
    t, 195
    beta, 201
    chi-square, 193
    double exponential, 189
    exponential, 180
    gamma, 200, 364

lognormal, 198
normal, 191, 266
pf(), 471
properties, 181
uniform, 179
distribution, discrete
Bernoulli, 142
binomial, 149
construction, 142
Poisson, 154
DNA, 125
DOB, 298
domain, 128
dot charts, 275
dot-product, 145
double exponential, 189
application, 190
definition, 189
parameter estimation, 190
standard, 189
Down's syndrome, 517
drinking water, 272
drug, 316
dry weight, 275
DSN, 61
duality, 323

EDA, 251, 272, 275
box plots, 251
Chebyshev's rule, 251
correlation, 251
empirical rule, 251
graphical methods, 252
histograms, 251
lattice plots, 252
mean, 253
Q-Q plots, 251
run-sequence plots, 252
scatter plots, 252
education, 242, 514
efficiency, 286
Efron and Tibshirani, 93, 235
elections, 46
element, 14, 18
elementary event, 105
elementary events, 112
elements, 98
elk, 190
emergency room, 239, 314, 363
emergency service, 542
emergency services, 547
emission, 154

empirical density, 140, 160, 239, 266, 544
empirical distribution, 191
empirical probabilities, 523
empirical rule, 251, 268
empirical sampling density, 304
empty set, 98
encounter rate, 380
engineering, 153
England, 121
enrollment, 288
enumerated types, 45
EPA, 347
epidemiology, 304
equidistant, 84
erectile dysfunction, 342
error
Type I, 316
Type II, 316
error term, 501
escape characters, 84
escape key, 258
estimate, 283
interval, 294
estimated distribution, 191
estimator
asymptotic properties, 286
best, 286
consistency, 286
efficiency, 286
precision, 285
unbiased, 285, 293
variance, 286
ethnicity, 463, 478, 482
EU, 464, 496
Euclidean plane, 271
Euclidean product, 271
Europe, 275, 374
European Commission, 353
evaluate, 88
Evans et al., 2000, 137
Evans, 2000, 104
event, 104, 140
compound, 177
dependent, 116
elementary, 105, 107
independent, 116
simple, 138
space, 138
event space, 104
events, 227
dependent, 120
independent, 120, 154

events (*continued*)
    rare, 154
evidence, 324
evolution, 97
exact, 302
exact method, 327, 363
exact methods, 242
Excel, 85
exceptional residuals, 533
excessive force, 542
exclosures, 492
execution time, 230
exhaustive hypotheses, 314
expected value, 183
    binomial, 151
    continuous density, 183
    discrete rv, 145
    exponential, 183
    geometric, 146
    Poisson, 156
    uniform, 183
experiment, 140
Exploratory Data Analysis, 251
explosion, 553
exponential, 200, 225, 229, 260, 292
    expected value, 183
    gamma, 200
    random, 188
    variance, 185
exponential density, 141, 181
    histogram, 184
exponential distribution, 180
exposure studies, 511
expression, 11
`expression()`, 467
extended real line, 128
extract elements, 20
extreme values, 257

factor, 45, 47, 89, 159, 230, 272
    ordered, 463
factorial, 122
factors, 463
    ordered, 48
faculty, 255, 260, 288
Fallujah, 548
false positive, 114
family, 284
feeding, 182, 186
females, 313
fetus, 517
finite variance, 232

First Chechen War, 542
First Intifada, 542
fish, 235, 328, 347, 511
fish meals, 348
Fisher's exact test, 386
fisheries, 357
fit, 161
fixed-effects, 463, 488
floor, 143
Florida, 46, 347
flu, 407
fluctuations, 556
formula, 89, 275
formula, $\sim$, 221
Fox, 2002, 3
France, 121, 507
frequentist, 111
function, 4, 30
    argument, 7
    argument order, 32
    code, 39
    optional argument, 31
    required argument, 31

G, 125
galaxies, 153
Gallup Poll, 375
gamma, 200, 364
    α, 200
    σ, 200
    $\hat{\alpha}$, 200
    $\hat{\sigma}$, 200
    applications, 200
    beta density, 200
    expectation, 200
    exponential, 200
    lifetime, 200
    scale parameter, 200
    shape parameter, 200
    variance, 200
gamma density, 193
gamma function, 167, 193, 200
Gelman, 95, 202
gender, 81, 496
generalized likelihood ratio, 516, 524, 531
generate levels, 230
genes, 123
Genmany, 465
genome, 125
genotypes, 164
geographic location, 543
geometric

density, 138
  expected value, 146
  variance, 147
geometric density, 138, 140
Germany, 507
Gini Coefficient, 260
global minimum, 293
goodness of fit, 525
goodness-of-fit, 195
Gore, 46
graduation, 76
grand jury, 324
graphical methods, 252
graphics device, 34, 139
graphics driver, 34
grasses, 313
Green Bay Packers, 119
ground water, 272
group, 469
group adjusted, 528
growth rate, 267, 276
guanine, 125
guilty, 314

Ha'aretz, 559
habitat, 164, 511
habitat type, 512
half-life, 290
Hamas, 183, 292, 552, 561
Hardy-Weinberg, 164
heart attack, 373
heavy tails, 289
height, 255
Help, Console, 8
herbivores, 492
hierarchical, 501
hinge, 276
histogram, 240
  exponential density, 184
histogram density, 140
histograms, 81, 251
homeless, 259
hospital, 314
household income, 478, 483
HSD, 483
Html help, 5
hummingbirds, 120
hypergeometric, 169
  applications, 170
  mean, 170
  variance, 170
hypertension, 373

hypothesis, 313
hypothesis testing, 314
  arbitrary density, small sample, 329
  arbitrary parameter, small sample, 329
  binomial, small sample, 327
  critical value, 322
  intensities, 313, 324
  intensities, small sample, 328
  large sample, 313, 318
  lower-tailed, 326, 328
  mean, 318
  mean, small sample, 326
  means, 313
  means two small samples, 384
  Poisson, 332
  Poisson, small sample, 328
  proportions, large sample, 323
  proportions, small sample, 327
  ratios, 313
  small sample, 313, 326
  two samples, 371
  two samples, intensities, 379
  two samples, mean, 370
  two samples, proportions, 375
  two-tailed, 326, 328
  upper-tailed, 326
  variance, 382

identically distributed, 232
IIEF, 342
immigration, 167
implementation, 304
import, 212, 264
income, 463, 482
incomplete beta, 202
incomplete beta function, 198
incomplete gamma function, 193
independent events, 116
independent sample, 227
independent variable, 463
index vector, 18, 51, 160, 264
induction, 251
inference, 521
inferential statistics, 313
infimum, 355
infinity, 23
influence, 533
infrastructure, 552
initial guess, 293
initial guesse, 293
injuries, 541
injuries per month, 547

inmates, 321, 378
innocent, 314, 316
insect-eating, 304
Insectivora, 304
installation directory, 37
integers, 137
integration, 185
intensities, 239, 240, 313, 324, 328
intensity, 153, 239, 291, 303, 318, 356
interaction effect, 501
interquartile range, 221, 265
interval, 159
interval estimate, 283, 294
introduction to R, 5
inverse distribution, 187
inverse logit, 513
invest, 314
Iowa, 327
IQ, 325
IQR, 265, 267
Iraq, 233, 252, 541
Irish Sea, 278
Israel, 353, 541, 554
Israelis, 140, 552
italic, 139
Italy, 465, 502

Jaynes, 2003, 111
Jerusalem Post, 559
Johnson et al., 1995, 137
Johnson, 94, 192
joint probability, 164
judge, 324
Julian, 233
Julian day, 67, 141, 542
jurors, 112

K-S test, 223
Keeling 2003, 87
Kolmogoroff, 1956, 111
Kolmogorov-Smirnov, 223
Kotz et al., 2000, 137
Krebs, 1989, 167
Kruger, 214
Kruskal-Wallis, 488
Kruskal-Wallis multiple comparisons, 491
Kurds, 548
kurtosis, 192

L-moments, 284
Lader and Meltzer, 2002, 151, 217
lag, 277
lag plots, 275, 276

lags, 543
Lake of the Woods, 328, 332
Laplace, 189
large sample, 295, 313, 318, 324
large samples, 234
Latitude, 465
lattice plots, 88, 252
law of large numbers, 111
lead, 373
lead in gasoline, 373
level, 89
levels, 47
leverage, 533
lifetime, 200
light gray, 80
likelihood, 519
likelihood function, 161, 285, 513, 520
limiting density, 232
line, best-fit, 220
linear contrasts, 484
linear model, 221, 514
linear regression, 463
lion, 388
lions, 163
list, 52, 226, 376
    length, 53
lists, 26
litter size, 313
Liverpool Bay, 278
lm(), 475
location, 192, 206, 252
location invariance, 370
log likelihood, 285, 520
log odds, 516, 522
log-likelihood, 293
logical, 13
logical value, 22
logical vector, 78, 372
logistic regression, 522
logistic transformation, 513, 522
logit transformation, 513
lognormal, 198
    μ, 199
    σ^2, 199
    $\widehat{\mu}$, 199
    $\widehat{\sigma}^2$, 199
    applications, 199
    standard, 199
London, 291
London Times, 553
longitude, 465
loop, 240

Lost Angeles Times, 553
lower quartile, 265
lower-tailed, 318, 326, 328, 346, 359, 371
LSD, 477
lung cancer, 290
lung damage, 290

Ma'ariv, 559
Madrid, 464
main effect, 496
male, 236
mammalian, 306
management, 314
Mann-Whitney U, 389
MANOVA, 495
mapping, 128
maps, 464
marital status, 496
mark recapture, 235
maternal age, 517
matrices, 51
matrix construction, 52
maximize, 520
maximum likelihood, 162, 284
McLaughlin99, 1999, 137
mean, 232, 234, 253, 254, 301, 318, 326
Mean squared error, 285
mean
    income, 259
    MLE, 254
    population, 254
    sample, 254
    trimmed, 260
means, 313
median, 221, 258, 276, 365, 388
    income, 259
    U.S. income, 259
medical facilities, 552
mercury, 347
metabolic rate, 304
Mexican American, 481, 483
MFA, 2004, 140
mice, 394
militant organizations, 542
military, 233
military successes, 550
minimize, 293
minimum detectable difference, 401
Minitab, 60
Minneapolis, 327
Minnesota, 328, 511
Minnesota Vikings, 119

miscarriage, 517
miscarriages, 517
missing values, 264
mist nets, 107
mixed-effects, 507
MLE, 161, 162, 195, 284, 293, 519, 524
    binomial, 163
    multinomial, 164
    numerical estimate, 293
    Poisson, 163
MLE estimator, 162
MLE Poisson, 162
MO, 542
mode, 15, 260
    character, 15
    numeric, 15
moments, 242, 284
monotonic, 161
Monte Carlo simulation, 188
mortality, 262, 264
Mosul, 548
Mozart, 147
MSE, 285
Multi-Racial, 481
multinomial, 163
    applications, 164
    MLE, 164
multiplicative rule, 117
multivariate, 55, 163
murder, 375
murder rate, 240
murder rates, 239
mutation, 412
mutations, 154, 156
mutually exclusive, 314
MySQL, 61

NA, 372
named arguments, 10
Nashville warbler, 81
NATO, 542
negative binomial, 166, 252, 541, 545, 552
    applications, 167
    mean, 166
    variance, 166
negative relationship, 271
nested, 496
nested hypothesis, 524
Nested models, 524
neuron, 202
neurons, 154

neuroscience, 154
neurotoxin, 373
New York, 56
New York Times, 553
News media, 548
no relationship, 271
Non-Hispanic Black, 481
Non-Hispanic white, 483
noncountable, 137
nonparametric, 388, 392, 488
nonparametric statistics, 195
normal, 205, 266, 296, 301, 313
    μ, 192
    σ^2, 192
    $\widehat{\gamma}_1$, 192
    $\widehat{\gamma}_2$, 192
    $\widehat{\mu}$, 192
    $\widehat{\sigma}^2$, 192
    approximation, 239
    approximation to the binomial, 299
    approximation, binomial, 215
    approximation, discrete, 214
    approximation, Poisson, 218
    area, 186
    confidence interval, 296
    density, 191
    distribution, 191
    fit, 214
    kurtosis, 192
    location, 206
    Poisson, approximation, 300
    scale, 206
    scores, 220
    skewness, 192
    standard, 206, 207, 220, 301
    testing, 220
normal approximation, 359
normal curve, 12
normalization, 58
North Sea, 278
Northern Europe, 264
Not a Number, 23
Not Available, 22
nuclear power, 356
nucleotide, 125
null hypothesis, 314
null set, 98
numeric, 15
numerical, 49, 77
numerical optimization, 162
numerical techniques, 521

object, 37

Octave, 60
ODBC, 60
ODBC driver, 61
odds ratio, 512
one-way, 463
One-way ANOVA, 468
one-way random-effects, 492
operator
    *, 20
    +, 20
    -, 20
    /, 20
    ^, 20
order, 306
ordered, 48, 463
ordinal, 48
ornithology, 117
Other Hispanic, 481
Otter Tail, 512, 529
outcome, 104
outlier, 374
outliers, 257, 274, 284
over dispersion, 545, 558
overdispersion, 167
ozone, 56

packages, 33
paired, 384, 393
paired comparison, 385
Paired comparisons, 477
paired design, 385
paired interactions, 88
paired signed rank, 388
pairwise, 369
Palestinian, 541, 552
Palestinian Liberation Organization, 561
Palestinians, 236
Papoulis, 1965, 104
parameter, 283
    population, 283
parameter space, 293
parameters, 161
Park, 324
particles, 154
patients, 239, 316
PD, 542
peace, 353
Pearson χ^2 statistic, 526
Pearson residual, 526, 533
Pearson type III, 242
Pearson's population ρ, 272
Pearson's sample R, 271

periodicities, 543
permutation, 122
permutations, 124
physics, 153
Plaice, 278
plaintiff, 324
plant communities, 492
plant growth, 83
plants, 300
plausible, 313
plot margins, 275
plots, 300
plotting
    tick labels, 467
    tick marks, 467
PMO, 552
point estimate, 325
Poisson, 154, 205, 218, 239, 240, 252, 288, 291, 300, 301, 303, 313, 324, 328, 332, 363, 380, 411, 545, 558
    confidence interval, 300
    expected value, 156
    mean, 163
    MLE, 162, 163
    mutations, 156
    normal approximation, 324
    power, 356
    sample size, 358
    variance, 157, 163
Poisson, two samples, 411
pollutant, 369
pollution, 469, 496
polygon, 186, 297
polygon
    vertex, 186
pooled comparison, 385
pooled variance, 380, 477
poor, 259
population, 155, 205, 283
    parameter, 283
population covariance, 271
population mean, 254, 318
population variance, 212, 261
positive relationship, 271
possibility space, 104
post-hoc, 475
postgresql, 61
power, 341, 344, 359, 401
    detectable difference, 342
    lower-tailed, 346, 359
    mean, for, 342
    Poisson, 356

Poisson, two samples, 411
profile, 402
proportions, 352, 407
small sample, intensities, 363
small sample, mean, 359
small sample, proportions, 361
two means, 401
two-tailed, 346, 360
upper-tailed, 346, 360
power profile, 344
power set, 98
precision, 285
predictions, 522, 556
premature death, 373
presence, 511
prey, 163
probability, 97, 111, 217
    addition rule, 119
    axioms, 111
    Bayesian, 113
    conditional, 113, 115, 120
    coverage, 294
    definition, 108
    joint, 115
    left-tail, 217
    multiplication rule, 120
    properties, 111
    rejection, 320
    right-tail, 217
probability densities, 542
probability plots, 221
profile, 402
projects, 37
properties
    continuous densities, 182
    continuous distributions, 181
proportion, 235, 318
proportions, 299, 323, 327, 352, 407
prosecutor, 314
prostate cancer, 342
prostatectomy, 342
protein, 125
pseudo-random, 109
public opinion, 318, 353

Q-Q plot, 221, 503
Q-Q plots, 251
quantile, 187, 189
quantile-quantile plot, 221
quantiles, 221, 242, 297, 329
quartile, 276

Ra224, 272

Ra226, 272
Ra228, 272
race, 321
radioactive, 290
radionucleotides, 272
Radon-222, 290
Ramadi, 548
random
    exponential, 188
random errors, 468
random events, 553
random mating, 122, 124
random numbers, 7, 109
random sample, 257
random variables, 127
random-effects, 469
range, 128, 261, 404
rank, 274
rank sum, 388
rank-sum, 488
rate, 154
rates, 239
ratio, 236, 298, 382
ratios, 202, 313
real line, 127
real number, 128
real numbers, 137
reasonable doubt, 314
recycle, 82
reelection, 318
region, 264
regions, 465
rejection probability, 320
relationship
    negative, 271
    no, 271
    positive, 271
relative frequency, 111
repeat measurements, 492
repetitions, 305
Residual analysis, 552
residuals, 543, 545
response variable, 469, 511
reward, 317
rich, 259
risk, 538
RNA, 125
RNA Codon Table, 125
robust, 284
ROC curve, 529
Rodentia, 304
rodents, 304

Roma, 464
Ross, 1993, 137
row bind, 52
run sequences, 543
run-sequence, 289
run-sequence plots, 252
rural, 505
Russia, 542
rv, 127

S, 60
sage sparrow, 301
Samarra, 548
sample, 205, 227, 283
    bias, 284
    covariance, 271
    density, 229
    independent, 227
    intensity, 205
    large, 318
    mean, 192, 205
    proportion, 205
    size, 341, 349, 401
    size profile, 352
    size, intensity, 356
    size, lower-tailed, 349
    size, Poisson, 358
    size, proportion, 409
    size, proportions, 355
    size, two means, 404
    size, two samples Poisson, 414
    size, two-tailed, 350
    size, upper-tailed, 349
    small, 301, 326, 359
    small binomial, 301, 302
    small normal, 301
    small Poisson, 301, 303
    small size for intensities, 364
    small size for proportions, 362
    small unknown density, 301
    small, power, 359
    small, size, 360
    space, 104
    standard, 212
    standard deviation, 318
    statistic, 313
    two small, intensities, 387
    two small, proportions, 386
    two small, unknown density, 388
    two, bootstrap, 394
    two, small, 380
    two, variance estimate, 380
    variance, 212, 261, 285

sample size, 360
    mean, 342
sampling, 118, 283
    densities, 205, 283
    density, 197, 228, 234, 283
    rule of thumb, 119
    space, 205
    with replacement, 118, 121, 304
    without replacement, 118
sampling density, 301, 318, 329, 345, 375
sampling space, 314
SAS, 60, 477
Saudi Arabia, 121
scale, 192, 206, 252
scale parameter, 200
scatter plot, 35, 86
scatter plots, 252
Scheffé, 487
scores, 220
script, 11, 231
SE, 318, 373
Second Chechen War, 542, 552
Second Intifada, 140, 541, 552
security guards, 555
seed bank, 115
semi-autonomy, 548
semicolon, 11
sensitivity, 529
sentencing, 298, 372
sequence, 6, 160
set, 98
    countable, 137
    noncountable, 137
sets
    associativity, 99
    commutativity, 99
    complements, 102
    difference, 102
    disjoint, 101
    distribution, 101
    equality, 99
    intersection, 100
    mutually exclusive, 101
    sum, 103
    transitivity, 98
    union, 99
shape parameter, 200
Shapiro, 65, 223
Shapiro-Wilk, 225
Shapiro-Wilk test, 223
sharks, 347
sheep, 167

SI, 552
side effects, 316
signal processing, 529
signed rank, 392
significance, 323
significance level, 317, 328, 341
significance, common sense, 325
Sildenafil citrate, 342
simple events, 138
simple random sample, 228
simulated annealing, 293
simulation, 188
    Monte Carlo, 188
size, 341, 349, 401
skewness, 192
skin color, 372, 514
small, 301, 326
small sample, 301, 313, 326
small samples, 359
smoking, 151, 217
smoking mother, 538
smoothing argument, 530
SO_2, 465, 466, 469, 496
song birds, 120, 182, 186
sort, 389
South Africa, 214
Southern Blight, 277
Southern US, 239
space, 98
Spain, 465
sparrows, 120
spatial, 153, 547
Spearman's rank correlation coefficient, 275
special values, 20, 22
species, 380
specificity, 529
Spotfin shiner, 529
SPSS, 60
SQL, 62
standard deviation, 147, 185, 208, 272, 318
standard deviations, 262
standard error, 233, 235, 241, 286, 321, 342, 371, 375, 521
standard lognormal, 199
standard normal, 12, 206, 207, 220, 301, 402
standard population, 291
standard uniform, 179
standardized deviance residuals, 534
Standardized Pearson residual, 534
stars, 154
Stata, 60, 212
state of nature, 316

station code, 497
Statistic, 284
statistic, 228, 229, 242, 304, 329, 369
Stephens, 74, 223
strategy, 561
Student t, 195
students, 299
subset, 98
subset, data, 18
subset, extraction, 18
subspecies, 301
suburban, 505
suicide bomber, 177
Sum of Squares, 470
`summary()`, 480
surge, 556
survey, 318, 353
Swain v. Alabama, 324
Swiss, 242
symmetric, 323
synchronization, 550
synchronizations, 548
synchrony, 561
Systat, 60

tables, 55
tag return, 236
Taji, 548
tall, 227
`tapply()`, 470
temporal, 541
terrorists, 121
test
    F-test, 469
    t vs. rank sum, 392
    t-test, 469
    Anderson-Darling, 223
    arbitrary parameter, small sample, 329
    binomial exact, 388
    Bonferroni, 477, 482
    conservative, 482
    exact method, 327
    Fisher's exact, 386, 388
    generalized likelihood ratio, 516
    HSD, 483
    intensities, large sample, 324
    intensities, small sample, 328
    K-S, 223
    Kruskal-Wallis, 488
    lower-tailed, 318, 326
    lower-tailed, two samples, 371
    LSD, 477, 481
    Mann-Whitney U, 389
    multiple pairs, 483
    nonparametric, 388, 392
    paired signed rank, 388
    paired, for means, 384
    paired, signed rank, 392
    proportions, large sample, 324
    proportions, small sample, 327
    rank-sum, 388, 488
    Scheffé, 487
    Shapiro-Wilk, 223, 225
    two samples, two-tailed, 371
    two samples, upper-tailed, 371
    two-tailed, 321, 326
    upper-tailed, 320, 326
    Wilcoxon rank sum, 389
    Wilson method, 327
test statistic, 370
tests of normality, 223
The R team, 3
theoretical density, 239
tick labels, 467
tick marks, 467
ticks, 167
time between attacks, 184
Time Series, 543
time series, 88, 252, 276
time-lags, 554
toxic, 511
toxicity, 373
transformations, 226
transpose, 80
trauma-treatment, 542
treatment, 394
tree diagram, 106
trial, 105
trim, 374
trimmed mean, 260, 289
trisomic fetuses, 517
trisomy, 517
Tukey, 77, 242, 251
tumor, 329
two samples, 371, 373
two-tailed, 321, 326, 328, 346, 360, 371
two-way, 463, 495
Type I, 316
Type I error, 401
type I error, 323, 341
Type II , 316
Type II error, 401
type II error, 341, 408

U, 125
U.S., 541

ubalanced design, 471
UK, 151, 217, 290
unbalanced design, 464
Unbiased, 285
unbiased, 293
uncertainty, 316
unform, standard, 179
uniform
    density, 180
    distribution, 179
    expected value, 183
    variance, 185
union, 314
unit effort, 357
units, 147, 262
univariate, 55
Universe, 154
universe, 98
unleaded gasoline, 373
unnamed argument, 9, 297
unpaired, 384
upper case, 466
upper hinge, 276
upper quartile, 265
upper whisker, 276
upper-tailed, 320, 326, 346, 360, 371
uracil, 125
uranium-238, 290
urban, 505
urologist, 342
UTF-8, 68

Validation, 533, 536
variance, 146, 185, 229, 241, 242, 285, 286, 292
    binomial, 151
    discrete rv, 146
    geometric, 147
    Poisson, 157
    pooled, 477
    population, 261
    ratio, 382
    sample, 261
    uniform, 185
vector, 6, 233
    element, 18
    index, 18

vector index, 252
Velleman and Hoaglin, 1981, 251
Venables and Ripley, 2003, 3
Venables et al, 2003, 4
Venables, 1994, 3
Venn diagram, 115
Venn diagrams, 98
Viagra, 342
visitors, 324
volume, 239
vote, 235

Wald-Z, 523
wall, 553
walleye, 389
war, 318
war on terrorism, 542
warning message, 294
Washington Post, 553
water hole, 214
waterfowl, 317
weighted average, 348, 521
West Bank, 236
Western Africa, 264
whisker, 276
whites, 372, 402
Whiting, 278
WHO, 57, 262
WI, 542
Wilcoxon rank sum, 389
Wildlife, 154
Wilson method, 302, 327
wing-chord, 301
with replacement, 304
Within MS, 470
Within SS, 470
within-subject, 492
wolves, 313
women, 553
workspace, 38
World Bank, 373
wrapper, 475

Yates' continuity correction, 378
Yediot Aharonot, 559
Yellow Medicine, 512, 529
yellow-headed blackbird, 327
Yellowstone, 146

Printed and bound in the UK by
CPI Antony Rowe, Eastbourne